Algebra

An incremental approach

L 24 last part
skip 19 20
#13
28.1 only #13
29 skip
skip 9-11 only 30

Algebra

An incremental approach

volume one

JOHN H. SAXON, JR.

Oscar Rose Junior College

PRENTICE-HALL, INC., Englewood Cliffs, New Jersey 07632

Library of Congress Cataloging in Publication Data

Saxon, John H (date)
Algebra, an incremental approach.

Includes index.
1. Algebra. I. Title.
QA152.2.S39 512.9 78-24371
ISBN 0-13-021600-3 (v. 1)

Printed in the United States of America

20 19 18 17 16 15 14 13 12 11

Editorial/production supervision by Zita de Schauensee
Cover and chapter-opening design by Judy Winthrop
Manufacturing buyers: Phil Galea and Edmund Leone

Prentice-Hall International, Inc., *London*
Prentice-Hall of Australia Pty. Limited, *Sydney*
Prentice-Hall of Canada, Ltd., *Toronto*
Prentice-Hall of India Private Limited, *New Delhi*
Prentice-Hall of Japan, Inc., *Tokyo*
Prentice-Hall of Southeast Asia Pte. Ltd., *Singapore*
Whitehall Books Limited, *Wellington, New Zealand*

To Mimmy
whose love, understanding, and approval
made this book possible

Contents

lesson one

lesson two

lesson three

lesson four

lesson five

lesson six

lesson seven

lesson eight

lesson nine

lesson ten

lesson eleven

lesson twelve

lesson thirteen

lesson fourteen

lesson fifteen

lesson sixteen

lesson seventeen

lesson eighteen

lesson nineteen

lesson twenty

lesson twenty-one

lesson twenty-two

lesson twenty-three

lesson twenty-four

lesson twenty-five

lesson twenty-six

lesson twenty-seven

lesson twenty-eight

lesson twenty-nine

lesson thirty

lesson thirty-one

lesson thirty-two

lesson thirty-three

lesson thirty-four

lesson thirty-five

lesson thirty-six

lesson thirty-seven

lesson thirty-eight

lesson thirty-nine

lesson forty

lesson forty-one

Preface

We believe that students who use this book will score higher on a departmental final examination or on a standardized subject matter test than students who have had the same teacher but have used another book. This claim is substantiated by subjective reports from teachers who have used the book in preliminary form and by results of tests conducted at the high school, junior college, and major university level on volume 2 of this two-volume edition.

The reason for the difference in scores is that most students have difficulty in absorbing algebraic abstractions in the short time permitted by the standard development of topics by chapters. Therefore, we have written the book in lessons so that an increment of a particular topic can be presented in one lesson and practiced in the homework before the next increment of the same topic is presented. Since this method is so successful and is unique, we give it a name. We call it an incremental continuous presentation.

This type of development requires that several disjoint topics be presented in each lesson. This disjointness turns out to be an unexpected bonus because students seem to be able to handle a little of three different topics more easily than they can handle three increments of one topic in depth. Their understanding of the lecture is enhanced by the understanding achieved in doing the homework, because every problem set touches on every topic presented previously. The problem sets contain no advanced problems of any kind—just variations of the problems discussed in the text. Thus, the successful completion of every problem set is possible. The problem sets define the task for each student as a function of his ability. Slower students require more time than more gifted students need to complete the homework, but they need more practice because they are slower. The additional time they spend on homework will permit them to write test papers equal to those written by more gifted students.

Because many students have difficulty in reading algebra books, we use more words in the developments than is customary. We tell what we are going to do and why; then we tell what we do as we do it; and finally, we review what we have done. Whenever possible, we begin a development with a concrete example and refer the abstract development to the concrete example, for we have found that many students project from the concrete to the abstract rather than the other way around. This presentation allows the book to be used in conjunction with lectures or as a text for individualized instruction.

We teach the solution to two equations in two unknowns early so that this tool can be used in the solution of word problems. The early word problems are restricted to percent problems and to simple value problems. Simple uniform motion word problems are discussed in later sections of the book.

We discuss the vocabulary of algebra—closure, commutativity, and the like—but avoid these words when explaining other topics. Thus, the student is given an adequate algebraic vocabulary but is not confused by the new words.

I would like to thank the following professors for their assistance in reviewing the manuscript: Jerald T. Ball, Chabot College; Suzanna Hessong, San Antonio College; Richard Pellerin, Northern Virginia Community College; Howard L. Penn and Richard D. Long, Mountain View College; and Richard Spangler, Tacoma Community College.

I thank Robert Park, my division chairman, for his backing and encouragement. I thank Carolyn Thomson and other students for help in various forms. Lastly, I thank Detia Roe, graduate physicist and outstanding typist, whose help made many tasks a joy rather than a chore.

Norman, Oklahoma

John H. Saxon, Jr.

lesson one

Numerals and numbers
Natural numbers • counting numbers • real numbers
Absolute value
Number line and negative numbers
Addition • subtraction • multiplication • division
Algebraic addition using directed numbers
Commutative property of addition

1.1 numerals and numbers

A **NUMBER** is an **IDEA**. A **NUMERAL** is a single **SYMBOL or a collection of symbols** that we use to express the **IDEA** of a particular **number**.

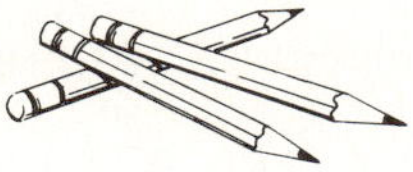

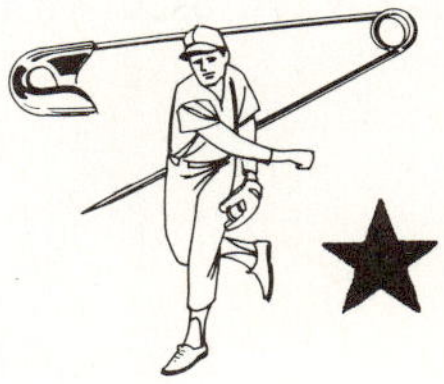

The three drawings above all have the quality of *threeness.* The three children and the three pencils both bring to mind the idea of *three.* The drawing at the right also brings to mind the idea of *three,* although all the things in the drawing are not of the same kind.

If we wish to use a symbol to designate the idea of 3, we could write any of the following:

$$\text{III},\quad 3,\quad \frac{30}{10},\quad \frac{27}{9},\quad \frac{33}{11},\quad 2+1,\quad 6 \div 2,\quad 11-8$$

Each of these is a symbolic representation for the **idea** of 3. Throughout the book when we use the word **number**, we are describing the **idea**, and we will use **numerals** to help us designate the **numbers**. But we will remember that none of the marks we make on paper are numbers because

A NUMBER IS AN IDEA!

Since the symbols

$$3 \quad \text{and} \quad \frac{30}{10}$$

are both **numerals** that represent the same **number**, we say that they have the same VALUE. Thus, the **value** of a **numeral** is the **number** represented by the **numeral**, and we see that the words **value** and **number** mean the same thing.

1.2 natural or counting numbers

The system of numeration that we use to designate numbers is called the decimal system. It was invented by the Hindus of India, passed to their Arab neighbors, and finally transmitted to Europe circa 1200 A.D. The decimal system uses 10 symbols that we call DIGITS. These digits are

$$0,\quad 1,\quad 2,\quad 3,\quad 4,\quad 5,\quad 6,\quad 7,\quad 8,\quad 9$$

We use these digits by themselves or in combination with one another to form the numerals that we use to designate decimal numbers.

We call the numbers that we use to count objects or things the **NATURAL NUMBERS** or the **COUNTING NUMBERS**. When we begin counting, we always begin with the number 1 and follow it with the number 2, etc.

$$1,\quad 2,\quad 3,\quad 4,\quad 5,\quad 6,\quad 7,\quad 8,\quad 9,\quad 10,\quad 11,\quad 12,\quad 13,\quad \ldots$$

It would not be natural to try to count by using numbers such as $\frac{1}{2}$ or 0 or $\frac{3}{4}$ so these numbers are not called **natural** or **counting numbers**. We designate the **natural** or **counting numbers** with the listing above. The three dots after the number 13 indicate that this listing continues without end.

1.3 real numbers

Any number that can be used to describe a physical distance is called a **POSITIVE REAL NUMBER**. Thus all the numbers shown here

$$\frac{3}{4}\quad .000163\quad 363\quad 3\frac{3}{8}\quad 46\quad \frac{11}{7}\quad 400.1623232323$$

are **positive real numbers**, for all of them can be used to describe a physical distance when used with descriptive units such as feet, yards, etc.

$$\frac{3}{4}\text{ mile}\qquad .000163\text{ yard}\qquad 363\text{ feet}\qquad 3\frac{3}{8}\text{ meters}$$

$$46\text{ inches}\qquad \frac{11}{7}\text{ kilometers}\qquad 400.16232323\text{ centimeters}$$

The number **ZERO** is not a positive number, but it can be used to describe a physical distance of no magnitude, and we also call **zero a real number**. In algebra we also use numbers that we call **NEGATIVE NUMBERS** and these numbers are also called real numbers. The ancients did not understand or use negative numbers. A man could not own negative 10 sheep. If he owned any sheep at all, the number of sheep had to be designated by a number greater than zero. The ancients could subtract 4 from 6 and get 2, but they felt that it was impossible to subtract 6 from 4 because that would result in a number that was less than zero itself. To their way of thinking, this was clearly impossible.

While some might tend to agree with the Ancients, to the modern mathematician, physicist, or chemist, the idea or concept of negative numbers does exist and it is a useful concept. **We say that every positive real number has a negative counterpart, and we call these numbers the negative real numbers.** The rest of this lesson will be devoted to the topics of **positive** and **negative numbers** and the use of these numbers in the operations of addition and subtraction. When we use the words **SIGNED NUMBERS**, we indicate that we are considering the non-zero real numbers, for each of these numbers is either a positive or a negative number. As we shall see, the use of signed numbers will allow us to lump the operations of addition and subtraction into one operation that we call **ALGEBRAIC ADDITION**.

1.4 number lines

In the 1950's the so-called new math appeared and among other things it introduced the **NUMBER LINE** at the elementary algebra level. The number line can be used as a graphic aid when discussing signed numbers, and it is especially useful when discussing the addition and subtraction of signed numbers.

To construct a number line, we first draw a line and divide it into equal units of length. The units may be any length as long as they are all the same length (Figure 1.1).

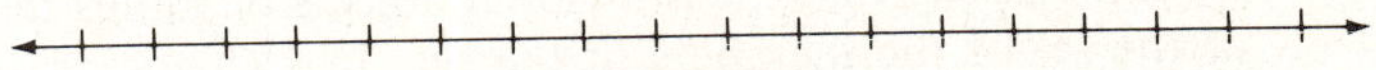

Figure 1.1

Now we choose a point on the line as our base point. We call this base point the **ORIGIN**, and we associate the number zero with this point (Figure 1.2).

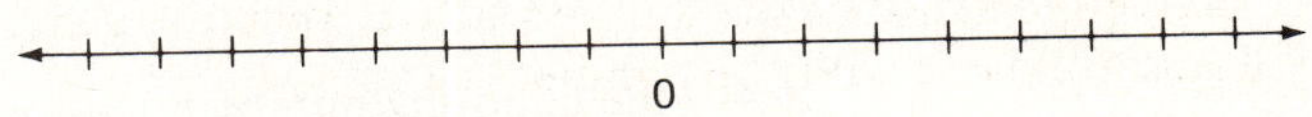

Figure 1.2

Then we associate the positive real numbers with the points to the right of the origin and the negative real numbers with the points to the left of the origin.

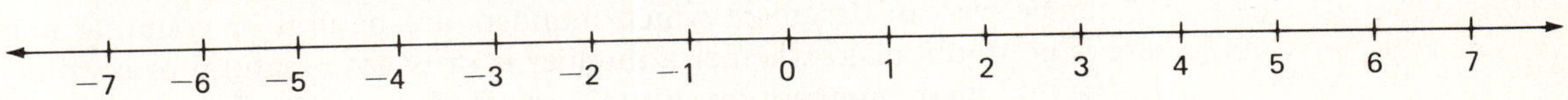

Figure 1.3

On the number line of Figure 1.3 we indicate the location of zero, the counting numbers and the negative counterpart of each counting number. As required, we can indicate the position of any real number by locating it in relation to the numbers shown. For example, on the number line of Figure 1.4 we indicate the position of $+\frac{3}{4}$, $-1\frac{1}{2}$, and $+2.6$ by placing a dot at the approximate location of these numbers.

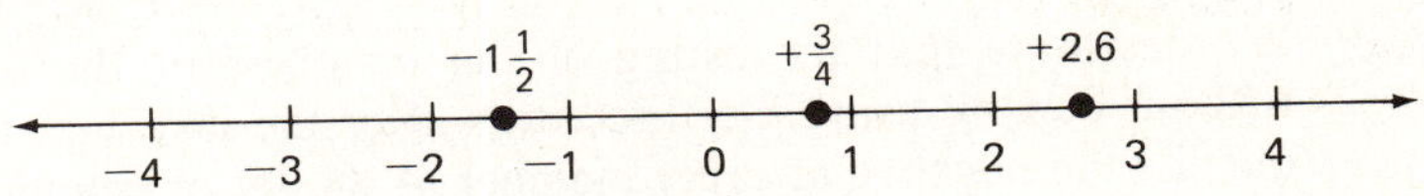

Figure 1.4

When we place a **DOT** on the number line to indicate the location of a number, we say that we have **GRAPHED** the number and that **the dot is the graph of the number**. Conversely, the number is said to be the **COORDINATE** of the point that we have graphed. We use the number line to tell if one number is **GREATER THAN** another number by saying that a number is **greater than** another number if its graph lies **TO THE RIGHT OF** the graph of the other number.

As we have said, every number that can be used to describe a physical distance is called a positive real number and is associated with a unique (one and only one) point on the right side of the number line. The number zero is called a real number and is neither positive nor negative. Equidistant from zero in the other direction lies the negative counterpart of each positive real number and these numbers are called the negative real numbers.

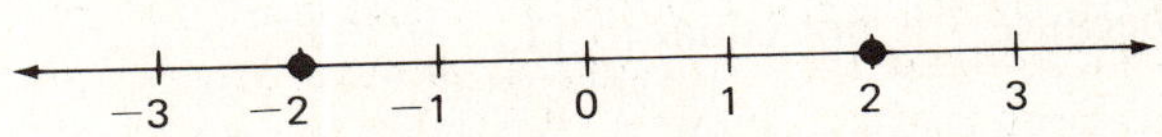

Figure 1.5

For example, in Figure 1.5 we have graphed $+2$ and -2. The number $+2$ (usually the $+$ sign is omitted) lies two units to the right of the origin and the number -2 lies two units to the left of the origin. Since the graphs of these numbers are equidistant from the origin but in opposite directions, it is sometimes helpful to think of each of these numbers as being the **OPPOSITE** of the other number. In this example, we say that -2 is the **opposite of** 2 and that 2 is the **opposite of** -2.

1.5 sets

We use the word **SET** to designate a well-defined collection of numbers, objects, or things. We say that the individual objects or things that make up the set are the elements of the set or the members of the set. It is customary to designate a set by enclosing the members of the set within braces.

$$A = \{1, 2, 3, 4, 5\}$$

We have designated set A as the set whose members are the counting numbers 1 through 5 inclusive. We could also designate this set by placing a verbal phrase within the braces as

$$A = \{\text{the counting numbers 1 through 5 inclusive}\}$$

or we could designate the members of this set by writing an English sentence:

"Set A is the set whose members are the counting numbers one through five inclusive."

Since we can designate which numbers are natural or counting numbers so that there is no doubt as to whether a number is or is not a natural or counting number, we can say that these numbers constitute a set. Both natural and counting are names for the same set and we normally use one name or the other. Thus we say

$$\text{Natural numbers} = \{1, 2, 3, 4, 5, \cdots\}$$

The three dots indicate that the listing of numbers continues without end.

If we include the number zero with the set of natural numbers, we say that we have designated the set of whole numbers.

$$\text{Whole numbers} = \{0, 1, 2, 3, 4, 5, \cdots\}$$

We designate that the listing of the members of the set of whole numbers continues without end by using the three dots after the last digit recorded. Now if we include in our list the **negative of every member of the set of counting numbers**, we have designated the set that we call the **INTEGERS.**

$$\text{Integers} = \{\cdots, -3, -2, -1, 0, 1, 2, 3, \cdots\}$$

The three dots on each end indicate that the listing continues without end in both directions.

1.6 symbols of equality and inequality

We use the equals $(=)$ sign to designate that two quantities are equal. Thus we can write

$$5 + 2 = 7$$

because the number represented by the notation $5 + 2$ is the same number as that represented by the numeral 7. In the same way we use the symbol, $\neq$, to designate that two quantities are not equal. Thus we can write that

$$5 + 2 \neq 11$$

because 7 is not equal to 11.

1.7 basic operations

The four basic operations of arithmetic are also the basic operations of algebra. The operations are addition, subtraction, multiplication, and division. We will review these operations here and will restrict our discussion to the numbers of arithmetic which are the positve real numbers and zero.

addition When we wish to add two numbers to get a result, we use the plus sign (+) to indicate the operation of addition. We call each of the numbers an **ADDEND**, and we call the result a **SUM**.

$$2 + 3 = 5$$

In this example, we use the plus sign to indicate addition; we say that the numbers 2 and 3 are addends, and we say that 5 is the sum.

We note that the sum of zero and any particular real number is the particular real number itself. This property or peculiarity is true of no other number and thus we say that this property is unique to the number zero. We call this property the **ADDITIVE PROPERTY OF THE NUMBER ZERO.**

$$4 + 0 = 4 \quad \text{and} \quad 15 + 0 = 15$$

subtraction When we wish to subtract one number from another number, we use the minus sign (−) to indicate the operation of subtraction. We call the first number the **MINUEND**; the second number, the **SUBTRAHEND**; and the result, the **DIFFERENCE**.

$$9 - 5 = 4$$

In this example, 9 is the minuend, 5 is the subtrahend, and 4 is the difference.

multiplication If two numbers are to be multiplied to achieve a result, each of the numbers is called a **FACTOR** and the result is called a **PRODUCT**. There are several ways to indicate the operation of multiplication.

$$4 \cdot 3 = 12, \quad 4(3) = 12, \quad (4) \cdot (3) = 12, \quad (4)(3) = 12, \quad 4 \times 3 = 12$$

In each of the five examples shown here, the notation indicates that 4 is to be multiplied by 3 and that the result is 12. In each case, we call 4 and 3 factors, and we say that 12 is the product.

We note that the product of a particular real number and the number one is the particular real number. We say that this property or peculiarity is *unique* to the number one because it is true for no other real number. We call this property the **multiplicative property of the number one.**

$$4 \cdot 1 = 4 \quad \text{and} \quad 15 \cdot 1 = 15$$

The number zero also has a unique multiplicative property. **The product of a particular number and the number zero is the number zero.** This property is called the **multiplicative property of the number zero.**

$$4 \cdot 0 = 0 \quad \text{and} \quad 15 \cdot 0 = 0$$

division If one number is to be divided by another number to achieve a result, the first number is called the **DIVIDEND**, the second number is called the **DIVISOR**, and the result is called the **QUOTIENT**.

$$\frac{10}{5} = 2 \qquad 10 \div 5 = 2$$

Both of the notations shown here indicate that 10 is to be divided by 5 and that the result is 2. We call 10 the dividend, call 5 the divisor, and say that the quotient is 2. When the indicated division is expressed in the form of a fraction such as $\frac{10}{5}$, we say that 10 is the **NUMERATOR** of the fraction and that 5 is the **DENOMINATOR** of the fraction.

1.8 absolute value

Every real number except zero has two parts. One of the parts is the plus or minus sign that indicates whether the number is a positive number or a negative number. The other part is the numerical part. We can think of the numerical part of the number as defining the quality of "bigness" of the number, and we use the words **ABSOLUTE VALUE** to describe this quality.† If we look at the two numbers,

$$+7 \quad \text{and} \quad -7$$

We see that one number is a positive number and one is a negative number, and that we can think of both of them as having the same degree of bigness or absolute value, which is 7. Since we normally use positive numbers to designate degrees of bigness, **we will use positive numbers to designate the absolute value of all real numbers except zero.** Since zero has no bigness, we say that the **absolute value of zero is zero.** If we enclose a number within vertical lines, we are designating the absolute value of the number. We will demonstrate this notation by designating the absolute value of zero, the absolute value of positive 7, and the absolute value of negative 7.

$|0| = 0$ read the absolute value of zero equals zero.
$|+7| = +7$ read the absolute value of 7 equals 7.
$|-7| = +7$ read the absolute value of -7 equals 7.

Since the plus sign in front of a positive number is customarily omitted, the above can be written without recording the plus signs:

$|7| = 7$ read the absolute value of 7 equals 7.
$|-7| = 7$ read the absolute value of -7 equals 7.

Thus we have two rules for stating the absolute value of a real number.

1. **The absolute value of zero is zero.**
2. **The absolute value of any non-zero real number is a positive number‡**

Here we designate the absolute value of zero and several other real numbers.

(a) $|0| = 0$ (b) $|-7.12| = 7.12$ (c) $|7.12| = 7.12$
(d) $|-5| = 5$ (e) $|5| = 5$ (f) $\left|\frac{3}{4}\right| = \frac{3}{4}$

example 1.8.1 Simplify (a) $|-5|$, (b) $|11 - 2|$, (c) $-|20 - 2|$.

solution (a) The absolute value of -5 is 5.

$$|-5| = 5$$

(b) First we simplify within the vertical lines:

$$|11 - 2| = |9|$$

and the absolute value of 9 is 9.

$$|9| = 9$$

(c) Again we simplify within the vertical lines:

$$-|20 - 2| = -|18|$$

The absolute value of 18 is 18, but we want the negative of this, so our answer is -18.

$$-|18| = -18$$

---

†Absolute value will be discussed again in Lesson 27 where we will present the formal definition, which uses no words—only symbols.

‡We *cannot* say that the absolute value of a number is the positive *value* of a number because numbers do not have two values. A non-zero real number is either a positive number or a negative number.

Test Your Understanding

Simplify:

(a) $|-42|$ (b) $|+15|$ (c) $|18-3|$

1.9 addition of signed numbers

In arithmetic the minus sign always means to subtract, but in algebra we use the same sign to designate that a number is a negative number. This can be confusing at first as we see if we look at the expression

$$7-4$$

and ask if the minus sign means to subtract or if it means that the 4 is a negative number. It turns out that we will find the same answer with either thought process, but in algebra we normally prefer the second process in which we think of the negative sign as designating that -4 is a negative number.

To help us with this discussion we will use a ploy that is used by many authors of recently published pre-algebra books. For a portion of the book the author will use raised signs such as $^+2$ and $^-2$ when he wishes the reader to think of the signed numbers positive 2 and negative 2 and will use regular signs such as $+2$ and -2 when he wishes to indicate the operations of addition and subtraction. If we use this raised sign notation, the numbers on the number line will have raised signs as shown in Figure 1.6 because the signs indicate that the numbers are positive or negative numbers and do not indicate the operations of addition and subtraction. **Note that the number zero is neither positive nor negative.**

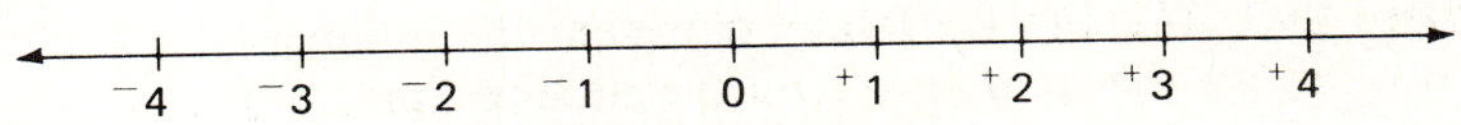

Figure 1.6

The pre-algebra books also use the concept of directed numbers to give the student a graphical or pictoral way to interpret the algebraic addition of signed numbers. The student is asked to think of **a positive number as an arrow that points to the right** whose length in units is the same as the absolute value of the number and **a negative number as an arrow that points to the left** whose length also corresponds to the absolute value of the number. Using this thought process a positive number, say $^+3$, would be designated by an arrow 3 units long that points to the right and a negative number, say $^-2$, by an arrow 2 units long that points to the left.

Thus to use directed numbers and the number line to add $^+3$ and $^-2$, we begin at the origin and draw an arrow 3 units long that points to the right to represent the number $^+3$ (Figure 1.7).

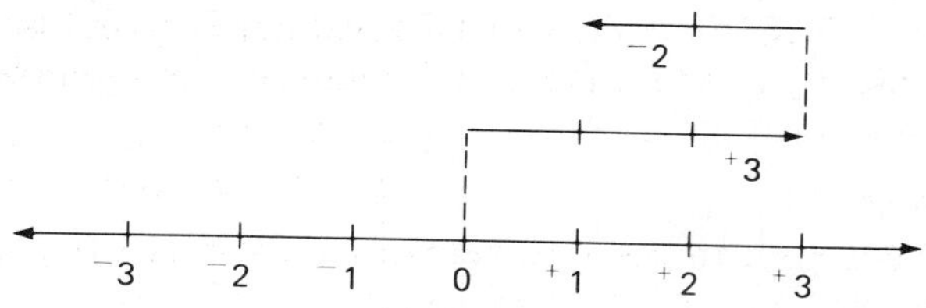

Figure 1.7

Then from the head of this arrow (Figure 1.7), we draw an arrow 2 units long that points to the left to represent the number $^-2$. The head of the last arrow is at $^+1$ and this gives us a graphical solution of the problem.

$$^+3 + {}^-2 = {}^+1$$

Please note that we obtain the same answer when we add signed numbers algebraically as we do when we use only the positive numbers of arithmetic and subtract!

$$3 - 2 = 1$$

It may seem that we are trying to turn an easy problem into a difficult problem but such is not the case. **In algebra the operations of addition and subtraction are lumped together in the one operation of ALGEBRAIC ADDITION** and this enables a straightforward solution to problems that would be very confusing if the concepts of signed numbers and algebraic addition are not used.

example 1.9.1 Use directed numbers and the number line to add $^+3$ and $^-5$.

solution

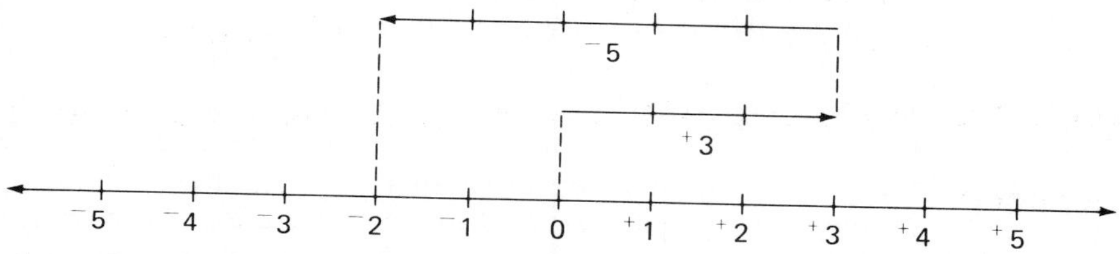

We begin at the origin and draw an arrow 3 units long that points to the right to represent the number $^+3$. From the head of this arrow we draw an arrow 5 units long that points to the left to represent the number $^-5$. The head of the second arrow is just above the number $^-2$ on the number line. Thus, our graphical solution to the problem of algebraically adding the numbers $^+3$ and $^-5$ is the number $^-2$, or

$$^+3 + {}^-5 = {}^-2$$

example 1.9.2 Use directed numbers and the number line to add $^-5$ and $^+3$.

solution

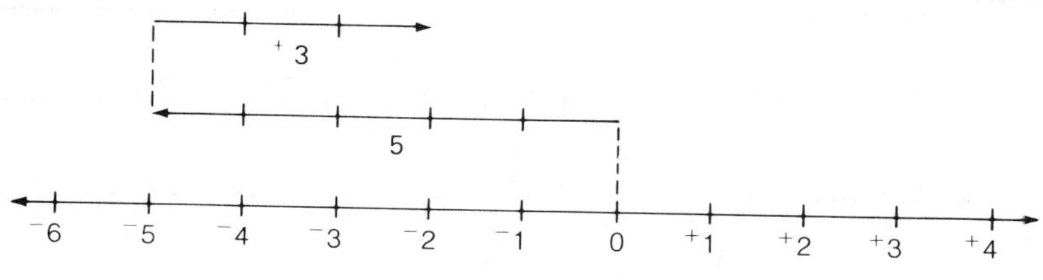

We begin at the origin and draw an arrow 5 units long pointing to the left to represent the number $^-5$, and from the head of this arrow we draw an arrow 3 units long pointing to the right to represent the number $^+3$. The head of this last arrow is above the number $^-2$ on the number line and our graphical solution to the problem of adding $^-5$ and $^+3$ is $^-2$. This problem is the same as the preceding problem except that in this problem we drew the $^-5$ arrow first and then drew the $^+3$ arrow. We note that the answer is $^-2$ for both problems. **This is a demonstration of the fact that the order in which two signed numbers are added does not affect the sum.**

This is true for **ANY TWO** real numbers and we give it the name of the commutative property of addition for real numbers. For example,

$$^{-}5 + {}^{+}3 = {}^{-}2 \quad \text{and} \quad {}^{+}3 + {}^{-}5 = {}^{-}2$$

also

$$4 + 3 = 7 \quad \text{and} \quad 3 + 4 = 7$$

To make a formal statement of the commutative property, we will use the letters a and b to designate any two real numbers.

> COMMUTATIVE PROPERTY OF ADDITION
>
> For any two real numbers a and b,
>
> $$a + b = b + a$$

Of course, if the two signed numbers have the same sign, the directed number arrows will point in the same direction as shown in the next two examples.

example 1.9.3 Use directed numbers and the number line to add $^{+}2$ and $^{+}1$.

solution

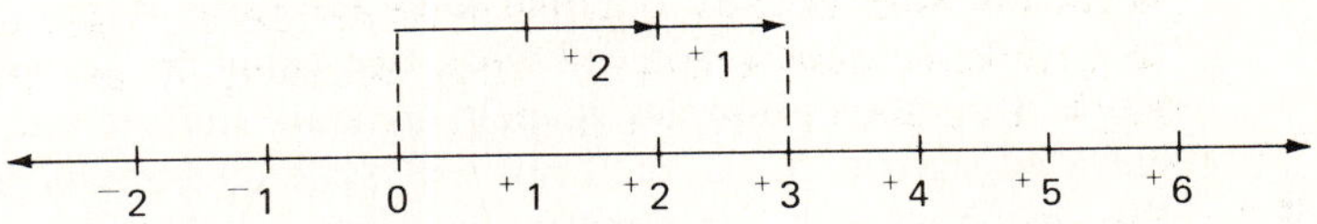

We see from the graph that the solution is $^{+}3$.

$$^{+}2 + {}^{+}1 = {}^{+}3$$

example 1.9.4 Use directed numbers and the number line to add $^{-}2$ and $^{-}1$.

solution

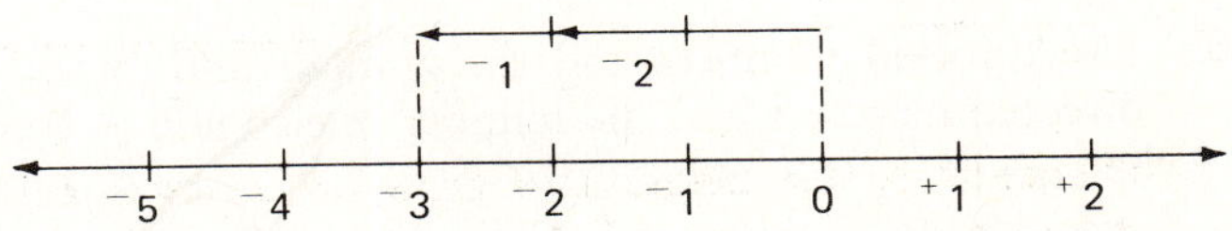

We see from the graph that the solution is $^{-}3$.

$$^{-}2 + {}^{-}1 = {}^{-}3$$

example 1.9.5 Use directed numbers and the number line to add $^{+}7$ and $^{-}4$.

solution

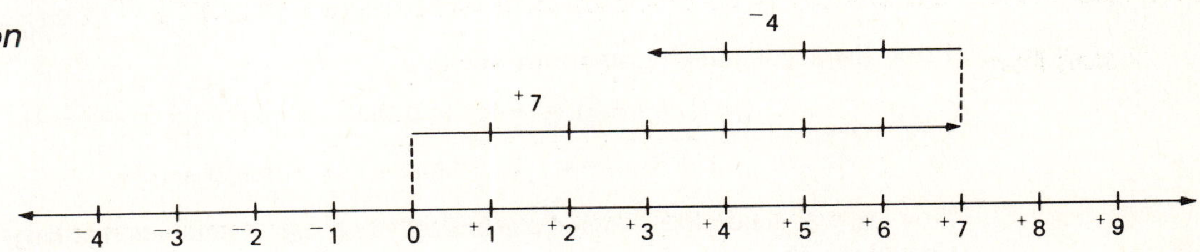

We see that the head of the $^{-}4$ arrow is just above the number $^{+}3$ on the number line so the graphical solution to the problem is $^{+}3$. Now our original problem was to decide whether the notation

$$7 - 4 = 3$$

meant to subtract a positive 4 from 7 and get 3 or whether we should think of the problem as indicating that we should add a negative 4 to positive 7 and get 3,

$$^{+}7 + {}^{-}4 = {}^{+}3$$

The answers are the same but in the first case we use the operation of subtraction and in the second case we use the concepts of signed numbers and algebraic addition.

TEST YOUR UNDERSTANDING

(d) Use directed numbers and the number line to add $^{+}3$ and $^{-}5$.

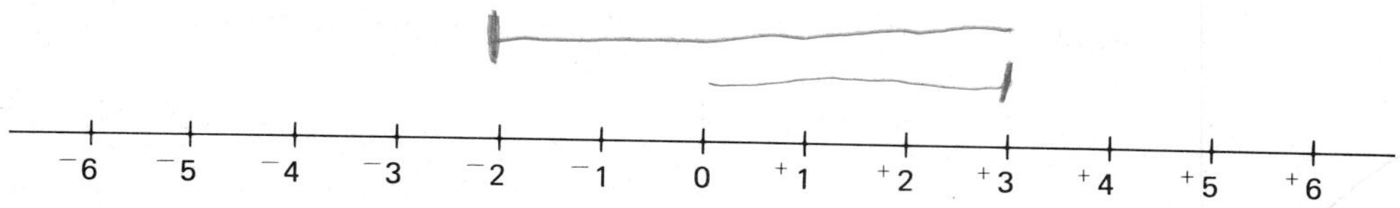

Henceforth we will use only the concept of algebraic addition, but we can't go through all algebra drawing number lines and arrows so we must develop rules that will allow us to do algebraic addition quickly and accurately. We need two rules—one to use when the numbers to be added have the same signs and one to use when the numbers have different signs. We will work two examples to help us state the first rule and then work two more examples to help us state the second rule. We will no longer use the elevated signs ($^{+}5$, $^{-}3$, etc.) but will write all signs in their normal position. **We will let the sign preceding the number designate whether the number is a positive number or a negative number, and we will mentally insert a plus sign in front of each number to designate algebraic addition.** If we use this process,

$4 - 3 + 2$ can be read as $(+4) + (-3) + (+2)$

and $-6 - 3 - 2 + 5$ can be read as $(-6) + (-3) + (-2) + (+5)$

example 1.9.6 Use directed numbers and the number line to add $+1$ and $+3$ algebraically and use directed numbers and the number line to add -1 and -3 algebraically.

solution $+1 + 3$ can be written as $(+1) + (+3)$ and $-1 - 3$ can be written as $(-1) + (-3)$.

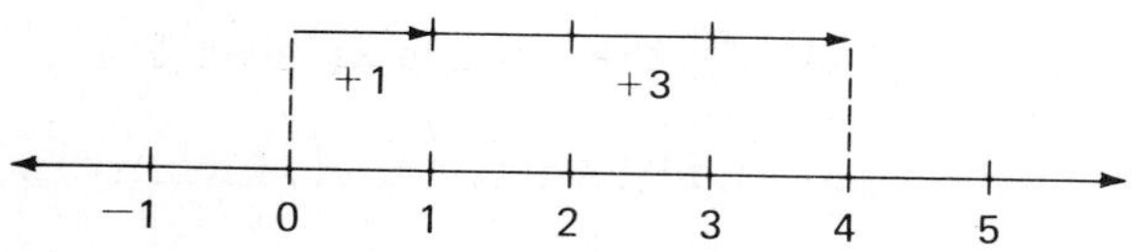

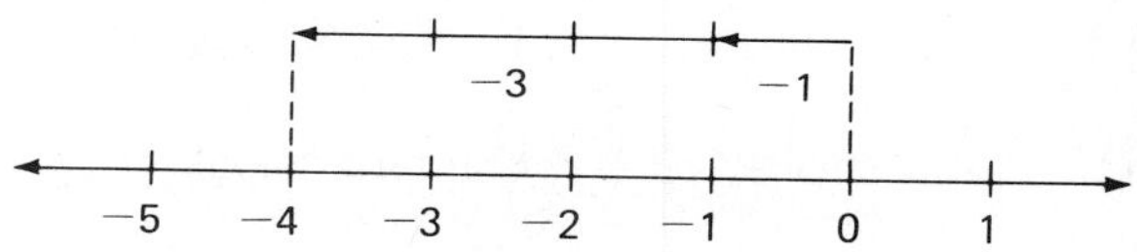

From the accompanying figure we see that

$(+1) + (+3) = +4$ and that $(-1) + (-3) = (-4)$

so $+1 + 3 = +4$ and $-1 - 3 = -4$

Now we will generalize. **To add algebraically two signed numbers that have the same sign, we add the absolute values of the numbers and give the result the same sign as the sign of the numbers.**

EXAMPLES

(a) $4 + 4$ means to think $(+4) + (+4)$, which equals $+8$.

(b) $-3 - 2$ means to think $(-3) + (-2)$, which equals -5.

(c) $-4 - 10$ means to think $(-4) + (-10)$, which equals -14.

(d) $7 + 11$ means to think $(+7) + (+11)$, which equals $+18$.

Now we will use two problems in which numbers with different signs are added algebraically to help us state the second rule.

example 1.9.7 Use directed numbers and the number line to add -2 and $+5$ algebraically and use directed numbers and the number line to add $+2$ and -5 algebraically.

solution We will insert + signs between the numbers so that the process of algebraic addition will be emphasized.

$$-2 + 5 \text{ can be written as } (-2) + (+5)$$

and

$$2 - 5 \text{ can be written as } (+2) + (-5).$$

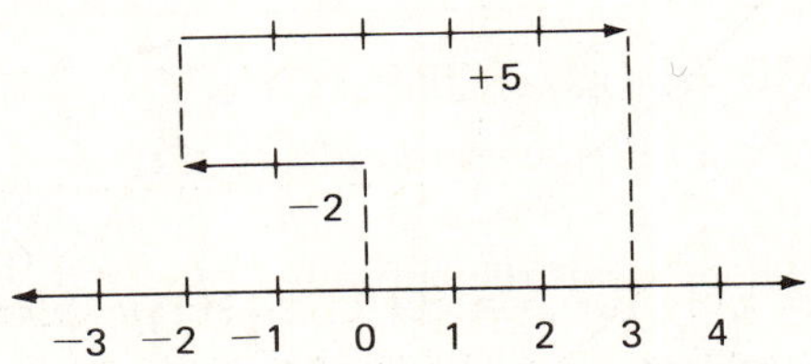

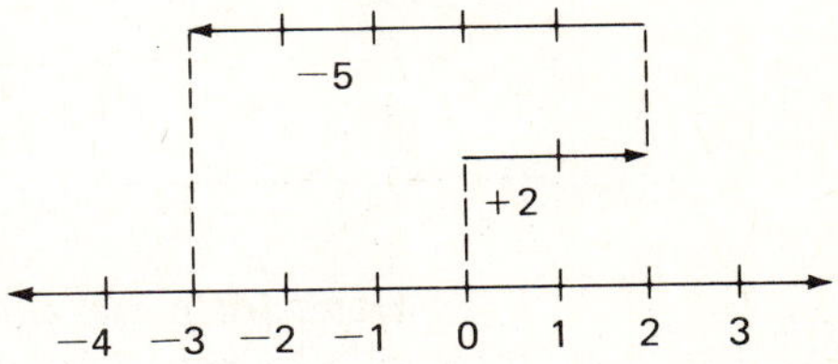

From the figure we see that the absolute value of each answer is 3 but that one of the answers is $+3$ and that one of the answers is -3.

$$(-2) + (+5) = +3 \qquad (+2) + (-5) = -3$$

In the first case the number $+5$ had the larger absolute value and thus the sign of the result was positive. In the second case the number -5 had the larger absolute value and thus the sign of the result was negative. In both cases the absolute value of the answer was the difference in the absolute values of the numbers.

Now we will generalize. **To add algebraically two signed numbers that have opposite signs, we take the difference in the absolute values of the numbers and give to this result the sign of the original number whose absolute value is the greatest.**

EXAMPLES

(a)	$-4 + 1$	means to think	$(-4) + (+1)$,	which equals -3.
(b)	$4 - 1$	means to think	$(+4) + (-1)$,	which equals $+3$.
(c)	$7 - 11$	means to think	$(+7) + (-11)$,	which equals -4.
(d)	$-7 + 11$	means to think	$(-7) + (+11)$,	which equals $+4$.

TEST YOUR UNDERSTANDING

Simplify without using a number line:

(e) $(-13) + (+6)$ (f) $-13 + 6$ (g) $-30 + 40$ (h) $30 - 40$

−13+6 −7 10 −10
−7

1.10 algebraic subtraction

In the foregoing paragraphs we have developed a procedure whose use allows us to delete the operation of subtraction from our algebraic vocabulary. We let the given signs denote whether the numbers are positive numbers or negative numbers and we mentally insert a plus sign between the numbers. Thus the subtraction problem shown on the left can be turned into the algebraic addition problem on the right.

$$7 - 4 = 3 \qquad 7 + (-4) = 3$$

However, we have not done away with subtraction because a definition of algebraic subtraction does exist and many authors prefer to use algebraic subtraction rather than the process described above. The result is exactly the same, but the definition uses the word subtraction. **To subtract algebraically, we change the sign of the subtrahend and add.**

$$7 - 4 = 3 \qquad 7 + (-4) = 3$$

The formal definition of the operation of algebraic subtraction is as follows.

ALGEBRAIC SUBTRACTION

If a and b are real numbers, then

$$a - b = a + (-b)$$

where $-b$ is the opposite of b.

1.11 addition of more than two signed numbers

Now to finish the topic of algebraic addition of signed numbers, we will use the number line to add the numbers $-4 + 2 - 1 + 5$ and then use the number line to add the numbers $-1 + 2 - 4 + 5$, which are the same numbers but are written in a different order. We will generalize the results of these two problems to the addition of any number of signed numbers.

example 1.11.1 Use directed numbers and the number line to find a graphical solution to the problem $-4 + 2 - 1 + 5$.

solution We will no longer insert the extra + signs but will remember to think of the problem as an addition problem since

$$-4 + 2 - 1 + 5 \quad \text{can be read as} \quad (-4) + (+2) + (-1) + (+5)$$

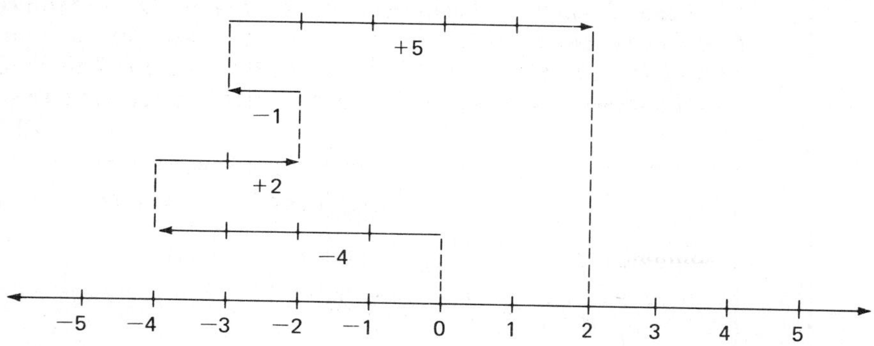

We began at the origin and moved 4 units to the left for -4, then 2 units to the right for $+2$, then 1 unit to the left for -1, and finally 5 units to the right for $+5$ and find that we end up directly above the number $+2$ on the number line. Thus

$$-4 + 2 - 1 + 5 = 2$$

Now we will change the order of the signed numbers and work the problem again.

example 1.11.2 Use directed numbers and the number line to find a graphical solution to the problem $-1 + 2 - 4 + 5$.

solution

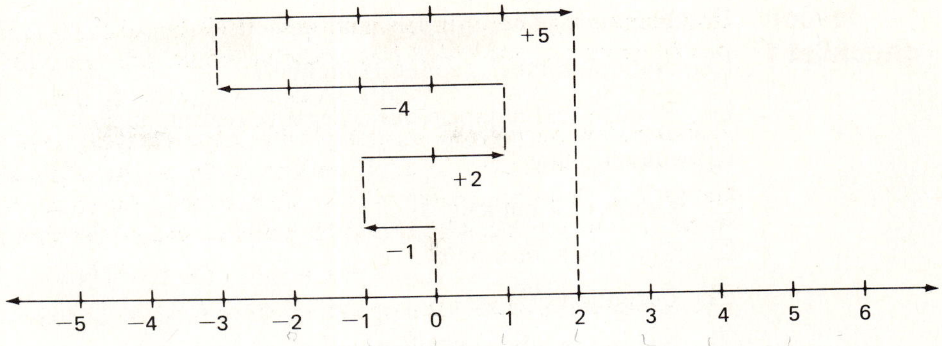

In the accompanying figure we use arrows representing directed numbers to find a graphical solution. We used four arrows, representing the four directed numbers. Although the order in which we used the arrows is different, the final result is the same as in the figure in Example 1.11.1. This is a demonstration of the fact that **The order in which signed numbers are added does not affect the value of the sum.**

It is interesting to note that algebraic addition is a binary operation.† Only two numbers can be added in one step. For instance, if we wish to find the sum of $2 + 3 + 4$, we can add 2 and 3 and then add 4; or add 4 and 3 and then add 2, or add 2 and 4 and then add 3. But we cannot add three numbers in one step because addition is a binary operation.

example 1.11.3 Add $-4 + 3 + 2 + 7$.

solution

(a) $-4 + 3 + 2 + 7$

$-1 + 9$

$\mathbf{+8}$

(b) $-4 + 3 + 2 + 7$

$-4 + 5 + 7$

$-4 + 5 + 7$

$-4 + 12$

$\mathbf{+8}$

(c) $-4 + 7 + 3 + 2$

$+3 + 5$

$\mathbf{+8}$

We found the sum three different ways. In (a) we added -4 and 3 and added 2 and 7 and then added the -1 and $+9$. In (b) we chose to add the two middle numbers to get 5 and then added 7 and added -4 as a last step. In (c) we changed the order of the addition and added the first two and the last two and then summed the results. In each case, however, we found that the sum of the four numbers was 8.

TEST YOUR UNDERSTANDING

Simplify:

(i) $-4 - 3 - 2 - 8$ (j) $-4 + 7 - 3 - 7$ (k) $+8 - 11 + 7$

---

†*Bi* is a prefix that comes from the Latin word *bis*, which means two. (A bicycle has two wheels; bimonthly means every two months, etc.)

lesson checklist 1

- ☐ Numbers, ideas, symbols, and numerals
- ☐ Natural numbers/counting numbers
- ☐ Positive real numbers, zero, negative real numbers
- ☐ Number line
- ☐ Graph of a number
- ☐ Coordinate of a point
- ☐ Definition of a set
- ☐ Natural numbers, whole numbers, and integers
- ☐ Symbols of equality and inequality
- ☐ Basic operations
- ☐ Addition, sum, addends, subtraction, difference, minuend, subtrahend, multiplication, product, factors
- ☐ Multiplicative property of the number one
- ☐ Additive and multiplicative properties of the number zero
- ☐ Division, quotient, dividend, divisor, numerator, denominator
- ☐ Absolute value
- ☐ Directed numbers
- ☐ Algebraic addition
- ☐ Commutative property of addition
- ☐ Three rules for adding signed numbers
- ☐ Algebraic subtraction
- ☐ Addition of more than two signed numbers

Answers to Test Your Understanding

(a)	42	(b)	15	(c)	15
(e)	−7	(f)	−7	(g)	10
(h)	−10	(i)	−17		
(j)	−7	(k)	4		

(d)

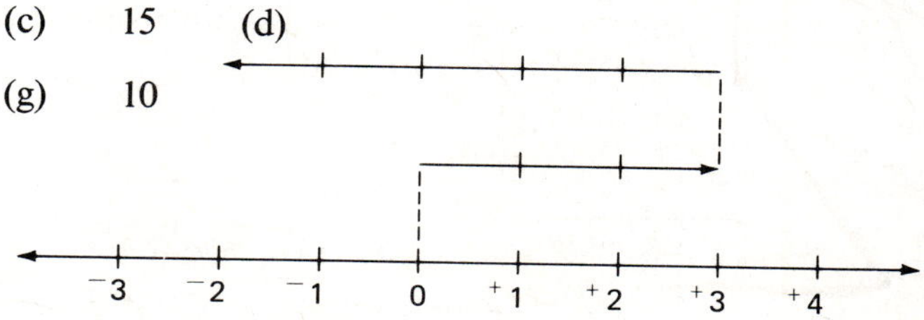

problem set 1

Note: The homework problem sets in this book are repetitive. Problems that appear in one problem set are repeated in later problem sets with little or no alternation in form. Students have found it helpful to do the problems on lined loose-leaf notebook paper and to keep the problems in a ring notebook reserved for this purpose. They find that these worked-out problems can be used for reference later in the course and are also helpful in mathematics and science courses that follow.

1. Is 143.26 a number or a numeral? Why?
2. Any number that can be used to describe a positive length is a real number. What other numbers are also real numbers?
3. Which real numbers *cannot* be paired with a point on the number line?
4. Designate the set of counting numbers.
5. Designate the set of integers.
6. Designate the set of whole numbers.
7. Define the word quotient.
8. Define the word sum.
9. What is a factor?
10. The number zero has two unique properties. List them and give a numerical example of each one.
11. The number one has one unique property. Give the name of this property and give a numerical example.
12. What is a product?

Simplify the following expressions:

*13. $|-5|$

*14. $|11 - 2|$

15. $|5 - 30|$

*16. $-|20 - 2|$

Use the number line to add:

17. $-2 + 1 + 4$

18. $-3 + 4 - 2$

Add algebraically without using the number line:

19. $(-4) + (+3) + (-2) + (+5)$

20. $(-8) + (7) + (-5) + (3)$

21. $(-2) + (3) + (-5) + (-10)$

22. $(-11) + (-15) + (12) + (-4)$

23. $(-17) + (6) + (-4) + (-3)$

24. $(-18) + (-11) + (4) + (20)$

Add algebraically without using the number line (remember to insert plus signs mentally between each of the numbers):

25. $-11 - 3 - 5$

26. $+21 - 3 - 8$

27. $-5 - 4 + 3 - 11 + 7$

28. $-14 + 3 - 8 + 7 + 4$

29. $-8 + 11 + 7 + 14 - 3 - 6$

30. $4 - 10 + 21 - 42$

31. $19 - 14 + 33 + 4 - 6$

32. $-8 - 3 - 6 - 7 - 4$

33. $8 + 4 + 13 - 15 + 7$

34. $-17 + 6 - 11 + 5 + 7 - 3$

35. $-8 - 11 - 14 + 22 - 3$

36. $-4 - 3 - 7 + 5 - 2 + 41$

37. $-8 + 5 - 17 + 6 - 8 - 11$

38. $-7 + 17 - 5 - 3 + 9 - 4$

39. $-15 - 5 + 10 + 15 - 11 + 20$

40. $-8 - 3 + 14 + 11 - 5 - 2$

41. $+20 - 5 - 8 - 7 + 3 - 8$

42. $-11 - 5 - 3 + 7 + 4 - 2$

---

*The problems marked with an asterisk were worked out in this lesson.

lesson two

Algebraic addition
Multiplication and division of signed numbers
Order of operations

2.1 the opposite of a number

In this lesson we will discuss the multiplication and division of signed numbers. For this discussion the number line will be helpful and so will the concept of the **OPPOSITE OF A NUMBER**, which we introduced in Lesson 1.

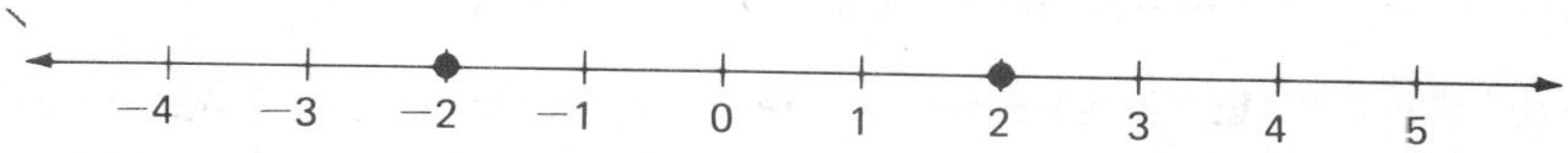

Figure 2.1

In Figure 2.1 we have graphed the numbers $+2$ and -2. From the figure we see that the number -2 lies on the **opposite** side of the origin from 2 and is exactly the same distance from the origin. Thus we can think of $+2$ as being the **opposite of** -2 and -2 as being the **opposite of** $+2$. Every real number except zero has a counterpart that is equidistant from zero and opposite in sign and we call these numbers **opposites**.

It is helpful sometimes to read the numeral -2 as the **opposite of 2**. If we use this wording, it is easy to locate $-(-2)$, for we read this as the opposite of the opposite of 2. Well, the opposite of 2 is -2 so the opposite of that must be 2 itself.

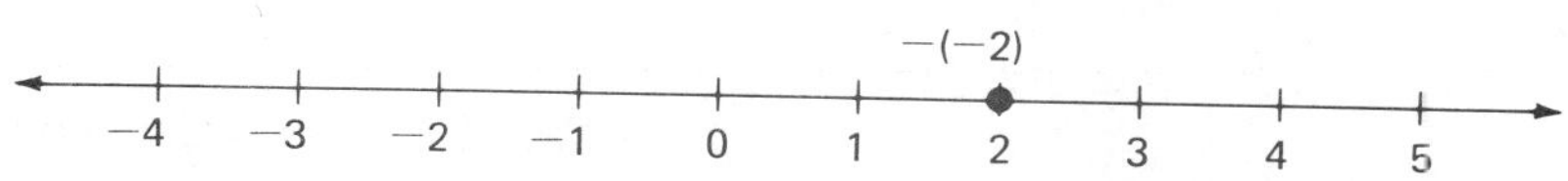

Figure 2.2

In Figure 2.2 we show the location of 2. On the opposite side of the origin lies -2, which is the opposite of 2. Now $-(-2)$, which is the opposite of the opposite of 2, is back on the right at the same point as 2. In section 1.1, we said that a numeral is a symbol that we use to represent the idea of a particular number. Thus 2 and $-(-2)$ are different numerals or symbols for the same number!

Now we can locate 2 and -2, which is the opposite of 2, and we can locate $-(-2)$, which is the opposite of the opposite of 2, but then where does $-[-(-2)]$ lie?

This number is the opposite of the opposite of the opposite of 2 (Figure 2.3).

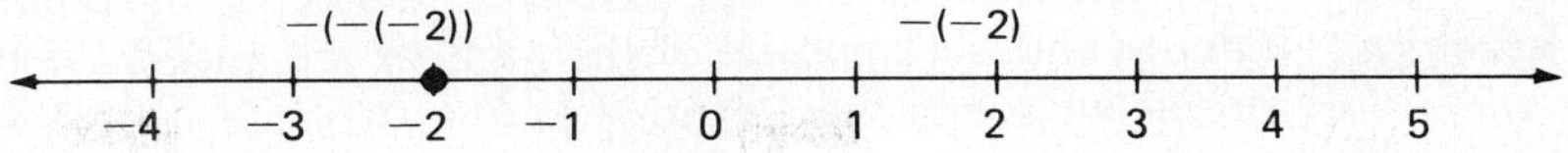

Figure 2.3

If we begin at 2, we can find the opposite of 2 at -2. Then the opposite of the opposite of 2 is back on the right side so the opposite of the opposite of the opposite of 2 is on the left and is another way to write -2. Of course, we could go on forever with this process —but we won't.

example 2.1.1 Simplify (use number line if necessary)

(a) $-[-(-4)]$ (b) $-(-4)$ (c) $-(+4)$

solution (a) **ans = -4** (b) **ans = 4** (c) **ans = -4**

2.2 algebraic addition

The concept of the opposite of a number can be useful in simplifying some indicated algebraic sums that may otherwise be rather confusing. For example, the expression

$$14 + (-2) - 2 - (+6)$$

can be written in a simpler form. The notation $+(-2)$ is read as **plus the opposite of 2,** which can be written simply as -2.

$$14 - 2 - 2 - (+6)$$

The last entry is the notation $-(+6)$, which is read as **the opposite of $+6$** and this can be written as -6. Now we add algebraically.

$$14 - 2 - 2 - 6 = \mathbf{4}$$

example 2.2.1 Simplify $-(-4) + (-2) - [-(-6)]$.

solution The first numeral is read as the opposite of the opposite of 4, which equals 4. The second numeral is read as plus the opposite of 2, which can be written as -2, while the third numeral is the opposite of (once) the opposite of (twice) the opposite of (three times) 6. With the aid of a mental number line we realize that this is another way of writing -6. Now we have

$$4 - 2 - 6 = \mathbf{-4}$$

example 2.2.2 Simplify $-(+4) - (-5) + 5 - (-3) + (-6)$.

solution Again as the first step we use the concept of the **opposite of** to simplify the signs. Then we add.

$$-4 + 5 + 5 + 3 - 6 = \mathbf{3}$$

TEST YOUR UNDERSTANDING

Simplify:

(a) $-(-3) + (-5) + 5 - 4 - (-6) + (-4)$

2.3 additive inverse

Another name for the opposite of a number is the **additive inverse** of the number. If a number and the opposite of the number are added algebraically, the sum is the number zero as we see in these examples.

$$4 + (-4) = 0 \qquad -5 + (5) = 0 \qquad -31.2 + (31.2) = 0$$

As we saw on the number line—every number except zero has an opposite or an additive inverse. This fact is very important and will be used extensively in the solutions of equations. **The algebraic sum of a number and the additive inverse of the same number is the number zero.**

2.4 multiplication of signed numbers

Multiplication is a shorthand method of repeated addition. The product of 3 and 2 is 6.

$$3 \cdot 2 = 6$$

We find the same result if we express the problem as a sum,

$$2 + 2 + 2 = 6 \quad \text{or} \quad 3 + 3 = 6$$

Three times 2 is a shorthand way of expressing the sum of three 2's or the sum of two 3's.

Thus to multiply $+3$ times $+2$ using directed numbers, we can show the result as the sum of two $+3$'s or as the sum of three $+2$'s. Either way, we find a product of $+6$ (Figure 2.4).

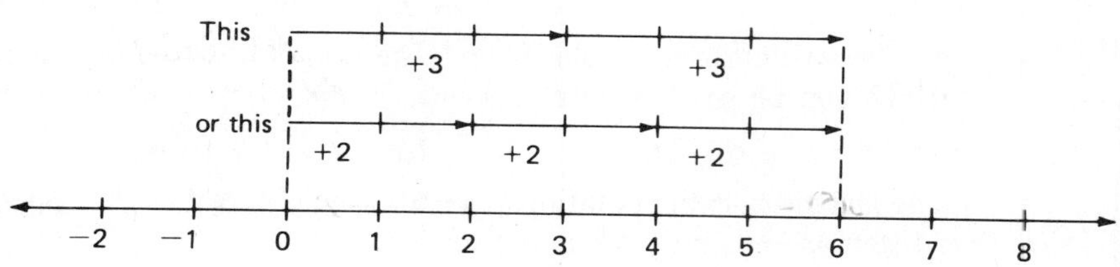

Figure 2.4

Now let's find the product of 2 and -3 on the number line. We can obtain the same answer by adding two -3's or computing the algebraic sum of $-3 - 3 = ?$

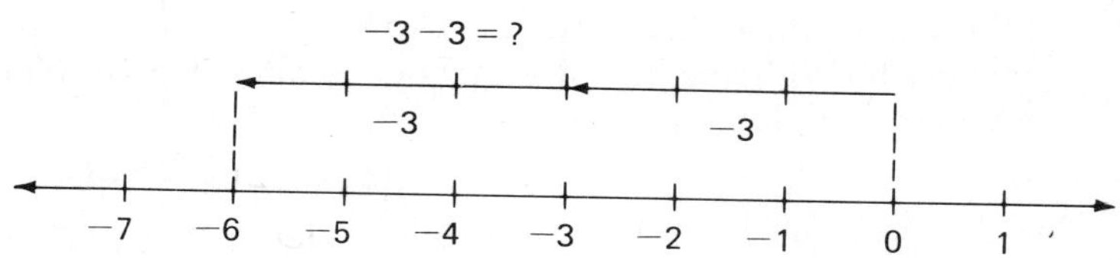

Figure 2.5

Thus from Figure 2.5 we see that

$$-3 - 3 = -6 \quad \text{so} \quad 2 \text{ times } -3 = -6$$

But now if we attempt to use the number line to show -3 times 2 by trying to draw -3 arrows that are $+2$ units long, we find that the task is impossible because we don't know how to draw -3 arrows, for any number of arrows we draw will be a number equal to or greater than one.

The number line was a useful graphic aid in understanding the concept of signed numbers and the algebraic addition of signed numbers but is of less help when discussing the multiplication of signed numbers (and also the division of signed numbers), so we will not try to use it further for this purpose.

We could now just give the rules for multiplication and be done with it, but let's try to get some feeling for why the rules are as they are.

We need three rules for multiplication:

1. A rule for use when both factors are positive.
2. A rule for use when one factor is negative and the other factor is positive.
3. A rule for use when both factors are negative.

2.5 multiplication with both factors positive

We developed the reasoning behind this rule for two positive factors in section 2.4 and in Figure 2.4.

The product of two positive real numbers is a positive real number whose absolute value is the product of the absolute value of the numbers.

EXAMPLES

(a) $(+4)(+5) = 20$ (b) $4(3) = 12$ (c) $2 \cdot 9 = 18$

2.6 multiplication with one negative factor and one positive factor

Let's use the example of Figure 2.5. We want to multiply $+2$ by -3 and find the magnitude and sign of the product. Now 3 times 2 equals 6.

$$3(2) = 6$$

So it seems reasonable that the product of the opposite of 3 times 2 would be the opposite of 6, or -6. It is!

$$(-3)(2) = -6 \qquad \text{and also} \qquad (3)(-2) = -6$$

The product of two signed real numbers of opposite sign is a negative real number whose absolute value is the product of the absolute values of the numbers.

EXAMPLES

(a) $-3(5) = -15$ (b) $(4)(-2) = -8$ (c) $(-2)(6) = -12$

2.7 multiplication with both factors negative

Three times 2 equals 6

$$3 \cdot 2 = 6$$

and in the last paragraph we said that 3 times the opposite of 2 equals the opposite of 6.

$$3(-2) = -6$$

Now it seems reasonable to guess that the product of the opposite of 3 times the opposite of 2 would be the opposite of the opposite of 6 which is 6 itself. It is!

$$(-3)(-2) = +6$$

The product of two negative real numbers is a positive real number whose absolute value is the product of the absolute values of the two numbers.

EXAMPLES

(a) $-3(-2) = 6$ (b) $(-5)(-4) = 20$ (c) $-5(-3) = 15$

TEST YOUR UNDERSTANDING

Simplify:

(b) $(-4)(2)$ (c) $(-4)(-5)$ (d) $(4)(-3)$

-8 20 -12

2.8 inverse operations

If one operation will *undo* another operation, the two operations are called inverse operations. If we take a particular number and then add and subtract the same number, the result is the particular number itself. For example, if we begin with the number 7 and add 3 and then subtract 3, the result is 7.

$$7 + 3 - 3 = 7^{\dagger}$$

Thus addition and subtraction are inverse operations.

Multiplication and division are also inverse operations. If we multiply 7 by 2 and then divide by 2, the result is 7.

$$\frac{7 \cdot 2}{2} = 7$$

There has been no change since dividing by 2 *undoes* the effect of multiplying by 2.

Since multiplication and division are inverse operations, our rules for the multiplication and division of signed numbers must be stated in such a way that these operations are inverse operations.

2.9 division of one positive number by another positive number

In section 2.5 we said that the product of two positive real numbers is a positive real number. For example, 4 times 3 equals 12.

$$4 \cdot 3 = 12$$

Now division must be defined in such a way that it will *undo* the multiplication, so

$$\frac{12}{3} = 4 \quad \text{and} \quad \frac{12}{4} = 3$$

If one positive real number is divided by another positive real number, the quotient is a positive real number whose absolute value is the quotient of the absolute values of the original numbers.

EXAMPLES

(a) $\frac{+12}{+2} = +6$ (b) $\frac{24}{8} = 3$ (c) $\frac{18}{3} = 6$

2.10 division of negative numbers

In section 2.6 we said that the product of two real numbers with opposite signs is a negative real number. Therefore, we must define the division of two numbers of opposite signs so that the division will be an inverse operation. Hence 2 times the opposite of 3 is the opposite of 6.

$$2(-3) = -6$$

Division must be defined in such a way that it will *undo* the multiplication.

$$\frac{-6}{-3} = 2 \quad \text{and} \quad \frac{-6}{2} = -3$$

Thus we see that we will have to define division so that the quotient of two negative real numbers is a positive real number and that the quotient of a negative real number divided by a positive real number is a negative real number.

If one negative real number is divided by another negative real number, the quotient is a positive real number whose absolute value is the quotient of the absolute values of the original numbers.

---

†Now we think of -3 as adding a negative 3 using the process of algebraic addition.

If a negative real number is divided by a positive real number, the quotient is a negative real number whose absolute value is the quotient of the absolute values of the original numbers.

We need to cover one more possibility. In section 2.7 we found that the product of the opposite of 2 times the opposite of 3 equals 6.

$$(-2)(-3) = 6$$

Thus, since division is the inverse operation of multiplication and must *undo* the multiplication, the following must be true.

$$\frac{6}{-2} = -3 \quad \text{and also} \quad \frac{6}{-3} = -2$$

This gives us the last case. The quotient of a positive real number divided by a negative real number is a negative real number.

If a positive real number is divided by a negative real number, the quotient is a negative real number whose absolute value is the quotient of the absolute values of the original numbers.

The following examples demonstrate all the rules for the division of signed numbers.

(a) $\frac{-12}{4} = -3$ (b) $\frac{-10}{-2} = 5$ (c) $\frac{20}{-2} = -10$ (d) $\frac{20}{2} = +10$

TEST YOUR UNDERSTANDING

Simplify:

(e) $\frac{-25}{5}$ (f) $\frac{-20}{-2}$ (g) $\frac{100}{-5}$

2.11 rules for the division and multiplication of signed real numbers

We can consolidate all we have learned about the multiplication and division of signed real numbers into two rules.

like signs

The product or the quotient of two real signed numbers that have the same sign is a positive real number whose absolute value is the absolute value of the product or the quotient of the absolute value of the original numbers.

unlike signs

The product or the quotient of two real signed numbers that have opposite signs is a negative real number whose absolute value is the absolute value of the product or the quotient of the absolute value of the original numbers.

We can state the rules above in a less rigorous but more easily remembered way if we say

> IN BOTH MULTIPLICATION AND DIVISION
>
> 1. *Like* signs $\xrightarrow{\text{yield}}$ a positive number
>
> 2. *Unlike* signs $\xrightarrow{\text{yield}}$ a negative number

2.12 division by zero

The operation of division is the inverse operation of the operation of multiplication, for division is defined as the process that will *undo* multiplication. Thus if $3 \times 2 = 6$, it is necessary that $\frac{6}{2} = 3$ and $\frac{6}{3} = 2$. We will now use the same thought process to try to decide what the result will be if we divide a non-zero number by zero. We will use the example of 6 divided by 0.

$$? = \frac{6}{0}$$

Since we say that division *undoes* multiplication, the multiplication that is to be undone by the above division must be

$$? \cdot 0 = 6$$

But the product of zero and any real number is zero—it is not 6. There is no number that we can substitute for ? so that the product of ? and 0 equals 6. Therefore, we say that since the multiplication does not exist that is to be undone by the division, the expression $\frac{6}{0}$ has no meaning or is undefined. A similar reasoning process is used to show that we can't divide zero by zero, and we say that zero divided by zero is indeterminate rather than saying it is undefined. Thus indicated divisions such as

$$\frac{0}{0} \qquad \frac{142}{0} \qquad \frac{6}{0} \qquad \frac{-5}{0}$$

have no meaningful simplifications and must be avoided.

2.13 commutative property of multiplication

In section 1.8 we stated the commutative property of addition for real numbers as

> COMMUTATIVE PROPERTY OF ADDITION
>
> For any two real numbers a and b,
>
> $$a + b = b + a$$

For example, $4 + 3 = 7$ and $3 + 4 = 7$

The statement for the commutative property of multiplication of real numbers is very similar.

> COMMUTATIVE PROPERTY OF MULTIPLICATION
>
> For any two real numbers a and b,
>
> $$ab = ba$$

For example, $4 \cdot 3 = 12$ and $3 \cdot 4 = 12$

The commutative property is a formal statement of the observation that the order of the factors does not affect the product of two real numbers.

2.14 order of multiplication

In section 1.9 we noted that algebraic addition is a binary operation in that only two numbers can be added at one time.

$$4 + 3 + 2$$

It is impossible to add these three numbers in one step or in one thought. We must add any two of the numbers and to this sum add the third number.

Multiplication is also a binary operation and the same restrictions apply

$$4 \cdot 3 \cdot 2$$

The product of these numbers cannot be determined in one step.

(a) $\underbrace{4 \cdot 3} \cdot 2$ → $12 \cdot 2$ → 24

(b) $4 \cdot \underbrace{3 \cdot 2}$ → $4 \cdot 6$ → 24

(c) $\underbrace{2 \cdot 4} \cdot 3$ → $8 \cdot 3$ → 24

In (a) we multiplied 4 by 3 and then multiplied by 2. In (b) we multiplied 3 by 2 and then multiplied by 4. In (c) we changed the order of the indicated multiplication, found the product of 2 and 4 and then multiplied by 3. In all three examples we found that the product was 24.

The order in which signed numbers are multiplied does not affect the value of the product!

example 2.14.1 Find the product of $(-4)(3)(-6)(-2)$.

solution

$$(-4)(3)(-6)(-2)$$
$$(-12)(-6)(-2)$$
$$(72)(-2)$$
$$\mathbf{-144}$$

In the first step we multiplied -4 by 3 and found -12. In the second step, we multiplied -12 by -6 and got $+72$, which we multiplied by -2 to get the final result of -144.

example 2.14.2 Find the product of $(-6)(-2)(3)(-4)$.

solution

$(-6)(-2)(3)(-4)$	given
$12(3)(-4)$	multiplied -6 by -2
$36(-4)$	multiplied 12 by 3
$\mathbf{-144}$	multiplied 36 and -4

example 2.14.3 Find the product of $(-6)(-4)(3)(-2)$.

solution

$(-6)(-4)(3)(-2)$	given
$24(3)(-2)$	multiplied -6 by -4
$72(-2)$	multiplied 24 by 3
$\mathbf{-144}$	multiplied 72 by -2

In each of the three problems above the same factors were multiplied, but the order of multiplication was different. The product was -144, however, regardless of the order of the factors.

TEST YOUR UNDERSTANDING

Simplify:

(h) $(-4)(2)(-3)(6)$

(i) $(-2)(4)(-1)(-5)$

2.15 multiplicative inverse or reciprocal

If one fraction is the inverted form of another fraction, each of the fractions is said to be the **RECIPROCAL** of the other fraction.

$$\frac{2}{3} \quad \text{is the reciprocal of} \quad \frac{3}{2}$$

$$\frac{3}{2} \quad \text{is the reciprocal of} \quad \frac{2}{3}$$

$$-\frac{4}{11} \quad \text{is the reciprocal of} \quad -\frac{11}{4}$$

$$-\frac{11}{4} \quad \text{is the reciprocal of} \quad -\frac{4}{11}$$

Since numbers such as 4 can also be written in a form such as $\frac{4}{1}$, these numbers also have a reciprocal.

$$\frac{1}{4} \quad \text{is the reciprocal of} \quad 4$$

$$4 \quad \text{is the reciprocal of} \quad \frac{1}{4}$$

$$-5 \quad \text{is the reciprocal of} \quad -\frac{1}{5}$$

$$-\frac{1}{5} \quad \text{is the reciprocal of} \quad -5$$

The **reciprocal** of a number is often called the **MULTIPLICATIVE INVERSE** of the number.

DEFINITION OF RECIPROCAL OR MULTIPLICATIVE INVERSE

For any non-zero real number a, the reciprocal or multiplicative inverse of the number is $\frac{1}{a}$.

If a number is multiplied by its multiplicative inverse or its reciprocal, the product is the number one. Thus

$$4 \cdot \frac{1}{4} = 1, \qquad -5 \cdot \frac{1}{-5} = 1, \quad \text{and} \quad -\frac{1}{13} \cdot (-13) = 1$$

This simple fact is of great importance and will be very useful in the solution of equations that will be discussed later in the book.

2.16 order of operations

If we wish to compute the value of

$$4 + 3 \cdot 2$$

we have a problem. It appears that there are two possible solutions.

(a) $4 + 3 \cdot 2$

Here we will first add 4 and 3 to get 7

$$7 \cdot 2$$

and multiply to get

$$7 \cdot 2 = \mathbf{14}$$

(b) $4 + 3 \cdot 2$

Here we will first multiply 3 by 2 to get 6

$$4 + 6$$

and then add to get

$$4 + 6 = \mathbf{10}$$

We worked the problem two ways and got two different answers. Neither way was basically more correct than the other, but since there are two possible ways to work the problem, mathematicians have found it necessary to agree that the **multiplication will be done before the addition** so that everyone will find the same result.

To simplify an expression such as

$$4 \cdot 3 + 5 - 6 + 4 - 3 \cdot 5 + 6 - 4 \cdot 2$$

we agree to use a two-step process. First we will perform all the multiplications from *left to right*, and we get

$$12 + 5 - 6 + 4 - 15 + 6 - 8$$

Now we again move from *left to right* and perform the algebraic additions as we come to them.

$$17 - 6 + 4 - 15 + 6 - 8$$
$$11 + 4 - 15 + 6 - 8$$
$$15 - 15 + 6 - 8$$
$$0 + 6 - 8$$
$$\mathbf{-2}$$

example 2.16.1 Simplify $4 \cdot 3 + 2 + (-3)5$.

solution We must perform all multiplications before we add and so we multiply $4 \cdot 3$ and (-3) times 5 as indicated above and get

$$12 + 2 - 15$$

which we now add algebraically.

$$12 + 2 - 15 = \mathbf{-1}$$

example 2.16.2 Simplify $-5(2) - 3 + 6(3)$.

solution We perform the two multiplications first

$$-10 - 3 + 18$$

and now add to get

$$\mathbf{5}$$

example 2.16.3 Simplify $4 \cdot 3 - 2 \cdot 5 + 6 - 5 \cdot 2$.

solution

$$12 - 10 + 6 - 10 \quad \text{performed multiplications}$$
$$\mathbf{-2} \quad \text{added algebraically}$$

2.17 identifying multiplication and addition

When confronted with an expression such as

$$4 - 3(5) - 7(-6) - (4)(-5)$$

the beginner often has difficulty telling whether quantities are to be added or multiplied. There is a simple way to identify indicated multiplication. **If there is no + or − sign between symbols, multiplication is indicated.** Let's simplify the expression above from left to right by first performing the indicated multiplications. The first place where there is no sign between the symbols is between the 3 and the parentheses enclosing the 5. The second place is between the 7 and the parentheses enclosing the −6 and the third is between the parentheses enclosing both the 4 and the −5.

$$4 - 3\underset{\uparrow}{\,}(5) - 7\underset{\uparrow}{\,}(-6) - (4)\underset{\uparrow}{\,}(-5)$$

The places where multiplication is indicated are designated by arrows. If we perform the indicated multiplications, we have

$$4 - 15 - (-42) - (-20)$$

Now we can simplify this expression and add.

$$4 - 15 + 42 + 20 = \mathbf{51}$$

> **Test Your Understanding**
>
> Simplify:
>
> (j) $-6 - 5(-2) + (-3)5$ (k) $2 + (-2)(-5) - 3$

lesson checklist 2

- ☐ Opposite of a number
- ☐ Additive inverse
- ☐ Multiplication as a shorthand for addition
- ☐ Rules for multiplication of signed numbers
- ☐ Inverse operations
- ☐ Rules for division of signed numbers
- ☐ Commutative property of multiplication
- ☐ Binary operations
- ☐ Multiplicative inverse or reciprocal
- ☐ Order of operations
- ☐ Identifying multiplication and addition

ANSWERS TO TEST YOUR UNDERSTANDING

(a) 1	(b) -8	(c) 20	(d) -12	(e) -5	(f) 10
(g) -20	(h) 144	(i) -40	(j) -11	(k) 9	

problem set 2

1. What is another name for the opposite of a number?
2. What is the product of a number and the reciprocal of the same number?
3. What is another name for the reciprocal of a number?
4. What is the difference between a number and a numeral?
5. Designate the set of whole numbers.
6. What is a quotient?
7. What is a factor?
8. What is the additive inverse of -8?
9. What is a binary operation?
10. Use real numbers to demonstrate the commutative property of addition.
11. Use real numbers to demonstrate the commutative property of multiplication.

Simplify:

***12.** $\dfrac{-12}{4}$ ***13.** $\dfrac{-10}{-2}$ ***14.** $\dfrac{20}{-2}$

***15.** $\dfrac{20}{2}$ ***16.** $-3(5)$ ***17.** $(4)(-2)$

***18.** $(-2)(6)$ ***19.** $-3(-2)$ ***20.** $(-5)(-4)$

***21.** $-5(-3)$ ***22.** $(-4)(3)(-6)(-2)$ ***23.** $(-2)(4)(-3)(5)$

***24.** $4 + 3 \cdot 2$ ***25.** $4 \cdot 3 + 5 - 6 + 4 - 3 \cdot 5 + 6 - 4 \cdot 2$

***26.** $4 \cdot 3 + 2 + (-3)5$ **27.** $(-4)(-10) - (-4)$ ***28.** $-5(2) - 3 + 6(3)$

***29.** $4 - 3(5) - 7(-6) - (4)(-5)$ **30.** $-(-5) + (-2)$

31. $-(+5) + 3 - (+2)$ **32.** $-5 - [-(-2)]$ **33.** $-(-3) - 2$

34. $-7 + (-14) - (-3)$ **35.** $-(-3) - (+2) + 5$ **36.** $(-4)(-3)$

37. $-(-3)(-2)$ **38.** $(-2)[-(-3)]$ **39.** $-3 + 2(-6)$

40. $-3 \cdot 2 + (-6)$ **41.** $-15 - (-5)(2)$

42. $-13 - (+10) - 3(-3)$ **43.** $-3 \cdot (-2) + 3 - (-2)$

44. $-5 - (5) - 3(-5)$ **45.** $(-2)(-1) - 2 - 1$ **46.** $-\{-[-(-4)]\}$

47. $-|18 - 4|$ **48.** $|-23 + 2 + 4|$

---

*The problems marked with an asterisk were worked out in this lesson.

lesson three

Symbols of inclusion
More on order of operations
Products of signed numbers

3.1 symbols of inclusion

In Lesson 2 we found that the proper result of simplifying the notation

$$4 + 3 \cdot 2$$

is 10 because we have agreed to do the multiplication first and then do the addition. So,

$$4 + 3 \cdot 2 = 4 + 6 = \mathbf{10}.$$

Parentheses, brackets, braces, and bars are called **SYMBOLS OF INCLUSION**, and can be used to help us emphasize the meaning of our notation. Using these symbols, the notation above could be written in any of the following ways:

(a) $4 + (3 \cdot 2)$ (b) $4 + [3 \cdot 2]$ (c) $4 + \{3 \cdot 2\}$ (d) $4 + \overline{3 \cdot 2}$

Each of the notations emphasizes that 3 is to be multiplied by 2 and that 4 is to be added to this product. A further benefit of the use of symbols of inclusion is that a non-standard order of operations can be indicated. For example, we can use parentheses to indicate that 4 is to be added to 3 and the result multiplied by 2 by writing

$$2(3 + 4) \quad \text{or} \quad (3 + 4)2$$

While bars and braces can be used as indicated above, we normally reserve the use of braces to indicate a set and bars are most often used as fraction lines as we show here.

$$\frac{4 + (3 \cdot 2)}{5(2 - 3)} = \frac{4 + 6}{-5} = \frac{10}{-5} = \mathbf{-2}$$

The parentheses in the numerator are used to emphasize that 4 is to be added to the product of 3 and 2 and the parentheses in the denominator are used to designate that 5 is to be multiplied by the algebraic sum of 2 and -3.

3.2 order of operations

To simplify a problem of the form of $\frac{-5+[7-(3-2)+5]}{4-(2+6)}$, begin in both the numerator and the denominator, and

(a) Simplify the expressions within the symbols of inclusion beginning with the innermost symbol of inclusion.
(b) Perform the indicated multiplications in order from left to right.
(c) Perform the indicated algebraic additions in order from left to right.
(d) Divide the denominator into the numerator or leave the expression in fractional form.

example 3.2.1 Simplify $4(3+2)-5(6-3)$.

solution

(a)	$4(5)-5(3)$	simplified parentheses
(b)	$20-15$	multiplied
(c)	**5**	added algebraically
(d)	not necessary as the denominator is 1	

example 3.2.2 Simplify $-3(2-3+5)-6(4+2)-3$.

solution

(a)	$-3(4)-6(6)-3$	simplified parentheses
(b)	$-12-36-3$	multiplied
(c)	**−51**	added algebraically
(d)	not necessary	

example 3.2.3 Simplify $-5+[7-(3-2)+5]$.

solution

(a)	$-5+[7-(1)+5]$	simplified innermost parentheses
(b)	$-5+11$	simplified bracket
(c)	**6**	added algebraically
(d)	not necessary	

example 3.2.4 Simplify $-5-[-5(-4)]+3\cdot 4$.

solution

(a)	$-5-(20)+3\cdot 4$	simplified inside bracket
(b)	$-5-20+12$	multiplied
(c)	**−13**	added algebraically
(d)	not necessary	

example 3.2.5 Simplify $\frac{5(-2)-(-7)(4)+2}{2-(2-5)-(-5)}$.

solution

(a)	$\frac{5(-2)+7(4)+2}{2+3+5}$	simplified parentheses
(b)	$\frac{-10+28+2}{2+3+5}$	multiplied

(c) $\dfrac{20}{10}$ added algebraically

(d) **2** divided

example 3.2.6 Simplify $\dfrac{-3(4-2)-(-5)}{4-(3)(-3)}$.

solution

(a) $\dfrac{-3(2)+5}{4-(3)(-3)}$ simplified parentheses

(b) $\dfrac{-6+5}{4+9}$ multiplied

(c) $\dfrac{-1}{13}$ added algebraically

(d) $-\dfrac{1}{\mathbf{13}}$ this doesn't divide evenly and we will leave it in fractional form

example 3.2.7 Simplify $\dfrac{-4-(-3)+7-6(2)}{7-(3)(-2)}$.

solution

$$\frac{-4+3+7-12}{7-(-6)} = \frac{-4+3+7-12}{7+6} = -\frac{6}{\mathbf{13}}$$

Test Your Understanding

Simplify:

(a) $\dfrac{-6[-8+(-3)]}{-2(4-15)}$ (b) $\dfrac{4[6-(-5)(-2)]}{-3(-3)}$

3.3 more on order of operations

In the discussion of the order of operations in section 2.16 we said that mathematicians have agreed that when they write

$$4 \cdot 3 + 2$$

the **multiplication should be done first** and then the addition.

$4 \cdot 3 + 2$

$12 + 2$ multiplied

14 added

We did not discuss division because if symbols of inclusion are properly used, the order in which division is to be performed is apparent.

If we write

$$\frac{4 \cdot 3 + 2}{-7 + 5}$$

we find the value of the numerator and the value of the denominator and then divide.

$$\frac{14}{-2} = -7$$

If the following problem is encountered, however,

$$4 + \frac{14}{2} - 3 \cdot 6$$

the notation clearly indicates that only 14 is to be divided by 2, and if we do this first, we get

$$4 + 7 - 3 \cdot 6$$

Now we do the multiplication and conclude with algebraic addition.

$$4 + 7 - 18$$

$$11 - 18$$

$$\mathbf{-7}$$

At this point in an algebra book, however, it is customary to give a rule for finding the number represented by

$$6 + 3 \cdot 6 \div 2 - 6 \cdot 2$$

The rule is to perform the operations from left to right in the following order.

1. Multiplication and division
2. Algebraic addition

First we will go through the problem from **left to right** performing the multiplications and divisions **in the order in which they are encountered**.

$6 + 3 \cdot 6 \div 2 - 6 \cdot 2$	original problem
$6 + 18 \div 2 - 6 \cdot 2$	multiplied 3 times 6
$6 + 9 - 6 \cdot 2$	divided 18 by 2
$6 + 9 - 12$	multiplied 6 times 2

Now we go through the problem again from left to right performing the algebraic additions as they are encountered.

$6 + 9 - 12$	
$15 - 12$	added 6 and 9
$\mathbf{3}$	added 15 and -12

If symbols of inclusion had been properly used, instead of stating the problem as

$$6 + 3 \cdot 6 \div 2 - 6 \cdot 2$$

the problem would have been written thusly,

$$6 + \frac{(3 \cdot 6)}{2} - (6 \cdot 2)$$

and here the method of solution is clearly indicated. We simplify within the parentheses as the first step.

$$6 + \frac{18}{2} - 12 \qquad \text{simplified within parentheses}$$

$$6 + 9 - 12 \quad \text{divided 18 by 2}$$
$$15 - 12 \quad \text{added 6 and 9}$$
$$\mathbf{3} \quad \text{added 15 and } -12$$

We will use symbols of inclusion to include the use of a bar as a fraction line when stating problems. Thus problems such as the one just discussed will not be encountered again in this book.

3.4 products of signed numbers

Let's review the concept of the opposite of a number by watching the pattern that develops here:

	READ AS	WHICH IS
2	2	2
-2	the opposite of 2	-2
$-(-2)$	the opposite of the opposite of 2	2
$-[-(-2)]$	opposite of opposite of opposite of 2	-2
$-\{-[-(-2)]\}$	opposite of opposite of opposite of opposite of 2	2

The expressions in the left-hand column are all equivalent expressions for 2 or for -2. If we look at the right-hand column, we see that every time an additional $(-)$ is included in the left-hand expression, the right-hand expression changes sign.

A similar alternation in sign occurs whenever a particular number is multiplied by a negative number. For instance,

$$-2 = -2$$
$$(-2)(-2) = +4$$
$$(-2)(-2)(-2) = -8$$
$$(-2)(-2)(-2)(-2) = +16$$
$$(-2)(-2)(-2)(-2)(-2) = -32$$

The numbers on the right have different absolute values, but they ALTERNATE IN SIGN. We note that

The product of **TWO** negative factors is **POSITIVE.**

The product of **THREE** negative factors is **NEGATIVE.**

The product of **FOUR** negative factors is **POSITIVE.**

The product of **FIVE** negative factors is **NEGATIVE.**

Without proof we will generalize these observations.

1. **The product of an even number of negative real numbers is a positive real number.**
2. **The product of an odd number of negative real numbers is a negative real number.**

We can use these observations to determine the sign of the product of several signed numbers. Let's consider

$$(4)(-3)(-4)(-2)(11) = ?$$

Here we have +4 and +11 as two of the five factors. Since multiplication by a positive number does not affect the sign of the product, we will not consider these numbers. The other three factors are negative. We can look at the rules we developed above and see that the sign of the product of three negative numbers is negative. Thus our answer can be expressed as

$$-(4 \cdot 3 \cdot 4 \cdot 2 \cdot 11) = \mathbf{-1056}$$

example 3.4.1 Determine the sign of the following products and give the reason. Do *not* do the multiplication.

solution

	SIGN	REASON
(a) $(-4)(-3)(2)(+5)(+6)$	positive	Even number of negative factors
(b) $(3)(+2)(6)$	positive	No negative factors
(c) $(-3)(-2)(6)(4)(-2)$	negative	Odd number of negative factors
(d) $(-3)(-2)(-5)(-7)(-2)$	negative	Odd number of negative factors
(e) $(-3)(-4)(-2) + 2(-3)$	?	Rule does not apply as this is an indicated *sum.* We will do the problem in three steps.
(f) $(-3)(-4)(-2)$	negative	Odd number of negative factors
(g) $2(-3)$	negative	Odd number of negative factors
(h) $-(3)(4)(2) - (2)(3)$	negative	Algebraic sum of two negative numbers is a negative number.

TEST YOUR UNDERSTANDING

Tell whether the *sign* of the following products is positive or negative. Do not do the multiplication.

(c) $(-3)(-4)(-7)(+5)(3)(2)(-4)(-4)(-27)(-815)$

(d) $(-3)(4)(176)(93)(+45)(-81)(-76)(-342)$

lesson checklist 3

- ☐ Symbols of inclusion
- ☐ Order of operations with parentheses
- ☐ Sign of the product of negative numbers

ANSWERS TO TEST YOUR UNDERSTANDING

(a) 3 (b) $\frac{-16}{9}$ (c) negative (d) positive

problem set 3

1. Is the product of 33 negative numbers and 2 positive numbers a positive number or a negative number?
2. Is the product of 33 negative numbers and 3 positive numbers a positive number or a negative number?
3. What is another name for the multiplicative inverse of a number?
4. Designate the set of integers.
5. Are addition, multiplication, and division all binary operations?
6. Are integers real numbers?
7. What is the multiplicative inverse of $-\frac{2}{3}$? of -4?

Simplify:

8. $4 + 2 - 5 - 4 + 3$
9. $5 - 3 + 2 - 5 - 7$
10. $-4 \cdot 3 + 5 - 6 + 2 \cdot 3$
11. $+3 + 2 \cdot 5 - 7 + 8$

*12. $4(3 + 2) - 5(6 - 3)$

*13. $-3(2 - 3 + 5) - 6(4 + 2) - 3$

*14. $-5 + [7 - (3 - 2) + 5]$

*15. $-5 - [-5(-4)] + 3 \cdot 4$

*16. $\dfrac{5(-2) - (-7)(4) + 2}{2 - (2 - 5) - (-5)}$

*17. $\dfrac{-3(4 - 2) - (-5)}{4 - (3)(-3)}$

*18. $\dfrac{-4 - (-3) + 7 - 6(2)}{7 - (3)(-2)}$

19. $-|4 \cdot 3 + 5|$
20. $|4(-3) + 5|$
21. $-[-(-4)]$
22. $-\{-[-(-15)]\}$
23. $4 \cdot (3 + 5) - 8$
24. $-8(-2)(-1) + 5 \cdot 3$
25. $-4[6 + (5 - 3)2] + 6$
26. $-2(4 - 2) + 5$
27. $-8[7 + 3(2 - 5) + 7]$
28. $\dfrac{-8}{2}$
29. $\dfrac{-14}{-2}$
30. $-5[-4(3 - 2)]$
31. $|-13|$
32. $-|-42 + 10|$
33. $-5(-6 - 2)$
34. $5(-6) - (-5 - 6)$
35. $-5[-4(-3 + 8)]$
36. $-4[(-5 - 3)(-6)] + 11$
37. $-3\{5[7 - 4(-3)] + 2\}$
38. $2\{-3[5 - 7(2 - 7)] + 6\}$
39. $-\{-[-5(-3 + 2)7]\}$
40. $-[-4(|-5| - 8) - 2]$
41. $3[11(-7 - 4)2] + 5$
42. $-7\left\{\dfrac{-14}{-2}[-7 + (-3)]\right\}$
43. $-3[-2(6|-5|)(-3)]$
44. $2 + [5 - (-3 - 6)]$

*The problems marked with an asterisk were worked out in this lesson.

lesson four

Variables and algebraic expressions
Evaluation of algebraic expressions

4.1 algebraic expressions

In section 1.1, we said that a **number** is an **idea** and that when we wish to write down something to represent this **idea**, we use a **numeral**. If we wish to bring to mind the number 7, we could write any of the following:

$$7, \qquad \frac{14}{2}, \qquad 4 + 3, \qquad \frac{-21}{-3}, \qquad 2 + 2 + 2 + 1$$

We call each of these notations a **NUMERICAL EXPRESSION** or just a **numeral**. Every numerical expression represents only one number and we call this number the **VALUE** of the expression. Each of the numerical expressions shown above has a **value** of 7.

In algebra we often use letters to represent numbers. When letters as well as numbers are used in an expression, we don't call the expression a **numerical expression** but call it by the more general name of **ALGEBRAIC EXPRESSION OR MATHEMATICAL EXPRESSION**. These words are used to describe expressions that contain only numbers or only letters or contain both numbers and letters.

If we write the algebraic expression

$$4 + x$$

the expression has a **value** that depends on the value that we assign to x. If we give x a value of 5, then the expression has a value of 9 because

$$4 + 5 = 9$$

If we give x a value of 11, then the expression has a value of 15 because

$$4 + 11 = 15$$

Thus the value that is assigned to x in this problem can change or be changed; or we can say that the value that is assigned to x can vary or be varied. For this reason we call the letters **VARIABLES**. We also call the letters **UNKNOWNS** as they represent unknown or unspecified numbers. The numeral 4 in this example does not change value and has a constant value of 4. For this reason the symbol that we use to denote a number is called a **CONSTANT**.

When we use variables in algebraic expressions, the notations that we use to indicate the operations of division and algebraic addition are the same as the notations that we use to indicate the division and algebraic addition of real numbers. The notation for the multiplication of variables is sometimes slightly different. We can denote that we wish to multiply 4 by the variable x by writing any of the following:

(a) $4x$ (b) $4(x)$ (c) $(4)(x)$ (d) $4 \cdot x$ (e) $4 \times x$

The last four notations are the same as the notations that we use for real number multiplication, but the notation shown in (a) is different from the real number notation.

$4x$ indicates that 4 is to be multiplied by the value of x

whereas

45 does not indicate that 4 is to be multiplied by 5 but instead is a numeral that represents the number 45.

Thus the expression xym indicates that the values of x, y, and m are to be multiplied. If we give x a value of 1, y a value of 2, and m a value of 3, the value of the expression xym can be found.

$$1 \cdot 2 \cdot 3 = \mathbf{6}$$

If we write the algebraic expression

$$4x + mx$$

we indicate that 4 is to be multiplied by the value of x and that the value of m is to be multiplied by the value of x and that the two products are to be summed. If we give x a value of 3 and m a value of 5, then we can find the value of the expression.

$$\begin{aligned} &4 \cdot 3 + 5 \cdot 3 \\ &12 + 15 \quad = \mathbf{27} \end{aligned}$$

Thus the value of the expression when x equals 3 and m equals 5 is 27.

If we give x the value of 2 and m the value of 6, then the expression will have a different value.

$$4x + mx = 4 \cdot 2 + 6 \cdot 2 = 8 + 12 = \mathbf{20}$$

In this case the value of the expression is 20. It is of **utmost importance** to note that in the first case when we gave x a value of 3, the value of x everywhere in the expression was 3. When we gave x a value of 2, the value of x everywhere in the expression was 2. **While the values assigned to variables may change or be changed, under any set of conditions the value assigned to a particular variable in an expression is the same value throughout the expression. Also when we begin solving equations and working problems, we must remember that the value assigned to any particular variable under any set of conditions must be the same value regardless of where the particular variable appears in the equation or the problem.**

example 4.1.1 Find the value of xpm if $x = 4$, $p = 2$, $m = 5$.

solution $xmp = 4 \cdot 5 \cdot 2 = 20 \cdot 2 = \mathbf{40}$

example 4.1.2 Evaluate (find the value of) $4yz - 5$ if $y = 2$ and $z = 10$.

solution $4yz - 5 = 4(2)(10) - 5 = 80 - 5 = \mathbf{75}$

Now we will investigate the most difficult problem of this type where the original expression contains minus signs and the variables are assigned the values of both positive and negative numbers.

example 4.1.3 Evaluate $y - z$ if $y = -2$ and $z = -6$.

solution The expression is read as y *minus* z and z has a value of *negative 6*. For this problem only we will use the raised sign of Lesson 1 to designate a negative number.

$$y - z = {}^-2 - {}^-6 = {}^-2 + {}^+6 = {}^+\mathbf{4}$$

We will normally write it as follows:

$$y - z = -2 - (-6)$$

and the $-(-6)$ is read as the opposite of the opposite of 6, which is 6 itself, so

$$y - z = -2 + 6 = \mathbf{4}$$

example 4.1.4 Evaluate $-x - (-a + b)$ if $x = 2$, $a = -4$, and $b = -6$.

solution It is often helpful to the beginner if he replaces each variable with parentheses and inserts the proper value for the unknown in the parentheses.

$$-(\quad) - [-(\quad) + (\quad)]$$
$$-(2) - [-(-4) + (-6)]$$

The first entry can be read as the opposite of 2 or -2. Inside the bracket we have $-(-4)$, read the opposite of the opposite of 4, which is 4 itself. The last entry is $+(-6)$, read plus the opposite of 6, which is the same as -6. Thus we have

$$-2 - (4 - 6) = -2 - (-2) = -2 + 2 = \mathbf{0}$$

example 4.1.5 Evaluate $-(-p + a) + [-a(p - a)]$ if $p = -2$ and $a = -4$.

solution We will use parentheses for each variable as we did in the last problem.

$$-[-(\quad) + (\quad)] + \{-(\quad)[(\quad) - (\quad)]\}$$

Now we will insert the proper value for the variables inside the parentheses.

$$-[-(-2) + (-4)] + \{-(-4)[(-2) - (-4)]\}$$
$$= -(2 - 4) + [4(-2 + 4)]$$
$$= -(-2) + [4(2)] = 2 + (8) = \mathbf{10}$$

example 4.1.6 Evaluate $x - y(-a + x)$ if $x = -2$, $y = +3$, and $a = -4$.

solution

$$(\quad) - (\quad)[-(\quad) + (\quad)]$$
$$(-2) - (+3)[-(-4) + (-2)]$$
$$= -2 - (3)(4 - 2) = -2 - (3)(2) = -2 - 6 = \mathbf{-8}$$

example 4.1.7 Evaluate $-(m + x)(-a + mx)$ if $m = 2$, $x = -3$, $a = -4$.

solution

$$-[(\quad) + (\quad)][-(\quad) + (\quad)(\quad)]$$
$$-[(2) + (-3)][-(-4) + (2)(-3)]$$
$$= -(2 - 3)[4 + (-6)] = -(-1)(4 - 6) = -(-1)(-2)$$
$$= -(2) = \mathbf{-2}$$

example 4.1.8 Evaluate $-4x(a - b) - xa$ if $x = -2$, $a = -4$, $b = 6$.

solution

$$\begin{aligned} &-4(\quad)[(\quad) - (\quad)] - (\quad)(\quad) \\ &= -4(-2)[(-4) - (6)] - (-2)(-4) \\ &= -4(-2)(-4 - 6) - (8) = -4(-2)(-10) - 8 \\ &= -4(20) - 8 = -80 - 8 = \mathbf{-88} \end{aligned}$$

example 4.1.9 Evaluate $-xa(x - a) + a$ if $a = -2$ and $x = 4$.

solution

$$\begin{aligned} &-(\quad)(\quad)[(\quad) - (\quad)] + (\quad) \\ &= -(4)(-2)[(4) - (-2)] + (-2) \\ &= -(-8)(4 + 2) - 2 = 8(6) - 2 = 48 - 2 = \mathbf{46} \end{aligned}$$

TEST YOUR UNDERSTANDING

Evaluate the following if $x = -2$, $y = +3$, and $a = -4$:

(a) $-2x(ay - x)$ (b) $[-ax(-3y + a)]$

lesson checklist 4

- ☐ Numerical expressions
- ☐ Variables or unknowns
- ☐ Algebraic expressions
- ☐ Value of an algebraic expression
- ☐ Evaluating algebraic expressions

ANSWERS TO TEST YOUR UNDERSTANDING

(a) -40 (b) 104

problem set 4

1. What is a variable?
2. Designate the set of whole numbers.
3. Define the word quotient.
4. Define the word factor.
5. Use the numbers 15 and 0 to demonstrate the two unique properties of the number zero.

6. Use the numbers 7 and 1 to demonstrate the unique property of the number one.
7. Define a binary operation.
8. Use numbers to demonstrate the commutative property of addition.
9. Use numbers to demonstrate the commutative property of multiplication.

Evaluate:

*10. $-x - (-a + b)$ if $x = 2$, $a = -4$, and $b = -6$

*11. $-(-p + a) + [-a(p - a)]$ if $p = -2$ and $a = -4$

*12. $x - y(-a + x)$ if $x = -2$, $y = +3$, and $a = -4$

*13. $-[m + x][-a + mx]$ if $m = 2$, $x = -3$, $a = -4$

*14. $-4x[a - b] - xa$ if $x = -2$, $a = -4$, $b = 6$

*15. $-xa[x - a] + a$ if $a = -2$ and $x = 4$

Simplify:

16. $-5 - [8 - (4 - 2) + 14]$
17. $-2\{-6[-4(-5 + 2)] + 3\}$
18. $\{6[-2(-5 - 3)]\} - 5$
19. $-8[4 - 3 \cdot 2(5 - 7)]$
20. $-14 - 11(-2) + 4 - 3 \cdot 6$
21. $-5[2 + 3(2 - 5) - 6]$
22. $-3\{7[-(-3 - 2)4] + 2\}$
23. $-8 \cdot 4(-3) + (-11) - (-5)$
24. $3 - \{-[-(-7)]\}$
25. $4[7 - (-4) - (-5)]$
26. $|4 - 3 - 11|$
27. $-|4 + 3|$
28. $\dfrac{-5(-3 - 6)}{-4 + (-7)}$
29. $\dfrac{-5[-8 - 3(6 - 4)]}{-2(5 - 8) - (-8)[6 - (-3)]}$
30. $\dfrac{5(-4 - 3) - 6}{-8 - 3 - (-7)}$
31. $\dfrac{-11(-2) - (-3) - 6}{4 - (-5 + 3) - 7}$
32. $-8 - (-3 + 7) - 4(-5 + 2)$
33. $-3 - (-2)(-5 - 3 - 2) - 6$
34. $4 - 3(-2) - 2 + (-3 - 5)$
35. $-2\{-6[-8(-4)]\}$
36. $2 - \{-3 + [-2(4 - 6)] - 7\} + 2$
37. $\{4 + [5(-2 - 8)] - 3\} + 6$
38. $-3(-5 - 6) + (-7 + 11)$
39. $-3 - 2(-4)(-6) + 8 - 3$
40. $-3[-2(-1 - 7)3] + 5$
41. $-3(-2)6 + (-3)(-2 - 8)(-4)$
42. $4 + 3(-6)[-6(2 - 3) - 5]$

---

*The problems marked with an asterisk were worked out in this lesson.

lesson five

Terms • factors and coefficients
The distributive property
Addition of like terms
Product theorem for exponents

5.1 terms

A **TERM** is an algebraic expression that

1. Consists of a single variable or constant.
2. Is the indicated product or quotient of variables and/or constants.
3. Is the indicated product or quotient of expressions that contain variables and/or constants.

$$4, \qquad x, \qquad 4x, \qquad \frac{4xy(a+b)}{p}, \qquad \frac{3x+2y}{m}$$

All the expressions above can be called **terms**. The first two are terms because they consist of a single symbol. The third is a term because it is an indicated product of symbols. The fourth and fifth are terms because they are considered to be indicated quotients although the numerator of the fourth term is an indicated product and the numerator of the fifth term is an indicated sum. **A term is thought of as a single entity that represents or has the value of one particular number**. The word term is very useful in allowing us to identify or talk about the parts of a larger expression. For instance, the expression

$$x + 4xym - \frac{6p}{y+2} - 8$$

is an expression that has four terms. We can speak of a particular term of this expression, say the third term, without having to write out the term in question. The terms of an expression are numbered from left to right beginning with the number 1. Thus for the expression above:

The first term is $+x$.

The second term is $+4xym$.

The third term is $-\dfrac{6p}{y+2}$.

The fourth term is -8.

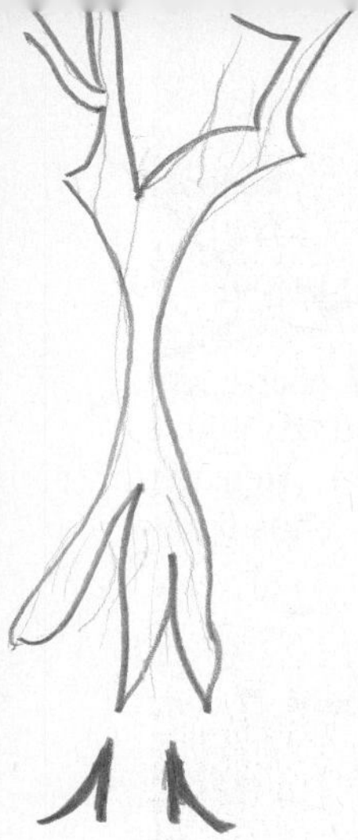

Note that the concept of signed numbers carries over to our discussion of terms. We do not think of the plus and minus signs in the expression as indicating addition and subtraction, but instead we use the thought of algebraic addition and think of the signs as designating the terms as positive or negative. But we must be careful. Let's look at the third term.

$$-\frac{6p}{y+2}$$

If p and y are given values such that $\frac{6p}{y+2}$ is a negative number, then $-\frac{6p}{y+2}$ will be positive. For example if p is equal to -4 and y is equal to 1, then

$$-\frac{6p}{y+2} = -\frac{6(-4)}{1+2} = -(-8) = +8$$

TEST YOUR UNDERSTANDING

(a) Evaluate the third term in the following expression if $x = 4$, $y = -3$, and $a = 2$.

$$x^2ay - 36xy - \frac{3ay - x}{4ax} - 6x + \frac{axy - ya}{4a - y}$$

5.2 factors and coefficients

If the form in which variables and constants are written in an expression indicates that the variables and constants are to be multiplied, we say that the expression is an **INDICATED PRODUCT**. If we write

$$4xy$$

we indicate that 4 is to be multiplied by the product of x and y. Each of the symbols is said to be a factor of the expression. Any one factor of an expression or product of factors of an expression can also be called the **COEFFICIENT** of the rest of the expression. Thus in the expression $4xy$ we can say that

(a) 4 is the coefficient of xy $\quad 4(xy)$
(b) x is the coefficient of $4y$ $\quad x(4y)$
(c) y is the coefficient of $4x$ $\quad y(4x)$
(d) xy is the coefficient of 4 $\quad xy(4)$
(e) $4y$ is the coefficient of x $\quad 4y(x)$
(f) $4x$ is the coefficient of y $\quad 4x(y)$

As we mentioned earlier, the value of a product is not affected by the order in which the multiplication is performed so we may arrange the factors in any order without affecting the value of the expression. You note that we change the order at will in (b) through (f) above.

If the coefficient is a number [as in (a) above], we call it a **numerical coefficient**, and if the coefficient consists entirely of variables or letters as in (b), (c), and (d) above, we call it a **literal coefficient**. We need to speak of numerical coefficients so often that we usually drop the adjective *numerical* and use the single word coefficient. Thus in the following expressions

$$4xy, \qquad -15pq, \qquad 81xmz$$

4 is the coefficient of xy, -15 is the coefficient of pq, and 81 is the coefficient of xmz.

5.3 the distributive property

The properties of real numbers that we have discussed thus far are listed here and two examples are shown, one using real numbers and one using variables for which any real numbers may be substituted.

1. Additive property of zero (section 1.7)
$$4 + 0 = 4 \qquad a + 0 = a$$
2. Multiplicative property of one (section 1.7)
$$4 \cdot 1 = 4 \qquad a \cdot 1 = a$$
3. Multiplicative property of zero (section 1.7)
$$4 \cdot 0 = 0 \qquad a \cdot 0 = 0$$
4. Commutative property of addition (section 1.9)
$$4 + 3 = 3 + 4 \qquad a + b = b + a$$
5. Commutative property of multiplication (section 2.13)
$$4 \cdot 3 = 3 \cdot 4 \qquad ab = ba$$

Now we will discuss another property of real numbers that is of considerable importance, the **DISTRIBUTIVE PROPERTY OF REAL NUMBERS**. If we write

$$4(5 - 3)$$

we indicate that we are to multiply 4 by the algebraic sum of the numbers 5 and -3. A property (or peculiarity) of real numbers permits the value of this product to be found two different ways.

$4(5 - 3)$	$4(5 - 3)$
$4(2)$	$4 \cdot 5 + 4(-3)$
8	$20 - 12$
	8

On the left we first added 5 and -3 to get 2 and then multiplied by 4 to get 8. On the right we first multiplied 4 by both 5 and -3 and then added the products 20 and -12 to get 8. Both methods of simplifying the expression led to the same result and **we call this property or peculiarity of real numbers the distributive property because we get the same result if we distribute the multiplication over the algebraic addition.**†

---

†Please note that while multiplication can be distributed over addition, the reverse is not true, for addition cannot be distributed over multiplication. For example,

$$2 + (3 \cdot 5) \neq (2 + 3) \cdot (2 + 5)$$

because
$$17 \neq 35$$

DISTRIBUTIVE PROPERTY

For any real numbers a, b, and c,

$$a(b + c) = ab + ac$$

It is possible to extend the distributive property so that the extension is applicable to the indicated product of a number or variable and the algebraic sum of any number of real numbers or variables.

EXTENSION OF THE DISTRIBUTIVE PROPERTY

For any real numbers $a, b, c, d, \cdots$

$$a(b + c + d + \cdots) = ab + ac + ad + \cdots$$

example 5.3.1 Use the distributive property to find the value of $4(6 - 2 + 5 - 7)$.

solution We begin by multiplying 4 by each of the terms within the parentheses and then we add the resultant products.

$$4(6 - 2 + 5 - 7) = 4(6) + 4(-2) + 4(5) + 4(-7)$$
$$= 24 - 8 + 20 - 28 = \mathbf{8}$$

example 5.3.2 Use the distributive property to expand $mn(x + y + 2p)$.

solution $mn(x + y + 2p) = \boldsymbol{mnx + mny + 2mnp}$

TEST YOUR UNDERSTANDING

Use the distributive property to evaluate (b) and to expand (c).

(b) $4(-5 + 3 - 8)$ (c) $4x(mn - 3)$

5.4 like terms

LIKE TERMS are terms that have the **same variables in the same form or in equivalent forms** so that the terms (excluding numerical coefficients) represent the same number regardless of the non-zero values assigned to the variables. Let's look at the indicated sum of terms

$$4xmp - 2pmx + 6mxp$$

Now whether terms are **like terms** or not does not depend on the sign of the terms or on the value of the numerical coefficients. So we won't consider the + and − signs or the numbers 4, 2, and 6. We just need to know if

$$xmp, \quad pmx, \quad \text{and} \quad mxp$$

are in the same form or equivalent forms and if each term (disregarding numerical coefficients) represents the same number regardless of the non-zero values that are assigned to the variables.

1. They are in equivalent forms, for they have the same variables in the form of an indicated product, and the order of multiplication of the factors does not affect the value of the product.
2. They do represent the same number regardless of the values assigned to the variables. We will not attempt to prove this but will demonstrate this with one set of values for the variables. If we let $x = 4$, $p = 6$, and $m = 2$, we have

xpm	pmx	mxp
$4 \cdot 6 \cdot 2$	$6 \cdot 2 \cdot 4$	$2 \cdot 4 \cdot 6$
48	**48**	**48**

Thus $4xmp$, $2pmx$, and $6mpx$ are like terms because the variables represent the same number regardless of the real number replacements used for the variables.

TEST YOUR UNDERSTANDING

(d) Which of the following are like terms?

$$-75xyp + 131qpx - 15xpa - 32xqp - 142pxm$$

5.5 addition of like terms

The extension of the distributive property of section 5.3 is restated as

$$a(b + c + d + \cdots) = ab + ac + ad + \cdots$$

The reverse of this property is also true.

$$ab + ac + ad + \cdots = a(b + c + d + \cdots)$$

We know that the multiplication of numbers or variables can be performed in any order without changing the product. This permits us to rearrange the order of the multiplications as we show here.

$$ba + ca + da + \cdots = (b + c + d + \cdots)a$$

We note that a is a common factor of each of the terms on the left and is written outside the parentheses on the right. If we look at the indicated sum of terms

$$4xpm - 2pmx + 6mxp$$

we see that the factor xpm is a COMMON FACTOR of all three terms and can be treated in the same manner as the a factor of the left side of the statement of the distributive property in the rearranged form.

Thus we can write the sum of three terms as a product of $(4 - 2 + 6)$ and pmx as we show here.

$$4xpm - 2pmx + 6mxp = (4 - 2 + 6)pmx = \mathbf{8pmx}$$

The factors of the three variables in the expression $8pmx$ can be written in any order without changing the *value* of the expression. Thus any of the following would be equally acceptable.

$$8pmx, \quad 8pxm, \quad 8xmp, \quad 8xpm, \quad 8mpx, \quad 8mxp$$

The above is a rather detailed approach to justify the following statement:

To add like terms we algebraically add the numerical coefficients.

Thus to add

$$4xpm - 2pmx + 6mxp$$

we simply add the numerical coefficients 4, −2, and +6 to get

$$4xpm - 2pmx + 6mxp = 8pxm$$

example 5.5.1 Simplify the following expression by adding like terms.

$$x - 3xy + yx + mx$$

solution There are only two terms here that are like terms, the second and the third. Since only like terms may be added, the only simplification possible is the algebraic addition of the second and third terms. So

$$x - 3xy + yx + mx = \mathbf{x - 2xy + mx}$$

example 5.5.2 Simplify by adding like terms.

$$xmy + 3ypx - xpy - 5mxy$$

solution The first and fourth terms are like terms and may be added. Also the second and third terms are like terms and may be added.

$$xmy + 3ypx - xpy - 5mxy = \mathbf{-4mxy + 2pxy}$$

example 5.5.3 Simplify $x + 3y - 6xy + 2x - 4 - y + 15 + xy$.

solution The value of the algebraic sum of a number of terms is not changed by changing the order of the indicated algebraic addition. It sometimes helps a little to rearrange the order of the terms first so that like terms are adjacent.

$$x + 2x + 3y - y - 6xy + xy - 4 + 15$$

Now we add like terms to complete the process of simplification.

$$\mathbf{3x + 2y - 5xy + 11}$$

No further simplification is possible because none of the remaining terms are like terms.

example 5.5.4 Simplify $-3xpy + ym - 4yxp - 3my + 2ypx - 5$.

solution We add like terms to get

$$\mathbf{-5pyx - 2ym - 5}$$

example 5.5.5 Simplify $4 - y - xy + 3yx + x - 4y + 5 - xy + 2$.

solution We add like terms to get

$$\mathbf{xy - 5y + x + 11}$$

example 5.5.6 Simplify $x + 3y - 5 + 2xy + 10 + 6mz - p$.

solution The only like terms are the third term and the fifth term. These terms may be added but no further simplification is possible.

$$\mathbf{x + 3y + 5 + 2xy + 6mz - p}$$

TEST YOUR UNDERSTANDING

Simplify by adding like terms:

(e) $3xym - 6px + 10mxy + 4xp$ (f) $3xym - 2mx + 5xm$

5.6 exponents

Often we find it necessary to indicate that a number is to be multiplied by itself a given number of times. For instance if we wish to indicate that 5 is to be used as a factor seven times, we could write $5 \cdot 5 \cdot 5 \cdot 5 \cdot 5 \cdot 5 \cdot 5$. This is a cumbersome expression and mathematicians have developed a sort of mathematical shorthand called **EXPONENTIAL NOTATION** that allows the expression to be written more concisely. The exponential notation for 5 used as a factor seven times is 5^7, read **five to the seventh**. The general form of the expression is x^n and indicates that x is to be used as a factor n times and is read **x to the n**.

DEFINITION OF EXPONENTIAL NOTATION†

$$\underbrace{x \cdot x \cdot x \cdot \cdot \cdot x}_{n \text{ factors}} = x^n$$

The symbol that is to be multiplied by itself is called the **BASE** and the symbol that indicates how many times it is to be used as a factor is called the **EXPONENT** or the **POWER**. For example,

$x^4 = x \cdot x \cdot x \cdot x$ The base is x and the exponent is 4.

$(-4)^3 = (-4)(-4)(-4)$ The base is (-4) and the exponent is 3.

$\left(\frac{1}{3}\right)^4 = \left(\frac{1}{3}\right)\left(\frac{1}{3}\right)\left(\frac{1}{3}\right)\left(\frac{1}{3}\right)$ The base is $\frac{1}{3}$ and the exponent is 4.

If we are given replacement values for the variables, we can find the **value** of an algebraic expression that contains exponents as we see in the following example problems.

example 5.6.1 Evaluate $x^2 + y^3$ if $x = 2$ and $y = -3$.

solution We replace x with 2 and y with 3. We use parentheses to help us avoid mistakes in sign.

$$(2)^2 + (-3)^3 = (2)(2) + (-3)(-3)(-3) = 4 + (-27) = 4 - 27 = \mathbf{-23}$$

example 5.6.2 Evaluate x^2z^3y if $x = 2$, $z = 3$, and $y = -2$.

solution We replace x with 2, z with 3, and y with -2.

$$(2)^2(3)^3(-2) = (4)(27)(-2) = \mathbf{-216}$$

---

†In this definition n is restricted to the set of positive integers.

example 5.6.3 Evaluate $pm^2 - z^3$ if $p = 1$, $m = -4$, and $z = -2$.

solution We replace p with 1, m with -4, and z with -2.

$$(1)(-4)^2 - (-2)^3 = (1)(16) - (-8) = 16 + 8 = \mathbf{24}$$

5.7 product theorem for exponents

From the definition of exponential notation, we know that

(a) x^5 means $x \cdot x \cdot x \cdot x \cdot x = x^5$

(b) $x^2 \cdot x^3$ means $(x \cdot x) \cdot (x \cdot x \cdot x) = x^5$

(c) $x^1 \cdot x^4$ means $(x) \cdot (x \cdot x \cdot x \cdot x) \longrightarrow x^5$

From the above we see that to multiply two exponential expressions that have the same base, we add the exponents. Thus

(a) $x^5 \cdot x^7 \cdot x^2 = \mathbf{x^{14}}$ (b) $5^2 \cdot 5^3 \cdot 5^2 = \mathbf{5^7}$

(c) $p^5 \cdot p^{12} = \mathbf{p^{17}}$ (d) $4^2 \cdot 4^3 \cdot 4^{25} = \mathbf{4^{30}}$

PRODUCT THEOREM FOR EXPONENTS

If m and n are real numbers and $x \neq 0$,

$$x^m \cdot x^n = x^{m+n}$$

example 5.7.1 Simplify $x^2y^2x^5y^3$.

solution Since rearranging the order of the factors of a product does not change the value of the product, we may write $x^2y^2x^5y^3$ as

$$x^2x^5y^2y^3 = \mathbf{x^7y^5}$$

example 5.7.2 Simplify $x^2y^3m^5x^3y^2$.

solution First we rearrange the factors and then we simplify.

$$x^2x^3y^2y^3m^5 = \mathbf{x^5y^5m^5}$$

Now we define the notation x^1.

DEFINITION

$$x^1 = x$$

This says that x means the same thing as x^1 and that x^1 means the same thing as x. **If any variable or constant is written without an exponent, it is understood to have an exponent of 1. Thus,**

5 equals 5^1, x^1 equals x, and 7^1 equals 7

example 5.7.3 Simplify xyy^2x^3.

solution

$$xx^3yy^2 = \mathbf{x^4y^3}$$

example 5.7.4 Simplify $m^3pmxm^2x^3p^5$.

solution We rearrange and simplify to get the result.

$$m^3mm^2xx^3pp^5 = \mathbf{m^6x^4p^6}$$

TEST YOUR UNDERSTANDING

Simplify:

(g) $x^2yy^4x^5$ (h) $mx^2y^3m^2xy^5$

Terms that contain exponents can be added if the terms are like terms. We will demonstrate this in the next three problems.

example 5.7.5 Simplify by adding like terms: $x^2yp^5 + xy^2p^5 + 3p^5x^2y + x^2y^5p$.

solution The first and third terms are like terms

$$x^2yp^5 = p^5x^2y$$

because the literal factors of the terms are the same factors. Thus these terms can be added.

$$\mathbf{4x^2yp^5 + xy^2p^5 + x^2y^5p}$$

example 5.7.6 Simplify by adding like terms.

$$2m^3xy^2p + 3pxy^2m^3 - 10xy^2m^3p + yx^2m^3p$$

solution The first three terms are like terms and may be added.

$$\mathbf{-5xy^2m^3p + x^2ym^3p}$$

example 5.7.7 Simplify by adding like terms.

$$2x^2y + 3yx^2 + x^2y^2 - x^2y - 4x^2y^2$$

solution The first, second, and fourth terms are like terms and so are the third and fifth terms. Thus we add the first, second, and fourth terms and the third and fifth terms to obtain

$$\mathbf{4x^2y - 3x^2y^2}$$

Test Your Understanding

Simplify by adding like terms:

(i) $3x^4ym^2z - 5zymx^4 + 8yzx^4m^2$

lesson checklist 5

- ☐ Term
- ☐ Indicated product
- ☐ Factors, numerical coefficient, literal coefficient
- ☐ Distributive property
- ☐ Like terms
- ☐ Rule for addition of like terms
- ☐ Exponential notation
- ☐ Base, exponent, and power
- ☐ Product theorem for exponents

Answers to Test Your Understanding

(a) $+\dfrac{11}{16}$	(b) -40	(c) $4mnx - 12x$
(d) $131qpx$ and $-32xqp$	(e) $13xym - 2px$	(f) $3xym + 3xm$
(g) x^7y^5	(h) $m^3x^3y^8$	(i) $11x^4ym^2z - 5zymx^4$

problem set 5

1. Use the letters a, b, and c to state the distributive property.
2. Use numbers to demonstrate that real numbers are commutative under addition.
3. Use numbers to demonstrate the two unique properties of zero.
4. In the expression $\dfrac{a + b + c}{m}$, what is the third term?
5. Which integers are not real numbers?
6. What is another name for the multiplicative inverse of a number?
7. Is the product of 142 negative numbers and 3 positive numbers a positive number or a negative number?
8. What is the reciprocal of $-\frac{4}{7}$?

9. What is a variable?

10. When can terms be called like terms?

Simplify by adding like terms:

***11.** $x - 3xy + yx + mx$

***12.** $xmy + 3ypx - xpy - 5mxy$

***13.** $x + 3y - 6xy + 2x - 4 - y + 15 + xy$

***14.** $-3xpy + ym - 4yxp - 3my + 2ypx - 5$

***15.** $4 - y - xy + 3yx + x - 4y + 5 - xy + 2$

***16.** $x + 3y - 5 + 2xy + 10 + 6mz - p$

***17.** $x^2yp^5 + xy^2p^5 + 3p^5x^2y + x^2y^5p$

***18.** $2m^3xy^2p + 3pxy^2m^3 - 10xy^2m^3p + yx^2m^3p$

***19.** $2x^2y + 3yx^2 + x^2y^2 - x^2y - 4x^2y^2$

Simplify:

***20.** $x^2y^2x^5y^3$

***21.** $x^2y^3m^5x^3y^2$

***22.** xyy^2x^3

***23.** $m^3pmxm^2x^3p^5$

Evaluate:

***24.** $x^2 + y^3$ if $x = 2$ and $y = -3$

***25.** x^2z^3y if $x = 2$, $z = 3$, and $y = -2$

***26.** $pm^2 - z^3$ if $p = 1$, $m = -4$, $z = -2$

Expand by using the distributive property:

27. $a(b + c)$

28. $xm(p - qc)$

29. $xy(a + b - c)$

30. $pxy(m - k - c)$

31. $k(mp - n - q)$

Evaluate:

32. $x(a - bc)$ if $x = -2$, $a = 4$, $b = 3$, $c = -3$

33. $a[-b(c - a) + 4]$ if $a = 2$, $b = -3$, $c = 5$

34. $-p[4(b - a)(a - b) + 2]$ if $a = 4$, $b = -2$, $p = 3$

35. $x - y(-a - x)$ if $x = -2$, $y = -4$, $a = 6$

36. $-x - 4[a(x - a)]$ if $a = -3$, $x = 6$

Simplify:

37. $\{4[-3(-6 - 2)]\}$

38. $-7[4 - 6 \cdot 2(4 - 7)]$

39. $-11 - (-2) + 6(-3)$

40. $|-13 - 11|$

41. $|41|$

42. $(-5)^2 + 5^3$

43. $(-5)^3 + (5)^2$

44. $-6[-3(2 - 11) + 8]$

45. $-3[-8 + (-5 - 3)] + 8$

46. $-14 + (-3)[-8(-4 + 11 - 2)]$

---

*The problems marked with an asterisk were worked out in this lesson.

lesson six

Statements • sentences • equations
Equivalent equations
Additive and multiplicative properties of equality
Solution of equations
Multiplication of exponentials

6.1 numerical and algebraic expressions (again)

Each of the following expressions

(a) 4 (b) $6 + 3$ (c) $4(2 + 3)$ (d) $\dfrac{7(8 + 4)}{5 + 2}$

is called a **numerical expression** because it consists of **a meaningful arrangement of numerals and symbols that designate specific operations**. Every numerical expression represents a particular number and we say that this number is the **value** of the expression. The **values** of expressions (a), (b), (c), and (d) are 4, 9, 20, and 12, respectively.

We use the words **algebraic expression** to describe **numerical expressions** and also to describe expressions that contain variables.

(e) 6 (f) $x + 4$ (g) $x^2 - 6$ (h) $x(x + 4)$

Each of the expressions shown here is an algebraic expression. The value of expression (e) is 6, but the values of expressions (f), (g), and (h) depend on the value that we assign to the variable x. If we assign to x the value of 3, then the values of expressions (f), (g), and (h) are 7, 3, and 21, respectively.

6.2 statements and sentences

If we wish to make a statement that certain quantities are equal or are not equal, we can do so by writing a grammatical **SENTENCE** in English.

(a) The number of peaches equals the number of apples.
(b) The number of peaches does not equal the number of apples.
(c) The number of peaches is greater than the number of apples.
(d) The number of peaches is less than the number of apples.

If we use the variables

$$N_p \quad \text{and} \quad N_a$$

to represent the number of peaches and the number of apples, and if we use the symbols

$=$	to mean	equal
$\neq$	to mean	not equal
$>$	to mean	greater than (read left to right)
$<$	to mean	less than (read left to right)

we can make the same statements, by writing algebraic sentences.

(a) $N_p = N_a$

(b) $N_p \neq N_a$

(c) $N_p > N_a$

(d) $N_p < N_a$

We see that all four are called statements or sentences but only the first one, (a), uses the equals sign. This algebraic statement is called an equation. The other three statements, (b), (c), and (d), do not use the equals sign and are called *inequalities*. We will discuss equations here and hold the topic of inequalities for a later lesson.

6.3 equations

An equation is an algebraic statement consisting of two algebraic expressions connected by an equals sign. Thus all the following are statements and all are also equations.

(a) $4 = 3 + 1$ (b) $4 + x = 2 + 2 + x$ (c) $4 = 6$

(d) $4 + x = 6 + x$ (e) $x + 4 = 8$

Equations do not always make true statements as we see here. Two of these equations make true statements, two make false statements, and the truth or falsity of one of them, (e), depends on the number we use as a replacement for x in the equation. We will discuss each of these equations beginning with equation (a).

(a) This is a true statement and a true equation because 4 does equal the sum of 3 and 1.

(b) This is a true statement and a true equation regardless of the number we use as a replacement for x. We will demonstrate this by replacing x, once with -3 and another time with $+7$.

WITH (-3)	WITH $(+7)$
$4 + (-3) = 2 + 2 + (-3)$	$4 + (+7) = 2 + 2 + (+7)$
$4 - 3 = 4 - 3$	$4 + 7 = 4 + 7$
$1 = 1$ True	$11 = 11$ True

(c) This is a false statement and a false equation because 4 is not equal to 6.

(d) This is a false statement and a false equation regardless of the number we use as a replacement for x. We will demonstrate this by replacing x, once with -3 and another time with $+7$.

WITH (-3)	WITH $(+7)$
$4 + (-3) = 6 + (-3)$	$4 + (+7) = 6 + (+7)$

$$4 - 3 = 6 - 3 \qquad 4 + 7 = 6 + 7$$

$$1 = 3 \quad \text{False} \qquad 11 = 13 \quad \text{False}$$

(e) The truth or falsity of this algebraic statement or equation depends on the number used as a replacement for x. If we use the number 4 as the replacement for x, we get a true statement or true equation:

$$(4) + 4 = 8$$
$$8 = 8 \quad \text{True}$$

and if we use any other real number, say -2, as a replacement for x, we get a false statement or false equation.

$$(-2) + 4 = 8$$
$$-2 + 4 = 8$$
$$2 = 8 \quad \text{False}$$

6.4 conditional equations and statements

We see that an algebraic statement of equality is also called an equation and that these equations can be true equations or false equations or have their truth or falsity depend on the number used to replace the variable in the equation.

Some mathematicians prefer to use the word **statement** to describe statements of equality or inequality that are always true or always false and to use the words **CONDITIONAL EQUATION** or just **EQUATION** to describe statements of equality whose truth or falsity depends on the replacement value of the variable.

STATEMENTS	EQUATIONS
(a) $4 = 2 + 2$	(e) $3x + 2 = 5$
(b) $3 \cdot 4 = 6 + 6$	(f) $x = 4$
(c) $3 + x = 6 - 3 + x$	(g) $4x + 5 = 0$
(d) $4x = x + x + x + x$	(h) $-3x = -2$

6.5 solutions and roots

If we begin with the equation

$$x + 2 = 6$$

and replace x in this equation with the number 4, we get the true statement

$$4 + 2 = 6$$
$$6 = 6 \quad \text{True}$$

Replacement values of the variable that turn the equation into a true statement are called solutions of the equation or roots of the equation and are said to satisfy the equation. Thus in the equation

$$x + 2 = 6$$

we say that the number 4 is a **SOLUTION** or **ROOT** of the equation, and we also say that the number 4 **SATISFIES** the equation.

6.6 equivalent equations

Two equations are said to be **EQUIVALENT EQUATIONS** if *every* solution of one of the equations is also a solution of the other equation.

(a) $x + 6 = 9$ (b) $x + 10 = 13$

The two equations shown are equivalent equations, for the number 3 will satisfy both equations and 3 is the only number that will satisfy either equation.

6.7 additive property of equality

$a = b$

$a + c = b + c$

If we begin with the true statement that

$$2 = 2$$

and add +4 to both sides of the equality, we get the true statement that 6 equals 6.

$$2 + 4 = 2 + 4 \qquad\qquad 4 + 2 = 4 + 2$$

$$6 = 6 \qquad\qquad 6 = 6$$

On the left we placed the 4's after the 2's and on the right we placed the 4's before the 2's. We note that both procedures yield the same result. We **emphasize** this fact in the formal definition given in the box **by writing the definition twice**—the second time with the order of the addends reversed.

> ADDITIVE PROPERTY OF EQUALITY
>
> If a, b, and c are any real numbers and if $a = b$, then
>
> $$a + c = b + c \quad \text{and also} \quad c + a = c + b$$

The additive property of equality can be used to find the solution of conditional equations such as

$$x + 4 = 6$$

This equation is a conditional equation and is neither true nor false because no value has been asigned to x so the additive property of equality does not apply. Thus we must hedge a little. We assume that some real number exists that when substituted for x will make the equation a true statement. We further assume that x in the equation represents this number. Now we have assumed that

$$x + 4 \text{ equals } 6$$

is a true statement and thus we can use the additive property of equality.

Our task is to isolate x on one side of the equation (either side) and have everything else on the other side of the equation. For no particular reason† we decide to isolate x on the left side of the equation. Thus we must find some way to dispose of the +4 that is now on the left side with x. We remember that the sum of a number and the additive inverse of the same number is zero, and we know that the additive property of equality permits us to add the same quantity to both sides of the equation. We will add −4 to both sides of the equation.

$$\begin{array}{rcrl} x + 4 & = & 6 & \\ -4 & & -4 & \quad \text{add } -4 \text{ to both sides} \\ \hline x + 0 & = & 2 & \\ x & = & 2 & \quad x \text{ plus 0 equals } x \end{array}$$

Now since we made some assumptions to permit the use of the additive property of equality, we must check our solution in the original equation.

$$\begin{array}{rl} x + 4 = 6 & \quad \text{original equation} \\ (2) + 4 = 6 & \quad \text{substitute 2 for } x \\ 6 = 6 & \quad \text{True} \end{array}$$

---

†Here we can isolate x on either side of the equation, and we have no reason for choosing the left side instead of the right side. We use the phrase *for no particular reason* to indicate that we chose the left side only because we had to choose one side or the other.

Since using the number 2 for x in the equation makes the equation a true statement, we say that the number 2 is a root or solution of the equation and that the number 2 satisfies the equation. It can be shown that the use of the additive property of equality will not change the numbers that satisfy the equation so we say that the use of the additive property of equality results in an equation that is an **equivalent equation** to the original equation.

> If the same quantity is added to both sides of an equation, the resulting equation will be an **equivalent equation** to the original equation and thus every solution of one of these equations will also be a solution of the other equation.

The rule says that we may add the same quantity to both sides of an equation, but it does not specify a particular format to be used and one format is usually just as acceptable as another. Here we show three possible formats for the problem worked above.

(a)
$$\begin{array}{rcr} x + 4 & = & 6 \\ -4 & = & -4 \\ \hline x & = & 2 \end{array}$$

(b)
$$\begin{array}{rcr} x + 4 & = & 6 \\ -4 & & -4 \\ \hline x & = & 2 \end{array}$$

(c)
$$\begin{aligned} x + 4 + (-4) &= 6 + (-4) \\ x + 0 &= 2 \\ x &= 2 \end{aligned}$$

In (a) we add -4 to both sides of the equation and place an equals sign between the -4's to *emphasize* that they are equal. In (b) we added -4 to both sides but omitted the equals sign as it really isn't necessary. In (c) we added the -4's on the same line with the rest of the numbers and variables. This form is adequate for very simple problems such as this one but is less desirable for more complicated problems. Hopefully, by the end of the course, the student will be able to perform this calculation mentally, without writing anything down, so that none of these formats will be necessary.

> TEST YOUR UNDERSTANDING
>
> Solve the following equations by adding the same quantity to both sides of the equation.
>
> (a) $x - 15 = 30$ (b) $x + 7 = -4$

6.8 multiplicative property of equality

To demonstrate the **multiplicative property of equality** we will begin with the true statement that

$$2 = 2$$

and multiply both sides of the equation by 3

$$\begin{aligned} 3 \cdot 2 &= 3 \cdot 2 \\ 6 &= 6 \end{aligned}$$

The result is the *true statement* that 6 equals 6.

In the formal statement of the multiplicative property of equality in the box below, note that we make the statement twice to **emphasize** that the order of the factors does not affect the product.

> MULTIPLICATIVE PROPERTY OF EQUALITY
>
> If a, b, and c are any real numbers
>
> and if $a = b$, then
>
> $$ca = cb \quad \text{and also} \quad ac = bc$$

We can use the multiplicative property of equality to enable the solution of equations such as

$$4x = 12$$

if we make some assumptions. The equation as it is now written is neither true nor false for no value has been assigned to x. So again we assume that some value of x exists that will make the equation a true statement and that x represents this number. Now we have mentally converted the equation into a true statement of equality and thus can use the multiplicative property of equality.

Again our problem is to **isolate *x* on one side of the equation** and have everything else on the other side of the equation. We can do this if we multiply both sides of the equation by the **RECIPROCAL** or **MULTIPLICATIVE INVERSE** of 4, which is $\frac{1}{4}$.

$$\frac{1}{4}(4x) = 12 \cdot \frac{1}{4}$$

The left side of the equation can be thought of as the product of three factors

$$\left(\frac{1}{4}\right)(4)(x)$$

If we now multiply $\frac{1}{4}$ by 4, the result is 1 because $\frac{1}{4}$ and 4 are reciprocals or each is the multiplicative inverse of the other. Now we have on the left side

$$1 \cdot x$$

and we know that $1 \cdot x$ equals x. On the right side we have $12 \cdot \frac{1}{4}$, which equals $\frac{12}{4}$ or 3. Thus

$$\frac{1}{4}(4x) = 12 \cdot \frac{1}{4} \longrightarrow 1 \cdot x = \frac{12}{4} \longrightarrow \boldsymbol{x = 3}^{\dagger}$$

Since it was necessary to make an assumption in order to use the multiplicative property of equality, we will check our solution in the original equation.

$$4x = 12 \longrightarrow 4(3) = 12 \longrightarrow \mathbf{12 = 12} \quad \text{True}$$

Note that in each of the examples shown below we are able to isolate x by multiplying the x term by the reciprocal of the coefficient of x.

	PROBLEM	SOLUTION		
(a)	$3x = 6$	$\frac{1}{3}(3x) = \frac{1}{3}(6)$	so	$x = 2$
(b)	$\frac{x}{3} = 9$	$3\left(\frac{x}{3}\right) = 3(9)$	so	$x = 27$
(c)	$\frac{2x}{5} = 7$	$\frac{5}{2}\left(\frac{2x}{5}\right) = \frac{5}{2}(7)$	so	$x = \frac{35}{2}$
(d)	$\frac{7x}{3} = 4$	$\frac{3}{7} \cdot \frac{7x}{3} = \frac{3}{7}(4)$	so	$x = \frac{12}{7}$

† When convenient we will use arrows to point the way to the next step in the solution.

TEST YOUR UNDERSTANDING

Solve the following equations by multiplying every term on both sides by the reciprocal of the coefficient of the variable.

(c) $4p = 12$ (d) $\frac{2}{3}x = 20$ (e) $\frac{m}{5} = -30$

It can be shown that the use of the multiplicative property of equality produces an equivalent equation to the original equation.

> If **every term** on both sides of an equation is multiplied by the same non-zero quantity, the resulting equation will be an equivalent equation to the original equation, and thus every solution of one of these equations will be a solution of the other equation.

It is helpful here if we note that we obtain the same result by dividing by a number as we get by multiplying by the reciprocal of the same number. If we write the number 12 down twice

$$12 \qquad\qquad 12$$

and multiply the left number by $\frac{1}{4}$ and divide the right number by 4, we get the same result!

$$12 \cdot \frac{1}{4} = 3 \qquad\qquad \frac{12}{4} = 3$$

We can use this fact to help us extend the use of the multiplicative property of equality by saying

> If **every term** on both sides of an equation is divided by the same non-zero quantity, the resulting equation will be an equivalent equation to the original equation, and thus every solution of one of these equations will also be a solution of the other equation.

Thus we can find the solution to the equation

$$4x = 12$$

by multiplying every term on both sides of the equation by $\frac{1}{4}$ or by dividing every term on both sides of the equation by 4.

MULTIPLYING BY $\frac{1}{4}$	DIVIDING BY 4
$\frac{1}{4}(4x) = \frac{1}{4}(12)$	$\frac{4x}{4} = \frac{12}{4}$
$\mathbf{x = 3}$	$\mathbf{x = 3}$

To recapitulate:

1. If the same quantity is added to both sides of an equation, the resulting equation will be an equivalent equation to the original equation, and thus every solution of one of these equations will be a solution to the other equation.
2. If every term on both sides of an equation is multiplied or divided by the same non-zero quantity, the resulting equation will be an equivalent equation to the original equation, and thus every solution of one of these equations will be a solution to the other equation.

6.9 solution of equations

We will begin by solving a problem that requires the use of the additive property of equality and also requires the use of the multiplicative property of equality.

example 6.9.1 Solve $3x - 2 = 10$.

solution **Our task is to isolate x on one side** (either side) **of the equation** with everything else on the other side. For no particular reason, we decide to isolate x on the left side of the equation, and thus we must somehow dispose of the 3 and the -2 that are now on the left side. We will begin by disposing of the -2 by adding $+2$ to both sides of the equation.

$$\begin{array}{rcl} 3x - 2 &=& 10 \\ +2 && +2 \\ \hline 3x &=& 12 \end{array} \qquad \text{add } +2 \text{ to both sides}$$

Now if we can get rid of the 3 on the left side, we will have isolated x. We can eliminate the 3 if we multiply both sides of the equation by $\frac{1}{3}$ or divide both sides by 3 because $\frac{1}{3}$ times 3 equals 1 as does 3 divided by 3.

MULTIPLY BY $\frac{1}{3}$	DIVIDE BY 3
$\frac{1}{3}(3x) = \frac{1}{3}(12)$	$\frac{3x}{3} = \frac{12}{3}$
$\boldsymbol{x = 4}$	$\boldsymbol{x = 4}$

example 6.9.2 Solve $3x + 2 - x + 4 = -5 - x - 4$.

solution **Again our task is to isolate x on one side of the equation.** We will begin the process by adding like terms where possible on both sides of the equation.

$$2x + 6 = -9 - x \qquad \text{added like terms}$$

Now we must use the additive property of equality to put all x's on one side (either side) of the equation and to put everything else on the other side of the equation. Let's put the x's on the right side. We can do this by adding $-2x$ to both sides and this procedure will eliminate the $2x$ on the left side.

$$\begin{array}{rcll} 2x+6 &=& -9-x & \\ -2x & & -2x & \text{add } -2x \text{ to both sides} \\ \hline 6 &=& -9-3x & \end{array}$$

Now we have all x's on the right but still have a -9 and -3 to dispose of so that x can be isolated on the right side of the equation. We begin by adding $+9$ to both sides to eliminate the -9.

$$\begin{array}{rcll} 6 &=& -9-3x & \\ +9 & & +9 & \text{add } +9 \text{ to both sides} \\ \hline 15 &=& -3x & \end{array}$$

Now we need to get rid of the -3 on the right side so that $+x$ can be isolated on this side of the equation. We can do this in either of two ways. On the left we divide both sides by -3.

$$\frac{15}{-3} = \frac{-3x}{-3} \qquad\qquad (-1)15 = -3x(-1)$$

$$-15 = 3x$$

$$\mathbf{-5 = x} \qquad\qquad \frac{-15}{3} = \frac{3x}{3}$$

$$\mathbf{-5 = x}$$

On the right we first multiply both sides by -1 (permitted by the multiplicative property of equality) to change the sign of $-3x$ to $+3x$. Then the last step is to divide by $+3$. Of course, both procedures give the same final result.

example 6.9.3 Solve $k + 3 - 4k + 7 = 2k - 5$.

solution

$$\begin{array}{rcll} -3k+10 &=& 2k-5 & \text{added like terms} \\ +3k & & +3k & \text{add } 3k \text{ to both sides} \\ \hline 10 &=& 5k-5 & \\ +5 & & +5 & \text{add } +5 \text{ to both sides} \\ \hline 15 &=& 5k & \\ \frac{15}{5} &=& \frac{5k}{5} & \text{divide both sides by 5} \\ \mathbf{3} &=& \mathbf{k} & \end{array}$$

example 6.9.4 Solve $-7n + 3 + 2n = 4n - 5 + n$

solution

$$
\begin{array}{rrcl}
-5n & +3 & = & 5n - 5 \\
-5n & & & -5n \\
\hline
-10n & +3 & = & -5 \\
 & -3 & & -3 \\
\hline
-10n & & = & -8
\end{array}
$$

simplified both sides; add $-5n$ to both sides; add -3 to both sides

$$(-1)(-10n) = (-1)(-8)$$

multiply both sides by (-1) to change signs

$$10n = 8$$

$$\frac{10n}{10} = \frac{8}{10} \rightarrow \boldsymbol{n} = \mathbf{\frac{4}{5}}$$

divide each side by 10 and simplify answer

example 6.9.5 Solve $2e - 5 + 7e = 5 + 3e + 10$.

solution

$$
\begin{array}{rcl}
9e - 5 & = & 3e + 15 \\
-3e & & -3e \\
\hline
6e - 5 & = & 15 \\
+5 & & +5 \\
\hline
6e & = & 20
\end{array}
$$

simplified; add $-3e$ to both sides; add $+5$ to both sides

$$\frac{6e}{6} = \frac{20}{6}$$

divide both sides by 6 and simplify answer

$$\boldsymbol{e} = \mathbf{\frac{10}{3}}$$

Many beginning algebra students would prefer to write the answer above as the mixed number $3\frac{1}{3}$. Both forms of the answer are equally correct but we prefer the improper fraction $\frac{10}{3}$ to the mixed number $3\frac{1}{3}$ for the reason that the improper fraction is easier to use. Suppose the instructions in this problem had been to, say, solve and then multiply the answer by $\frac{11}{4}$. If we had written the answer as the mixed number $3\frac{1}{3}$, we would have to change it back to the improper fraction $\frac{10}{3}$ before the multiplication could be performed. Instructors at this level and in higher courses usually prefer the improper fraction to the mixed number.

example 6.9.6 Solve $3p + 7 - (-3) = p + (-2)$.

solution If the equation to be solved contains parentheses, the first step in the solution is to eliminate the parentheses and simplify where possible.

$$3p + 7 + 3 = p - 2 \qquad \text{clear parentheses}$$

$$3p + 10 = p - 2 \qquad \text{add like terms}$$

$$
\begin{array}{rcl}
-p & & -p \\
\hline
2p + 10 & = & -2 \\
-10 & & -10 \\
\hline
2p & = & -12
\end{array}
$$

add $-p$ to both sides; add -10 to both sides

$$\frac{2p}{2} = \frac{-12}{2} \rightarrow \boldsymbol{p} = \mathbf{-6} \qquad \text{divide both sides by 2}$$

We review our procedure for solving equations by listing four steps.

1. Clear parentheses and add like terms.
2. Use the additive property of equality by adding the same thing to both sides of the equation in such a way that all terms that contain the variable are on one side (either side) of the equation and all other terms are on the other side.
3. Use the multiplicative property of equality (if necessary) by multiplying each term on both sides of the equation by the reciprocal of the coefficient of the unknown or (the same thing) divide both sides of the equation by the coefficient of the unknown.
4. Simplify the result if possible.

TEST YOUR UNDERSTANDING

Solve the following equations:

(f) $6x - (-4) + x = 2x - 7$ (g) $-3x + 8 - 3 = -x - 7 + 2x$

6.10 multiplication of exponentials

Since variables are used to represent unspecified real numbers, the distributive property of real numbers is applicable to expressions that contain variables, as we demonstrate in the following example problems.

example 6.10.1 Expand $xy(y^2 - x^2z)$.

solution

$$xy(y^2 - x^2z) = xyy^2 - xyx^2z = \mathbf{xy^3 - x^3yz}$$

example 6.10.2 Expand $4x^2(ax - 2)$.

solution

$$4x^2(ax - 2) = 4x^2ax - 4x^2(2) = \mathbf{4ax^3 - 8x^2}$$

example 6.10.3 Expand $4xy^3(x^4y - 5x)$.

solution

$$4xy^3(x^4y - 5x) = 4y^4x^5 - 4xy^3 \cdot 5x = \mathbf{4x^5y^4 - 20y^3x^2}$$

example 6.10.4 Expand $4x^2y^3z(x - 2x^2yz + cz)$.

solution

$$4x^3y^3z - 8x^4y^4z^2 + 4cx^2y^3z^2$$

TEST YOUR UNDERSTANDING

Expand:

(h) $4x^2(a^2y - 3)$

(i) $-3x^2y^4(a^4xy - 2x^2y)$

lesson checklist 6

- ☐ Algebraic statements, algebraic sentences
- ☐ True and false statements, true and false sentences
- ☐ Conditional equations and statements
- ☐ Solution, roots, and satisfy
- ☐ Equivalent equations
- ☐ Additive property of equality
- ☐ Multiplicative property of equality
- ☐ To solve means to isolate
- ☐ Solution of equations
- ☐ Multiplication of exponentials

ANSWERS TO TEST YOUR UNDERSTANDING

(a) $x = 45$	(b) $x = -11$	(c) $p = 3$
(d) $x = 30$	(e) $m = -150$	(f) $x = -\frac{11}{5}$
(g) $x = 3$	(h) $4x^2a^2y - 12x^2$	(i) $-3x^3a^4y^5 + 6x^4y^5$

problem set 6

1. Is a numerical expression also an algebraic expression?
2. Use variables to state the additive property of equality.
3. Use variables to state the multiplicative property of equality.
4. Designate the set of integers.
5. Use the numerals 2 and 3 to demonstrate the commutative property of addition.
6. Which real number cannot be paired with a point on the number line?
7. Define the word quotient.

8. Give a numerical example of the multiplicative property of the number one.

9. Use the numerals 4 and 7 to demonstrate the commutative property of multiplication.

10. Use the letters a, b, and c to state the distributive property.

11. What is another name for the reciprocal of a number?

12. What is a variable?

Solve:

*13. $3x - 2 = 10$

*14. $3x + 2 - x + 4 = -5 - x - 4$

*15. $k + 3 - 4k + 7 = 2k - 5$

*16. $-7n + 3 + 2n = 4n - 5 + n$

*17. $2e - 5 + 7e = 5 + 3e + 10$

*18. $3p + 7 - (-3) = p + (-2)$

Expand by using the distributive property:

*19. $xy(y^2 - x^2z)$

*20. $4x^2(ax - 2)$

*21. $4xy^3(x^4y - 5x)$

*22. $4x^2y^3z(x - 2x^2yz + cz)$

Simplify:

23. $x^2y^5m^2y^2x$

24. $p^5x^4mxp^3$

25. $m^5x^2y^5mx^3y$

26. $x^4m^2xx^2ym^3$

Add like terms:

27. $5x^2pm - 3ap + 2x^2mp + 5pa$

28. $2x^2a - a^2x + ax^2 + 3xa^2$

29. $-7k^2pa + 5k^2p + 2pk^2a$

30. $4px^2k - 3x^2pk + 4kpx^2$

31. $mxyp + 2xmp^2 - 3pxym + 3p^2mx$

32. $4a^2x^3 + 6x^3a^2 - 3x^2a^3 + 5a^3x^2$

Simplify:

33. $-3[-2 - 5(9)]$

34. $2 - 3(-6)$

35. $-3 - (-4)(-2)[7 - (-5)]$

36. $-5[4 - 3(7 - 14) + 2]$

Evaluate:

37. $-a[a - (-b)] + b - a$ if $a = -4$, $b = 6$

38. $a[-b(b - a)]$; $a = 5$, $b = -10$

39. $(a - b)[b - (-a)]$; $a = 5$, $b = -3$

40. $a + (-b)(-a) - b$; $a = 2$, $b = -5$

41. $a(b\{b + [-(-a)]\})$; $a = 4$, $b = -3$

42. $|-15 - 3|$

43. $-|-(-15)|$

44. $-a^2 + (-b)$ if $a = -2$ and $b = 2$

45. $(-2)^2 - 3^2 + (-4)^3$

Solve:

*46. $\frac{x}{3} = 9$

*47. $\frac{2z}{5} = 7$

*48. $\frac{7p}{3} = 4$

Simplify:

49. $-3[-2(3 - 5 - 1)(-3)] + 8$

50. $2 + (-6)[-5 - (3 - 7) + 4]$

51. $-8[-5(-3 - 7) + 2]$

---

*The problems marked with an asterisk were worked out in this lesson.

lesson seven

Negative exponents
Product theorem for exponents
Zero exponents
Solution of equations

7.1 negative exponents

It is convenient to have a notation for the reciprocal of a number written in exponential notation. For instance, it is convenient to be able to write $\frac{1}{5^4}$ as a number whose denominator is 1 (understood). We therefore state that 5^{-4} is an alternate way to write $\frac{1}{5^4}$. In the formal definition we will use x and n to represent the base and the exponent.

> DEFINITION OF x^{-n}
>
> If n is any real number and x is any real number that is not zero,
>
> $$\frac{1}{x^n} = x^{-n}$$

(a) $\frac{1}{x^4} = x^{-4}$ (b) $x^{-4} = \frac{1}{x^4}$ (c) $\frac{1}{p^{-4}} = p^4$ (d) $m^{-a} = \frac{1}{m^a}$

In example (a) above we moved the x^4 from the denominator to the numerator **and changed the sign of the exponent** from plus to minus. In (b) we moved the x^{-4} from the numerator to the denominator **and changed the sign of the exponent** from minus to plus. In (c) we moved the p^{-4} from the denominator to the numerator **and changed the sign of the exponent** from minus to plus. In (d) we moved m^{-a} to the denominator and **changed the sign of the exponent**. The formal definition of x^{-n} is stated in the box above. We will now state the definition informally.

> A number or variable in exponential form can be written in reciprocal form if the sign of the exponent is changed.

If the exponent was positive, it is negative in the reciprocal form. If the exponent was negative, it is positive in the reciprocal form.

example 7.1.1 Simplify 3^{-2}.

solution The negative exponent is meaningless as an operation indicator. Thus the first step in the solution is to write 3^{-2} in reciprocal form and change the negative exponent to a positive exponent.

$$3^{-2} = \frac{1}{3^2}$$

Now we can complete the simplification because a positive exponent is an operation indicator because 3^2 means $3 \cdot 3$.

$$3^{-2} = \frac{1}{3^2} = \frac{1}{3 \cdot 3} = \frac{1}{9}$$

example 7.1.2 Simplify $\frac{1}{3^{-3}}$.

solution Again, as the first step we write the expression in reciprocal form so that the negative exponent can be eliminated from the problem.

$$\frac{1}{3^{-3}} = 3^3$$

Now 3^3 is meaningful as $3 \cdot 3 \cdot 3$ so 3^3 equals 27.

$$\frac{1}{3^{-3}} = 3^3 = 3 \cdot 3 \cdot 3 = \mathbf{27}$$

example 7.1.3 Simplify $(-3)^{-2}$.

solution We first eliminate the negative exponent by writing the expression in reciprocal form and changing the sign of the exponent.

$$(-3)^{-2} = \frac{1}{(-3)^2} = \frac{\mathbf{1}}{\mathbf{9}}$$

Test Your Understanding

Simplify:

(a) 4^{-2} (b) $\frac{1}{3^{-3}}$ (c) $\frac{1}{(-3)^{-3}}$

We have defined negative exponents so that their use will not conflict with the definition of the product theorem that we repeat here.

PRODUCT THEOREM FOR EXPONENTS

If m and n are real numbers and $x \neq 0$,

$$x^m \cdot x^n = x^{m+n}$$

We can write

$$x \cdot x \cdot x \cdot x \cdot x \text{ as } x^5 \quad \text{and we can write} \quad \frac{1}{x \cdot x} \text{ as } \frac{1}{x^2}$$

Thus if we multiply $x^5 \cdot \frac{1}{x^2}$, we can simplify as follows:

$$x^5 \cdot \frac{1}{x^2} = x \cdot x \cdot x \cdot x \cdot x \cdot \frac{1}{x \cdot x} = \frac{x \cdot x \cdot x \cdot x \cdot x}{x \cdot x} = \boxed{\frac{x \cdot x \cdot}{x \cdot x}}\ \frac{x \cdot x \cdot x}{}$$

which equals $x \cdot x \cdot x$ or x^3 because $\frac{x \cdot x}{x \cdot x}$ is an equivalent expression for 1 if $x \neq 0$.

Thus we see that

$$x^5 \cdot \frac{1}{x^2} = \mathbf{x^3}.$$

If we write $\frac{1}{x^2}$ as x^{-2} and then perform the multiplication using the product theorem for exponents, we have

$$x^5x^{-2} = x^{5-2} = \mathbf{x^3}. \qquad \text{The same result!}$$

example 7.1.4 Simplify $x^4m^2x^{-2}m^{-5}$.

solution We first change the order of multiplication

$$x^4x^{-2}m^2m^{-5}$$

and then multiply the exponentials that have the same bases to get

$$\mathbf{x^2m^{-3}}$$

Now this answer can be written in several forms by using the definition of x^{-n} as stated at the beginning of this lesson.

$$x^2m^{-3}, \qquad \frac{x^2}{m^3}, \qquad \frac{1}{x^{-2}m^3}, \qquad \frac{m^{-3}}{x^{-2}}$$

All the expressions above are equivalent expressions for all have the same value or represent the same number regardless of the non-zero replacement values used for x and m. Since they are equivalent expressions, one of them is no more correct than another and none of them is preferred. To give the student practice in writing exponentials, however, it is customary to have problems in elementary algebra books that specify the form of the answer.

example 7.1.5 Simplify the following exponential expressions and write the result using only positive exponents.

(a) $x^{-4}m^2x^2m$ (b) $p^{-5}x^2m(x^{-4}p^2)$ (c) $z^5m^4p^{-4}z^{-2}m$

solution (a) $x^{-4}m^2x^2m = x^{-2}m^3 = \dfrac{m^3}{x^2}$

(b) $p^{-5}x^2m(x^{-4}p^2) = x^{-2}p^{-3}m = \dfrac{m}{x^2p^3}$

(c) $z^5m^4p^{-4}z^{-2}m = m^5z^3p^{-4} = \dfrac{m^5z^3}{p^4}$

example 7.1.6 Write the answers to Example 7.1.5 using only negative exponents.

solution (a) $\dfrac{m^3}{x^2} = \dfrac{x^{-2}}{m^{-3}}$

(b) $\dfrac{m}{x^2p^3} = \dfrac{x^{-2}p^{-3}}{m^{-1}}$

(c) $\dfrac{m^5z^3}{p^4} = \dfrac{p^{-4}}{m^{-5}z^{-3}}$

example 7.1.7 Now write the same answers with all exponentials in the numerator.

solution (a) $\dfrac{m^3}{x^2} = m^3x^{-2}$

(b) $\dfrac{m}{x^2p^3} = mx^{-2}p^{-3}$

(c) $\dfrac{m^5z^3}{p^4} = m^5z^3p^{-4}$

> **TEST YOUR UNDERSTANDING**
>
> Simplify and express the result with all exponents positive:
>
> (d) $x^2y^{-2}x^{-4}y^5$ (e) $m^{12}x^5ym^{-4}x^{-3}y^{-4}$

7.2 zero exponents

We know that $\dfrac{4 \cdot 4}{4 \cdot 4}$ is the same as $\dfrac{16}{16}$, which is an equivalent expression for 1. In the same way,

(a) $\dfrac{x^3}{x^3}$ (b) $\dfrac{m^{15}}{m^{15}}$ $(x, m \neq 0)$

also have a value of 1 no matter what non-zero number is used as a replacement for the variables x and m. If we write expression (a) above with both exponentials in the numerator, we have

$$x^3 \cdot x^{-3}$$

and if we use the product theorem for exponents, we get

$$x^3 \cdot x^{-3} = x^{3-3} = x^0$$

We began by noting that $\dfrac{x^3}{x^3}$ had a value of 1, and we used the product theorem of exponents to change this expression to x^0. Therefore x^0 must have a value of 1! In the same way, any non-zero expression raised to the zero power must also have a value of 1.

$$(x + y + z^2)^0 = 1 \qquad (pm)^0 = 1 \qquad (px^{-4})^0 = 1$$

All the above have a value of 1 if the expression in parentheses does not have a value of zero.

> DEFINITION
>
> If x is any real number that is not zero,
>
> $$x^0 = 1$$

example 7.2.1 Simplify the following expressions: (a) $x^2y^5y^{-2}x^{-2}$ (b) $m^5b^2mb^{-2}$

solution (a) $x^2y^5y^{-2}x^{-2} = y^3x^0 = y^3$ (because $x^0 = 1$ if $x \neq 0$)

(b) $m^5b^2mb^{-2} = m^6b^0 = m^6$ (because $b^0 = 1$ if $b \neq 0$)

Since we say that we must not use the expression 0^0, it is necessary that in problems such as (a) and (b) above we assume that the variable with the zero exponent is not zero. **Further we will assume in the problem sets that no variable with a zero exponent has a value equal to zero.**

> TEST YOUR UNDERSTANDING
>
> Simplify:
>
> (f) $x^2y^5x^{-2}x^0$ (g) $y^{-5}y^0x^{-4}y^5p$

7.3 solution of equations

In Lesson 6 we learned how to use the additive and multiplicative properties of equality to solve first-degree equations in one unknown such as $2x + 4 = 10$.

$$2x + 4 = 10$$

$$-4 \quad -4 \qquad \text{add } -4 \text{ to both sides}$$

$$2x = 6$$

$$\frac{2x}{2} = \frac{6}{2} \qquad \text{divide both sides by 2}$$

$$x = 3$$

In some problems it is necessary to use the distributive property and then simplify before using the additive and multiplicative properties of equality.

example 7.3.1 Solve $2(3 - b) = -2(1 - 3b) + 5$.

solution As the first step we will use the distributive property on the left side to expand $2(3 - b)$ and on the right side to expand $-2(1 - 3b)$.

$$6 - 2b = -2 + 6b + 5 \quad \text{used distributive property}$$
$$6 - 2b = 3 + 6b \quad \text{added like terms}$$
$$+2b \qquad +2b \quad \text{add } 2b \text{ to both sides}$$
$$6 = 3 + 8b$$
$$-3 \qquad -3 \quad \text{add } -3 \text{ to both sides}$$
$$3 = 8b$$
$$\frac{3}{8} = b \quad \text{divided both sides by 8}$$

example 7.3.2 Solve $3(-4 + 2x + 5) + 7 = -4(x + 2)$.

solution As the first step we will simplify within the parentheses on the left. Then we will use the distributive property as required on both sides and solve.

$$3(1 + 2x) + 7 = -4(x + 2) \quad \text{simplify left side}$$
$$3 + 6x + 7 = -4x - 8 \quad \text{used distributive property}$$
$$10 + 6x = -4x - 8 \quad \text{combined like terms}$$
$$+4x \qquad +4x \quad \text{add } 4x \text{ to both sides}$$
$$10 + 10x = -8$$
$$-10 \qquad -10 \quad \text{add } -10 \text{ to both sides}$$
$$10x = -18$$
$$x = -\frac{18}{10} \longrightarrow x = -\frac{9}{5} \quad \text{divided by 10 and simplified}$$

example 7.3.3 Solve $15(4 - 5b) = 16(4 - 6b) + 10$.

solution As the first step we will use the distributive property as required on both sides of the equation.

$$60 - 75b = 64 - 96b + 10 \quad \text{used distributive property}$$
$$60 - 75b = 74 - 96b \quad \text{added like terms}$$
$$+96b \qquad +96b \quad \text{add } 96b \text{ to both sides}$$
$$60 + 21b = 74$$
$$-60 \qquad -60 \quad \text{add } -60 \text{ to both sides}$$
$$21b = 14$$
$$b = \frac{14}{21} \longrightarrow b = \frac{2}{3} \quad \text{divided both sides by 21 and simplified}$$

example 7.3.4 Solve $(3 - 5)m - 2m = 4(-2 + 5)$.

solution $-2m - 2m = 4(3) \longrightarrow -4m = 12 \longrightarrow m = -3$

example 7.3.5 Solve $12 - 5(2x - 3) = 5(5 - 2x) + x$.

solution $12 - 10x + 15 = 25 - 10x + x \longrightarrow -10x + 27 = 25 - 9x \longrightarrow \mathbf{2 = x}$

The parentheses in the problems we have just worked have been used to indicate the multiplication of the contents of parentheses by a variable or a constant. To solve the next two equations, we will need to have rules that tell how to eliminate parentheses when they are preceded by a plus sign or a minus sign.

1. **When parentheses are preceded by a plus sign, both the parentheses and the sign may be discarded as we demonstrate here.**

$$+(-4 + 3x) = -4 + 3x$$

2. **When parentheses are preceded by a minus sign, both the minus sign and the parentheses may be discarded if the signs of all terms within the parentheses are changed.** This is because the minus sign indicates the negative of or the opposite of the quantity within the parentheses.

$$-(x - 3y + 6 - k) = -x + 3y - 6 + k$$

example 7.3.6 Solve $12 - (2x + 5) = -2 + (x - 3)$.

solution As the first step we drop the parentheses, remembering to change the signs of all terms that are in parentheses that are preceded by a minus sign.

$$12 - 2x - 5 = -2 + x - 3$$

Now we simplify on both sides of the equation

$$7 - 2x = x - 5$$

and solve for x:

$$\begin{array}{rcl} 7-2x & = & x-5 \\ +5+2x & & +2x+5 \\ \hline 12 & = & 3x \end{array} \longrightarrow \frac{12}{3} = \frac{3x}{3} \longrightarrow \mathbf{4 = x}$$

add $+5 + 2x$ to both sides

example 7.3.7 Solve $-(4y - 17) + (-y) = (2y - 1) - (-y)$.

solution Again we remember that when we discard parentheses preceded by a minus sign, the signs of all terms within the parentheses are changed.

$$-4y + 17 - y = 2y - 1 + y$$

Simplifying and solving:

$$\begin{array}{rcl} -5y+17 & = & 3y \quad -1 \\ +5y+\ 1 & & +5y \quad +1 \\ \hline 18 & = & 8y \end{array} \longrightarrow \frac{18}{8} = \frac{8y}{8} \longrightarrow \mathbf{\frac{9}{4} = y}$$

add $+5y + 1$ to both sides

TEST YOUR UNDERSTANDING

Solve:

(h) $4(x - 3) - (-2x + 5) = 3 + (-2x + 4)$

lesson checklist 7

- ☐ Definition of negative exponents
- ☐ Product theorem for exponents
- ☐ Zero exponents
- ☐ Solution of equations

ANSWERS TO TEST YOUR UNDERSTANDING

(a) $\frac{1}{16}$	(b) 27	(c) -27	(d) $\dfrac{y^3}{x^2}$
(e) $\dfrac{m^8x^2}{y^3}$	(f) y^5	(g) $\dfrac{p}{x^4}$	(h) $x = 3$

problem set 7

1. Use the numbers 4 and 3 to demonstrate the commutative property of multiplication.
2. Designate the set of whole numbers.
3. What is the difference between a variable and an unknown?
4. Define the word quotient.
5. What is the difference between the set of counting numbers and the set of natural numbers?
6. Name and give an example of the unique property of the number one.
7. Name and give an example of the two unique properties of the number zero.
8. Define the word factor.

Simplify:

***9.** $x^{-4}m^2x^2m$ ***10.** $p^{-5}x^2m(x^{-4}p^2)$ ***11.** $z^5m^4p^{-4}z^{-2}m$

***12.** $x^2y^5x^{-2}y^{-2}$ ***13.** $m^5b^2mb^{-2}$

---

*The problems marked with an asterisk were worked out in this lesson.

Expand:

14. $x^0y^2(y^0x - y^{-2})$

15. $p^{-3}y^0(p^{-3}y^2 - y^3p^3)$

16. $x^5y^0z(p^{-3}z^0 - 4x^{-5}z^{-1})$

17. $x^2y^{-2}(x^{-2}y^2 + 4)$

18. $4x^0y^5(x^5y^0 - 4p^0 + 2y^{-5})$

19. $x^{-2}y^{-2}(x^2y^2 + x^4y^4)$

Solve:

20. $b - 3b + 3 - 2b = 4b - 6 + 2$

21. $x - 5 - 3x = 2x - 6 + 11$

22. $p + 5p - p - 11 = 2p$

23. $k - 5k + 3 + 2k = 3k + 2 - 5$

***24.** $2(3 - b) = -2(1 - 3b) + 5$

***25.** $3(-4 + 2x + 5) + 7 = -4(x + 2)$

***26.** $15(4 - 5b) = 16(4 - 6b) + 10$

27. $12 - 5(2x - 3) = 5(5 - 2x) + x$

Add like terms:

28. $3a^2x^4 - 2a^3x^3 + 5x^3a^3 - 2x^4a^2$

29. $4m^2p^2x + 6m^2xp^2 + 3m^2xp + 2mx^2p^2$

30. $4a^2x + 6xa - 3x^2a^2 + 2xa^2 - 4ax$

31. $-2a^2xp^2 + 3p^2xa^2 + 5xp^2a^2$

32. $3a^2xmz^4 + 7a^2z^4xm - 2a^2xz^4m$

Simplify:

33. $|-5 - 11(2)|$

34. $-(-|-11|)$

35. $-4 - (-3) + 7 - (6)2$

36. $-[5(-3)] - 6$

37. $-[-3(-5 + 2) - 6] + 7$

38. $4 - [(-3)(-2) + 5]$

39. $7[-7 + (-5 - 2) + 3]$

Find the value of:

40. $a^2[b - c^2(-a)]$ if $a = -2$, $b = 2$, $c = -3$

41. $3b(-c) - 4(b - 3c)$ if $b = 2$ and $c = -4$

42. $(x - a)(-3) - 2 + a(-2x)$ if $a = 3$ and $x = -3$

43. $-2a(-x) - 3(a + x)$ if $x = 2$, $a = -4$

Simplify:

44. $(-2)^2 + 4(-2)^3$

45. $(-3)^3(-2)^2$

46. 3^{-3}

47. $(-3)^{-2}$

Solve:

48. $3m = 20$

49. $\frac{m}{5} = 45$

50. $\frac{2p}{3} = \frac{7}{5}$

lesson eight

Prime numbers and greatest common factor
Factoring
Minus signs and negative exponents

8.1 prime numbers

A **PRIME NUMBER** is a positive integer greater than 1 whose only integral divisors† are 1 and the number itself. The number 17 is a **prime number** because the only integral divisors of 17 are 1 and 17. The number 23 is a **prime number** because the only integral divisors of 23 are 23 and 1. The first 10 **prime numbers** are 2, 3, 5, 7, 11, 13, 17, 19, 23, and 29.

A number that can be written as a **product of prime numbers** is called a **COMPOSITE NUMBER**. The number 12 is a **composite number** because it can be written as a product of **prime numbers** as

$$2 \cdot 2 \cdot 3$$

The number 12 can be written as a product of integral factors in four different ways.

(a) $12 \cdot 1$ (b) $4 \cdot 3$ (c) $2 \cdot 6$ (d) $2 \cdot 2 \cdot 3$

In (a), (b), and (c), one of the factors is not a prime number, but in (d) all three of the factors are prime numbers. A **PRIME FACTOR** is a factor that is a prime number.

example 8.1.1 Express 80 as a product of prime factors.

solution We will divide by prime numbers.

$$\frac{80}{2} = 40, \quad \frac{40}{2} = 20, \quad \frac{20}{2} = 10, \quad \frac{10}{2} = 5$$

†A divisor that is an integer.

Using the five factors we have found, we can express 80 as a product of prime factors as $\mathbf{2 \cdot 2 \cdot 2 \cdot 2 \cdot 5}$.

example 8.1.2 Express 147 as a product of prime factors.

solution 147 is not divisible by 2 or by 5, so let's try 3.

$$\frac{147}{3} = 49 \quad \text{and} \quad \frac{49}{7} = 7$$

So 147 expressed as a product of prime factors is $\mathbf{3 \cdot 7 \cdot 7}$.

TEST YOUR UNDERSTANDING

(a) Write 900 as a product of prime factors.

8.2 greatest common factor

The discussion of prime factors above dealt only with real numbers. If an algebraic expression has factors that are represented by letters, these factors are called **LITERAL FACTORS**. We use the words **ALGEBRAIC FACTOR** as a general term to cover factors that contain either numbers or letters or both numbers and letters. In later lessons we will discuss factors that are sums of numbers and/or letters, and we will also call these factors algebraic factors.

DEFINITION

The greatest common factor of two or more terms is the product of all prime algebraic factors common to every term, each to the highest power that it occurs in all of the terms.

The expression $6x^2y^2m^2 + 3xy^3m^2 + 3x^3y^2$ can be written as

$$2 \cdot 3 \cdot x \cdot x \cdot y \cdot y \cdot m \cdot m + 3 \cdot x \cdot y \cdot y \cdot y \cdot m \cdot m + 3 \cdot x \cdot x \cdot x \cdot y \cdot y.$$

Now only the first term has 2 as a factor, so 2 is not a factor of the greatest common factor. Each term has 3 as a factor at least once, so 3 is a factor of the greatest common factor of all the terms.

$$3$$

Each term has x as a factor at least once in every term so x is a factor of the GCF.

$$3x$$

In the same way, y is used as a factor at least twice in every term so the greatest common factor of the three given terms is

$$3xy^2$$

The variable m is not included because it is not a factor of the third term of the original expression.

example 8.2.1 Find the GCF of $8z^4m^2p - 12z^3m^4p^2$.

solution The greatest common factor of the terms is $4z^3m^2p$.

example 8.2.2 Find the greatest common factor of $4x^2y^3z - 8y^2xz^3$.

solution The GCF is $4xy^2z$.

example 8.2.3 Find the GCF of $16x^2yp^3 - 4x^3y^2p + 2x^2y^2p^{15}$.

solution The GCF is $2x^2yp$.

TEST YOUR UNDERSTANDING

(b) Find the GCF of $6x^2y^5m - 9x^5y^2m^4 + 12x^3y^4m^2$

8.3 factoring

As we have pointed out in past lessons, the distributive property permits multiplication to be distributed over addition as indicated by the statement

$$a(b + c) = ab + ac$$

If we use the distributive property, we can write the indicated product $xy^2(y^4 - 3x)$ as a sum.

$$xy^2(y^4 - 3x) = \mathbf{xy^6 - 3x^2y^2}$$

If we write the distributive property in reverse order

$$ab + ac = a(b + c)$$

we see that the distributive property will allow us to express an indicated sum as an indicated product if each term of the sum has a common factor. Thus we can write $xy^6 - 3x^2y^2$ as an indicated product, for each term has xy^2 as a factor.

$$xy^6 - 3x^2y^2 = xy^2(y^4 - 3x)$$

The process of writing an indicated sum as a product of factors is called FACTORING.

If we are asked to factor $xb + xc$, we are being asked to write $xb + xc$ in the form of an indicated product. To do this

$$(\ )(\ )$$

we first write down two sets of parentheses to form an indicated product. Now we look at the expression $xb + xc$ and note that each term has x as a factor and also note that x is the only factor that is common to both terms. Thus we call x the greatest common factor of both terms. We insert the greatest common factor of both terms in the first parentheses.

$$(x)(\ )$$

Now we must decide what to write in the second parentheses so that (x) multiplied by the expression in the second parentheses will equal

$$xb + xc$$

The proper entry for the second parentheses is $b + c$ because

$$\boldsymbol{x(b + c)} = xb + xc$$

example 8.3.1 Factor $a^3x^2m^2 + a^2xm^3 - a^4x^3m^2$.

solution We want to write this sum as a product. We begin by writing two sets of parentheses.

$$(\quad)(\quad)$$

In the first parentheses we want to write the greatest common factor of all three terms. To find this GCF, we will write the three terms as products of individual factors,

$$a^3x^2m^2 + a^2xm^3 - a^4x^3m^2$$
$$a \cdot a \cdot a \cdot x \cdot x \cdot m \cdot m + a \cdot a \cdot x \cdot m \cdot m \cdot m - a \cdot a \cdot a \cdot a \cdot x \cdot x \cdot x \cdot m \cdot m$$

Look at the a's. Each term has at least two a's so a^2 is part of the greatest common factor

$$(a^2 \quad)(\quad)$$

Each term has at least one x, so x is a part of the greatest common factor.

$$(a^2x \quad)(\quad)$$

Finally each term has at least 2 m's so m^2 is a part of the greatest common factor. No other factors are common to all 3 terms.

$$(a^2xm^2)(\quad)$$

Now the first entry in the second parentheses must be ax because a^2xm^2ax equals $a^3x^2m^2$, the first term of the original expression

$$(a^2xm^2)(ax \quad)$$

The second entry in the second parentheses must be m because $a^2xm^2(m) = a^2xm^3$, the second term of the original expression.

$$(a^2xm^2)(ax + m \quad)$$

The third entry must be $-a^2x^2$ because $a^2xm^2(-a^2x^2) = -a^4x^3m^2$, the last entry in the original expression. The desired factored expression is $a^2xm^2(ax + m - a^2x^2)$ because

$$a^2xm^2(ax + m - a^2x^2) = a^3x^2m^2 + a^2xm^3 - a^4x^3m^2$$

example 8.3.2 Factor $4a^3b^4z^3 + 2a^2bz^4$.

solution

$$2 \cdot 2 \cdot a \cdot a \cdot a \cdot b \cdot b \cdot b \cdot b \cdot z \cdot z \cdot z + 2 \cdot a \cdot a \cdot b \cdot z \cdot z \cdot z \cdot z$$

Each term has, at least one 2, two a's, one b, and three z's as factors. Thus the greatest common factor is $2a^2bz^3$, so we write

$$(2a^2bz^3)(\quad)$$

The first term in the second parentheses is $2ab^3$ because $2a^2bz^3(2ab^3) = 4a^3b^4z^3$, the first term in the original expression.

$$(2a^2bz^3)(2ab^3 \quad)$$

The second term in the second parentheses is z because $2a^2bz^3(z) = 2a^2bz^4$, the second term in the original expression. Now

$$(2a^2bz^3)(2ab^3 + z)$$

is our answer because

$$(2a^2bz^3)(2ab^3 + z) = 4a^3b^4z^3 + 2a^2bz^4$$

example 8.3.3 Factor $6a^2x^2 + 2a^3x^3 + 4a^4x^3$.

$$(\quad)(\quad)$$

solution The greatest common factor is $2a^2x^2$.

$$(2a^2x^2)(\quad)$$

The entry in the second parentheses is

$$(3 + ax + 2a^2x)$$

because

$$(2a^2x^2)(3 + ax + 2a^2x) = 6a^2x^2 + 2a^3x^3 + 4a^4x^3$$

example 8.3.4 Factor $3m^3xy^2 + m^2y$

solution $m^2y(3mxy + 1)$ is the answer because

$$m^2y(3mxy + 1) = 3m^3xy^2 + m^2y$$

TEST YOUR UNDERSTANDING

Factor:

(c) $a^{10}x^5 - 5ax^5$

(d) $a^2x^6 - a^2x + 2a^4x^8$

8.4 multiplication of fractions

Two fractions are multiplied by multiplying the numerators to form the new numerator and multiplying the denominators to form the new denominator. For example,

$$\frac{3}{2} \cdot \frac{5}{7} = \frac{15}{14}$$

Since variables stand for unspecified real numbers, all the rules for real numbers also apply to expressions that contain variables. Thus fractions that contain variables are multiplied using the rule given above.

$$\frac{mx}{4y} \cdot \frac{ax}{2y} = \frac{\mathbf{amx^2}}{\mathbf{8y^2}}$$

The distributive property of real numbers is also applicable to the multiplication of fractions that contain variables. Fractions of this type are often called **RATIONAL EXPRESSIONS.**

$$\frac{x}{y}\left(\frac{a}{y} - b\right) = \frac{\mathbf{xa}}{\mathbf{y^2}} - \frac{\mathbf{bx}}{\mathbf{y}}$$

example 8.4.1 Use the distributive property to permit the four multiplications indicated in (a), (b), (c), and (d).

	PROBLEM		SOLUTION
(a)	$\frac{x^2}{y}\left(\frac{x^2}{y} - \frac{3y^3}{m}\right)$	=	$\frac{x^4}{y^2} - \frac{3x^2y^3}{ym} = \mathbf{\frac{x^4}{y^2} - \frac{3x^2y^2}{m}}$
(b)	$\frac{m}{z}\left(\frac{axp}{mk} - 2m^4p^4\right)$	=	$\frac{maxp}{mzk} - \frac{2m^5p^4}{z} = \mathbf{\frac{axp}{zk} - \frac{2m^5p^4}{z}}$
(c)	$\frac{ab}{c^2}\left(\frac{xab}{c} + 2bx - \frac{4}{c^2}\right)$	=	$\mathbf{\frac{xa^2b^2}{c^3} + \frac{2ab^2x}{c^2} - \frac{4ab}{c^4}}$
(d)	$\frac{ax}{c^2}\left(\frac{b^4}{xk} - 2b^2\right)$	=	$\frac{axb^4}{c^2xk} - \frac{2axb^2}{c^2} = \mathbf{\frac{ab^4}{c^2k} - \frac{2axb^2}{c^2}}$

TEST YOUR UNDERSTANDING

Multiply and then simplify:

(e) $\frac{m^2x^4}{p}\left(\frac{p^5m}{x^6} + \frac{5p^2}{m^2x^4}\right)$

8.5 minus signs and negative exponents

Simplification of expressions that contain minus signs and negative exponents such as the following is difficult for the beginner in algebra. The minus sign in front of the expression indicates **the opposite of** while **the negative exponent does not indicate the opposite of**.

(a) 4^2 (b) 4^{-2} (c) -4^2 (d) $(-4)^2$ (e) -4^{-2} (f) $(-4)^{-2}$

The notations in (a) and (b) are easy to simplify because in each of these we have a positive number raised to a power.

(a) $4^2 = \mathbf{16}$ (b) $4^{-2} = \frac{1}{4^2} = \mathbf{\frac{1}{16}}$

The notations in (c) and (d) often give difficulty because of the difficulty in handling the minus sign in front of the 4. When the minus sign is not enclosed in parentheses as in (c).

$$-4^2$$

it is sometimes helpful to think of moving the minus sign far to the left and the 4^2 far to the right as

$$-\qquad\qquad 4^2$$
↖ sign

and then performing the operation of squaring 4 to get 16,

$$-\qquad\qquad 16$$
↖ sign

and then moving the sign back in front of the 16 to get the final answer of negative 16.

$$-16$$

From this we see that -4^2 is read as *the opposite of* four squared and is not read as *the opposite of 4* squared. The notation in (d) which is

$$(-4)^2$$

does mean that -4 is to be multiplied by itself.

$$(-4)^2 = (-4)(-4) = \mathbf{+16}$$

The notations in (e) and (f) are similar to the two we have just discussed. They are

$$\text{(e)}\quad -4^{-2} \qquad \text{and} \qquad \text{(f)}\quad (-4)^{-2}$$

In the expression on the left we may again think of moving the minus sign far to the left and the rest of the expression far to the right.

$$- \qquad\qquad\qquad 4^{-2}$$

↖ sign

and now we simplify 4^{-2} as

$$- \qquad 4^{-2} = \frac{1}{4^2} = \frac{1}{16}$$

↖ sign

and lastly we bring back the negative sign for our final answer of negative one-sixteenth.

$$-\mathbf{\frac{1}{16}}$$

In (f) the $-$ sign is inside the parentheses and cannot be separated from the rest of the problem. Thus our simplification is

$$(-4)^{-2} = \frac{1}{(-4)^2} = \frac{1}{(-4)(-4)} = \mathbf{\frac{1}{16}}$$

We will practice problems like these on every homework from here to the end of the book.

TEST YOUR UNDERSTANDING

Simplify:

(f) -3^{-2} (g) $(-3)^{-2}$ (h) $\dfrac{1}{-3^{-2}}$ (i) $\dfrac{1}{(-3)^2}$

lesson checklist 8

- ☐ Prime numbers and prime factors
- ☐ Greatest common factor
- ☐ Factoring
- ☐ Rule for multiplication of fractions
- ☐ Negative numbers raised to powers

Answers to Test Your Understanding

(a) $2 \cdot 2 \cdot 3 \cdot 3 \cdot 5 \cdot 5$	(b) $3x^2y^2m$	(c) $ax^5(a^9 - 5)$
(d) $a^2x(x^5 - 1 + 2a^2x^7)$	(e) $\frac{m^3p^4}{x^2} + 5p$	(f) $-\frac{1}{9}$
(g) $\frac{1}{9}$	(h) -9	(i) $\frac{1}{9}$

problem set 8

1. What is a prime number?
2. Is 14 a prime number? Why?
3. What is the greatest common factor of two or more terms?
4. Designate the set of integers.
5. Use letters to state the distributive property of real numbers.
6. Why do 4^{-2} and $\frac{1}{4^2}$ represent the same number?
7. Use letters to state the additive property of equality.
8. Is an algebraic expression also a numerical expression?
9. Define the word sum.
10. What is the absolute value of a number?

Express as a product of prime factors:

*11. 80 *12. 147 13. 200

14. 98 15. 36 16. 175

Factor:

17. $ab + ac$

*18. $a^3x^2m^2 + a^2xm^3 - a^4x^3m^2$

*19. $4a^3b^4z^3 + 2a^2bz^4$

20. $6a^2x^2 + 2a^3x^3 + 4a^4x^3$

*21. $3m^3xy^2 + m^2y$

22. $6a^3x^2m - 3m^2x^2a^2$

Multiply using the distributive property:

*23. $\frac{x}{y}\left(\frac{a}{y} - b\right)$

*24. $\frac{x^2}{y}\left(\frac{x^2}{y} - \frac{3y^3}{m}\right)$

25. $\frac{m}{z}\left(\frac{axp^2}{m} - 2m^4p^4\right)$

Simplify:

26. $m^0x^{-2}y^5(3x^0y^{-7}m - 6x^2y^{-5})$

27. $x^0x^{-3}y^0p^2(ax^{-2}y^2 - 3p^{-2} + 6x^3p^{-2})$

Solve:

28. $-7n - (3n)2 = 3n - (4 - 3)$

29. $-3p + 7 - (-3) = p + (-2)$

30. $4(7z - 9) - 7(4z + 3) = 6(9z - 10)$

*The problems marked with an asterisk were worked out in this lesson.

31. $3y - 2(2y - 7) = 2(3 + y) - 4$ **32.** $10 - 7x = 4(11 - 6x)$

33. $12 = -\{-3[4z - 2(z - 2)]\}$

Add like terms:

34. $4k^2x + 3ak^2x - 2xk^2 + axk^2$ **35.** $abc - 2a^2bc^2 - 3cba + 5bca$

36. $xmp + 3mpx - pxm$

Simplify:

37. $|-11 - 2|$ **38.** $|-11|$ **39.** $|-4 - 5 - 3|$

40. $-6[-2(-3 - 2)] + 5$ **41.** $-(-3 - 2) - 6$

Evaluate:

42. $-a(b + c) - c$ if $a = 1, b = -2, c = -3$

43. $-k + 2b - (-c)$ if $k = -3, b = 2, c = 1$

44. $-p(b - a) + p$, if $a = 2, p = -3, b = 4$

Simplify:

***45.** -4^2 ***46.** $(-4)^2$ ***47.** -4^{-2}

***48.** $(-4)^{-2}$

Evaluate:

49. $a^2 + b^2c$ if $a = -2, b = 2$, and $c = -3$

Solve:

50. $\frac{5}{11}x = 6$

lesson nine

Graphical solutions of inequalities
Quotient theorem for exponents
Addition of like terms

9.1 inequalities

We begin by reviewing the discussion of equations presented in Lesson 6. In that lesson we said that an equation is a statement of equality consisting of two algebraic expressions connected by an equals sign. Thus an equation is a symbolic statement that the two expressions are equal.

$$4 = 2 + 2 \quad \text{True} \qquad\qquad 5 = 11 \quad \text{False}$$

As we see here, equations can be true equations or false equations. The equation on the left is a true equation, for both 4 and 2 + 2 have the same value as both are numerals or symbols that make us think of the number 4. The equation on the right is a false equation because 5 is not equal to 11.

A statement of equality whose truth or falsity depends on the replacement values used for the variables is called a **conditional statement of equality** or a **conditional equation**. Usually, we don't write the word conditional and just use the one word **equation**. If in the equation

$$x + 6 = 10$$

we use the numbers 4 and 2 as replacements for the variable x;

USING 4	USING 2
$4 + 6 = 10$ True	$2 + 6 = 10$ False

we obtain a true statement when we use 4 and a false statement when we use 2.

If the number we use as a replacement for the variable makes the equation a true statement, we call the number a **solution** or **root** of the equation and say that **the number satisfies the equation.**

Algebraic statements may also state that quantities are not equal, and we call these statements **INEQUALITIES**. The symbol $<$ is a symbol of inequality. The small or pointed end points to the quantity that is designated as less than and the large or open end of the symbol points to the quantity that is designated as greater than. Thus the symbol above can be said to be a **greater than** symbol or a **less than** symbol. When reading the symbol in a stated inequality, **we read the part of the symbol that we come to first and do not read the other part**. For example, we read the following statement from left to right

$$4 < 10$$

as 4 is less than 10 because when we read from left to right, we come to the small or pointed end first. We read the same statement from right to left as 10 is greater than 4 because when we read from right to left, we come to the large or open end first. Many authors call the symbol $<$ a **less than symbol** because we normally read from left to right. If the symbol is reversed, $>$, they call it a **greater than symbol**.

The basic symbols of equality and inequality are the equals sign and the greater than/less than symbol and a combination of the two symbols that means equal to or greater than/less than.

$$= \qquad > \qquad \geq$$

A diagonal line can be used to negate the basic symbol. For example, the notations

$$4 \neq 6 \qquad x \neq 7$$

are read 4 **is not equal to** 6, and x **is not equal to** 7. The inequality

$$x \nless 6$$

is read from left to right as x **is not less than** 6 and from right to left as 6 **is not greater than** x.

The inequality

$$x \nleq 6$$

is read from left to right as x **is not less than or equal to** 6 and from right to left as 6 **is not greater than or equal to** x.

Inequalities that contain variables such as $x < 4$ can be true statements or false statements depending on the replacement value used for the variable. If a number that we use as a replacement for the variable makes the inequality a true statement, we say that the number is a **solution of the inequality** and say that the number **satisfies the inequality**. If the number that we use as a replacement for the variable makes the inequality a false statement, then the number is not a solution of the inequality and does not satisfy the inequality. Since more than one number will often satisfy a given inequality, there is often more than one solution to the inequality.

> We call the **set of numbers** that will satisfy a given equation or inequality the **solution set** of the equation or inequality.

9.2 greater than and less than

Zero is a real number. Also any other number that can be used to describe a physical distance is a real number and the opposite of each of these numbers is also a real number.

We can use the number line to give us a picture of the way real numbers are ordered. We know that the number 4 is greater than the number 2 and we can **see** this on the number line. We say that **a number is greater than a given number if its graph is farther to the right than the graph of the given number**.

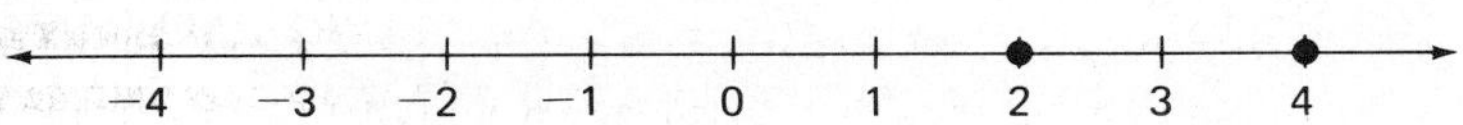

Figure 9.1

In Figure 9.1 we have graphed the numbers 2 and 4. The graph of 4 is farther to the right than the graph of 2 and by the definition above, we can **see** that 4 is greater than 2.

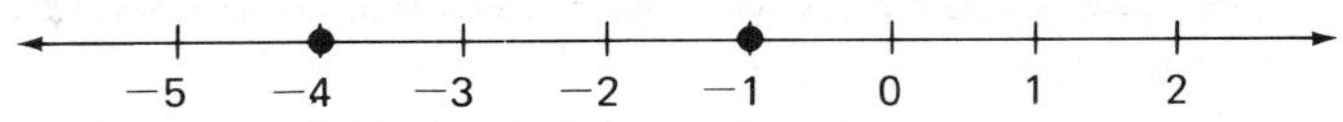

Figure 9.2

In Figure 9.2 we show the graphs of -1 and -4 and in this figure we can see that -1 is greater than -4 because the graph of -1 lies farther to the right than does the graph of -4.

9.3 graphical solutions of inequalities

We can use the number line to display the graph or the picture of the solution to many problems. For instance

$$x > 2$$

is read from left to right as x **is greater than** 2. If we are asked what replacement values for x would satisfy this inequality, we would reply that any real number that is greater than 2 will satisfy the inequality and is therefore a solution to the inequality. We can indicate the numbers that satisfy the inequality by using an arrow on the number line as shown in Figure 9.3.

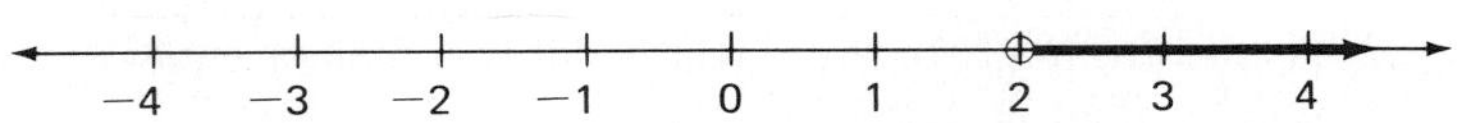

Figure 9.3

We draw a heavy line to indicate all numbers that are greater than 2. The open circle at 2 indicates that 2 is not a solution because 2 is not greater than 2. The arrowhead indicates that all numbers to the right of the arrowhead are also solutions.

example 9.3.1 Graph the solution of $x \leq 2$.

solution This inequality is read from left to right as x **is less than or equal to** 2. Thus we are asked to show the location on the number line of all numbers that equal 2 or are less than 2.

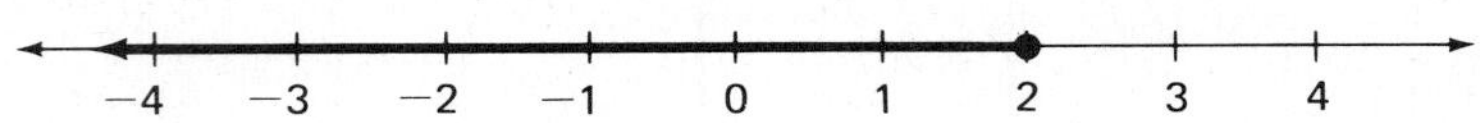

The locations of the numbers that satisfy the condition are indicated by the heavy line. The solid circle at 2 indicates that the number 2 is a part of the solution of $x \leq 2$.

example 9.3.2 Graph the solution of $x \nless 2$.

solution We read the inequality from left to right as x **is not less than** 2. Thus we are to indicate the location of all numbers that are not less than 2.

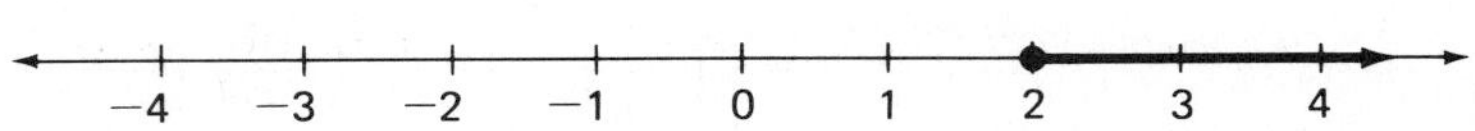

The solid circle at 2 indicates that 2 is a number that is not less than 2 and thus the number 2 also satisfies the stated condition.

example 9.3.3 Graph the solution of $x \ngeq 3$.

solution We are asked to indicate all numbers that **are not greater than or equal to** 3.

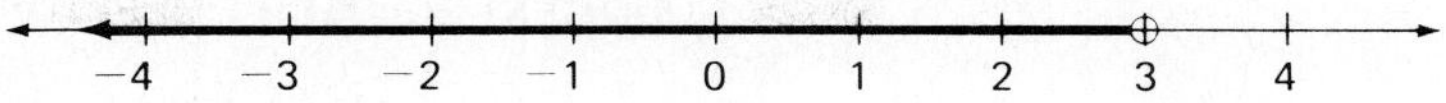

Here we graph all numbers that are not greater than or equal to 3. This is the set of all numbers that are less than 3. The open circle at 3 indicates that the number 3 does not satisfy the given condition.

TEST YOUR UNDERSTANDING

(a) Graph the solution to $x \nleq -2$.

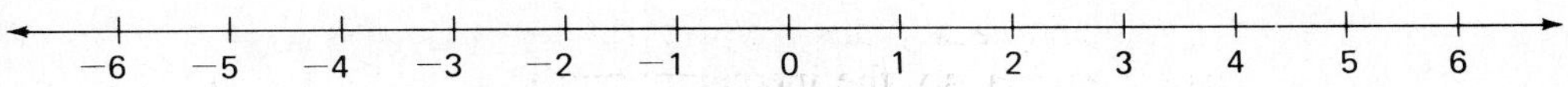

9.4 quotient theorem for exponents

Let's review our theorems and definitions for exponents.

DEFINITION:	$\underbrace{x \cdot x \cdot x \cdots x}_{n \text{ factors}} = x^n$
DEFINITION:	If x is not zero, $x^0 = 1$
DEFINITION:	$x^1 = x$
PRODUCT THEOREM:	If x is not zero, $x^m \cdot x^n = x^{m+n}$
DEFINITION:	If x is not zero, $x^{-n} = \dfrac{1}{x^n}$

The **QUOTIENT THEOREM FOR EXPONENTS** is really an extension of the last definition above that says $x^{-n} = \frac{1}{x^n}$. If we wish to multiply

$$x^5 \text{ times } \frac{1}{x^2}$$

we can use the definition of x^{-n} to write $\frac{1}{x^2}$ as x^{-2} and then multiply by using the product theorem

$$x^5 \cdot \frac{1}{x^2} = x^5 \cdot x^{-2} = x^{5-2} = x^3$$

The quotient theorem permits just the same procedure in just one step.

> QUOTIENT THEOREM FOR EXPONENTS
>
> If m and n are real numbers and $x \neq 0$,
>
> $$\frac{x^m}{x^n} = x^{m-n} = \frac{1}{x^{n-m}}$$

example 9.4.1 Simplify $\dfrac{x^6}{x^4}$.

solution We know that we can move the x^4 from the denominator to the numerator if we change the sign of the 4 from plus to minus.

$$\frac{x^6}{x^4} = x^6x^{-4} = x^{6-4} = \boldsymbol{x^2}$$

If we use the quotient theorem, we can omit the first step and write

$$\frac{x^6}{x^4} = x^{6-4} = \boldsymbol{x^2}$$

example 9.4.2 Simplify $\dfrac{x^6}{x^4}$ but this time write the exponential in the denominator. We will use the quotient theorem and write

solution

$$\frac{x^6}{x^4} = \frac{1}{x^{4-6}} = \boldsymbol{\frac{1}{x^{-2}}}$$

example 9.4.3 Simplify $\dfrac{x^6}{x^{-4}}$.

solution We will work the problem twice. The first time we will use the quotient theorem so that the final exponential is in the numerator.

(a) $$\frac{x^6}{x^{-4}} = x^{6+4} = x^{10}$$

(b) This time we will put the exponential in the denominator.

$$\frac{x^6}{x^{-4}} = \frac{1}{x^{-4-6}} = \frac{1}{x^{-10}}$$

example 9.4.4 Simplify $\dfrac{x^{-a}}{x^b}$.

solution We will use the quotient theorem to simplify with the final exponential being in the numerator.

(a) $$\frac{x^{-a}}{x^b} = x^{-a-b}$$

(b) And this time we will put the exponential in the denominator.

$$\frac{x^{-a}}{x^b} = \frac{1}{x^{b+a}}$$

The two answers shown in (a) and (b) are **equivalent expressions**, which means that they have the same **value** no matter which non-zero real number is used as a replacement for x.

example 9.4.5 Simplify $\dfrac{x^{-5}y^6z}{z^{-3}y^2x}$.

solution We will find four equivalent expressions for this expression.

(a) $x^{-6}y^4z^4$ (b) $\dfrac{1}{x^6y^{-4}z^{-4}}$ (c) $\dfrac{y^4z^4}{x^6}$ (d) $\dfrac{x^{-6}}{y^{-4}z^{-4}}$

Answer (a) is written with all exponentials in the numerator; (b) has all exponentials in the denominator; (c) has all exponents positive; and (d) has all exponents negative. No one of these forms is more correct than another and none is necessarily preferred. However many authors and teachers prefer final results wirtten with all exponents positive as in (c).

example 9.4.6 Multiply $x^{-2}y\left(\dfrac{y^4}{x^2} - \dfrac{x^4}{y^2}\right)$ and write the product with all exponentials in the numerator.

solution We begin by using the distributive property to multiply

$$x^{-2}y\left(\frac{y^4}{x^2} - \frac{x^4}{y^2}\right) = \frac{x^{-2}yy^4}{x^2} - \frac{x^{-2}yx^4}{y^2} = \frac{y^5}{x^4} - \frac{x^2}{y}$$

If we wish we can write this result with all exponentials in the numerator as

$$\mathbf{x^{-4}y^5 - x^2y^{-1}}$$

example 9.4.7 Multiply $\dfrac{k^2b}{p^{-2}}\left(\dfrac{ab^{-1}}{k^2} - \dfrac{4p^2}{b}\right)$ and write the product with all exponentials in the denominator.

solution Again we begin by using the distributive property to multiply

$$\frac{k^2b}{p^{-2}}\left(\frac{ab^{-1}}{k^2} - \frac{4p^2}{b}\right) = \frac{k^2bab^{-1}}{p^{-2}k^2} - \frac{4p^2k^2b}{p^{-2}b}$$

If we wish we can write this product with all exponentials in the denominator as

$$\mathbf{\frac{1}{a^{-1}p^{-2}} - \frac{4}{p^{-4}k^{-2}}}$$

TEST YOUR UNDERSTANDING

Simplify (write answers with all exponents negative):

(b) $\dfrac{x^{-5}y^6zx^0}{y^0y^4z^3}$ (c) $\dfrac{k^{-2}b^0}{x^2}(x^{-3} + k^5x^{-4})$

9.5 adding like terms

Sometimes it is difficult to determine if terms in an expression are like terms. For instance, if we look at the expression

$$\frac{bx}{x^3y^{-2}} - 3by^2x^{-2} + \frac{4y^2}{bx^2}$$

it is rather difficult to see that two of the terms are like terms and thus may be added. If each of the terms is written in the same form, however, it is easy to identify like terms. Let's write each of the three terms with all exponents positive.

$$\frac{by^2}{x^2} - \frac{3by^2}{x^2} + \frac{4y^2}{bx^2}$$

We see that the first two terms are like terms and may be added. The last term is different and thus cannot be added.

$$\frac{by^2}{x^2} - \frac{3by^2}{x^2} + \frac{4y^2}{bx^2} = -\frac{\mathbf{2by^2}}{\mathbf{x^2}} + \frac{\mathbf{4y^2}}{\mathbf{bx^2}}$$

Now we will work the problem again by first writing the original terms so that all exponents are negative.

$$\frac{x^{-2}}{b^{-1}y^{-2}} - \frac{3x^{-2}}{b^{-1}y^{-2}} + \frac{4b^{-1}x^{-2}}{y^{-2}} = -\frac{2x^{-2}}{b^{-1}y^{-2}} + \frac{4b^{-1}x^{-2}}{y^{-2}}$$

Again we see that the first two terms are like terms and may be added. To see if terms are like terms, we can put them in any form we wish as long as we use the same form for every term. However, most people feel more comfortable with the exponentials written with all positive exponents.

example 9.5.1 Add like terms.

$$\frac{bx}{x^3y^{-2}} - 3by^2x^{-2} + \frac{4y^2}{b^{-1}x^2}$$

solution To help us identify like terms, we will rewrite each term so that all exponents are positive.

$$\frac{by^2}{x^2} - 3\frac{by^2}{x^2} + \frac{4by^2}{x^2}$$

We see that all three terms are like terms and can be added by adding the numerical coefficients.

$$\frac{by^2}{x^2} - \frac{3by^2}{x^2} + \frac{4by^2}{x^2} = \frac{2by^2}{x^2}$$

example 9.5.2 Add like terms.

$$\frac{a^{-3}b}{b^{-3}} + \frac{2b^4}{a^3}$$

solution We begin by writing each term with all exponents positive.

$$\frac{b^4}{a^3} + \frac{2b^4}{a^3}$$

Now we see that the terms are like terms and may be added.

$$\frac{b^4}{a^3} + \frac{2b^4}{a^3} = \frac{3b^4}{a^3}$$

example 9.5.3 Add like terms.

$$\frac{7a^{-3}b^2}{c^{-1}} - \frac{5b^2}{a^3c^{-1}} + \frac{3b^2}{a^3c}$$

solution We begin by writing each term with all exponents positive.

$$7\frac{cb^2}{a^3} - 5\frac{cb^2}{a^3} + 3\frac{b^2}{a^3c}$$

Now we see that the first two terms are like terms and may be added but that the third term is different. Thus the simplification is

$$\frac{2cb^2}{a^3} + \frac{3b^2}{a^3c}$$

TEST YOUR UNDERSTANDING

(d) Add like terms: $4x^{-4}yp^{-3} - 2\frac{yp^3}{x^4} + 5\frac{y}{x^4p^3}$

lesson checklist 9

- ☐ Symbols of inequality
- ☐ Definition of *greater than*
- ☐ Graphical solution of inequalities
- ☐ Quotient theorem for exponents
- ☐ Equivalent expressions
- ☐ Writing expressions in the same form

ANSWERS TO TEST YOUR UNDERSTANDING

(a) 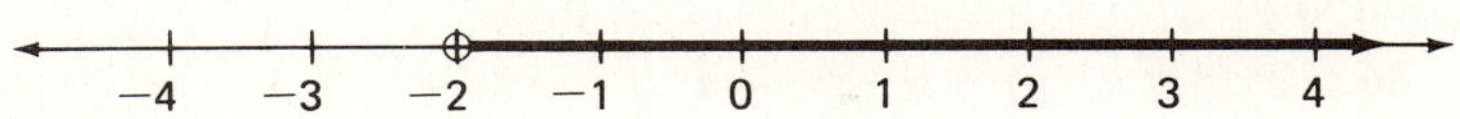

(b) $\frac{z^{-2}x^{-5}}{y^{-2}}$ (c) $k^{-2}x^{-5} + \frac{x^{-6}}{k^{-3}}$ (d) $\frac{9y}{x^4p^3} - \frac{2yp^3}{x^4}$

problem set 9

1. Is 23 a prime number? Why?
2. Use the numbers 7 and 11 to demonstrate the commutative property of addition.
3. Use the letters a, b, and c to demonstrate the distributive property.
4. What is another name for the multiplicative inverse of a number?

Graph on a number line:

***5.** $x \le 2$ ***6.** $x \nless 2$ ***7.** $x \ngtr 3$

8. $x < 3$ **9.** $x \ge -3$

Simplify:

***10.** $\dfrac{x^{-5}y^{6}z}{z^{-3}y^{2}x}$ **11.** $\dfrac{a^2b^{-2}c}{c^2ab}$ **12.** $\dfrac{m^7k^{-4}x^0p}{m^2k^6p^4}$

13. $\dfrac{z^4p^2pz}{z^{-2}p^2m^0}$ **14.** $\dfrac{15a^2x^4m^{-3}a^0}{3a^{-2}x^2m^{-4}}$

Add like terms:

***15.** $\dfrac{a^{-3}b}{b^{-3}} + \dfrac{2b^4}{a^3}$ **16.** $\dfrac{2x^2}{y} + \dfrac{6y^{-1}}{x^{-2}}$

17. $\dfrac{2ab}{c^{-2}} - 3b^2ac + 2c^2ab - 3b^2ac$ **18.** $\dfrac{2a^2bc}{x} - \dfrac{3ba^2}{c^{-1}x} + \dfrac{5bc}{xa^{-2}}$

Multiply:

***19.** $x^{-2}y\left(\dfrac{y^4}{x^2} - \dfrac{x^4}{y^2}\right)$ ***20.** $\dfrac{k^2b}{p^{-2}}\left(\dfrac{ab^{-1}}{k^2} - \dfrac{4p^2}{b}\right)$

21. $\dfrac{km^0}{p^2}\left(\dfrac{4k^{-2}}{p^{-4}} - \dfrac{5x^0p^2}{k}\right)$

Express as a product of prime factors:

22. 250 **23.** 525 **24.** 54

25. 585 **26.** 462

Factor:

27. $x^2yb^3 - 4x^2b^3y^2 - bx^2y$ **28.** $2m^2b^3x^5 + 2mb^3x^6 - 2mb^6x^3$

Simplify:

29. $|-14|$ **30.** $|14 - 28|$

Solve:

31. $k - 5 - 3k + 6 = -4(2 - 5)$ **32.** $2b - 2 + (b - 3) = 3(-2b + 5)$

33. $9 - 3(2x - 7) - 9x = 5x - 2[6x - (4 - x) - 20]$

34. $2z + 10 - 3z = 7 - 3z + 5$ **35.** $-3(2a - 2) + 6 = -3(-3) + 2$

Simplify:

36. $2x(3x^2 - x) - x^2(3x - 4)$

37. $4(2m - 3n) - \{5[4m - 2(m - 3n)] - 10n\}$

38. $3x(4x^2 - 2x - 3) - 2x(3x^2 - x + 1)$

39. $\dfrac{x^0y^0}{m^{-6}}$ **40.** $\dfrac{m^0x^0y^{-3}}{m^2x^{-2}y^{-5}}$

Evaluate:

41. $-a(b + c) - c$ when $a = 1, b = -2, c = -3$

42. $-k + 2b - (-c)$ when $k = -3, b = 2, c = 1$

---

*The problems marked with an asterisk were worked out in this lesson.

43. $\dfrac{(k)k + c}{-k + c}$ when $k = 1, c = -2$

44. $-p(b - a) + p$ when $a = 2, p = -3, b = 4$

45. $\dfrac{-2m(3p - n)}{(p + 2n)(-2)}$ when $m = 4, n = -2, p = 1$

46. $a^2 - ab^2$ if $a = -2, b = -4$

Simplify:

47. $\dfrac{1}{(-3)^{-2}}$

Solve:

48. $\dfrac{p}{3} = 16$

49. $\dfrac{2}{11}p = 41$

50. $\dfrac{2p}{11} = 41$

Simplify:

51. $-3 + (-5) - \{-4[3 - 7(4 - 3) + 2]\}$

52. $-6(-3\{-2 - 5 - 4\} - 2)$

lesson ten

Equations with more than one variable
Least common multiple
Denominator numerator same quantity theorem
Addition of fractions

10.1 equations with more than one variable

If we are asked to **SOLVE** the equation $x - 11 + 3x = -6 - x + 5$, we proceed as follows:

$x - 11 + 3x = -6 - x + 5$	original equation
$4x - 11 = -1 - x$	added like terms
$+x \qquad\qquad +x$	add $+x$ to both sides
$5x - 11 = -1$	
$+11 \qquad +11$	add $+11$ to both sides
$5x = 10$	
$\frac{5x}{5} = \frac{10}{5}$	divide both sides by 5
$\boldsymbol{x = 2}$	

The original equation had several terms that had x as a factor and also had terms that were constants. **When we were asked to solve the equation, we were asked to rearrange the equation so that x would be the sole member of one side of the equation (either side) and that the other side of the equation would consist of a single numeral.** We accomplished this in the last step when we wrote

$$x = 2$$

If an equation contains more than one variable and we are asked to solve the equation for one of the variables therein, our task is the same as that described above. **We are asked to rearrange the equation so that the designated variable is the sole member of one side of the equation (either side).** This time, however, the other side of the equation will contain variables as well as numerals.

example 10.1.1 Solve for y: $6y - x + z = 4$.

solution We will begin the process of **isolating** y on one side of the equation (either side) by moving terms that do not contain y to the other side of the equation.

$$\begin{array}{rl} 6y - x + z = 4 & \text{original equation} \\ +x - z \quad +x - z & \text{add } +x - z \text{ to both sides} \\ \hline 6y \quad = 4 + x - z & \end{array}$$

Now we complete the isolation by dividing every term by 6.

$$\frac{6y}{6} = \frac{4}{6} + \frac{x}{6} - \frac{z}{6} \longrightarrow y = \frac{2}{3} + \frac{x}{6} - \frac{z}{6}$$

example 10.1.2 Solve for y: $4x - 2y + 2 = y - 4$.

solution We begin by adding $2y$ to both sides.

$$\begin{array}{rl} 4x - 2y + 2 = \quad y - 4 & \text{original equation} \\ +2y \qquad +2y & \text{add } +2y \text{ to both sides} \\ \hline 4x \quad +2 = \quad 3y - 4 & \end{array}$$

Now we have all the y's on the right side. To **isolate** y on the right we must eliminate the -4 and the 3 that are on the right side. To eliminate the -4, we add $+4$ to both sides.

$$\begin{array}{rl} 4x + 2 = 3y - 4 & \\ +4 \qquad +4 & \text{add } +4 \text{ to both sides} \\ \hline 4x + 6 = 3y & \end{array}$$

Now we complete the isolation of y by dividing every term by 3.

$$\frac{4x}{3} + \frac{6}{3} = \frac{3y}{3}$$

$$\frac{4x}{3} + 2 = y \qquad \text{divided every term by 3}$$

example 10.1.3 Solve for p: $4p + 2a - 5 = 6a + p$.

solution For no particular reason we decide to place all p's on the left side so as a first step we add $-p$ to both sides.

$$\begin{array}{rl} 4p + 2a - 5 = 6a + p & \text{original equation} \\ -p \qquad\qquad -p & \text{add } -p \text{ to both sides} \\ \hline 3p + 2a - 5 = 6a & \end{array}$$

Now all p's are on the left side so we add $-2a + 5$ to both sides to position all terms that do not contain p on the right side.

$$\begin{array}{rl} 3p + 2a - 5 = \quad 6a & \\ -2a + 5 \quad -2a + 5 & \text{add } -2a + 5 \text{ to both sides} \\ \hline 3p \qquad = \quad 4a + 5 & \end{array}$$

As the final step we divide every term by 3 and get

$$p = \frac{4}{3}a + \frac{5}{3} \qquad \text{divided every term by 3}$$

example 10.1.4 Solve for x: $5y + x - 2y - 4 + 3x = 0$.

solution We will begin by adding like terms and then use the additive property of equality to place all terms that contain x on one side of the equation and all terms that do not contain x on the other side of the equation.

$$\begin{array}{rlr} 3y+4x-4= & 0 & \text{added like terms} \\ -3y \qquad +4= & -3y+4 & \text{add } -3y+4 \text{ to both sides} \\ \hline 4x \qquad = & -3y+4 & \end{array}$$

Now we complete the process of **isolating** x by dividing every term by 4.

$$\frac{4x}{4} = \frac{-3y}{4} + \frac{4}{4}$$

$$\boldsymbol{x = -\frac{3}{4}y + 1} \qquad \text{divided by 4}$$

example 10.1.5 Solve for n: $3m + 2n + a = 5m - 4n + 4$.

solution We decide to place n on the right side and everything else on the left side. We can do this in one step by adding $-2n - 5m - 4$ to both sides of the equation.

$$\begin{array}{rlr} 3m+2n+a = & 5m-4n+4 & \text{original equation} \\ -4-5m-2n \qquad & -5m-2n-4 & \text{add } -5m-2n-4 \text{ to both sides} \\ \hline -4-2m \qquad +a = & -6n & \end{array}$$

We want to end up with n as the sole member of the right side of the equation. We see that on the right side we now have $-6n$, so we need to eliminate both the negative sign and the 6. Both can be eliminated in one step by dividing both sides by -6, but many students at this level find a two-step method leads to fewer mistakes.

$$-2m + a - 4 = -6n$$

As the first step we will multiply every term on both sides of the equation by -1. **Multiplication by -1 will change every sign.** Thus we get

$$2m - a + 4 = 6n$$

Now we complete the problem by dividing every term by 6.

$$\frac{2m}{6} - \frac{a}{6} + \frac{4}{6} = \frac{6n}{6} \longrightarrow \boldsymbol{\frac{1}{3}m - \frac{1}{6}a + \frac{2}{3} = n}$$

example 10.1.6 Solve for y: $4y + 6x - 4 = 2$.

solution Since only one term contains a y, we begin by moving all other terms to the right side.

$$\begin{array}{rlr} 4y+6x-4= & 2 & \text{original equation} \\ -6x+4 \qquad & +4-6x & \text{add } +4-6x \text{ to both sides} \\ \hline 4y = & 6-6x & \\ \frac{4y}{4} = & \frac{6}{4} - \frac{6x}{4} & \text{divide by 4} \\ \boldsymbol{y} = & \boldsymbol{\frac{3}{2} - \frac{3}{2}x} & \end{array}$$

TEST YOUR UNDERSTANDING

(a) Solve for y: $2y + 3x - 5 = -4y + 3$

10.2 least common multiple

If we are given the numbers

$$4 \quad 5 \quad \text{and} \quad 8$$

and are asked to find the **smallest number that is evenly divisible by each of the numbers**, a reasonable guess would be the product of the numbers, which is 160, because we know that each of the numbers will divide 160 evenly.

$$\frac{160}{4} = 40 \qquad \frac{160}{5} = 32 \qquad \frac{160}{8} = 20$$

But 160 is not the smallest number that is evenly divisible by the three numbers. The number 40 is.

$$\frac{40}{4} = 10 \qquad \frac{40}{5} = 8 \qquad \frac{40}{8} = 5$$

We call the number that is the smallest number that can be divided evenly by each of a group of specified numbers the **LEAST COMMON MULTIPLE** of the specified numbers.

We can give a two-step procedure for finding the least common multiple.

1. Write each of the given numbers as a product of prime factors.
2. Write the least common multiple as a product of factors in which **each factor of the numbers appears as many times as it is used as a factor in any of the original numbers.**

Thus to find the least common multiple of 4, 5, and 8, we first write each of the numbers as a product of prime factors.

$$\overbrace{2 \cdot 2}^{4} \quad \overbrace{5}^{5} \quad \overbrace{2 \cdot 2 \cdot 2}^{8}$$

Now 2 is a factor of two of the numbers, and it appears three times as a factor of 8. Thus it must appear three times as a factor of the least common multiple.

$$2 \cdot 2 \cdot 2$$

Every factor of the original numbers must be a factor of the least common multiple so 5 must be factor. Thus the least common multiple of 4, 5, and 8 is 40 because

$$5 \cdot 2 \cdot 2 \cdot 2 = 40$$

example 10.2.1 Find the least common multiple of 100, 15, and 8.

solution We begin by writing each of the numbers as a product of prime numbers.

$$\begin{array}{ccc} 100 & 15 & 8 \\ 2 \cdot 2 \cdot \underbrace{\mathbf{5 \cdot 5}} & \underbrace{\mathbf{3}} \cdot 5 & \underbrace{\mathbf{2 \cdot 2 \cdot 2}} \end{array}$$

We note that 5 is a factor twice, 3 is a factor once, and 2 is a factor three times. Thus the LCM is

$$\underbrace{\mathbf{5 \cdot 5}} \cdot \underbrace{\mathbf{3}} \cdot \underbrace{\mathbf{2 \cdot 2 \cdot 2}} = 600$$

So 600 is the smallest number that is evenly divisible by each of the three numbers, 100, 15, and 8.

example 10.2.2 Find the LCM of 80 and 75 and 30.

solution We begin by writing each of the numbers as a product of prime numbers.

$$\begin{array}{ccc} 80 & 75 & 30 \\ \underbrace{\mathbf{2 \cdot 2 \cdot 2 \cdot 2}} \cdot \underbrace{5} & 3 \cdot \underbrace{\mathbf{5 \cdot 5}} & 2 \cdot 3 \cdot 5 \end{array}$$

So the LCM is $\underbrace{\mathbf{2 \cdot 2 \cdot 2 \cdot 2}} \cdot \underbrace{\mathbf{3}} \cdot \underbrace{\mathbf{5 \cdot 5}} = \underbrace{\mathbf{1200}}$

TEST YOUR UNDERSTANDING

(b) Find the LCM of 16, 12, and 50.

DEFINITION

The **least common multiple** of two or more terms is the product of all prime algebraic factors of the terms, each to the highest power that it occurs in any one of the terms.

In the following problems we will demonstrate how to use this definition.

example 10.2.3 Find the LCM of $15a^2b$ and $10ab^3$.

solution We begin by writing the expressions as products of factors whose exponents are one.

$$15a^2b \qquad\qquad 10ab^3$$

$$\underbrace{3} \cdot 5 \cdot \underbrace{a \cdot a} \cdot b \qquad\qquad \underbrace{2} \cdot \underbrace{5} \cdot a \cdot \underbrace{b \cdot b \cdot b}$$

The LCM is

$$2 \cdot 3 \cdot 5 \cdot a \cdot a \cdot b \cdot b \cdot b = \mathbf{30a^2b^3}$$

example 10.2.4 Find the LCM of $4x^2m$ and $6x^3m$.

solution We begin the same way.

$$4x^2m \qquad\qquad 6x^3m$$

$$\underbrace{2 \cdot 2} \cdot x \cdot x \cdot m \qquad\qquad \underbrace{3} \cdot 2 \cdot \underbrace{x \cdot x \cdot x} \cdot \underbrace{m}$$

The LCM is

$$2 \cdot 2 \cdot 3 \cdot x \cdot x \cdot x \cdot m \quad \text{or} \quad \mathbf{12x^3m}$$

TEST YOUR UNDERSTANDING

(c) Find the LCM of $15x^2ym$, $5xy^3m^2$, and $20x^3y^2m^5$.

10.3 denominator numerator same quantity theorem

If we divide a non-zero real number by itself, we have an equivalent expression for 1.

$$\frac{5}{5} \quad \frac{4}{4} \quad \frac{16}{16}$$

Thus all the expressions above represent the number 1. Since the product of a number and 1 is the number itself, we do not change the value of a number if we multiply the number by another number and also divide by the other number. This is a useful tool in changing the form of a particular number. For instance, if we wish to express 10 as a fraction that has a denominator of 17, we can multiply 10 by 17 over 17 and get $\frac{170}{17}$, which is another numerical expression for 10

$$10 \cdot \frac{17}{17} = \mathbf{\frac{170}{17}}$$

What we really did was to multiply 10 by 1 for $\frac{17}{17}$ equals 1. If we wish to write $\frac{5}{8}$ as a fraction that has a denominator of 40, we can multiply $\frac{5}{8}$ by 5 over 5.

$$\frac{5}{8} \cdot \frac{5}{5} = \mathbf{\frac{25}{40}}$$

We did not change the value of $\frac{5}{8}$ because we multiplied by $\frac{5}{5}$, which is really multiplying by 1. The expressions $\frac{5}{8}$ and $\frac{25}{40}$ are equivalent expressions for the same number. The theorem that describes this phenomenon is normally called **THE FUNDAMENTAL THEOREM OF FRACTIONS** and since another name for a fraction is a **RATIONAL EXPRESSION**, this theorem is also called **THE FUNDAMENTAL THEOREM OF RATIONAL EXPRESSIONS.** We will use non-standard terminology and call it the **DENOMINATOR NUMERATOR SAME QUANTITY THEOREM** because we will use this theorem often and this name helps us to remember what the theorem is.

> DENOMINATOR NUMERATOR SAME QUANTITY THEOREM
>
> The denominator and the numerator of a fraction may be multiplied by the same non-zero quantity without changing the value of the fraction.

10.4 addition of fractions

The sum of

$$\frac{1}{11} + \frac{3}{11}$$

is $\frac{4}{11}$ because the procedure for adding fractions that have the same denominator is to add the numerators algebraically and write the sum of the numerators over a single denominator. Thus

$$\frac{2}{11} - \frac{5}{11} + \frac{9}{11} = \frac{2 - 5 + 9}{11} = \mathbf{\frac{6}{11}}$$

The procedure is the same for fractions whose denominators contain variables. For example, if we wish to add the rational expressions

$$\frac{5}{a+6} + \frac{a+b}{a+6} + \frac{2}{a+6}$$

we see that the denominators are the same so we can add the numerators and record them over a single denominator.

$$\frac{5}{a+6} + \frac{a+b}{a+6} + \frac{2}{a+6} = \frac{5 + a + b + 2}{a+6} = \mathbf{\frac{7 + a + b}{a + 6}}$$

example 10.4.1 Add $\dfrac{4}{2x^2 + y} - \dfrac{6ax}{2x^2 + y}$.

solution The denominators are the same so we can add the numerators.

$$\frac{4}{2x^2 + y} - \frac{6ax}{2x^2 + y} = \mathbf{\frac{4 - 6ax}{2x^2 + y}}$$

example 10.4.2 Add $\dfrac{5}{a^2 + 7y} - \dfrac{3}{a^2 + 7y} + \dfrac{z}{a^2 + 7y}$.

solution The denominators are the same so we can add the numerators.

$$\frac{5}{a^2 + 7y} - \frac{3}{a^2 + 7y} + \frac{z}{a^2 + 7y} = \frac{5 - 3 + z}{a^2 + 7y} = \mathbf{\frac{2 + z}{a^2 + 7y}}$$

TEST YOUR UNDERSTANDING

(d) Add: $\dfrac{4x - 3}{x^2 + b^2} + \dfrac{3x + 5}{x^2 + b^2} - \dfrac{5x}{x^2 + b^2}$

We cannot find the sum of

$$\frac{1}{4} + \frac{1}{2}$$

in this form because the denominators are not the same. But if we use the **denominator numerator same quantity theorem** and multiply both the numerator and the denominator of $\frac{1}{2}$ by 2, we get $\frac{2}{4}$, which is an equivalent expression for $\frac{1}{2}$.

$$\frac{1 \cdot (2)}{2 \cdot (2)} = \frac{2}{4}$$

Now the fractions may be added, for they both have a denominator of 4.

$$\frac{1}{4} + \frac{1}{2} = \frac{1}{4} + \frac{1(2)}{2(2)} = \frac{1}{4} + \frac{2}{4} = \mathbf{\frac{3}{4}}$$

If the fractions to be added have different denominators, the procedure shown here can be used to rewrite the fractions as equivalent fractions with the same denominators.

example 10.4.3 Add $\dfrac{3}{4} + \dfrac{2}{b}$.

solution We will use the **denominator numerator same quantity theorem** and a three-step procedure to rewrite the fractions above as equivalent fractions that have the same denominator.

(a) As the first step we write the fraction lines with the proper sign between them:

$$\underline{\quad\quad} + \underline{\quad\quad}$$

(b) Now we write the least common multiple of the denominators as the new denominators. When the least common multiple is used in this fashion, we call it the **LEAST COMMON DENOMINATOR.**

$$\frac{}{4b} + \frac{}{4b}$$

(c) The first two steps were automatic. We used no theorems or rules. Now we use the **denominator numerator same quantity theorem**. We have multiplied the denominator, 4, of the first fraction by b to get $4b$ so we must also multiply the numerator, 3, by b and get $3b$.

$$\frac{3b}{4b} + \frac{}{4b}$$

We have multiplied the denominator, b, of the second fraction by 4 to get $4b$ so we must also multiply the numerator, 2, by 4 to get the new numerator of 8.

$$\frac{3b}{4b} + \frac{8}{4b}$$

Now the fractions have the same denominator and can be added.

$$\frac{3b}{4b} + \frac{8}{4b} = \frac{3b + 8}{4b}$$

example 10.4.4 Add $\frac{4}{b} + \frac{1}{c} + \frac{1}{2}$.

solution

(a) $\underline{\quad} + \underline{\quad} + \underline{\quad}$ write the fraction lines

(b) $\frac{}{2bc} + \frac{}{2bc} + \frac{}{2bc}$ use the LCM as the denominator of every term

(c) $\frac{8c}{2bc} + \frac{2b}{2bc} + \frac{bc}{2bc} = \frac{\mathbf{8c + 2b + bc}}{\mathbf{2bc}}$ find the new numerators and add

example 10.4.5 Add $\frac{m}{c^3} + \frac{4}{c^2} - \frac{1}{2}$.

solution

(a) $\underline{\quad} + \underline{\quad} - \underline{\quad}$ write the fraction lines

(b) $\frac{}{2c^3} + \frac{}{2c^3} - \frac{}{2c^3}$ use the LCM as the denominator of every term

(c) $\frac{2m}{2c^3} + \frac{8c}{2c^3} - \frac{c^3}{2c^3} = \frac{\mathbf{2m + 8c - c^3}}{\mathbf{2c^3}}$ find the new numerators and add

example 10.4.6 Add $\frac{p}{4} - \frac{a}{2} + \frac{c}{b}$.

solution

(a) $\underline{\quad} - \underline{\quad} + \underline{\quad}$ write the fraction lines

(b) $\frac{}{4b} - \frac{}{4b} + \frac{}{4b}$ use the LCM as the denominator of every term

(c) $\frac{pb}{4b} - \frac{2ab}{4b} + \frac{4c}{4b} = \frac{\mathbf{pb - 2ab + 4c}}{\mathbf{4b}}$ find the new numerators and add

example 10.4.7 Add $\frac{a}{c^2} + \frac{3}{4c^3} + \frac{m}{3c^4}$

solution (a) $\_\_\_ + \_\_\_ + \_\_\_$ write the fraction lines

(b) $\frac{}{12c^4} + \frac{}{12c^4} + \frac{}{12c^4}$ use the LCM as the denominator of every term

(c) $\frac{12ac^2}{12c^4} + \frac{9c}{12c^4} + \frac{4m}{12c^4} = \mathbf{\frac{12ac^2 + 9c + 4m}{12c^4}}$ find the new numerators and add

> **TEST YOUR UNDERSTANDING**
>
> (e) Add: $\frac{4}{b^3} + \frac{k}{b^2c} - \frac{x}{b^2c^2}$

lesson checklist 10

- ☐ Solution of equations that contain more than one variable
- ☐ Least common multiple
- ☐ Denominator numerator same quantity theorem
- ☐ Addition of fractions
- ☐ Least common denominator

> **ANSWERS TO TEST YOUR UNDERSTANDING**
>
> (a) $y = -\frac{1}{2}x + \frac{4}{3}$ (b) $2 \cdot 2 \cdot 2 \cdot 2 \cdot 3 \cdot 5 \cdot 5 = 1200$ (c) $60x^3y^3m^5$
>
> (d) $\frac{2x + 2}{x^2 + b^2}$ (e) $\frac{4c^2 + kbc - bx}{b^3c^2}$

problem set 10

*1. $6y - x + z = 4$, solve for y.

*2. $4x - 2y + 2 = y - 4$, solve for y.

*3. $4p + 2a - 5 = 6a + p$, solve for p.

*4. $5y + x - 2y - 4 + 3x = 0$, solve for x.

*5. $3m + 2n + a = 5m - 4n + 4$, solve for n.

6. $4y + 6x - 4x = 2$, solve for y.

Add:

7. $\frac{3}{4} + \frac{2}{b}$

*8. $\frac{4}{b} + \frac{1}{c} + \frac{1}{2}$

*9. $\frac{m}{c^3} + \frac{4}{c^2} - \frac{1}{2}$

*10. $\frac{p}{4} - \frac{a}{2} + \frac{c}{b}$

*11. $\frac{a}{c^2} + \frac{3}{4c^3} + \frac{m}{3c^4}$

---

*The problems marked with an asterisk were worked out in this lesson.

Graph on a number line:

12. $x \geq 5$ **13.** $x > 5$ **14.** $x \not\geq 5$ **15.** $x \not> 5$

Simplify:

16. $\dfrac{m^5b^2}{m^2b^{-5}}$ **17.** $\dfrac{m^2 4^2 a^{-4}}{4a^{-8}m}$ **18.** $\dfrac{(-2)^2}{4^{-2}}$

19. $\dfrac{a^0x^2y^{-4}a^6}{a^6x^{-2}y^{-8}}$ **20.** $\dfrac{p^5zx^0y^2}{y^{-3}p^{-2}z^{-5}}$ **21.** $\dfrac{(-2)^{-2}}{x^{-8}}$

22. $\dfrac{x^5y^{-2}m^2x^{-2}}{x^{-3}y^4m^3}$ **23.** $\dfrac{k^0x^2}{m^{-2}y^{-4}}$ **24.** $\dfrac{abc^2d^{-2}}{d^{-4}b^2c^{-2}}$

25. $\dfrac{am^2p^{-4}}{p^{-6}am^0}$

Add like terms:

26. $x^5y - 2xy^5 + 3x^5y - 7yx^5$ **27.** $\dfrac{2a^2x}{m^5} + 3a^2xm^{-5} - \dfrac{8m^{-5}x}{a^{-2}}$

Multiply:

28. $\dfrac{m^{-3}p}{x^2}\left(m^3x^{-2} + \dfrac{4}{m^{-3}x^2}\right)$

Find the LCM of:

***29.** 100, 15, 8 ***30.** 80, 75, 30

Solve:

31. $2(k - 3) + 4 - k = 2[7 + 3(k + 6)]$

32. $4m(4 - 6) + 3 = -2(m + 4) - 6$

33. $3x - 3 + (x - 4) = 3(-2x + 7)$

34. $5p - 4(-p + 2) + 5 = 8(p + 4) - 3$

Evaluate:

35. $x - 4[x + (-y)]$ if $x = 2, y = -3$

36. $-b(-a)(4 + a)$ if $b = 2, a = -3$

37. $-m(n - 3) + m(-n + 3)$ if $m = -2, n = -4$

38. $x[y - b(-k)]$ if $k = 2, b = -4, y = 3, x = 2$

Factor:

39. $xc^2 - bxc$ **40.** $a^2b^3c^4 - 2ab^2c^2 + 3a^2b^2c^3$

Simplify:

41. $|42 + 3|$ **42.** $|-42 - 3|$ **43.** $(-5)^{-2}$ **44.** 5^{-2}

Evaluate:

45. $a - a^2b$ if $a = -2, b = -5$

Solve:

46. $\dfrac{p}{3} = 4$ **47.** $\dfrac{5}{3}p = 9$ **48.** $\dfrac{5p}{3} = 9$

Simplify:

49. $-3 - 6[-5 - 2(6 - 3)(-2)]$

50. $-5[-3(6 - 4)(-2)] + 3[-5(-2 - 1)]$

lesson eleven

Monomials • binomials • trinomials • polynomials

Addition and multiplication of polynomials

11.1 polynomials

Thus far we have encountered expressions such as

(a) $\dfrac{x^{-3} + m}{x}$ (b) $\dfrac{2y}{x^3}$ (c) $4a^{-2}x + m^3$

(d) rt^{n-2} (e) 4 (f) $-4a^2$

(g) $7x^2 + 2$ (h) $4x^2y$ (i) $7y^3 + 3y + 2$

All the expressions shown here are called **algebraic expressions** or **mathematical expressions** and are individual terms or indicated sums of terms. The more complicated expressions as shown in (a) through (d) have no special names and are just called **terms** or **expressions.** The simple expressions as shown in (e) through (i) occur so often and are so useful that we give these expressions a special name: **POLYNOMIAL.**†

A polynomial in one variable is one term or a sum of individual terms each of which has the form

$$ax^n$$

where ***a* is a real number** and ***n* is a whole number,** such as the following:

(a) $4x^2$ (b) $-x^3$ (c) $-1.414x^{32}$

(d) $2x^4$ (e) -7

Each of the five expressions meets all three of the requirements for being called a polynomial:

---

†It is unfortunate that we use such an intimidating word to describe the simplest type of expression.

1. Each expression is in the **form ax^n.**
2. The numerical coefficient of each expression, (a), is a **real number.**
3. The exponent of the variable, (n), is a **positive integer or is the number zero.**

The last polynomial shown, -7, can be thought of as being $-7x^0$, which is the same as -7 if x has a value other than zero because **any non-zero quantity raised to the zero power has a value of one!**

We refresh our memory with the following examples.

$$(-15)^0 = 1 \qquad \left(\frac{75}{14}\right)^0 = 1 \qquad x^0 = 1 \qquad (x+y)^0 = 1 \qquad (x, x+y \neq 0)$$

Thus, since x^0 equals 1 if x is not zero, we can write

$$-7x^0 = -7(1) = -7$$

A **POLYNOMIAL** of one term is called a **MONOMIAL**, so each of the five expressions above can be called a polynomial or it may be described by using the more restrictive name of monomial.

A **POLYNOMIAL OF TWO TERMS** is also called a **BINOMIAL** and a **POLYNOMIAL OF THREE TERMS** is also called a **TRINOMIAL.** Thus each of the following

(a) $4 + x$ (b) $p^{15} - 4p$ (c) $-y^{10} + 3y$

can be described by using either the word **polynomial** or the word **binomial** and these expressions

(a) $x^2 + 2x + 4$ (b) $y^{14} - 1.6y^2 + 4$ (c) $m^4 + 2m - 1.6$

can be called either **polynomials** or **trinomials**. Indicated sums of more than three monomial terms have no special names and are just called polynomials. Thus we can call any of the following expressions a polynomial.

(a) -14.2 (b) $\frac{7}{2}$ (c) $4x^2$ (d) $6x^2 + 4$

(e) $x^4 - 3x^2 + 2x$ (f) $-7x^{15} + 2x^3 - 5x^2 + 6x + 4$

The first two expressions are monomials or polynomials because the exponents of the understood variable is the number zero.

$$-14.2m^0 = -14.2 \qquad \frac{7}{2}y^0 = \frac{7}{2}$$

The third, fourth, and fifth expressions are all polynomials or can be described in order as being a monomial, a binomial and a trinomial. The sixth expression, (f), is a polynomial because each term in the indicated sum is a monomial. This expression does not have a more restrictive name, for we usually use no special names for polynomials of four or more terms.

> DEFINITION OF A POLYNOMIAL IN ONE UNKNOWN
>
> A polynomial in one unknown is an algebraic expression of the form
>
> $$a_nx^n + a_{n-1}x^{n-1} + \cdots + a_1x + a_0$$
>
> where n is a whole number, x is a variable, and the coefficients $a_n, a_{n-1}, \ldots, a_1, a_0$ are real numbers.

Thus none of the following expressions is a polynomial.

(a) $4x^{-3}$ (b) $\frac{-6x + y}{z}$ (c) $-15y^{-5}$

The expressions (a) and (c) have real number coefficients but the exponent of the variable is not a positive integer or zero. Expression (b) is not a polynomial because it is not in the required form of ax^n.

11.2 polynomials, general

> DEFINITION OF A POLYNOMIAL
>
> A polynomial in one or more unknowns is an algebraic expression having only terms of the form $ax^ny^mz^p \cdots$ where the coefficient a is a real number and the exponents $n, m, p, \cdots$ are whole numbers.

Thus the general definition of a polynomial has the same restrictions that are given by the definition of a polynomial in one unknown, namely:

1. The numerical coefficients of the individual terms must be **REAL NUMBERS.**
2. The exponents of the variables must be **WHOLE NUMBERS.**

The following can therefore be called polynomials in more than one variable.

(a) xyz^2m (b) $4x^{15}ym^3 + pq^5$ (c) $-11x^2p^4 + 2$

11.3 degree

The **DEGREE OF A TERM** of a polynomial is the sum of the exponents of the variables in the term. Thus

$4x^3, 6xym, 2x^2y$ are third-degree terms.

$4x^2m^3, 3y^5, 2xypmz$ are fifth-degree terms.

The **DEGREE OF A POLYNOMIAL** is the same as the degree of its highest degree term.

$3x^2 + xyz + m$ is a third-degree polynomial because the degree of its highest degree term (xyz) is three.

$4x^5 + yx^3 + 2x^2 + 2$ is a fifth-degree polynomial because the degree of its highest degree term ($4x^5$) is five.

> TEST YOUR UNDERSTANDING
>
> What is the degree of each of the following polynomials?
>
> (a) $xyzm + 6$ (b) $x^2 + xym^2$ (c) $x^2y^5 + z^3$

Polynomials are usually written in descending powers of one of the variables. The polynomials

$$x^5 - 3x^4 + 2x^2 - x + 5$$

$$x^4m + x^3m^2 - 2xm^5 - 6$$

$$-2xm^5 + x^3m^2 + x^4m - 6$$

are written in descending powers of a particular variable. The first two polynomials are written in descending powers of x. The third polynomial is the same polynomial as the second but is written in descending powers of m instead of descending powers of x.

11.4 addition of polynomials

Since polynomials are composed of individual terms, the rule for adding polynomials is the same rule that we use for adding terms—**like terms may be added.**

example 11.4.1 Add $(x^3 + 3x^2 + 2) + (2x^3 + 4)$.

solution We remember that we can discard parentheses preceded by a plus sign without changing the signs of the terms therein.

$$(x^3 + 3x^2 + 2) + (2x^3 + 4) = x^3 + 3x^2 + 2 + 2x^3 + 4$$
$$= \mathbf{3x^3 + 3x^2 + 6}$$

example 11.4.2 Add $(3x^4 - 2x^2 + 3) - (x^4 - 2x^3 + x^2)$.

solution When we discard the parentheses in this problem, we remember to **change the sign of every term in the second parentheses** because the second parentheses are preceded by a minus sign.

$$3x^4 - 2x^2 + 3 - x^4 + 2x^3 - x^2 = \mathbf{2x^4 + 2x^3 - 3x^2 + 3}$$

example 11.4.3 Add $(3x^3 + 2x^2 - x + 4) - (x^2 - 7x - 5)$.

solution The first parentheses can be removed with no change to the terms inside, but because the second parentheses are preceded by a minus sign, when we remove these parentheses, we must **change the sign of all terms therein.**

$$(3x^3 + 2x^2 - x + 4) - (x^2 - 7x - 5) = 3x^3 + 2x^2 - x + 4 - x^2 + 7x + 5$$
$$= \mathbf{3x^3 + x^2 + 6x + 9}$$

TEST YOUR UNDERSTANDING

(d) Add: $(4x^3 - 2x) - (3x^3 + 2x^2 + 5x - 3)$

(e) Add: $-(x^4 + 2x^3 + 3) + (-2x^4 - 5)$

11.5 multiplication of polynomials

To perform the multiplication of the monomial $-2x^2$ by the binomial $4x^3 - 5$

$$-2x^2(4x^3 - 5)$$

we remember that the distributive property of real numbers permits the multiplication of $-2x^2$ by each of the terms of the binomial $4x^3 - 5$ and indicates that these products are to be expressed as an indicated sum. We remember that when we multiply exponentials that have the same base, we multiply the numerical coefficients and add the exponents of like variables.

$$-2x^2(4x^3 - 5) = (-2x^2)(4x^3) + (-2x^2)(-5) = \mathbf{-8x^5 + 10x^2}$$

To develop a procedure for multiplying a binomial by a binomial, we will use the distributive property to multiply

$$(a + b)(c + d)$$

The notation $(a + b)(c + d)$ tells us that $(a + b)$ is to be multiplied by c and that $(a + b)$ is also to be multiplied by d and that the two products are to be summed.

$$(a + b)(c + d) = (a + b)c + (a + b)d$$

Now we again use the distributive property to multiply $(a + b)$ by c

$$(a + b)c = ac + bc$$

and to multiply $(a + b)$ by d

$$(a + b)d = ad + bd$$

and the products are summed

$$(a + b)c + (a + b)d = ac + bc + ad + bd$$

The final result consists of four terms which are the sums of the product of each member of the first binomial with each member of the second binomial.

example 11.5.1 Multiply $(x + y)(a + b)$.

solution From the discussion above, we realize that the notation of this problem indicates that x is to be multiplied by both a and b and that y is to be multiplied by both a and b and that the products are to be summed.

$$(x + y)(a + b) = xa + xb + ya + yb$$

example 11.5.2 Multiply $(4x + 5)(3x - 2)$.

solution The notation indicates that $4x$ is to be multiplied by both $3x$ and -2

$$4x(3x - 2) = \underbrace{12x^2 - 8x}$$

and that $+5$ is to be multiplied by both $3x$ and -2

$$+5(3x - 2) = \underbrace{15x - 10}$$

and that the products are to be added algebraically.

$$(4x + 5)(3x - 2) = \underbrace{12x^2 - 8x} + \underbrace{15x - 10} = \mathbf{12x^2 + 7x - 10}$$

example 11.5.3 Multiply $(4x^2 + 3x)(2x^2 - x)$.

solution

$$(4x^2 + 3x)(2x^2 - x) = 8x^4 - 4x^3 + 6x^3 - 3x^2 = \mathbf{8x^4 + 2x^3 - 3x^2}$$

An extension of the procedure described above can be used to perform the multiplication of polynomials of any number of terms. **Each term in one expression** (either expression) **is to be multiplied by each term in the other expression and the products are to be summed.** Thus we can perform the multiplication indicated by

$$(4x - 2)(x^2 + x + 4)$$

by multiplying each term in one expression (say the first expression) by each term in the second expression and summing the products of these multiplications:

$$\begin{aligned}(4x - 2)(x^2 + x + 4) &= 4x(x^2 + x + 4) - 2(x^2 + x + 4)\\ &= 4x^3 + 4x^2 + 16x - 2x^2 - 2x - 8\\ &= \mathbf{4x^3 + 2x^2 + 14x - 8}\end{aligned}$$

Much of this book will deal with the products of first-degree binomials such as

$$(x + 2)(x - 3).$$

We customarily perform this multiplication by using either of two different formats. The first format is called horizontal multiplication and is performed in the manner just described.

$$(4x + 2)(x - 5) = 4x^2 - 20x + 2x - 10 = \mathbf{4x^2 - 18x - 10}$$

The second format is to write the two polynomials vertically with either polynomial on top.

$$\begin{array}{r} 4x + 2 \\ x - 5 \\ \hline \end{array}$$

Now, the x of $x - 5$ is multiplied by both terms of $4x + 2$ and the products recorded.

$$\begin{array}{r} 4x + 2 \\ x - 5 \\ \hline 4x^2 + 2x \end{array}$$

Now the -5 of $x - 5$ is multiplied by both terms of $4x + 2$ and the products are recorded so that like terms (if any) are recorded below like terms to facilitate addition.

$$\begin{array}{lll} & 4x & +2 \\ & x & -5 \\ \hline 4x^2 & +2x & \\ & -20x & -10 \\ \hline 4x^2 & -18x & -10 \end{array}$$

The product $-20x$ was recorded below the term $+2x$. There was no constant in the first product so -10 was recorded out to the right—by itself.

example 11.5.4 Multiply $(4x + 2)(3x - 5)$.

solution We will use the vertical format to multiply.

$$\begin{array}{rrrl} & 4x & + 2 & \\ & 3x & - 5 & \\ \hline 12x^2 & + 6x & & \text{product of } 3x \text{ and } (4x+2) \\ & -20x & - 10 & \text{product of } -5 \text{ and } (4x+2) \\ \hline 12x^2 & - 14x & - 10 & \text{sum of the products} \end{array}$$

example 11.5.5 Multiply $(x - 5)(7x - 4)$.

solution Again we choose the vertical format.

$$\begin{array}{rl} x - 5 & \\ 7x - 4 & \\ \hline 7x^2 - 35x & \text{product of } 7x \text{ and } (x-5) \\ - 4x + 20 & \text{product of } -4 \text{ and } (x-5) \\ \hline 7x^2 - 39x + 20 & \text{sum of the products} \end{array}$$

Test Your Understanding

Multiply:

(f) $(4x + 5)(3x - 4)$ (g) $(5y - 7)(-3y + 5)$

lesson checklist 11

- ☐ Definition of a polynomial
- ☐ Monomial, binomial, trinomial
- ☐ Degree of a polynomial
- ☐ Descending powers of the variable
- ☐ Addition of polynomials
- ☐ Multiplication of polynomials

Answers to Test Your Understanding

(a) fourth degree (b) fourth degree (c) seventh degree

(d) $x^3 - 2x^2 - 7x + 3$ (e) $-3x^4 - 2x^3 - 8$ (f) $12x^2 - x - 20$

(g) $-15y^2 + 46y - 35$

problem set 11

Simplify:

*1. $(x^3 + 3x^2 + 2) + (2x^3 + 4)$ *2. $(3x^4 - 2x^2 + 3) - (x^4 - 2x^3 + x^2)$

Multiply:

*3. $(4x + 5)(3x - 2)$ *4. $(4x - 2)(x^2 + x + 4)$ *5. $(4x + 2)(3x - 5)$

*The problems marked with an asterisk were worked out in the lesson.

Solve for y:

6. $4x + 2y - 5 = 0$

7. $2x + 3y = y - 2$

8. $ax + m - 4 = 3y + 2$

Add:

9. $\frac{b}{c} + \frac{x}{mc^2}$

10. $\frac{1}{c} + \frac{x}{mc^2} + d$

11. $\frac{a}{mx} + \frac{b}{c}$

12. $a + \frac{b}{m} + \frac{d}{mc^2}$

Graph on a number line:

13. $x \not< -2$

14. $x \not> -2$

15. $x > 2$

16. $x \not\leq 4$

Simplify:

17. $\frac{3^4}{3^2}$

18. $\frac{a^4x^3b^{-2}c}{c^{-3}x^2b}$

19. $\frac{a^{-2}x^2p^{-4}k}{m^3xa^2p^{-2}}$

20. $\frac{m^{-4}x^2p^0}{m^{-2}x^{-2}p^3}$

21. $\frac{m^4x^2p^{-6}}{m^{-2}x^2p^3}$

22. $\frac{p^{-3}x^0m^{-2}}{p^2x^{-2}m^4}$

23. $\frac{x^0py^{-2}}{x^{-3}yp^{-3}}$

24. $\frac{x^4m^{-2}y}{y^3mx^4}$

25. $\frac{p^{-2}x^4m^2}{m^0x^2}$

Add like terms:

26. $x^5y^{-2} + 3y^{-2}x^5 - 4x^{-2}y^5$

27. $\frac{p^5x^2}{y} - 3x^2p^5y^{-1} + 2x^2p^5y$

Multiply:

28. $ab^2c^0\left(\frac{mb}{c} - \frac{xb^{-2}}{p}\right)$

Find the LCM of:

29. 50, 36, 24

30. 75, 42, 18

Solve:

31. $x + 4x - 5 + (-4x) = 7 - 2$

32. $4y + y - 2 = (-2)(-y + 4)$

33. $p + 7 - 4p + 7 - p(-3) = 2(-5 + p)$

34. $k(-4 + 7) - 2k - 5 = -3(-k + 2)$

Evaluate:

35. $4c(-b) - 3(c - 2b)$ if $b = 2$, $c = -4$

36. $(x - y)(-4) - 3 + y(-2a)$ if $a = 4$, $x = -1$, and $y = 2$

37. $-3p(-m) - 3(p + m)$ if $p = 2$, $m = -3$

38. $x[-k(k - x)]$ if $x = 5$, $k = -4$

Factor:

39. $m^2x^4y^6 - 3m^3x^2y^3$

40. $x^2y^3z^5 + mx^2y^2z^3$

Simplify:

41. $|-23 + 2|$

42. $|7|$

43. $(-7)^{-2}$

44. -7^{-2}

Evaluate:

45. $a^2 - b^3x$ if $a = -2$, $b = -2$, $x = 3$

Solve:

46. $\frac{p}{7} = 14$

47. $\frac{2p}{7} = 14$

48. $\frac{3}{7}p = 14$

Simplify:

49. $-2[-5 + 3(-5 - 4)(2) - 8]$

50. $-3\{7[4 - 3(-2 - 3)(2)]\}$

lesson twelve

First-degree equations in two unknowns
Cartesian coordinate system
Graphing linear equations
Addition of rational expressions

12.1 first-degree equations in two unknowns

In section 11.3 we said that the **degree of a term** of a polynomial is the sum of the exponents of the variables in the term. Thus

$2x^3$	is a third-degree term
x	is a first-degree term
xyz	is a third-degree term

Also we said that the **degree of a polynomial** is the same as the degree of its highest degree term. Thus

$x^3 + xy + m$	is a third-degree polynomial
$2x + 4y$	is a first-degree polynomial
$2x$	is a first-degree polynomial

If two polynomial expressions are connected by an equals sign, we call the equation a POLYNOMIAL EQUATION. **The degree of a polynomial equation is the same as the degree of its highest degree term.** Thus

$x^4 - 3x^2 + 2 = 0$	is a fourth-degree polynomial equation
$xyz + y = 4$	is a third-degree polynomial equation
$y = 2x + 4$	is a first-degree polynomial equation

There is an infinite number of pairs of values of x and y that are solutions to any first-degree polynomial equation in two variables. We will use the equation

$$y = 2x + 4$$

to investigate. If we assign a value to x, the equation will then indicate the value of y

that is paired with the assigned value of x. For instance, if we assign to x a value of 2, then

$$y = 2(2) + 4 \longrightarrow y = 8$$

the paired value of y is 8. If we give x a value of 2 and y a value of 8 in the original equation, we find that these values of x and y satisfy the equation and are solutions to the equation because the replacement of the variables by these numbers makes the equation a true statement.

$$\begin{aligned} y &= 2x + 4 \\ 8 &= 2(2) + 4 \\ 8 &= 4 + 4 \\ 8 &= 8 \quad \text{True!} \end{aligned}$$

If, in the original equation, we give x a value of -5, then

$$y = 2(-5) + 4 \longrightarrow y = -10 + 4 \longrightarrow \mathbf{y = -6}$$

We find that the paired value of y is -6. Thus the pair of values $x = -5$ and $y = -6$ will also satisfy the equation for the use of both of these numbers in the original equation in place of x and y will cause the equation to be a true statement, as we show here.

$$\begin{aligned} y &= 2x + 4 \\ -6 &= 2(-5) + 4 \\ -6 &= -10 + 4 \\ -6 &= -6 \quad \text{True!} \end{aligned}$$

We can assign to x the value of any real number and use the equation to find the value of y that, when used with the assigned value of x, will make the equation a true statement.

In both examples above we **assigned** a value to the variable x. The variable x, to which we assign a value, is called the **INDEPENDENT VARIABLE**. We see that, in each case, the value of y **depended** on the value that we assigned to x. Therefore, in our examples, we call the variable y the **DEPENDENT VARIABLE**. We could have assigned a value to y and then used the equation to find the corresponding value of x, in which case y would be the independent variable and x would be the dependent variable. **It is customary, however, to avoid confusion, to use the letter x to designate the independent variable and to use the letter y to designate the dependent variable. We will follow this custom in this book.**

12.2 ordered pairs

In the section above we found that, given the equation,

$$y = 2x + 4$$

if we let $x = 2$ and $y = 8$, this pair of values of x and y would make the equation a true equation. Also the pair of values $x = -5$ and $y = -6$ will make the equation a true equation. Since writing $x = 2$ and $y = 8$ and writing $x = -5$ and $y = -6$ is rather cumbersome, it is customary to write just (2, 8) and $(-5, -6)$, **with the x value always designated by the first number in a parentheses and the y value always designated by the second number**. Since the numbers are written in order with x first and y second, we designate this notation as an **ORDERED PAIR** of x and y. The general form of an ordered pair of x and y is (x, y). If two other variables are used instead of x and y, it is necessary to designate which variable will be represented by each of the entries in the parentheses. If the variables m and p are to be used and we wish to write the m value first, we could designate this at the outset by making a statement about the ordered pair (m, p).

It is important to remember that in ordered pairs of x and y, the first number will always designate the value of x and the second number will always designate the value of y.

> TEST YOUR UNDERSTANDING
>
> Write the given values of x and y as ordered pairs:
>
> (a) $y = -4, x = -2$ (b) $y = -7, x = 3$

12.3 cartesian coordinate system

In Lesson 9 we learned that the solution of an equation or inequality in one variable can be presented in graphical form on a single number line.

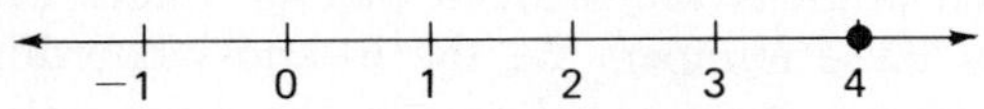

$x = 4$
graphical solution

$x \geq -2$
graphical solution

The graphical solution to equations or inequalities that contain two variables cannot be displayed on a single number line. We must have one number line for one of the variables and another number line for the other variable. It is customary to use the variables x and y and draw the x number line horizontally, the y number line vertically and to let the number lines intersect at the origin of both number lines. The positive values of x are located to the right of the origin on the horizontal or x number line and the positive values of y are located above the origin on the vertical or y number line. The x number line is called the **X-AXIS** or **HORIZONTAL AXIS** and the y number line is called the **Y-AXIS** or **VERTICAL AXIS**.

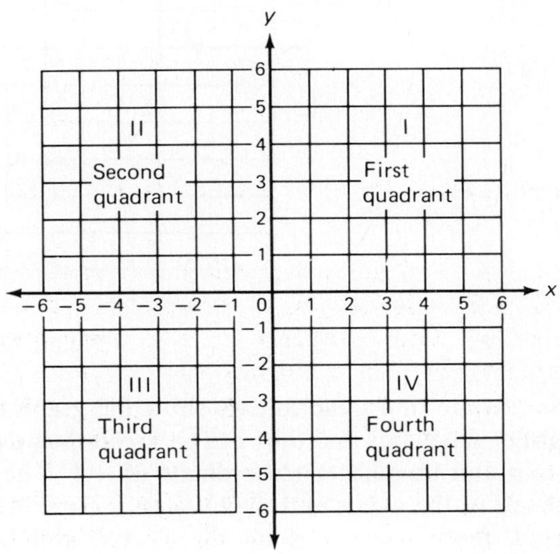

Figure 12.1

Figure 12.1 shows that the two number lines divide the plane into four QUARTERS or QUADRANTS. The quadrants are named the **FIRST QUADRANT, SECOND QUADRANT, THIRD QUADRANT**, and **FOURTH QUADRANT** as shown. Figure 12.1 in its entirety is called a system of **CARTESIAN COORDINATES** or a **CARTESIAN COORDINATE SYSTEM** after the famous seventeenth-century French philosopher and mathematician René Descartes. It is also called a **RECTANGULAR COORDINATE SYSTEM** and is sometimes called a **COORDINATE PLANE.**

When we use a single number line to graph the solution to an equation in one variable such as $x = 2$ (Figure 12.2), we call the mark we make on the number line the graph of the point, and conversely we call the number the coordinate of the point on the line designated by the graph. The point is defined to be without size and thus the graph is not the point but denotes the location of the point.

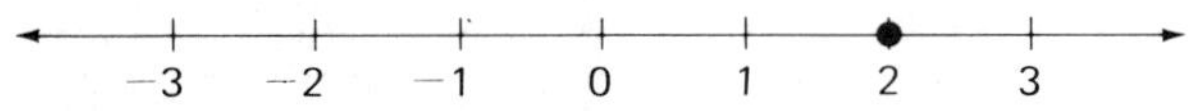

Figure 12.2

Since a rectangular coordinate system has two number lines, **it is necessary to associate with every point on a rectangular coordinate system two numbers or coordinates.** These numbers designate the location of the point. In Figure 12.3 we show the graphs of four points. Written by each point is the ordered pair of values of x and y that we associate with the point and these numbers are the x- and y-coordinates of the point. **The number written first is always the x-coordinate** of the point and is also called the **ABSCISSA** of the point. This number denotes the measure of the distance of the point to the right (+) or left (−) of the vertical axis. **The number written second is always the y-coordinate** of the point and is also called the **ORDINATE** of the point. This number denotes the measure of the distance of the point above (+) or below (−) the horizontal axis.

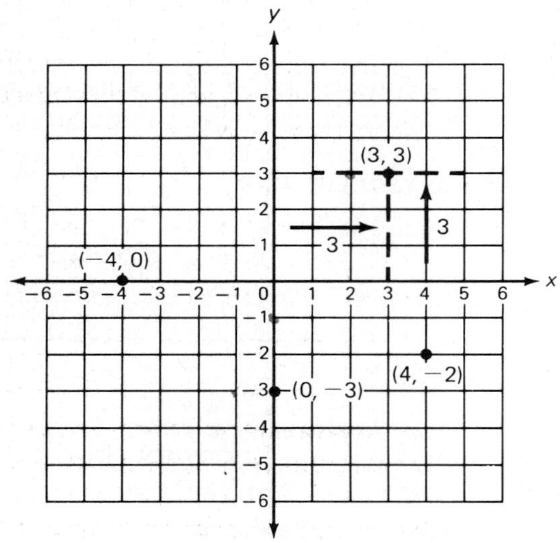

Figure 12.3

In the first quadrant we show the graph of the point (3, 3), which is **3 units to the right** of the y-axis and thus has an **x-coordinate of +3**. The point is also **3 units above** the x-axis and thus has a **y-coordinate of +3**. The next point, (−4, 0), is located 4 units to the left of the y-axis and thus has an x-coordinate of −4. The point is no units above or below the *x-axis* (it is on the x-axis) and thus has a y-coordinate of 0. The exact positions of the other two points shown are designated by the ordered pairs (0, −3) and (4, −2). Thus we see that the location of every point on the coordinate plane can be designated by starting the x- and y-coordinates of the point.

We have stated that a point on a number line is without size and that the graph of a point is not the point but only denotes the location of the point. In the same way, a point in a rectangular coordinate system is without size and the graph of the point is not the point but designates the location of the point.

TEST YOUR UNDERSTANDING

Graph the points:

(c) $(-3, 4)$

(d) $(4, -2)$

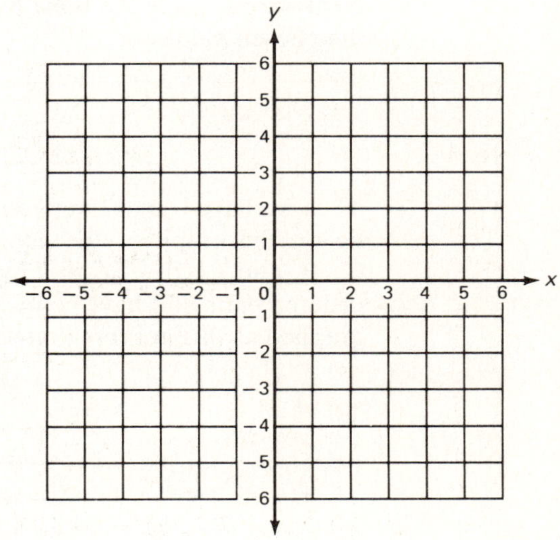

12.4 graphs of linear equations

If we use a rectangular coordinate system to graph the ordered pairs that satisfy a first-degree polynomial equation in two unknowns, we find that the graph is the graph of a straight line. For this reason, we call a first-degree polynomial equation in one or more variables a **LINEAR EQUATION**. The following equations are examples of linear equations in two unknowns,

$$y - 2x + 1 = 0 \qquad x - 4 = -2y \qquad x + y = 0$$

for the graph of each of the equations is a straight line.

To graph a linear equation in two unknowns, we need to know only two ordered pairs of x and y that satisfy the equation and thus lie on the line that is the graph of the equation, since two points adequately determine the graph of a line. But since the topic is a new one, we will learn how to find three or more ordered pairs of x and y that lie on the line. To graph the equation

$$y - 2x + 1 = 0$$

we first solve the equation for y by using the additive and multiplicative properties of equality as required.

$$\begin{array}{lcl} y - 2x + 1 = 0 & & \text{original equation} \\ \quad +2x - 1 \quad +2x - 1 & & \text{add } 2x - 1 \text{ to both sides} \\ \hline y \qquad = \quad 2x - 1 & & \end{array}$$

Next we make a table and choose convenient values for the independent variable x. The values chosen for x should not be too close together and should not be so large that they will graph off our coordinate system. Numbers such as 0, 2, -2, 3, and -3 usually work well.

x	0	3	-3	2	-2
y					

Now we will use the numbers 0, 3, -3, 2, and -2 one at a time as replacement values for x in the equation $y = 2x - 1$ to find the paired values of y.

when $x = 0$	when $x = 3$	when $x = -3$	when $x = 2$	when $x = -2$
$y = 2(0) - 1$	$y = 2(3) - 1$	$y = 2(-3) - 1$	$y = 2(2) - 1$	$y = 2(-2) - 1$
$y = -1$	$y = 5$	$y = -7$	$y = 3$	$y = -5$

Next we complete the table by entering the values of y that the equation has paired with the chosen values of x.

x	0	3	-3	2	-2
y	-1	5	-7	3	-5

Thus we have found five ordered pairs of x and y that satisfy the equation and therefore lie on the graph of the equation. These ordered pairs are $(0, -1)$, $(3, 5)$, $(-3, -7)$, $(2, 3)$, and $(-2, -5)$. In Figure 12.4 we have graphed the points designated by these ordered pairs and have connected them with a straight line. The point $(-3, -7)$ was not graphed as this point fell outside the borders of our coordinate system.

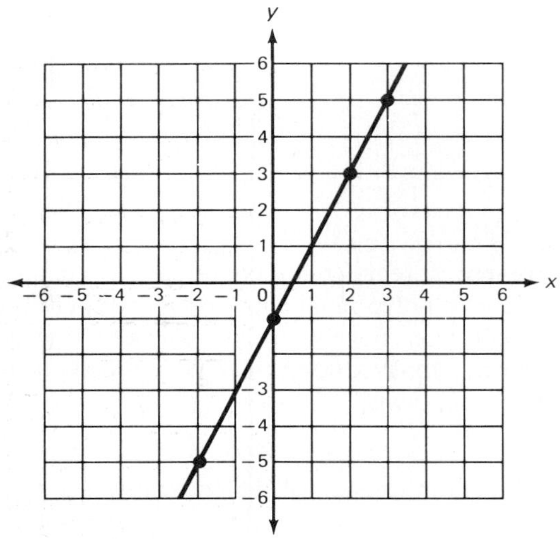

Figure 12.4

Since a line (straight line) in mathematics is defined to be infinite in length, we have graphed only a segment of the line or a **LINE SEGMENT**. Also we say that a mathematical line has no width and the line we have drawn has a width that can be measured so it is not a mathematical line but is a **graph of a mathematical line** and indicates the location of the mathematical line in question.

example 12.4.1 Graph the equation $x + 2y - 4 = 0$.

solution (a) Write the equation of the line in the form $y = mx + b$ by solving for y.

$$\begin{array}{lll} x + 2y - 4 = & 0 & \text{original equation} \\ -x \qquad\;\; + 4 & -x + 4 & \text{add } -x + 4 \text{ to both sides} \\ \hline 2y \;\;= & -x + 4 & \\ y \;\;= & -\frac{1}{2}x + 2 & \text{divided each term by 2} \end{array}$$

(b) Make a table and choose convenient values for x.

x	0	2	− 2
y			

(c) Find the values of y that the equation $y = -\frac{1}{2}x + 2$ pairs with the chosen values for x.

when $x = 0$	when $x = 2$	when $x = -2$
$y = -\frac{1}{2}(0) + 2$	$y = -\frac{1}{2}(2) + 2$	$y = -\frac{1}{2}(-2) + 2$
$y = 2$	$y = 1$	$y = 3$

(d) Insert the y values to complete the table.

x	0	2	− 2
y	2	1	3

(e) Then we graph the points and draw the line through the points.

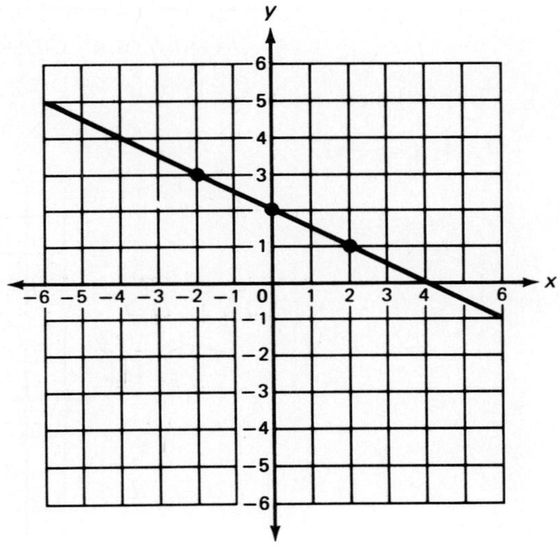

example 12.4.2 Graph the equation $x + y = 0$.

solution (a) Solve for y. $x + y = 0 \rightarrow y = -x$

(b) Make a table.

x	0	3	− 3
y			

(c) Find the values of y for each value of x.

when $x = 0$	when $x = 3$	when $x = -3$
$y = -(0)$	$y = -(3)$	$y = -(-3)$
$y = 0$	$y = -3$	$y = 3$

(d) Complete the table.

x	0	3	− 3
y	0	−3	3

(e) Graph the points and draw the line.

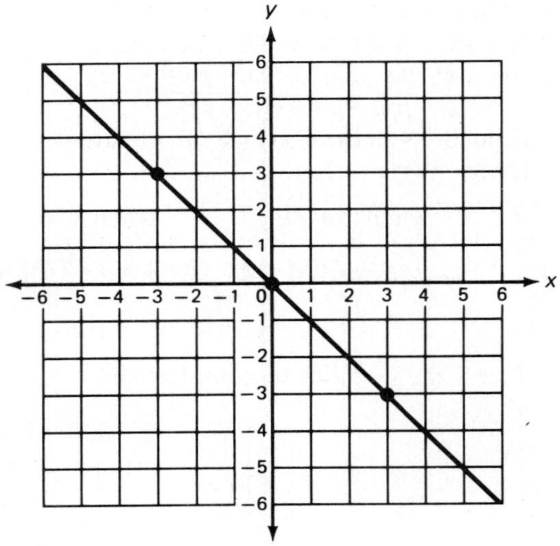

TEST YOUR UNDERSTANDING

(e) Graph the equation

$4y + x = -12$

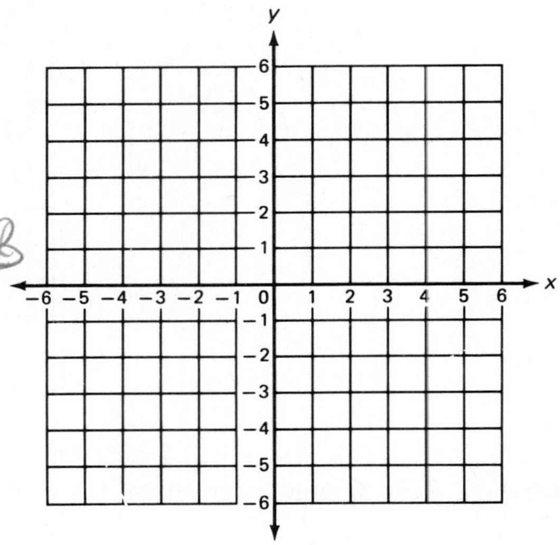

12.5 vertical and horizontal lines

In the last paragraph we drew the graph for the equations

$$y - 2x + 1 = 0, \qquad x + 2y - 4 = 0, \qquad x + y = 0$$

We note that each of these equations has both an x term and a y term. We also note that none of the graphs was a vertical line or a horizontal line. **Some equations of a straight line contain either an x term or a y term but not both an x term and a y term, and the graph of these equations is either a vertical line or a horizontal line.**

example 12.5.1 Graph $y = 4$.

solution This equation places no restriction on the x-coordinates of the points on the line. It tells us that the y-coordinate of every point on the line is 4, regardless of the x-coordinate of the point. The line that satisfies these conditions is shown in the accompanying figure. The y-coordinate of every point on the line is 4, regardless of the value of the x-coordinate.

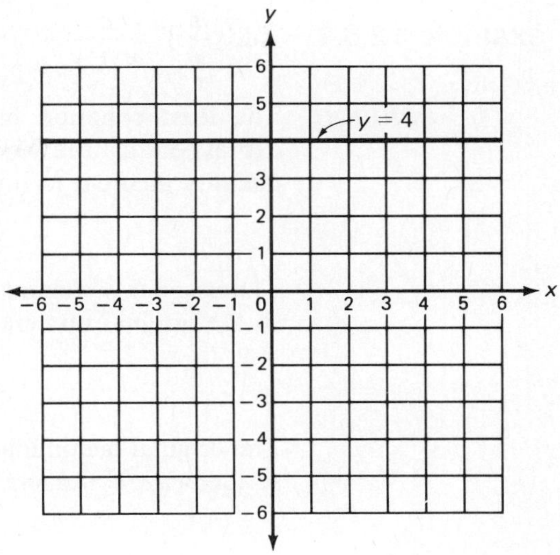

example 12.5.2 Graph $x = -2$.

solution This equation places no restriction on the y-coordinates of the points on the line. It tells us that the x-coordinate of every point on the line is -2 regardless of the y-coordinate of the point. The line that satisfies these conditions is shown in the accompanying figure. The x-coordinate of every point on the line is -2 regardless of the value of the y-coordinate.

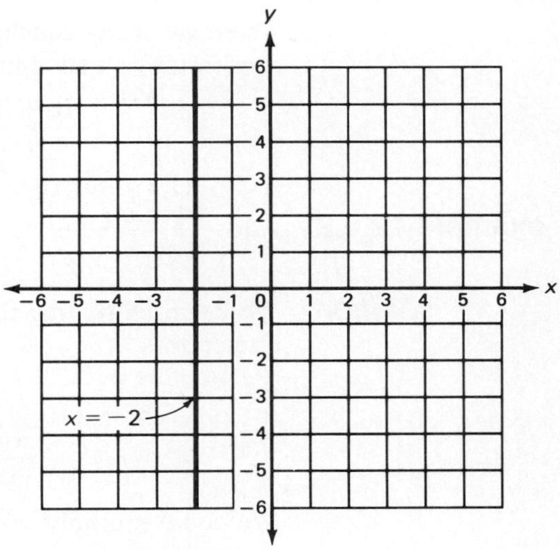

We will remember that equations of the form

$$y = \pm k \quad \text{or} \quad x = \pm k$$

such as

$$y = 2, \qquad y = -4, \qquad x = 5, \qquad x = -4$$

are special forms of the equation of a straight line. Equations of the form

$$y = \pm k$$

are horizontal lines while equations of the form

$$x = \pm k$$

are vertical lines.

12.6 addition of rational expressions

If the denominator of a term in an addition problem is in the form of a sum, the sum must be a factor of the least common multiple of the denominators.

example 12.6.1 Add $\frac{a}{x} + \frac{b}{x + y}$.

solution The least common multiple of the denominators is $x(x + y)$ and we will use this expression as the new denominator. We will use the **denominator numerator same quantity theorem** to find the new numerators.

$$\frac{\quad}{x(x+y)} + \frac{\quad}{x(x+y)}$$

The original denominator of the first term was x, and it has been multiplied by $(x + y)$, so the original numerator, a, must also be multiplied by $x + y$.

$$\frac{a(x+y)}{x(x+y)} + \frac{\quad}{x(x+y)}$$

The original denominator of the second term was $x + y$ and it has been multiplied by x, so the original numerator, b, must also be multiplied by x. Now we add the numerators.

$$\frac{a(x+y)}{x(x+y)} + \frac{xb}{x(x+y)} = \mathbf{\frac{a(x+y)+xb}{x(x+y)}}$$

There are many equally correct forms of the answer. We may multiply out in either the numerator or denominator or both. Thus all three forms shown below are also correct.

$$\frac{ax + ay + xb}{x(x+y)}, \qquad \frac{a(x+y)+xb}{x^2+xy}, \qquad \frac{ax+ay+xb}{x^2+xy}$$

example 12.6.2 Add $\frac{4x + a}{a + b} + \frac{c}{x}$.

solution We begin by writing the fraction lines and the new denominators and then find the new numerators.

$$\frac{\quad}{x(a+b)} + \frac{\quad}{x(a+b)}$$

$$\frac{x(4x+a)}{x(a+b)} + \frac{c(a+b)}{x(a+b)} = \mathbf{\frac{4x^2+ax+ca+cb}{x(a+b)}}$$

We could multiply $x(a + b)$ in the denominator, but we decide to leave it in the more concise form.

example 12.6.3 Add $\frac{a + b}{x} + \frac{c}{m} + d$.

solution We begin by writing the fraction lines with the new denominators and then find the new numerators.

$$\frac{\quad}{xm} + \frac{\quad}{xm} + \frac{\quad}{xm}$$

$$\frac{m(a+b)}{mx} + \frac{cx}{mx} + \frac{dxm}{mx} = \mathbf{\frac{m(a+b)+cx+dxm}{xm}}$$

We could multiply $m(a + b)$ in the numerator, but we decide to leave it as it is.

example 12.6.4 Add $\frac{a}{b + c} - x + \frac{d + m}{k}$.

solution The first step is to write the new denominators and then we find the new numerators.

$$\frac{\quad}{k(b+c)} - \frac{\quad}{k(b+c)} + \frac{\quad}{k(b+c)}$$

$$\frac{ka}{k(b+c)} - \frac{xk(b+c)}{k(b+c)} + \frac{(d+m)(b+c)}{k(b+c)} = \mathbf{\frac{ka - xk(b+c) + (d+m)(b+c)}{k(b+c)}}$$

Again we choose not to use the distributive property, and we leave the answer in the more concise form.

TEST YOUR UNDERSTANDING

(f) Add: $\frac{a}{x} + m + \frac{c}{x+y}$

lesson checklist 12

- ☐ Degree of a polynomial equation
- ☐ Dependent and independent variables
- ☐ Ordered pairs
- ☐ Cartesian coordinate system
- ☐ Coordinates of a point
- ☐ Graph of a point
- ☐ Abscissa and ordinate
- ☐ Graph of a linear equation
- ☐ Line segment
- ☐ Graph of constant linear equations
- ☐ Addition of rational expressions

ANSWERS TO TEST YOUR UNDERSTANDING

(a) $(-2, -4)$

(b) $(3, -7)$

(f) $\frac{a(x+y) + mx(x+y) + cx}{x(x+y)}$

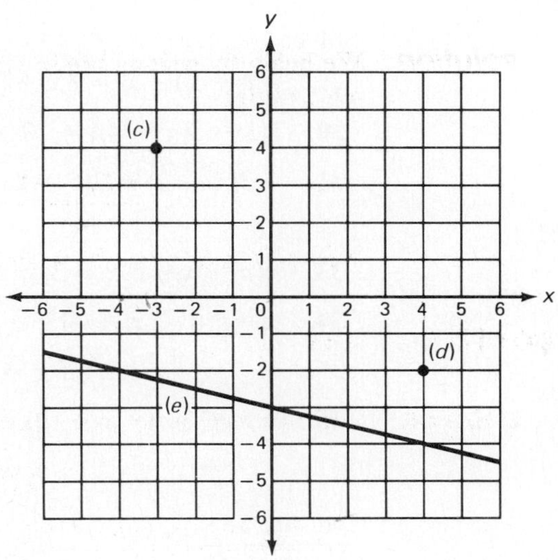

problem set 12

Graph on a rectangular coordinate system:

*1. $y - 2x + 1 = 0$ *2. $x + 2y - 4 = 0$ *3. $x + y = 0$

*4. $x = -2$ *5. $y = 4$

Add:

6. $(4x^2 + 2x - 5) - (x^2 - x + 6)$ 7. $\frac{1}{a} + \frac{2}{ab} + \frac{3}{ax}$

8. $\frac{m}{a} + \frac{bx}{a^2y} + \frac{c}{m}$ 9. $\frac{a}{mx} + \frac{c}{mx^2} + 4$ 10. $\frac{x^2}{2} + \frac{y}{a} + \frac{m}{z}$

Solve for y:

11. $4x + 5y - 6 = 0$ 12. $7x + 2a - 3y + 7 = 0$

13. $3x + 2y - 4 = x + y - 3$

Multiply:

14. $(3x + 2)(x - 4)$ 15. $(4 + 2x)(x^2 + x - 5)$

Add:

*16. $\frac{a}{x} + \frac{b}{x + y}$ *17. $\frac{4x + a}{a + b} + \frac{c}{x}$

*18. $\frac{a + b}{x} + \frac{c}{m} + d$ *19. $\frac{a}{b + c} - x + \frac{d + m}{k}$

Graph on a number line:

20. $x \leq 5$ 21. $x > -3$ 22. $x \not\leq 2$

Simplify:

23. $\frac{p^{-2}x^4m^2}{m^0x^2}$ 24. $\frac{p^{-2}xp^2}{x}$ 25. $\frac{m^{-4}a^{-2}xp^0}{p^2m^2}$

Add like terms:

26. $\frac{2a^2bc}{x} - \frac{3ba^2}{c^{-1}x} + \frac{5bc}{xa^{-2}}$ 27. $\frac{-3m^2}{x} + \frac{2m^{-2}}{x} + \frac{4x^{-1}}{m^{-2}} + \frac{3}{m^2x}$

Multiply:

28. $\frac{a^{-2}x}{p}(-a + 2b^2p^{-1})$

Solve:

29. $4(2 - x) - 6 = 8x + [-3(3x - 11) + 20]$

30. $2[3 - 5(x - 4)] = 10 - 5x$

31. $-3(x - 2) = 3(5 - 2x)$ 32. $20 = 18 - \{-2[3z - 2(z - 1)]\}$

Evaluate:

33. $x(a - xy)$ if $x = 3, a = 2, y = -2$

34. $a[-c(c - a) + 4]$ if $a = 2, c = -3$

35. $x - y(-a + x)$ if $x = 2, a = -1, y = 3$

36. $-x - 3[a(-x + a)]$ if $x = 6, a = -3$

*The problems marked with an asterisk were worked out in the lesson.

Factor:

37. $mx^2y - 3m^2x^2y^2z$

38. $4p^4x^3 - 3p^2x^2m$

Simplify:

39. $|5 - 2|$

40. $|2 - 5|$

Express as a product of prime factors:

41. 99

42. 168

Simplify:

43. -4^2

44. $(-4)^2$

45. $(-4)^{-2}$

46. -4^{-2}

Solve:

47. $\frac{2p}{3} = 4$

48. $-\frac{2p}{3} = 4$

49. $\frac{2p}{3} = -4$

Simplify:

50. $-4\{-3[-2 - 6(4 - 2) + 3]\}$

lesson thirteen

Power theorem for exponents
Solving simultaneous equations by substitution

13.1 power theorem for exponents

The last eight problem sets have contained problems involving exponents whose solution required the use of one or more of the following definitions and theorems.

DEFINITION:	$x \cdot x \cdot x \cdot x \cdots x = x^m$	(m is a whole number)
DEFINITION:	$x^0 = 1$	$(x \neq 0)$
DEFINITION:	$x^1 = x$	
DEFINITION:	$x^{-n} = \dfrac{1}{x^n}$	$(x \neq 0)$
PRODUCT THEOREM:	$x^m \cdot x^n = x^{m+n}$	$(x \neq 0)$
QUOTIENT THEOREM:	$\dfrac{x^m}{x^n} = x^{m-n}$ or $\dfrac{1}{x^{n-m}}$	$(x \neq 0)$

To complete the list of rules for exponents, we will introduce the **POWER THEOREM FOR EXPONENTS**, which is a logical extension of the first definition listed above.

In the definition of exponential notation above

$$x \cdot x \cdot x \cdot x \cdots x = x^m$$

we say that we can indicate that x is to be used as a factor m times by writing x^m. Thus if we wish to indicate the following multiplication in a more compact form,

(a) $$x^5 \cdot x^5 \cdot x^5$$

and by the definition of exponential notation that is stated above, we can write

(b) $$x^5 \cdot x^5 \cdot x^5 = (x^5)^3$$

The first product (a), $x^5 \cdot x^5 \cdot x^5$, by the product theorem could be written as x^{5+5+5} or x^{15}. In order to get the same result from

$$(x^5)^3$$

we must specify that, **when an exponential is raised to a power, the proper result is obtained by multiplying the exponents.**

(b) $$(x^5)^3 = x^{15}$$

This leads to the *power theorem for exponents.*

> POWER THEOREM FOR EXPONENTS
>
> If m and n and x are real numbers
>
> $$(x^m)^n = x^{mn}$$

Examples:

(a) $(a^5)^2 = a^{10}$ (b) $(x^{-4})^{-2} = x^8$ (c) $(m^{-2})^4 = m^{-8}$

(d) $(p^{-7})^{-7} = p^{49}$ (e) $(x^3)^5 = x^{15}$ (f) $(m^k)^2 = m^{2k}$

The use of the power theorem in expressions such as the above usually gives little trouble, but the use of this theorem to simplify expressions such as $(2x^5y^2z)^3$ is more difficult. We know that the notation indicates that $2x^5y^2z$ is to be used as a factor three times because 3 is the exponent of the whole expression.

$$(2x^5y^2z)(2x^5y^2z)(2x^5y^2z)$$

Since the order of multiplication of factors of a product does not affect the value of the product, we will rearrange the order of multiplication to get

$$2 \cdot 2 \cdot 2 \cdot x^5 \cdot x^5 \cdot x^5 \cdot y^2 \cdot y^2 \cdot y^2 \cdot z \cdot z \cdot z$$

which could be written by use of the product theorem as

$$2^3x^{15}y^6z^3 \qquad \text{which equals} \qquad \mathbf{8x^{15}y^6z^3}$$

To obtain the same result by using the power theorem

$$(2x^5y^2z)^3$$

we would have to multiply the exponent of each factor of the given term by 3 or raise each factor of the given term to the third power.

$$(2x^5y^2z)^3 = (2)^3(x^5)^3(y^2)^3(z)^3 = \mathbf{8x^{15}y^6z^3}$$

This is clear to the mathematician from the notation that $(x^m)^n = x^{mn}$, but it is sometimes not clear to the student at this level. Therefore, we will state it as follows:

To raise a term that contains no indicated additions to a given power, the exponent indicating the power is multiplied by the exponent of every factor of the numerator and the denominator (if there is one) **of the term.** For example,

(a) $\left(\dfrac{x^{-2}y^5}{z^4}\right)^{-2} = \dfrac{x^4y^{-10}}{z^{-8}}$ (b) $(a^{-2}b^2z^{-10})^{-5} = a^{10}b^{-10}z^{50}$

(c) $\left(\dfrac{4xy}{m^{-2}}\right)^2 = \dfrac{4^2x^2y^2}{m^{-4}} = \dfrac{16x^2y^2}{m^{-4}}$ (d) $\left(\dfrac{3xy^{-2}}{p^5}\right)^{-2} = \dfrac{3^{-2}x^{-2}y^4}{p^{-10}} = \dfrac{x^{-2}y^4}{9p^{-10}}$

(e) $\left(\frac{3x^0y}{k^4}\right)^3 = \frac{27y^3}{k^{12}}$ (f) $\left(\frac{2^0x^{-2}y^4}{z}\right)^{15} = \frac{x^{-30}y^{60}}{z^{15}}$

TEST YOUR UNDERSTANDING

Simplify:

(a) $(x^{-3})^2$ (b) $(2x^4y^5)^3$ (c) $(x^{-4}y^2p)^{-3}$ (d) $\left(\frac{p^0x^3}{y^{-2}}\right)^3$

13.2 substitution axiom

Since Lesson 4 we have been finding the value of a particular expression that contains variables by assigning a value to each variable. Thus the value of

$$\frac{x^2y}{p} \quad \text{when } x = 2, y = 4, \text{and } p = -1 \text{ is} \quad \frac{(2)^2(4)}{(-1)} = \frac{16}{-1} = \mathbf{-16}$$

We have been using what is usually called the **SUBSTITUTION AXIOM.**

The substitution axiom is stated in different ways by different authors. Three statements of this axiom that are frequently used follow:

SUBSTITUTION AXIOM

1. Changing the numeral by which a number is named in an expression does not change the value of the expression.
2. For any numbers a and b, if $a = b$, then a and b may be substituted for each other.
3. If $a = b$, then a may replace b or b replace a in any statement without changing the truth or falsity of the statement.

The definition given in (1) seems to apply only to individual expressions. Definition (2) is general enough but not sufficiently specific. Definition (3) seems to apply only to statements and not to individual expressions. We will use the definition below to state formally and exactly the thought that **if two expressions have equal value, it is permissible to use either expression.**

SUBSTITUTION AXIOM

If two expressions a and b are of equal value, $a = b$, then a may replace b or b may replace a in another expression without changing the value of the expression. Also a may replace b or b may replace a in any statement without changing the truth or falsity of the statement. Also a may replace b or b replace a in any equation or inequality without changing the solution set of the equation or inequality.

Thus the substitution axiom applies to expressions, equations, and inequalities. We have already been using this axiom to evaluate expressions as shown in the problem worked out at the beginning of this section. Now we will use the axiom to solve a system of first-degree linear equations in two unknowns.

13.3 simultaneous equations

If we are given the two equations

$$\text{(a)}\quad 2x - y = 1 \qquad \text{and} \qquad \text{(b)}\quad x = -3y + 11$$

we see that (3, 5) is a solution to equation (a) and that (5, 2) is a solution to equation (b).

(a) $2x - y = 1$	(b) $x = -3y + 11$
$2(3) - (5) = 1$	$(5) = -3(2) + 11$
$6 - 5 = 1$	$5 = -6 + 11$
$1 = 1$ True	$5 = 5$ True

But neither (3, 5) nor (5, 2) is a solution to both equations at the same time or simultaneously. If we need one solution that will satisfy two or more equations, we call the equations a **system of simultaneous equations**. We can designate that two or more equations form a system of equations by so stating in words or by using a brace:

$$\begin{cases} 2x - y = 1 \\ x = -3y + 11 \end{cases}$$

13.4 solution of simultaneous equations by substitution

We can find the common solution to a system of two first-degree simultaneous equations in two unknowns by using the substitution axiom.

example 13.4.1 Solve

$$\begin{matrix} \text{(a)} \\ \text{(b)} \end{matrix} \begin{cases} 2x - y = 1 \\ x = -3y + 11 \end{cases}$$

solution First we must **assume** that an ordered pair of real numbers exists which will satisfy both of these equations and that x and y in the equations represent these real numbers. If our assumption is correct, we can find the value of the members of this ordered pair. If our assumption is incorrect, the attempted solution will degenerate into an expression involving real numbers such as $1 = 2$ or $4 + 2 = 6$ or $0 = 0$.[†]

Now since both equations are assumed to be true equations and also since x in both equations stands for the number that is the x value of the ordered pair that will satisfy both equations, we can **replace the variable x in the equation (a) with the equivalent expression for x given by equation (b).**

$2x - y = 1$	equation (a)
$2(-3y + 11) - y = 1$	replaced x with $-3y + 11$
$-6y + 22 - y = 1$	distributive property
$-7y + 22 = 1$	added like terms
$-7y = -21$	added -22 to both sides
$y = 3$	divided both sides by -7

We have found that the y value of the desired ordered pair is 3. Now we may find the value of x by substituting the number 3 for y in either of the original equations.

[†]To be discussed in a later lesson.

IN EQUATION (a)	IN EQUATION (b)
$2x - y = 1$	$x = -3y + 11$
$2x - (3) = 1$	$x = -3(3) + 11$
$2x - 3 = 1$	$x = -9 + 11$
$2x = 4$	$\mathbf{x = 2}$
$\mathbf{x = 2}$	

Thus the ordered pair of x and y that will satisfy both equations is **(2, 3).**

example 13.4.2 Solve $\begin{cases} 2x + 3y = -13 \\ y = x - 6 \end{cases}$

solution The second equation states that y equals $x - 6$. Therefore, in the first equation we will replace y with $x - 6$ and solve for x.

$$2x + 3y = -13 \longrightarrow 2x + 3(x - 6) = -13$$
$$\longrightarrow 2x + 3x - 18 = -13 \longrightarrow 5x = 5 \longrightarrow \mathbf{x = 1}$$

Now the paired value for y may be found by substituting 1 for x in either of the original equations.

FIRST EQUATION	SECOND EQUATION
$2x + 3y = -13$	$y = x - 6$
$2(1) + 3y = -13$	$y = (1) - 6$
$2 + 3y = -13$	$y = 1 - 6$
$3y = -15$	$\mathbf{y = -5}$
$\mathbf{y = -5}$	

Thus the solution is the ordered pair **(1, −5)**.

example 13.4.3 Solve $\begin{cases} 3x + 2y = 3 \\ x = 3y - 10 \end{cases}$

solution We will replace x in the first equation with its equivalent $(3y - 10)$ from the second equation.

$3x + 2y = 3$	first equation
$3(3y - 10) + 2y = 3$	replaced x with $(3y - 10)$
$9y - 30 + 2y = 3$	multiplied
$11y = 33$	simplified
$\mathbf{y = 3}$	divided by 11

Now the paired value for x may be found by replacing y with 3 in either of the original equations.

FIRST EQUATION	SECOND EQUATION
$3x + 2(3) = 3$	$x = 3(3) - 10$
$3x + 6 = 3$	$x = 9 - 10$
$3x = -3$	$\mathbf{x = -1}$
$\mathbf{x = -1}$	

Thus the solution is the ordered pair **(−1, 3)**.

example 13.4.4 Solve $\begin{cases} -x - 2y = 4 \\ y = -3x + 8 \end{cases}$.

solution We will replace y in the top equation by its equivalent $(-3x + 8)$ from the lower equation.

$$-x - 2y = 4 \longrightarrow -x - 2(-3x + 8) = 4 \longrightarrow -x + 6x - 16 = 4$$
$$\longrightarrow 5x - 16 = 4 \longrightarrow 5x = 20 \longrightarrow \mathbf{x = 4}$$

Now we can replace x with 4 in either of the original equations to find the value of y.

FIRST EQUATION	SECOND EQUATION
$-x - 2y = 4$	$y = -3x + 8$
$-(4) - 2y = 4$	$y = -3(4) + 8$
$-4 - 2y = 4$	$y = -12 + 8$
$-2y = 8$	$\mathbf{y = -4}$
$\mathbf{y = -4}$	

Thus the solution is the ordered pair **(4, −4)**.

example 13.4.5 Solve $\begin{cases} 2x + 3y = 5 \\ x = y \end{cases}$

solution In the top equation we replace x with its equivalent (y) from the bottom equation.

$$2x + 3y = 5 \longrightarrow 2(y) + 3y = 5 \longrightarrow 2y + 3y = 5$$
$$\longrightarrow 5y = 5 \longrightarrow \mathbf{y = 1}$$

and since $x = y$, $\mathbf{x = 1}$

Thus the solution is the ordered pair **(1, 1)**.

TEST YOUR UNDERSTANDING

(e) Solve: $\begin{cases} 3x - 5y = -19 \\ x = 2y - 8 \end{cases}$

lesson checklist 13

- ☐ Power theorem for exponents
- ☐ Substitution axiom
- ☐ Simultaneous equations
- ☐ Solution of simultaneous equations by substitution

ANSWERS TO TEST YOUR UNDERSTANDING

(a) x^{-6} (b) $8x^{12}y^{15}$ (c) $x^{12}y^{-6}p^{-3}$ (d) $\dfrac{x^9}{y^{-6}}$ (e) (2, 5)

problem set 13

Simplify:

*1. $\left(\frac{x^{-2}y^5}{z^4}\right)^{-2}$ *2. $(a^{-2}b^2z^{-10})^{-5}$ *3. $\left(\frac{4xy}{m^{-2}}\right)^2$

*4. $\left(\frac{3xy^{-2}}{p^5}\right)^{-2}$ *5. $\left(\frac{2^0x^{-2}y^4}{z}\right)^{15}$

Use substitution to solve:

*6. $\begin{cases} 2x - y = 1 \\ x = -3y + 11 \end{cases}$ *7. $\begin{cases} 2x + 3y = -13 \\ y = x - 6 \end{cases}$ *8. $\begin{cases} 3x + 2y = 3 \\ x = 3y - 10 \end{cases}$

*9. $\begin{cases} -x - 2y = 4 \\ y = -3x + 8 \end{cases}$ *10. $\begin{cases} 2x + 3y = 5 \\ x = y \end{cases}$

Add:

11. $(x + 2) - (x^2 - 3x - 4)$ 12. $(x^2 - 3x + 2) - (-2x - 5)$

13. $\frac{a}{b} - \frac{c}{x}$ 14. $\frac{a}{b} - c + \frac{m}{x}$

15. $\frac{4 + x}{a} - \frac{m}{2}$ 16. $\frac{x}{m - n} - \frac{6}{m} + k$

Graph on a rectangular coordinate system:

17. $2y - x - 4 = 0$ 18. $x - y + 4 = 4$ 19. $y + 2x - 2 = 0$

20. $-3 = y$ 21. $x = 2$

Graph on a number line:

22. $x \not> 2$ 23. $x > -4$ 24. $x \not\geq -4$

Simplify:

25. $\frac{(x^2a)^0x^0}{(x^{-2})^2}$ 26. $\frac{(a^{-3})^0x^2m^{-2}}{xm^{-3}}$ 27. $\frac{b^2cp^{-4}}{c^4bp^2}$

Add like terms:

28. $4x^2y^3 - 3y^3x^2 + \frac{2}{x^{-2}y^3}$ 29. $\frac{x^5yp}{m} - \frac{4y^5xp}{m} + \frac{2yp}{x^{-5}m}$

Multiply:

30. $p^{-2}x\left(\frac{p^2}{x} - \frac{2pa}{m}\right)$

Solve:

31. $6(3m - 5) = 3[4(1 - m) - 7]$

32. $5x - 7(4 - 2x) + 8 = 10 - 9(11 - x)$

33. $3[-53 - 7(4x - 9) + 18x] = 0$

34. $5p + 7 - 8(-8 + p) = (-4)(p + 7)$

Evaluate:

35. $x[-6(y - x)]$ if $x = 2, y = -3$

36. $(x - y)[y - (-x)]$ if $x = 7, y = -4$

37. $p + (-p)(-m)$ if $p = 2, m = -3$

---

*The problems marked with an asterisk were worked out in this lesson.

38. $a(b + c) - (-a)$ if $a = -3, b = 2, c = -1$

Factor:

39. $kmx^4p^2 - 3m^2xp^3$

40. $x^5yz^4 - xy^2z^3p$

Simplify:

41. $|-42 + 10|$

42. $|15|$

Express as a product of prime factors:

43. 30

44. 152

Simplify:

45. $(-3)^{-2} + \frac{1}{9}$

46. 3^{-2}

Evaluate:

47. $-ab^2(-c)$ if $a = 2, b = -2, c = -3$.

Solve:

48. $-\frac{2}{5}x = 13$

49. $\frac{-2x}{5} = 13$

Simplify:

50. $-4 - (-3)[-6(-2 + 4)(-2)]$

lesson fourteen

Properties of zero and one
Simplification of quotients of fractions
Finite and infinite sets
Membership in a set
Percent problems

14.1 properties of one and zero

The numbers one (1) and zero (0) have special properties that no other numbers have. We discussed these properties briefly in Lesson 1 and because these properties are important, we will go over them again.

14.2 unique properties of the number zero

The product of the number zero and any real number is zero. For example,

$$0 \cdot 4 = 0 \qquad 0 \cdot (-3) = 0 \qquad 0 \cdot \frac{3}{2} = 0$$

MULTIPLICATIVE PROPERTY OF ZERO
For any real number a, $a \cdot 0 = 0$ and $0 \cdot a = 0$

The property is shown two ways to emphasize the fact that the same result accrues whether 0 is the first factor of the indicated product or the second factor of the indicated product.

The sum of the number zero and any real number is the real number. For example,

$$0 + 4 = 4 \qquad -3 + 0 = -3 \qquad 0 + \frac{3}{2} = \frac{3}{2}$$

ADDITIVE PROPERTY OF ZERO

For any real number a,

$$a + 0 = a \quad \text{and} \quad 0 + a = a$$

The property is shown two ways to emphasize the fact that the addend of 0 can be the first addend or the second addend without changing the resultant sum.

14.3 unique properties of the number one

The product of the number one and a real number is the real number. For example,

$$4 \cdot 1 = 4 \qquad (-3) \cdot 1 = -3 \qquad \frac{3}{2} \cdot 1 = \frac{3}{2}$$

MULTIPLICATIVE PROPERTY OF ONE

For any real number a,

$$1 \cdot a = a \quad \text{and} \quad a \cdot 1 = a$$

The statement emphasizes the fact that the number 1 may be the first factor or the second factor without altering the resultant product.

DIVISIVE PROPERTY OF ONE

We won't put all this property in a box because it is an extension of the multiplicative property of 1 and some books omit it completely. We remember from Lesson 2 that the operation of division is defined to be the inverse operation of the operation of multiplication. Since $2 \cdot 3 = 6$, we say that

$$\frac{6}{2} = 3 \quad \text{or} \quad \frac{6}{3} = 2$$

for division must *undo* multiplication. Thus since any number multiplied by 1 equals the number,

$$1 \cdot a = a$$

we must define the inverse operation, division, as follows:

$$\frac{a}{1} = a$$

The quotient of any real number divided by the number 1 is the real number.

$$\frac{a}{1} = a$$

14.4 complex fractions

The multiplicative property of 1 states that the product of the number 1 and any real number is the real number. Since any non-zero number divided by itself is an equivalent expression for 1, any particular number or expression can be multiplied by another number **and** divided by the other number without changing the value of the particular number or expression. For example, if we multiply the number 5 by (2 divided by 2), the product is 10 divided by 2 which is another way to write the number 5.

$$5 \cdot \left(\frac{2}{2}\right) = \frac{10}{2}$$

In Lesson 10 we gave this thought the name **denominator numerator same quantity theorem**. We will use this theorem to simplify complex fractions such as

$$\frac{\frac{a}{b}}{\frac{c}{d}} \qquad \begin{bmatrix} b \neq 0 \\ c \neq 0 \\ d \neq 0 \end{bmatrix}^{\dagger}$$

If we multiply the denominator of this complex fraction by the reciprocal of the denominator, we obtain a product of 1 because we remember that **the product of a number and the reciprocal of the same number is the number one.**

$$\frac{c}{d} \cdot \frac{d}{c} = \frac{cd}{dc} = 1$$

But if we wish to multiply the denominator of the fraction by its reciprocal, which is $\frac{d}{c}$, we must also multiply the numerator by $\frac{d}{c}$ so that the value of the original expression will not be changed.

$$\frac{\frac{a}{b}}{\frac{c}{d}} = \frac{\frac{a}{b} \cdot \frac{d}{c}}{\frac{c}{d} \cdot \frac{d}{c}} = \frac{\frac{ad}{bc}}{\frac{cd}{dc}} = \frac{\frac{ad}{bc}}{1} = \frac{ad}{bc}$$

We have simplified the original fraction by multiplying both the numerator and denominator by the **RECIPROCAL OF THE DENOMINATOR.**

example 14.4.1 Simplify $\frac{\frac{a}{b}}{c}$ $(b, c \neq 0)$.

solution We multiply both the numerator and the denominator by $\frac{1}{c}$, which is the reciprocal of the denominator c.

$$\frac{\frac{a}{b}}{c} = \frac{\frac{a}{b} \cdot \frac{1}{c}}{\frac{c}{1} \cdot \frac{1}{c}} = \frac{\frac{a}{bc}}{\frac{c}{c}} = \frac{\frac{a}{bc}}{1} = \boldsymbol{\frac{a}{bc}}$$

example 14.4.2 Simplify $\frac{\frac{1}{c}}{\frac{1}{b}}$ $(b, c \neq 0)$.

solution We multiply both the numerator and the denominator by b, which is the reciprocal of $\frac{1}{b}$.

---

† To ensure that we are not implying that division by zero is permissible.

$$\frac{\frac{1}{c}}{\frac{1}{b}} = \frac{\frac{1}{c}\cdot\frac{b}{1}}{\frac{1}{b}\cdot\frac{b}{1}} = \frac{\frac{b}{c}}{1} = \boldsymbol{\frac{b}{c}}$$

example 14.4.3 Simplify $\frac{a}{\frac{1}{c}}$ $(c \neq 0)$.

solution We multiply both the numerator and the denominator by c, which is the reciprocal of $\frac{1}{c}$.

$$\frac{a}{\frac{1}{c}} = \frac{\frac{a}{1}\cdot\frac{c}{1}}{\frac{1}{c}\cdot\frac{c}{1}} = \frac{\frac{ac}{1}}{1} = \boldsymbol{ac}$$

example 14.4.4 Simplify $\frac{\frac{a}{x}}{b}$ $(x, b \neq 0)$.

solution We multiply both the numerator and the denominator by $\frac{1}{b}$, which is the reciprocal of b.

$$\frac{\frac{a}{x}}{b} = \frac{\frac{a}{x}\cdot\frac{1}{b}}{\frac{b}{1}\cdot\frac{1}{b}} = \boldsymbol{\frac{a}{xb}}$$

The rule for dividing fractions is sometimes stated as follows: **To divide one fraction by another fraction, invert the fraction in the denominator and multiply**. If we use this rule to simplify

$$\frac{\frac{m}{n}}{\frac{x}{y}} \qquad (n, y \neq 0)$$

the solution is $\frac{m}{n}\cdot\frac{y}{x} = \frac{my}{nx}$.

If we use the **numerator denominator same quantity theorem** to perform the same simplification, we obtain

$$\frac{\frac{m}{n}}{\frac{x}{y}} = \frac{\frac{m}{n}\cdot\frac{y}{x}}{\frac{x}{y}\cdot\frac{y}{x}} \longrightarrow \boldsymbol{\frac{my}{nx}}$$

This procedure yields the same result as that obtained by using the rule, but hopefully we have some understanding of what we did and can justify our procedure.

Many algebraic manipulations can be justified by one of the following:

1. **The denominator numerator same quantity theorem.**
2. **The multiplicative property of equality.**
3. **The additive property of equality.**

The reader is encouraged to justify his manipulations by one of these three rules whenever possible.

TEST YOUR UNDERSTANDING

Simplify:

(a) $\dfrac{\frac{a}{b}}{\frac{x}{y}}$ (b) $\dfrac{a}{\frac{c}{d}}$ (c) $\dfrac{\frac{a}{x}}{y}$

14.5 finite and infinite sets

The words **FINITE** and **INFINITE** are basic words and are difficult to define. The word finite implies the thought of bounded or limited while the word infinite implies the thought of unbounded or without limit. Thus when we say that a set has a finite number of members, we are describing a set such as

$$\{6, 7, 8, 9, 10\}$$

in which the listing of the members has an end. **A set with a finite number of members is called a finite set.**

When we say that a set has an infinite number of members, we are describing a set such as

$$\{1, 2, 3, 4, 5, \ldots\}$$

in which the listing of the members of the set continues without end. **Sets that have an infinite number of members are called infinite sets.**

Some authors define a finite set as a set whose members could be counted if we could live the number of lifetimes necessary to do the counting. Thus the set that has

$$63{,}072{,}000{,}000{,}000{,}000$$

members is a finite set because we could count the members of this set if we counted for 1 billion years at two counts per second.

Using the same definition the set

$$\{1, 2, 3, 4, \ldots\}$$

is an infinite set because we could never count the number of members of this set as the listing has no end.

TEST YOUR UNDERSTANDING

Tell whether or not the following sets are finite or infinite:

(d) $\{3, 4, 5, \ldots, 15, 16, 17\}$ (e) $\{0, 1, 2, 3, \ldots\}$ (f) {the real numbers}

14.6 membership in a set

If we have the numbers

$$0, 0, 0, 0, 0, 0, 1, 1, 1, 1, 2, 2, 2, 2$$

and wish to designate these numbers as members of a set B, we would write

$$B = \{0, 1, 2\}$$

which is read *B is the set whose members are the numbers 0, 1, and 2.* **Note that each of the numbers is only listed once.** Thus, if we have the set,

$$D = \{0, 1, 2, 3, 4\}$$

we have said that the set consists of 0, or any number of zeroes; also the set consists of 1's, 2's, 3's, and 4's, but not necessarily just one 1 and one 2 and one 3 and one 4.

example 14.6.1 Represent the following numbers as being members of set K: 1, 0, 2, 1, 0, 5, 7, 4, 5, 7.

solution We list each number only once. The order in which the numbers are listed is unimportant.

$$\mathbf{K = \{0, 1, 2, 4, 5, 7\}}$$

example 14.6.2 Represent the following numbers as constituting set L:

$$7, \quad 15, \quad 0, \quad 1, \quad 15, \quad 0, \quad 8, \quad -13, \quad 42$$

solution We list each number only one time. The order in which the numbers are listed is unimportant.

$$L = \{7, 15, 0, 1, 8, -13, 42\}$$

We use the symbols

$$\in \quad \text{and} \quad \notin$$

to designate that a particular symbol or number is a member of a given set or is not a member of the given set.

$$0 \in K \qquad 15 \notin K$$

We would read the above as *zero is a member of set K* and *15 is not a member of set K.*

example 14.6.3 Given the sets $A = \{0, 1, 3, 5\}$, $B = \{0, 4, 6, 7\}$, $C = \{1, 2, 3, 5, 7\}$, which of the following is true? (a) $5 \in A$ (b) $4 \in C$ (c) $5 \notin B$

solution

(a) $5 \in A$ True, because 5 is a member of set A.

(b) $4 \in C$ False, because 4 is not a member of set C.

(c) $5 \notin B$ True, because 5 is not a member of set B.

example 14.6.4 Given the sets

$$M = \{0, 1, 2, 3\}, \qquad L = \{5, 6, 7\}, \qquad N = \{0, 1\}$$

which of the following statements is true?

solution

(a) $6 \in M$ False, because 6 is not a member of set M.

(b) $0 \in N$ True, because 0 is a member of set N.

TEST YOUR UNDERSTANDING

Given the sets $A = \{1, 2, 3\}$, $B = \{1, 3, 7\}$, $C = \{3, 5, 9\}$, which statements are true and which are false?

(g) $5 \in A$ (h) $1 \in A$ (i) $3 \in C$

14.7 percent

The word percent comes from the Latin word *per*, which means by, and the Latin word *centum*, which means hundred. Thus percent literally means *by the hundred.*

When we say 6 percent or 15 percent or 133 percent, we are designating the numerator of a fraction whose denominator is 100. For example,

$$\begin{array}{lccc} 6 \text{ percent} & \text{means} & \dfrac{6}{100} & \text{or} & .06 \\ 15 \text{ percent} & \text{means} & \dfrac{15}{100} & \text{or} & .15 \\ 133 \text{ percent} & \text{means} & \dfrac{133}{100} & \text{or} & 1.33 \\ k \text{ percent} & \text{means} & \dfrac{k}{100} & \text{or} & (.01)k \end{array}$$

The denominator of a fraction tells us into how many parts a whole has been divided and the numerator tells us how many of these parts we are considering. If we speak of three-fourths of Jim's farm, we indicate that we have divided Jim's farm into four parts and are considering three of the parts.

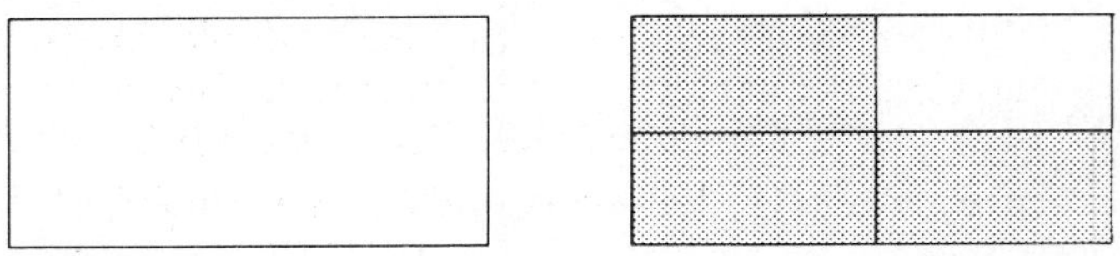

(a) Jim's farm (b) $\frac{3}{4}$ of Jim's farm

Figure 14.1

On the left of Figure 14.1 we show a rectangle that represents Jim's farm and on the right we have divided the rectangle into four parts and have shaded three of these four parts to represent three-fourths of Jim's farm. We can speak of the same part of Jim's farm by using the phrase seventy-five percent of Jim's farm. The word percent tells us that we have divided Jim's farm into one hundred parts and are considering 75 of these 100 parts.

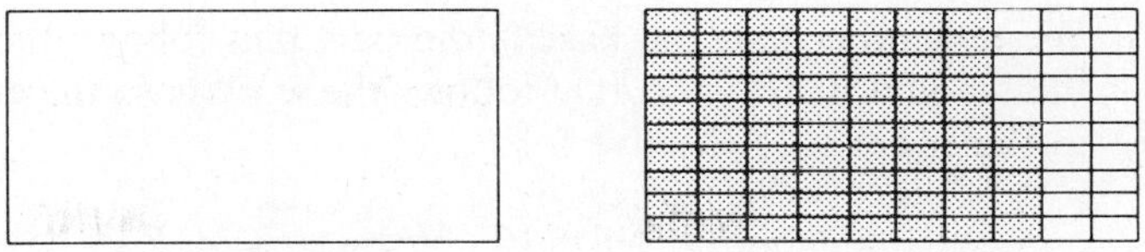

(a) Jim's farm (b) 75 percent of Jim's farm

Figure 14.2

On the left of Figure 14.2 we show Jim's farm and the shaded area on the right indicates 75 percent of Jim's farm.

Thus we see that when we speak of 37 percent of "something," we have mentally divided the something into 100 parts and are considering 37 of the parts. If we speak of 133 percent of something, we have again divided the something into 100 parts and are considering 133 of these parts. Of course, 133 parts are more parts than we began with as we see in Figure 14.3.

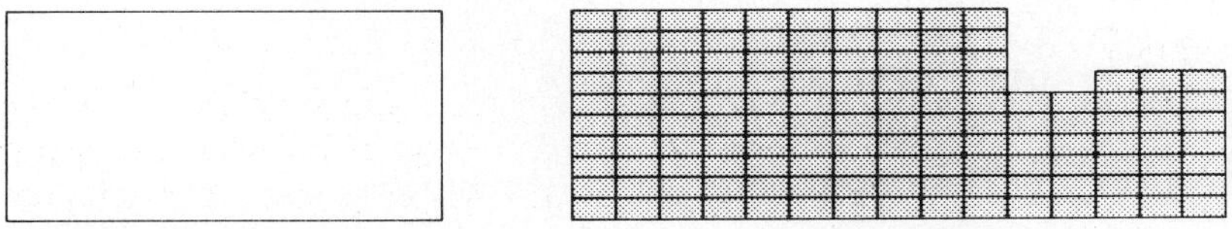

(a) Jim's farm (b) 133 percent of Jim's farm

Figure 14.3

Thus basic percent problems involve multiplying some number by a fraction whose denominator is 100.

(a) 75 percent of 40 is 30 means $\frac{75}{100} \times 40 = 30.$

(b) 30 percent of 12 is 3.6 means $\frac{30}{100} \times 12 = 3.6.$

(c) 60 percent of 4 is 2.4 means $\frac{60}{100} \times 4 = 2.4.$

Since percent problems have a fixed format, we find it convenient to name the parts. Of course, the numerator of the fraction is called the **PERCENT**,

$$\frac{\textbf{PERCENT}}{100} \times \textbf{BASE} = \textbf{AMOUNT}$$

the number by which it is multiplied is called the **BASE**, and the result of the multiplication is called the **AMOUNT**. In percent problems, we are given the values for two of these parts and asked to find the value of the third. Our problem is to identify the two parts that are given. Fortunately, there is an easy way.

1. The percent is identified by the word **percent** or the symbol %.
2. The base is the part that usually follows the word **of**.
3. The amount is the part that follows the word **is**.

example 14.7.1 What percent of 50 is 10?

solution We begin by writing the percent equation.

$$\frac{\text{Percent}}{100} \times \text{base} = \text{amount}$$

We remember that the base is the part that follows the word **of** so the base is 50. Also the amount is the word that follows the word **is** so the amount is 10. We put these values in the equation and solve for P.

$$\frac{P}{100} \times 50 = 10 \longrightarrow \frac{50}{100} \times P = 10 \longrightarrow .5P = 10$$

$$\longrightarrow \frac{.5P}{.5} = \frac{10}{.5} \longrightarrow \mathbf{P = 20\ percent}$$

Some people have difficulty remembering whether the **base** or the **amount** is identified by the words **of** or **is**, and they write the percent equation using **of** and **is** instead of **base** and **amount**.

$$\left(\frac{P}{100}\right)(\text{of}) = (\text{is})$$

If we do this, we place the number identified by **of** in the **of parentheses** and the number identified by **is** in the **is parentheses**.

example 14.7.2 952 is what percent of 5600?

solution We begin with three sets of parentheses

$$\left(\frac{P}{100}\right)(\text{of}) = (\text{is})$$

and note the three key phrases in the statement of the problem.

What percent **of** 5600 **is** 952?

We replace **of** with 5600 and **is** with 952.

$$\left(\frac{P}{100}\right)(5600) = 952$$

We remove the parentheses and write

$$\frac{5600}{100}P = 952 \longrightarrow 56P = 952$$

and now we complete the solution by dividing both sides of the equation by 56.

$$56P = 952 \longrightarrow \frac{56P}{56} = \frac{952}{56} \longrightarrow \mathbf{P = 17\ percent}$$

example 14.7.3 Ten is 20 percent of what number?

solution We will use the same format to solve.

$$\left(\frac{P}{100}\right)(\text{of}) = (\text{is})$$

The key phrases in the statement of the problem are

20 percent **of** what number **is** 10?

$$\left(\frac{20}{100}\right)(WN) = 10$$

We remove the parentheses and solve by multiplying both sides by $\frac{100}{20}$.

$$\frac{20}{100}WN = 10 \longrightarrow \frac{100}{20}\frac{20}{100}WN = \frac{100}{20}(10) \longrightarrow \mathbf{WN = 50}$$

TEST YOUR UNDERSTANDING

(j) 43.47 is 23 percent of what number?

Unfortunately, the problems that we encounter in everyday life are not stated in a nice neat form such as

20 percent of what number is 10?

where the words **is**, **percent**, and **of** can be used to identify the basic parts of the problem. Most of the time the given problem must be reworded so that it fits our format.

example 14.7.4 Kathy, John, and Susie have only 20 chickens left. If they began with 80 chickens, what percent of the original flock remains?

solution We reword the problem as follows:

What percent of 80 is 20?

Now we can use our three-part format to solve

$$\left(\frac{P}{100}\right)(\text{of}) = (\text{is}) \longrightarrow \left(\frac{WP}{100}\right)(80) = 20 \longrightarrow .8\,WP = 20$$

$$\longrightarrow \frac{.8\,WP}{.8} = \frac{20}{.8} \longrightarrow \boldsymbol{WP = 25 \text{ percent}}$$

example 14.7.5 Meme and Jim have 75 thingamabobs. They want to give Hal 14 percent of them. How many thingamabobs do they give Hal?

solution We reword the problem into our format.

14 percent of 75 is what number?

and now we write the equation and solve

$$\left(\frac{14}{100}\right)(75) = WN \longrightarrow \frac{(14)(75)}{100} = WN \longrightarrow \boldsymbol{10.5 = WN}$$

example 14.7.6 Beau has only 48 raisins while Christy has 64 raisins. What percent of Christy's raisins does Beau have?

solution We reword as follows:

What percent of 64 is 48?

and now we use our three-part format to solve.

$$\left(\frac{WP}{100}\right)(64) = 48 \longrightarrow \frac{64}{100}\,WP = 48 \longrightarrow \frac{100}{64}\cdot\frac{64}{100}\,WP = \frac{100}{64}(48)$$

$$\longrightarrow \boldsymbol{WP = 75 \text{ percent}}$$

TEST YOUR UNDERSTANDING

(k) Lizzabel and Eva reduced the price to $208. If this was 80 percent of the original price, what was the original price?

lesson checklist 14

- ☐ Multiplicative property of zero
- ☐ Additive property of zero
- ☐ Multiplicative property of one
- ☐ Divisive property of one
- ☐ Simplification of a quotient of fractions
- ☐ Finite and infinite sets
- ☐ Membership in a set
- ☐ Definition of percent
- ☐ The form of the percentage equation

ANSWERS TO TEST YOUR UNDERSTANDING

(a) $\frac{ay}{bx}$	(b) $\frac{da}{c}$	(c) $\frac{a}{xy}$	(d) finite
(e) infinite	(f) infinite	(g) false	(h) true
(i) true	(j) 189	(k) $260	

problem set 14

*1. Meme and Jim have 75 thingamabobs. They want to give Hal 14 percent of them. How many thingamabobs do they give Hal?

*2. Kathy, John, and Susie have only 20 chickens left. If they began with 80 chickens, what percent of the original flock remains?

*3. Beau has only 48 raisins while Christy has 64 raisins. What percent of Christy's raisins does Beau have?

4. Sixty-eight is 4 percent of what number?

*The problems marked with an asterisk were worked out in this lesson.

Simplify:

5. $\dfrac{(x^{-2})y^{-4}}{y^{-4}(y^{15})^0}$

6. $\left(\dfrac{3x^{-2}y^4}{z^{-1}}\right)^{-2}$

7. $\dfrac{(x^{-2})^{-3}(y^2z)^{-1}}{xyz^{-4}}$

8. $\left(\dfrac{p^{-2}}{xp^{-3}}\right)\left(\dfrac{x^2}{p^3}\right)$

Use substitution to solve:

9. $\begin{cases} x = 5 - 2y \\ 2x + 3y = 7 \end{cases}$

10. $\begin{cases} y = -3x + 6 \\ 2x + 2y = 0 \end{cases}$

11. $\begin{cases} 7x - 2y = 32 \\ x = 3y + 10 \end{cases}$

Add:

12. $(3 - 2x + x^2) - (3x^2 + 4x - 5)$

13. $\dfrac{a}{b} + \dfrac{m}{c^2}$

14. $\dfrac{a}{b + c} - m + \dfrac{1}{k}$

15. $\dfrac{a}{b} - c$

16. $\dfrac{a}{b + c} + \dfrac{m}{c^2}$

Graph on a rectangular coordinate system:

17. $2y - x + 6 = 0$

18. $2y - 3x + 2 = 0$

19. $x = -6$

20. $y = -2$

Graph on a number line:

21. $x \not> 4$

22. $x < 2$

23. $x \leq -3$

Multiply:

24. $(4x + 2)(x - 3)$

Add like terms:

25. $x^4 - 3x^3zp^{-1} + \dfrac{1}{x^{-4}} + \dfrac{2z}{x^{-3}p}$

26. $4x^2ym^{-3} + \dfrac{2y^2}{x^{-2}m} - \dfrac{y}{x^2m^3}$

27. $ax^2m + 3x^2ma - \dfrac{2x^2m}{a^{-1}}$

Multiply:

28. $p^{-2}x\left(\dfrac{p^2x^{-1}}{z^4} + \dfrac{5}{x}\right)$

29. $\dfrac{x^2p^{-2}}{y}\left(\dfrac{y^{-4}}{x^{-2}p} + 6y\right)$

Simplify:

30. $4 + (-3)[-5 + 7(-2)]$

31. $-[-(-3 + 5)]$

Solve:

32. $4(x + 2) - 3[-(-5)] = 7$

33. $4\{3x - 2[-x - 3(-2)]\} = -3(x + 2)$

34. $25 - 3[4(x - 2) - 12x + 4] = 17x + 33$

35. $9 - 3(m - 2) = 2(5 - m)$

Evaluate:

36. $y[-a + (-b)]$ if $a = -2, b = 3, y = -2$

37. $(a - b)[b - (-a)]$ if $a = -3, b = 4$

38. $a^2[-c^2(b^3 - a)]$ if $a = 3, c = -2, b = 4$

39. $m + n(-n) - [m + (-n)]$ if $m = 2, n = -3$

Factor:

40. $mp^2x - m^2zpx + 4m^2p^3x^4$

41. $a^2bc - ab^2c + 3za^2b^2c$

Simplify:

42. $|-62 + |-2||$

43. $|-4 - |-2||$

Express as a product of prime factors:

44. 36

45. 175

Simplify:

***46.** $\dfrac{\frac{a}{b}}{\frac{c}{d}}$

***47.** $\dfrac{\frac{a}{b}}{c}$

48. $\dfrac{\frac{1}{c}}{\frac{1}{b}}$

49. $\dfrac{a}{\frac{1}{c}}$

50. $\dfrac{\frac{a}{x}}{b}$

lesson fifteen

Word problems
More on the substitution method
Subsets
Subsets of the set of real numbers

15.1 word problems Competency in writing mathematical statements that correspond to the written statements in word problems comes through practice. Thus word problems will be an integral part of each problem set for the remainder of the book. For students who have difficulty in writing the proper equations, the answers to these problems in the back of the book will contain both the required equations and the answers.

15.2 meaningful unknowns Instead of the customary variables, x, y, and z, we will use variables that are more meaningful in that we can easily relate the symbol that is used in the equation to the quantity in the word problem that the symbol represents. We will begin with problems that have only one unknown.

example 15.2.1 The sum of twice a number and 13 is 75. Find the number.

solution We will use the letter N to represent the unknown number. The statement of the problem speaks of the sum of twice the number, $2N$, and 13, which we write as $2N + 13$. This sum equals 75, so our equation is

$$2N + 13 = 75$$

To solve for N we will use the additive and multiplicative properties of equality.

$$\begin{aligned} 2N + 13 &= 75 \\ -13 & -13 \qquad \text{add } -13 \text{ to both sides} \\ \hline 2N &= 62 \\ N &= \mathbf{31} \qquad \text{divided both sides by 2} \end{aligned}$$

Check: $2(31) + 13 = 75 \longrightarrow 62 + 13 = 75 \longrightarrow 75 = 75$ Check

example 15.2.2 Find a number such that 13 less than twice the number equals 137.

solution We will use N to represent the unknown number. Then twice the unknown number is $2N$ and 13 less than that is $2N - 13$.

$$\begin{array}{rl} 2N - 13 = & 137 \\ +13 \quad & +13 \qquad \text{add } +13 \text{ to both sides} \\ \hline 2N \quad = & 150 \\ N = & \mathbf{75} \qquad \text{divided both sides by 2} \end{array}$$

Check: $2(75) - 13 = 137 \longrightarrow 150 - 13 = 137 \longrightarrow 137 = 137$ Check

example 15.2.3 Six more than the sum of four times a number and five times the number equals 123. Find the number.

solution We use N to represent the number, $4N$ to represent four times the number and $5N$ to represent five times the number.

$$\begin{array}{rl} 4N + 5N + 6 = 123 & \\ 9N + 6 = 123 & \text{added like terms} \\ -6 \quad -6 & \text{add } -6 \text{ to both sides} \\ \hline 9N \quad = 117 & \\ N = \mathbf{13} & \text{divided both sides by 9} \end{array}$$

Check: $4(13) + 5(13) + 6 = 123 \longrightarrow 52 + 65 + 6 = 123 \longrightarrow 123 = 123$ Check

example 15.2.4 Find a number which equals its opposite increased by 14.

solution We will use N to represent the number and $-N$ to represent its opposite. Thus the opposite of the number increased by 14 can be written as $-N + 14$.

$$\begin{array}{rl} N = -N + 14 & \\ +N \quad +N & \text{add } N \text{ to both sides} \\ \hline 2N = 14 & \\ N = \mathbf{7} & \text{divided both sides by 2} \end{array}$$

Check: $(7) = -(7) + 14 \longrightarrow 7 = -7 + 14 \longrightarrow 7 = 7$ Check

example 15.2.5 We get the same result if we multiply a number by 3 *or* if we multiply the number by 5 and then add 2. Find the number.

solution The statement of the problem leads to the following equation.

$$\begin{array}{rl} 3N = \quad 5N + 2 & \\ -5N \quad -5N & \text{add } -5N \text{ to both sides} \\ \hline -2N = \qquad 2 & \\ N = \quad -1 & \text{divided both sides by } -2 \end{array}$$

Check: $3(-1) = 5(-1) + 2 \longrightarrow -3 = -5 + 2 \longrightarrow -3 = -3$ Check

example 15.2.6 Five times a number is 15 less than twice the number. What is the number?

solution We write five times a number as $5N$, and since this is 15 less than $2N$, we subtract 15 from $2N$ to get a statement of equality.

$$\begin{array}{rrl} 5N= & 2N-15 & \\ -2N & -2N & \text{add } -2N \text{ to both sides} \\ \hline 3N= & -15 & \\ N= & \mathbf{-5} & \text{divided both sides by 3} \end{array}$$

Check: $5(-5) = 2(-5) - 15 \longrightarrow -25 = -10 - 15 \longrightarrow -25 = -25$ Check

example 15.2.7 Find the number which equals the opposite of the same number increased by 22.

solution

$$\begin{array}{rrl} N= & -N+22 & \\ N & N & \text{add } N \text{ to both sides} \\ \hline 2N= & 22 & \\ N= & \mathbf{11} & \text{divided both sides by 2} \end{array}$$

Check: $(11) = -(11) + 22 \longrightarrow 11 = -11 + 22 \longrightarrow 11 = 11$ Check

---

TEST YOUR UNDERSTANDING

(a) Find the number that multiplied by 5 gives the same result as multiplying the sum of the number and 28 by -2.

15.3 substitution

In Lesson 13 we learned how to use **substitution** to find the ordered pair of x and y that will satisfy a system of two first-degree equations in two unknowns. We will review the process by finding the ordered pair that is a solution to the system given here.

example 15.3.1 Use substitution to solve the system of equations shown here. $\begin{cases} 2x + 3y = 12 \\ x = 4y - 5 \end{cases}$

solution

$$\begin{array}{rl} 2(4y - 5) + 3y = 12 & \text{substitute } (4y - 5) \text{ for } x \\ 8y - 10 + 3y = 12 & \text{expand} \\ 11y = 22 & \text{simplify} \\ \mathbf{y = 2} & \text{divide by 11} \end{array}$$

Now $x = 4y - 5 \longrightarrow$ so $x = 4(2) - 5 \longrightarrow \mathbf{x = 3}$.

Thus the solution is the ordered pair (3, 2).

example 15.3.2 Use substitution to solve the system $\begin{matrix} (a) \\ (b) \end{matrix} \begin{cases} x - 2y = -1 \\ 2x - 3y = 4 \end{cases}$.

solution To use substitution to solve this system, it is necessary to rearrange one of the equations by using the additive and/or multiplicative property of equality so that one variable is expressed in terms of the other variable and numbers. We choose to solve for x in the top equation because the x term in this equation has a coefficient of 1 and thus we can solve this equation for x in just one step.

$$\begin{array}{rcl} x - 2y & = & -1 \\ 2y & & 2y \\ \hline x \quad & = & 2y - 1 \end{array} \qquad \text{add } 2y \text{ to both sides}$$

Now we can substitute the expression $2y - 1$ for x in the second equation and complete the solution.

$$2x - 3y = 4 \longrightarrow 2(2y - 1) - 3y = 4 \longrightarrow 4y - 2 - 3y = 4$$
$$\longrightarrow y - 2 = 4 \longrightarrow \mathbf{y = 6}$$

We can find the value of x by replacing the variable y with the number 6 in either of the original equations. We will use both of the original equations to demonstrate that either one can be used to find x.

USING EQUATION (a)	USING EQUATION (b)
$x - 2y = -1$	$2x - 3y = 4$
$x - 2(6) = -1$	$2x - 3(6) = 4$
$x - 12 = -1$	$2x - 18 = 4$
$\mathbf{x = 11}$	$\mathbf{x = 11}$

Thus the ordered pair of x and y that satisfies both equations is (11, 6).

example 15.3.3 Solve
$$\begin{array}{l} (a) \\ (b) \end{array} \begin{cases} 2x - y = 10 \\ 4x - 3y = 16 \end{cases}$$

solution We will first solve equation (a) for y and then substitute the resulting expression for y in equation (b).

$$2x - y = 10 \longrightarrow -y = 10 - 2x \longrightarrow \mathbf{y = -10 + 2x}$$

Now we substitute $-10 + 2x$ for y.

$$4x - 3y = 16 \longrightarrow 4x - 3(-10 + 2x) = 16 \longrightarrow 4x + 30 - 6x = 16$$
$$\longrightarrow -2x + 30 = 16 \longrightarrow -2x = -14 \longrightarrow \mathbf{x = 7}$$

To finish the solution we can use either of the original equations to solve for y.

USING EQUATION (a)	USING EQUATION (b)
$2x - y = 10$	$4x - 3y = 16$
$2(7) - y = 10$	$4(7) - 3y = 16$
$14 - y = 10$	$28 - 3y = 16$
$-y = -4$	$-3y = -12$
$\mathbf{y = 4}$	$\mathbf{y = 4}$

The solution is the ordered pair (7, 4).

example 15.3.4 Solve
$$\begin{array}{l} (a) \\ (b) \end{array} \begin{cases} 4x - 2y = 38 \\ 2x + y = 25 \end{cases}$$

solution We will first solve equation (a) for y and then substitute the resulting expression for y in equation (b).

$$2x + y = 25 \longrightarrow \mathbf{y = 25 - 2x}$$

Now we substitute $25 - 2x$ for y in equation (b).

$$4x - 2y = 38 \longrightarrow 4x - 2(25 - 2x) = 38 \longrightarrow 4x - 50 + 4x = 38$$

$$\longrightarrow 8x = 88 \longrightarrow \mathbf{x = 11}$$

Now we can use either of the original equations to solve for y.

USING EQUATION (a)	USING EQUATION (b)
$4x - 2y = 38$	$2x + y = 25$
$4(11) - 2y = 38$	$2(11) + y = 25$
$44 - 2y = 38$	$22 + y = 25$
$-2y = -6$	$\mathbf{y = 3}$
$\mathbf{y = 3}$	

The solution is the ordered pair (11, 3).

TEST YOUR UNDERSTANDING

(b) Use substitution to solve: $\begin{cases} 3x - 5y = 36 \\ x + 3y = -16 \end{cases}$

15.4 subsets

If all the members of one set are also members of a second set, the first set is said to be a subset of the second set. If we have the two sets

$$B = \{1, 2\} \qquad A = \{1, 2, 3\}$$

then we can say that set B is a subset of set A because all the members of set B are also members of set A. We use the symbol $\subset$ to mean **is a subset of**. Therefore, we can write

$$B \subset A$$

which is read as *set B is a subset of set A* or as *B is a subset of A*. If we also consider set C where

$$C = \{3, 2, 1\}$$

we could say that $C \subset A$, read *set C is a subset of set A* because all the members of set C are also members of set A. Sets A and C are said to be **EQUAL SETS** because they have the **SAME MEMBERS**. Set C is said to be an **IMPROPER SUBSET** of set A and set A is said to be an **improper subset** of set B because **equal sets are defined to be improper sets of each other.** Conversely, set B is a **PROPER SUBSET** of set A because every member of set B is also a member of set A and the sets are not equal sets.

The set that has no members is defined to be a **proper subset** of every set that has members and to be an **improper subset** of itself. This set is called the **EMPTY SET** or the **NULL SET** and can be designated by using either of the symbols shown here.

$\{\ \ \}$ is the empty set $\qquad \varnothing$ is the null set

Thus we can say that $\{\ \ \} \subset A$ or $\varnothing \subset A$, read the empty set is a subset of set A or the null set is a subset of set A because **this set is considered to be a subset of every set**. The slash can be used to negate the symbol $\subset$ as

$$A \not\subset B$$

read set A is not a subset of set B because all the members of set A are not members of set B.

example 15.4.1 Given the sets $D = \{0, 1, 2\}$, $E = \{1, 2, 3, \ldots\}$, $G = \{1, 3, 5\}$, tell which of the following assertions are true and which are false and why.

solution

(a)	$E \subset G$	False	All members of set E are not members of set G
(b)	$G \subset E$	True	All members of set G are members of set E
(c)	$D \subset E$	False	0 is not a member of set E

15.5 subsets of the set of real numbers

The set of **REAL NUMBERS** is said to constitute an **INFINITE SET** because there are an infinite number of members of this set. An infinite set has an infinite number of subsets, but we normally restrict our attention to five subsets of the set of real numbers. These are the sets of **natural numbers, whole numbers, integers, rational numbers**, and **irrational numbers**. It would seem reasonable to assume that all authorities use the same definitions for these sets of numbers, but unfortunately this is not the case. In this book we will use the definitions used by almost all authors of recent algebra textbooks. We define the set of natural numbers (counting numbers) as follows:

$$\text{Natural numbers} \qquad N = \{1, 2, 3, \ldots\}$$

If we list the number zero in addition to the set of natural numbers, we have designated the set of whole numbers.

$$\text{Whole numbers} \qquad W = \{0, 1, 2, 3, \ldots\}$$

Many students want to argue that numbers such as -3 and -15 should also be called whole numbers because they are not fractions and they are not decimal numbers. Maybe they should be called whole numbers but if we use the definitions given above we must call these numbers **INTEGERS** and we cannot call them whole numbers. We designate the set of integers by listing the set of whole numbers and including the opposite of every natural number.

$$\text{Integers} \qquad J = \{\cdots, -3, -2, -1, 0, 1, 2, 3, \cdots\}$$

Every number greater than zero that can be used to describe a physical distance is a positive real number and the set of all of these numbers is called the set of positive real numbers. Each of these numbers has an opposite and the set whose members are the opposites of every positive real number is called the set of negative real numbers. **If we take the set of positive real numbers together with the number zero and the set of negative real numbers, we have the set of real numbers.** Each of these numbers can be paired with a point on the number line and the numbers are called the coordinates of the points. Thus we can also describe the set of real numbers by saying that this set consists of the coordinates of every point on the number line.

Every real number is either a rational number or an irrational number. A rational number is a number that **can be** represented as a quotient† (fraction) of integers. Thus each of the following numbers is a rational number.

†A fraction of integers whose denominator is not zero.

(a) $\frac{1}{4}$ (b) $\frac{-3}{17}$ (c) -10 (d) .013

The numbers $\frac{1}{4}$ and $\frac{-3}{17}$ are already expressed as quotients of integers. The number -10 can be written as a quotient of integers in many ways such as

$$\frac{-100}{10} \quad \text{and} \quad \frac{3000}{-300} \quad \text{and} \quad \frac{270}{-27}$$

and the number .013 can be written as a quotient of integers in many ways, one of which is

$$\frac{13}{1000}$$

There are an infinite number of numbers such as the square root of 2 that cannot be written as a quotient of integers. These numbers constitute the set of **irrational numbers** and will be investigated further in later sections of this book. For now, we will remember that the set of real numbers can be divided into two disjoint subsets as we show here.

REAL NUMBERS

(The coordinates of all points on the number line)

Q = **RATIONAL NUMBERS**	P = **IRRATIONAL NUMBERS**
(All real numbers that **can** be expressed as quotients of integers)	(All real numbers that **cannot** be expressed as quotients of integers)

In this book when we ask to which sets a particular number belongs, we restrict the possible answers to the sets of natural numbers, whole numbers, integers, rational numbers, irrational numbers, and real numbers. Problems such as the following will give the reader practice in distinguishing between these sets.

example 15.5.1 $\frac{1}{2} \in$ {what sets of numbers?}

solution This asks that we identify the sets of which the number $\frac{1}{2}$ is a member. If we restrict our reply to the sets discussed above, the answer is the sets of **rationals,** and **reals.**

example15.5.2 $5 \in$ {what sets of numbers?}

solution The **naturals, wholes, integers, rationals,** and **reals.**

example 15.5.3 Tell whether the following statements are true or false and tell why.

solution

(a)	{Reals} $\subset$ {Integers}	False	The reals are not a subset of the integers. The integers are a subset of the reals.
(b)	{Irrationals} $\subset$ {Reals}	True	All irrational numbers are also real numbers.
(c)	{Irrationals} $\subset$ {Rationals}	False	The irrationals are not a subset of the rationals. In fact, no irrational number is also a rational number and vice versa.
(d)	{Whole} $\subset$ {Natural}	False	It's the other way around.
(e)	{Integers} $\subset$ {Reals}	True	All five sets just discussed are subsets of the real numbers.

Test Your Understanding

The following numbers are members of what sets of numbers?

(c) 5 (d) $\frac{1}{2}$ (e) -14

lesson checklist 15

- ☐ Meaningful unknowns
- ☐ Rearranging an equation before substitution
- ☐ Definition of a subset
- ☐ Proper and improper subsets
- ☐ Empty set and null set
- ☐ Major subsets of the set of real numbers
- ☐ Rational numbers and irrational numbers

Answers to Test Your Understanding

(a) $5N = (N + 28)(-2) \rightarrow N = -8$ (b) $(2, -6)$

(c) Natural, wholes, integers, rationals, reals

(d) Rationals, reals (e) Integers, rationals, reals

problem set 15

*1. The sum of twice a number and 13 is 75. Find the number.

*2. Find a number such that 13 less than twice the number equals 137.

*3. Six more than the sum of 4 times a number and 5 times the number equals 123. Find the number.

*4. Find a number which equals its opposite increased by 14.

*5. We get the same result if we multiply a number by 3 or if we multiply the number by 5 and then add 2. Find the number.

6. Four percent of the 3000 people living in Marshalville voted. How many people voted?

7. The list price of a camera is \$175. If the store owner gives a discount of \$35, what percent discount does he give?

8. Fifteen percent of a number is 120. What is the number?

Use substitution:

*9. $\begin{cases} 2x + 3y = 12 \\ x = 4y - 5 \end{cases}$ *10. $\begin{cases} 2x - y = 10 \\ 4x - 3y = 16 \end{cases}$ 11. $\begin{cases} 4x - 2y = 26 \\ 2x + y = 3 \end{cases}$

*The problems marked with an asterisk were worked out in this lesson.

Add:

12. $-(x^3 - 4x^2 - 5) + (x^2 - 5)$

13. $\frac{x}{y} - m + \frac{1}{b + c}$

14. $\frac{a + x}{m} - \frac{1}{c}$

15. $\frac{k}{m} + c - \frac{1}{d + x}$

16. $\frac{a}{x^3(y + c)} + \frac{m}{x^2} - d$

17. $\frac{x}{m^3c} - \frac{1}{c^3p}$

Graph on a rectangular coordinate system:

18. $3y - 4x + 6 = 0$ **19.** $y = -2$ **20.** $y = -2x$ **21.** $x = -y - 4$

Graph on a number line:

22. $x \leq 2$ **23.** $x \geq -4$ **24.** $x \nleq -2$

Multiply

25. $(x + 2)(3x - 4)$ **26.** $(x + 2)(x^2 + 3x + 4)$

Simplify:

27. $\frac{x^{-2}yp^3x^3}{x^2y^{-2}p}$ **28.** $\frac{-3xy}{2y^2x^{-2}}$ **29.** $\left(\frac{x^{-2}y^2}{x^{-3}}\right)^{-3}$ **30.** $\left(\frac{a^5b}{x^{-1}}\right)^0$

Add:

31. $1 + \frac{x}{y}$

Add like terms:

32. $\frac{3a^2x}{b^2} - 2a^2b^{-2}x + \frac{4b^2x}{a^2}$

33. $\frac{4a^2}{b} - \frac{3a}{b^2} + 2a^2b^{-1} - 5ab^{-2}$

Multiply:

34. $\frac{mx^2}{b}\left(\frac{2b}{m^{-1}x^{-2}} + \frac{4}{x^2}\right)$

Simplify:

35. $|-14|$

36. $-5[3 - (-5)] - 2$

Evaluate:

37. $y + (-x)(-y)$ if $x = -3, y = 2$

38. $-y(a - b) + (-b)$ if $y = -2, a = 3, b = -1$

39. $-x^2 - a(x - a^2)$, if $x = -2, a = 3$

Solve:

40. $-3(-x + 2) + 4 = -x(5 - 11)$

41. $5p - 3p + 4(-p) = 2(p - 6)$

42. $m - 3m + (-4)(-m) = 2(-m + 3)$

Simplify:

43. $\frac{\frac{a}{x}}{c}$ **44.** $\frac{\frac{a}{x}}{\frac{1}{c}}$ **45.** $\frac{\frac{a + b}{x}}{\frac{1}{x}}$ **46.** $\frac{\frac{a - b}{a + b}}{\frac{1}{a + b}}$ **47.** $\frac{-5}{5^{-2}}$

***48.** True or False: {Integers} $\subset$ {Natural}

***49.** $5 \in$ {what sets of numbers?}

Simplify:

50. $-6\{-3[2 - 7(4 - 2)(-3)]\}$

lesson sixteen

Roots of positive real numbers
Square roots of positive real numbers
Products of square roots theorem
Domain
Additive property of inequality

16.1 roots of a positive real number

The operations of addition and subtraction are inverse operations because subtraction can undo addition and addition can undo subtraction. For example, if we begin with a number, say 7, and add 2 and then subtract 2, the result is 7, the number with which we began.

$$7 + 2 - 2 = 7$$

The operations of multiplication and division are also inverse operations because multiplication can undo division and division can undo multiplication. If again we begin with the number 7 and then multiply and divide by 2, the result is 7, the number with which we began.

$$\frac{7 \cdot 2}{2} = 7$$

By the definition of exponential notation

$$2^2 = 2 \cdot 2 = 4$$

This operation is called raising to a power. **The inverse operation of raising to a power is called extracting a root or taking the root of.** Since

$$2^2 = (2)(2) = 4$$

we say that 2 is a square root of 4, and since

$$(-2)^2 = (-2)(-2) = 4,$$

we say that -2 is a square root of 4. To indicate the square root of 4, we use the expression $\sqrt{4}$. This expression is called a **RADICAL EXPRESSION**. The symbol $\sqrt{\ }$ is called a **RADICAL SIGN** and the number or expression under the radical sign is called the **RADICAND.**

We note that $2^2 = 4$ and also $(-2)^2 = 4$ so the number 4 has two square roots. In a like manner every positive number has two square roots, one positive and one negative.

The number zero has only one square root, which is zero itself. **We reserve the symbol $\sqrt{x}$ to indicate the positive square root or principal square root of the positive† number x.** Thus when we write $\sqrt{4}$, we are indicating the principal square root of 4, which is +2. To consolidate the above, we write

$$2^2 = 4 \quad \text{and} \quad (-2)^2 = 4 \quad \text{but} \quad \sqrt{4} = 2 \text{ only}$$

If we wish to indicate the negative square root of 4, we must write

$$-\sqrt{4} = -2$$

DEFINITION OF SQUARE ROOT

If x is greater than or equal to zero, then $\sqrt{x}$ is the unique **positive** real number such that

$$(\sqrt{x})^2 = x$$

Unfortunately, the square root of every positive real number is not an integer. In fact, the square roots of many numbers cannot even be represented exactly by using a decimal number that has a finite number of digits. The **decimal approximation** often used for the square root of 2 is 1.414.

$$\sqrt{2} \cong 1.414 \qquad (\cong \text{ is read } \textit{approximately equal to})$$

A more exact approximation is given by a particular pocket calculator as

$$\sqrt{2} \cong 1.414213562$$

and a table of square roots gives an approximation to 20 places as

$$\sqrt{2} \cong 1.41421356237309504880$$

The square root of 2 is a non-repeating decimal number of infinite length and thus cannot be expressed exactly as a decimal number.

In Lesson 15 we said that an irrational number was a number that cannot be written as a quotient‡ of integers. Another way to define an irrational number is to say that **an irrational number is a number whose decimal representation is a non-repeating decimal number of infinite length.** Thus the $\sqrt{2}$ is an irrational number. The square roots of many integers are also irrational numbers. For example, the following numbers are all irrational numbers.

$$\sqrt{2}, \quad \sqrt{3}, \quad \sqrt{5}, \quad \sqrt{6}, \quad \sqrt{7}, \quad \sqrt{8}, \quad \sqrt{10}, \quad \sqrt{11}, \quad \sqrt{12}, \quad \sqrt{13}$$

The easiest way to find an approximation for the square root of a positive real number is to use a table of square roots or a pocket calculator. If these are not available, some variation of a *cut-and-try* approach will give a good approximation. Let's cut-and-try to find an approximation for the square root of 5.

Now $2 \times 2 = 4$ and $3 \times 3 = 9$, so $\sqrt{5}$ lies between 2 and 3 and is closer to 2.

Try 2.1	$2.1 \times 2.1 = 4.41$	not large enough
Try 2.2	$2.2 \times 2.2 = 4.84$	still not large enough
Try 2.3	$2.3 \times 2.3 = 5.29$	too large

†In this course, we will not discuss square roots of negative numbers. This is a topic for a more advanced course in algebra.

‡Fraction.

Try 2.25	$2.25 \times 2.25 = 5.0625$	a little too large
Try 2.24	$2.24 \times 2.24 = 5.0176$	still too large
Try 2.23	$2.23 \times 2.23 = 4.9729$	now too small

Thus we see that $\sqrt{5}$ is a number somewhere between 2.23 and 2.24. We could continue this cut-and-try procedure to find an approximation of $\sqrt{5}$ to as many decimal places as we desire.

Although the decimal representation of $\sqrt{2}$ is a non-repeating decimal number of infinite length, **the $\sqrt{2}$ is a real number and can be associated with a specific point on the number line.** In Figure 16.1 we graph the $\sqrt{2}$ and also several other irrational numbers.

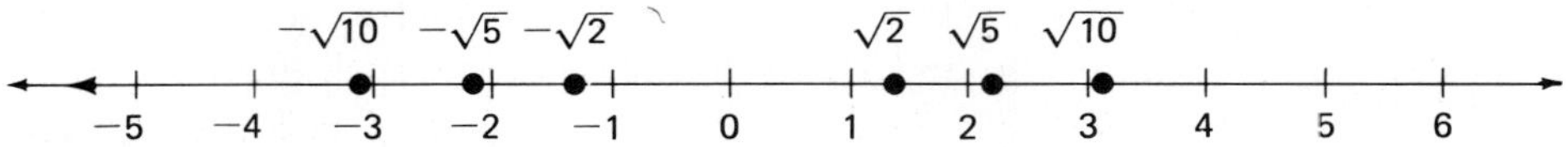

Figure 16.1

The definition of square root stated earlier in this lesson says that the square root of a given positive number is that positive number which multiplied by itself equals the given number.

(a) $\sqrt{2}\,\sqrt{2} = 2$ (b) $\sqrt{a}\,\sqrt{a} = a \quad (a > 0)$

(c) $\sqrt{2.42}\,\sqrt{2.42} = 2.42$ (d) $\sqrt{x}\,\sqrt{x} = x \quad (x > 0)$

Square roots of many numbers such as $\sqrt{50}$, $\sqrt{200}$, and $\sqrt{147}$ can be written in a simplified form. To write one of these numbers in a simplified form, we use the following theorem.

> PRODUCT OF SQUARE ROOTS THEOREM
>
> If m and n are non-negative real numbers, then
>
> $$\sqrt{m}\,\sqrt{n} = \sqrt{mn} \quad \text{and} \quad \sqrt{mn} = \sqrt{m}\,\sqrt{n}$$

This theorem can be generalized to the product of any number of factors and thus we say that the square root of any product may be written as the product of the square roots of the factors of the product. For example, $\sqrt{2 \cdot 5 \cdot 5}$ can be written as $\sqrt{2}\,\sqrt{5}\,\sqrt{5}$ and $\sqrt{3 \cdot 3 \cdot 3 \cdot 5}$ can be written as $\sqrt{3}\,\sqrt{3}\,\sqrt{3}\,\sqrt{5}$. We will use this theorem in the following problems to help us simplify the expressions $\sqrt{50}$, $\sqrt{200}$, $\sqrt{147}$ and $\sqrt{108}$.

example 16.1.1 Simplify $\sqrt{50}$.

solution We will first write 50 as a product of prime factors.

$$\sqrt{5 \cdot 5 \cdot 2}$$

Now we use the product of square roots theorem to write the square root of the product as a product of square roots.

$$\sqrt{5}\,\sqrt{5}\,\sqrt{2}$$

Now by definition, $\sqrt{5}\,\sqrt{5} = 5$ so we have

$$(\sqrt{5}\,\sqrt{5})\sqrt{2} = \mathbf{5\sqrt{2}}$$

example 16.1.2 Simplify $\sqrt{200}$.

solution First we write 200 as a product of prime factors

$$\sqrt{200} = \sqrt{2 \cdot 2 \cdot 2 \cdot 5 \cdot 5}$$

Now the square root of the product is written as the product of square roots.

$$\sqrt{2 \cdot 2 \cdot 2 \cdot 5 \cdot 5} = \sqrt{2}\,\sqrt{2}\,\sqrt{2}\,\sqrt{5}\,\sqrt{5}$$

Since by definition $\sqrt{2}\,\sqrt{2}$ equals 2 and $\sqrt{5}\,\sqrt{5} = 5$,

$$(\sqrt{2}\,\sqrt{2}\,)\sqrt{2}\,(\sqrt{5}\,\sqrt{5}\,) = 2\sqrt{2}\,5 = \mathbf{10\sqrt{2}}$$

example 16.1.3 Simplify $\sqrt{147}$.

solution

(a) $\sqrt{147} = \sqrt{3 \cdot 7 \cdot 7}$ — write as product of prime factors

(b) $\sqrt{3 \cdot 7 \cdot 7} = \sqrt{3}\,\sqrt{7}\,\sqrt{7}$ — root of product equals product of roots

(c) $\sqrt{3}\,(\sqrt{7}\,\sqrt{7}\,) = \mathbf{7\sqrt{3}}$ — definition of square root

example 16.1.4 Simplify $\sqrt{108}$.

solution

(a) $\sqrt{108} = \sqrt{2 \cdot 2 \cdot 3 \cdot 3 \cdot 3}$ — write as product of prime factors

(b) $\sqrt{2 \cdot 2 \cdot 3 \cdot 3 \cdot 3} = \sqrt{2}\,\sqrt{2}\,\sqrt{3}\,\sqrt{3}\,\sqrt{3}$ — root of product equals product of roots

(c) $(\sqrt{2}\,\sqrt{2}\,)(\sqrt{3}\,\sqrt{3}\,)\sqrt{3} = 2 \cdot 3\sqrt{3} = \mathbf{6\sqrt{3}}$ — definition of square root

TEST YOUR UNDERSTANDING

Simplify:

(a) $\sqrt{48}$ (b) $\sqrt{125}$

16.2 the domain

In problems in mathematics (and in physics, chemistry, and other mathematically based disciplines) the numbers that may be used as replacements for the variables are often restricted by the nature of the problem or by a restriction stated in the problem. For instance, if a girl goes to the store with 25¢ to buy eggs for her mother, and eggs cost 10¢ each, the total amount of money that she can spend on eggs can be represented by the equation.

$$\text{total cost} = 10N_E \text{ cents}$$

where N_E represents the number of eggs she buys. She may buy no eggs or one egg or two eggs. Thus the total cost of her eggs is as shown in (a), (b), or (c).

(a) cost $= 10N_E$

cost $= 10(0)$

cost $= 0¢$

(b) cost $= 10N_E$

cost $= 10(1)$

cost $= 10¢$

(c) cost $= 10N_E$

cost $= 10(2)$

cost $= 20¢$

In (a) we use 0 as the replacement for the variable and in this case the cost is 0¢. In (b), we use 1 as the replacement for the variable and find the cost to be 10¢. In (c) we use 2 as the replacement for the variable and find that in this case the cost is 20¢. We could not use 8 as a replacement for the variable for we were restricted in the possible replacement values for N_E because the statement of the problem said she only had 25¢. Neither could we use -4 as a replacement for the variable because buying -4 eggs makes no sense. We were restricted by the statement of the problem to using only the numbers

$$\{0, 1, 2\}$$

The set of numbers that constitutes the set of permissible replacement values for the variables in a particular equation or inequality is called the domain for that equation or inequality.

Since **every equation and every inequality have a domain**† and since this is an important concept, it is customary in courses in algebra to include problems in which the domain is specified. These problems should help the reader with the concept of domain. The domains for the problems in this book were chosen by the author, sometimes with a purpose and sometimes just arbitrarily so that the problems would have specified domains. We will use the capital letter D as the symbol for the domains in this book and will indicate the domains as sets by enclosing them with braces. For instance,

(a) $D = \{1, 2, 3\}$ (b) $D = \{\text{Reals}\}$ (c) $D = \{\text{Positive integers}\}$

The domains specified here are (a) the numbers 1, 2 and 3; (b) the set of real numbers; and (c) the set of positive integers.

example 16.2.1 Graph $x < 3$, $D = \{\text{Integers}\}$.

solution We are asked to indicate the **INTEGERS** that are less than 3.

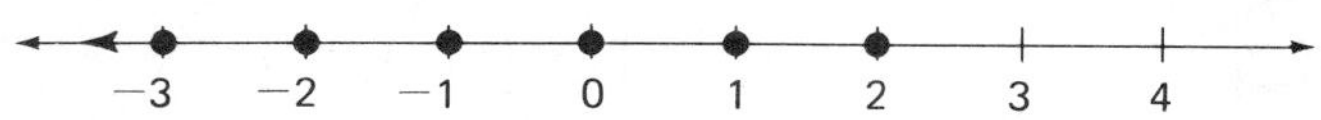

On the number line we have indicated the integers whose values are less than 3. Note that it was not necessary to place an open circle at 3. The arrow on the left indicates an infinite continuation.

example 16.2.2 Graph $x \geq -1$, $D = \{\text{Reals}\}$.

solution We are asked to indicate all **REAL NUMBERS** that are equal to or greater than -1.

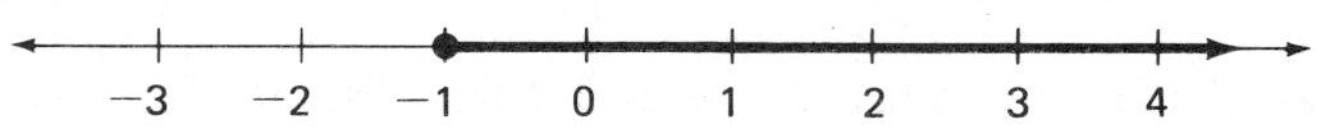

The graph indicates all real numbers that are greater than -1. The solid circle at -1 indicates that -1 is a member of the solution set.

example 16.2.3 Graph $x < -1$, $D = \{\text{Positive integers}\}$.

solution The solution is $\varnothing$, the null set, or { }, the empty set, because there are *no* positive integers that are less than -1.

---

†If the domain is not stated, it is implied.

example 16.2.4 Graph $x \geq -5$, $D = \{\text{Positive integers}\}$.

solution

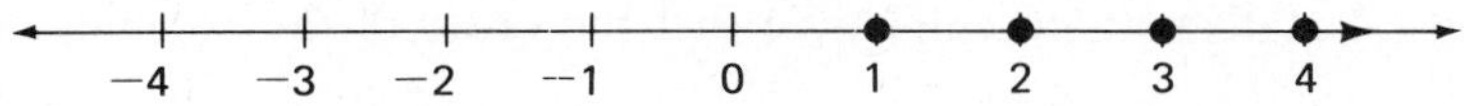

The graph indicates the numbers that are greater than or equal to -5 and that are also members of the set of positive integers.

TEST YOUR UNDERSTANDING

(c) Graph: $x \geq -3$, $D = \{\text{Negative integers}\}$

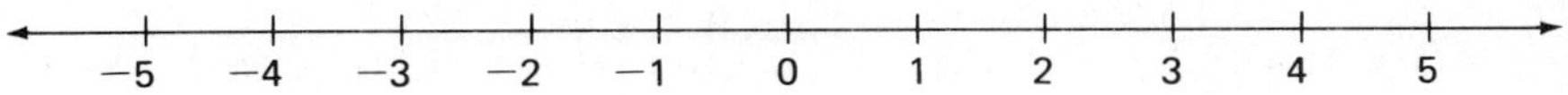

16.3 additive property of inequality

The **additive property of equality is restated.** Note that we write $a + c = b + c$ and also write $c + a = c + b$. We do this to emphasize that the order of addition of the addends does not affect the result.

ADDITIVE PROPERTY OF EQUALITY

If a, b, and c are any real numbers such that

$$a = b$$

then

$$a + c = b + c \quad \text{and} \quad c + a = c + b$$

In Lesson 6 we learned that if we make certain assumptions, we can use the additive property of equality to help us solve equations. To solve $x + 4 = 8$, we add -4 to both sides of the equation.

$$\begin{array}{r} x + 4 = 8 \\ -4 \quad -4 \\ \hline x = 4 \end{array}$$

The **additive property of inequality** is stated in the same way as the additive property for equality except that we use the $>$ symbol rather than the $=$ sign.

ADDITIVE PROPERTY OF INEQUALITY

If a, b, and c are any real numbers such that

$$a > b$$

then $a + c > b + c$ and also $c + a > c + b$

This statement implies that the same quantity may be added to both sides of a statement of inequality without changing the solution set of the inequality. Again we used two statements to emphasize that the order of the addends does not affect the result.

example 16.3.1 Graph $x + 2 < 0$, $D = \{\text{Reals}\}$.

solution We isolate x by adding -2 to both sides of the equation, as permitted by the additive property of inequality.

$$\begin{array}{rrl} x + 2 < & 0 & \text{given} \\ -2 & -2 & \text{added } -2 \text{ to both sides} \\ \hline x < & -2 & \end{array}$$

Now we graph the inequality $x < -2$.

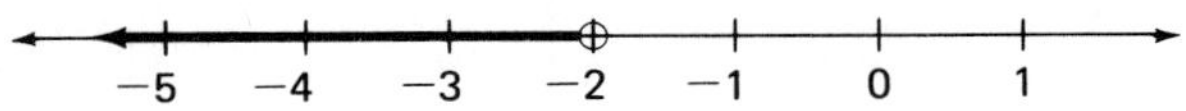

The open circle at -2 indicates that -2 is not a member of the solution set.

example 16.3.2 Graph $x - 3 \geq -5$, $D = \{\text{Integers}\}$.

solution We isolate x by adding $+3$ to both sides of the equation, as permitted by the additive property of inequality.

$$\begin{array}{rrl} x - 3 \geq & -5 & \text{given} \\ +3 & +3 & \text{add } +3 \text{ to both sides} \\ \hline x \quad \geq & -2 & \end{array}$$

Now we graph the solution $x \geq -2$ and remember that the domain is the set of integers.

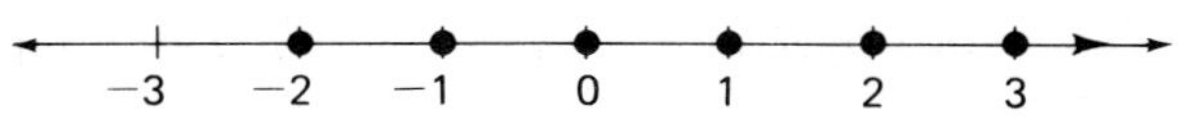

Note that -2 is a solution to the inequality $x \geq -2$.

TEST YOUR UNDERSTANDING

(d) Graph: $x + 4 > 6$, $D = \{\text{Reals}\}$

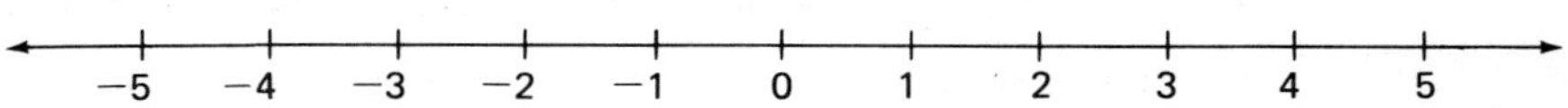

lesson checklist 16

- ☐ Raising to a power and taking the root of
- ☐ Radical expressions
- ☐ Definition of square root
- ☐ Non-repeating decimal numbers of infinite length

- ☐ Product of square roots theorem
- ☐ Definition of domain
- ☐ Additive property of inequality
- ☐ Graphs of inequalities with domains

ANSWERS TO TEST YOUR UNDERSTANDING

(a) $4\sqrt{3}$ (b) $5\sqrt{5}$

(c)

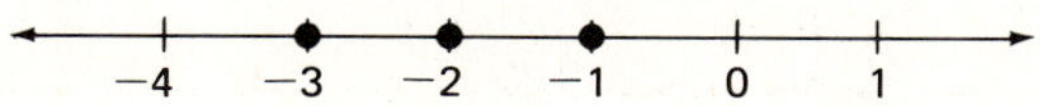

(d)

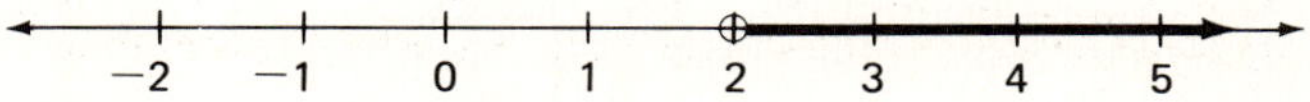

problem set 16

1. If 27 is added to twice a number, the sum is 51. What is the number?
2. If 11 is added to three times a number, the sum is −13. What is the number?
3. The sum of three times a number and 27 is 21. What is the number?
4. Find a number which equals its opposite decreased by 54.
5. What number is 5 percent of 20?
6. Seventy-five is 20 percent of what number?
7. Twenty-five is what percent of 200?

Simplify:

*8. $\sqrt{50}$ *9. $\sqrt{200}$

*10. $\sqrt{147}$ *11. $\sqrt{108}$

Graph on a number line:

*12. $x < 3$, $D = \{\text{Integers}\}$ *13. $x \geq -1$, $D = \{\text{Reals}\}$

*14. $x \geq -5$, $D = \{\text{Positive integers}\}$ *15. $x + 2 < 0$, $D = \{\text{Reals}\}$

Use substitution:

16. $\begin{cases} 5x - 3y = -5 \\ 2x + y = 9 \end{cases}$ 17. $\begin{cases} 3x - y = 1 \\ 2x + 3y = 8 \end{cases}$

18. $\begin{cases} 11x - 4y = -5 \\ 3x - y = 1 \end{cases}$ 19. $\begin{cases} 3x + 2y = -8 \\ x = -4 \end{cases}$

Multiply:

20. $(x - 3)(2x^2 + 3x + 2)$

*The problems marked with an asterisk were worked out in this lesson.

Add:

21. $-(x^2 - x - 3) + (2x^2 + x)$ **22.** $\dfrac{m + n}{a^2} + \dfrac{b}{a^3c}$

23. $\dfrac{m}{x} + c - \dfrac{m}{y}$ **24.** $\dfrac{x(a + b)}{c} - \dfrac{c}{x^2}$ **25.** $\dfrac{a}{b + d} - x - \dfrac{1}{d}$

Graph on a rectangular coordinate system:

26. $y - 3x = -2$ **27.** $y = -3$

Simplify:

28. $\sqrt{8}$ **29.** $\sqrt{32}$ **30.** $\sqrt{27}$

Add:

31. $1 + \dfrac{1}{y}$

Simplify:

32. $\dfrac{(y^{-2}z)^{-4}z^2}{y^4}$ **33.** $\dfrac{(x^{-2})^0y^{-2}}{(x^2)^{-3}y^{-4}}$ **34.** $\left(\dfrac{a^{-2}x^4}{x}\right)^{-2}$

35. $\dfrac{(-2)^3}{2^2}$ **36.** $\dfrac{m^{-3}x^0y^2}{(y^{-2})^{-2}m^{-3}}$

Add like terms:

37. $\dfrac{m^{-5}x^4}{y^{-2}} + \dfrac{3y^2x^4}{m^{-5}} - \dfrac{2x^4y^{-2}}{m^2}$ **38.** $\dfrac{4p^{-3}y^2}{m} - \dfrac{4x^2}{m^2} + \dfrac{m^{-2}}{x^{-2}} - \dfrac{y^2}{p^3m}$

Multiply:

39. $p^{-2}y^2\left(\dfrac{p^4y}{z^{-5}} - \dfrac{3p}{y^2}\right)$.

Simplify:

40. $-4 - (-3)[-6 + (-5)]$ **41.** $-2(-5 + 11)(-3)$

Evaluate:

42. $-a(-b)(-c + a)$ if $a = -2$, $b = 4$, $c = -1$

43. $-mx + y(-x)$ if $m = -3$, $y = -2$, $x = 1$

44. $-xy(-z) + [-y(-x)]$ if $x = -2$, $y = 3$, $z = 4$

Solve:

45. $x + 3(2x - 5) + 4 = 2x + 3(x + 2)$

46. $x + (-3)(3x - 2) = -5(x + 3)$

Simplify:

47. $\dfrac{\frac{a}{x + y}}{a}$ **48.** $\dfrac{\frac{m}{n}}{\frac{1}{xy}}$ **49.** $\dfrac{\frac{p}{q}}{xp}$

50. True or false: {Natural} $\subset$ {Reals}

51. $\sqrt{2} \in$ {what sets of numbers?}

lesson seventeen

Addition of radical expressions
Simplification of radical expressions
Solving simultaneous equations by elimination
Complex fractions

17.1 addition of radical expressions

If we use the commutative properties of addition and multiplication, we can rearrange the standard expression of the distributive property into two additional forms as we show here.

1. $a(b + c) = ab + ac$ standard form
2. $(b + c)a = ba + ca$ rearranged
3. $ba + ca = (b + c)a$ rearranged and reversed

In these expressions, a, b, and c represent unspecified real numbers. If in expression (3), we let a equal 4, b equal 3, and c equal 2, we find

$$3 \cdot 4 + 2 \cdot 4 = (3 + 2)4$$
$$12 + 8 = (5)4$$
$$20 = 20 \quad \text{True}$$

Since radical expressions such as the $\sqrt{2}$ designate real numbers, the distributive property in form (3) gives us the procedure for the addition of **LIKE RADICAL TERMS**.†

The indicated sum $4\sqrt{2} + 3\sqrt{2}$ is in the same form as $ba + ca$. From form 3 above, we see that $ba + ca$ can be written as $(b + c)a$. We can use the same procedure to write the sum of $4\sqrt{2} + 3\sqrt{2}$ as we show here.

$$4\sqrt{2} + 3\sqrt{2} = (4 + 3)\sqrt{2} \quad \text{or} \quad 7\sqrt{2}$$

This use of the distributive property of real numbers can be generalized to apply to the algebraic addition of any number of like radical terms for each of these terms represents the same real number.

$$ba + ca + da + ea + \cdots + ka = (b + c + d + e + \cdots + k)a$$

†Radical terms that have the same index and the same radicand.

example 17.1.1 Add $4\sqrt{2} - 5\sqrt{2} + 12\sqrt{2}$.

solution We use the distributive property to write

$$4\sqrt{2} - 5\sqrt{2} + 12\sqrt{2} = (4 - 5 + 12)\sqrt{2} = \mathbf{11\sqrt{2}}$$

example 17.1.2 Add $4\sqrt{3} + 3\sqrt{5} - 6\sqrt{3}$.

solution Only like radical terms may be added.

$$4\sqrt{3} - 6\sqrt{3} + 3\sqrt{5} = (4 - 6)\sqrt{3} + 3\sqrt{5} = \mathbf{-2\sqrt{3} + 3\sqrt{5}}$$

example 17.1.3 Add $-3\sqrt{2} + 5\sqrt{3} - 2\sqrt{2} + 8\sqrt{3}$.

solution We omit the intermediate step and write the answer directly by simply adding the coefficients of like radical terms.

$$-3\sqrt{2} + 5\sqrt{3} - 2\sqrt{2} + 8\sqrt{3} = \mathbf{-5\sqrt{2} + 13\sqrt{3}}$$

example 17.1.4 Add $4\sqrt{3} - 2\sqrt{2} + 6\sqrt{5}$.

solution No two of these radical terms are like radical terms so no addition is possible.

TEST YOUR UNDERSTANDING

(a) Add: $5\sqrt{3} + 7\sqrt{3} - 2\sqrt{3} + 5\sqrt{3}$

17.2 simplification of radical expressions

In Lesson 16 we learned to simplify expressions such as $\sqrt{50}$ by using the product of square roots theorem.

$$\sqrt{50} = \sqrt{5 \cdot 5 \cdot 2} = \sqrt{5}\,\sqrt{5}\,\sqrt{2} = \mathbf{5\sqrt{2}}$$

We can use the same procedure to simplify expressions such as

$$\sqrt{8} - \sqrt{50} + \sqrt{98}$$

if we first simplify each expression and then add the like radical terms. We begin by writing each radicand as a product of prime factors.

$$\sqrt{2 \cdot 2 \cdot 2} - \sqrt{5 \cdot 5 \cdot 2} + \sqrt{7 \cdot 7 \cdot 2}$$

Now we write the roots of products as products of roots.

$$\sqrt{2}\,\sqrt{2}\,\sqrt{2} - \sqrt{5}\,\sqrt{5}\,\sqrt{2} + \sqrt{7}\,\sqrt{7}\,\sqrt{2}$$

And to finish we simplify and add like radical terms

$$2\sqrt{2} - 5\sqrt{2} + 7\sqrt{2} = \mathbf{4\sqrt{2}}$$

example 17.2.1 Simplify $\sqrt{18} + \sqrt{8}$.

solution

$\sqrt{18} + \sqrt{8}$	given
$\sqrt{2 \cdot 3 \cdot 3} + \sqrt{2 \cdot 2 \cdot 2}$	write each radicand as a product of prime factors
$\sqrt{2}\,\sqrt{3}\,\sqrt{3} + \sqrt{2}\,\sqrt{2}\,\sqrt{2}$	write roots of products as products of roots
$3\sqrt{2} + 2\sqrt{2} = \mathbf{5\sqrt{2}}$	simplify and add like radical terms

example 17.2.2 Simplify $8\sqrt{27} - 3\sqrt{75}$.

solution The radicals in this problem have coefficients so we will use parentheses to help us prevent errors. We can simplify $\sqrt{27}$ and $\sqrt{75}$ as

$$\sqrt{27} = \sqrt{3 \cdot 3 \cdot 3} = \sqrt{3}\,\sqrt{3}\,\sqrt{3} = \mathbf{3\sqrt{3}}$$

and
$$\sqrt{75} = \sqrt{5 \cdot 5 \cdot 3} = \sqrt{5}\,\sqrt{5}\,\sqrt{3} = \mathbf{5\sqrt{3}}$$

and now we replace $\sqrt{27}$ with $3\sqrt{3}$ and replace $\sqrt{75}$ with $5\sqrt{3}$.

$$8\sqrt{27} - 3\sqrt{75} = 8(3\sqrt{3}) - 3(5\sqrt{3}) = 24\sqrt{3} - 15\sqrt{3} = \mathbf{9\sqrt{3}}$$

example 17.2.3 Simplify $\sqrt{27} - 3\sqrt{18} - 6\sqrt{45}$.

solution

$\sqrt{27} - 3\sqrt{18} - 6\sqrt{45}$	given
$\sqrt{3 \cdot 3 \cdot 3} - 3\sqrt{3 \cdot 3 \cdot 2} - 6\sqrt{3 \cdot 3 \cdot 5}$	products of prime factors
$\sqrt{3}\,\sqrt{3}\,\sqrt{3} - 3\sqrt{3}\,\sqrt{3}\,\sqrt{2} - 6\sqrt{3}\,\sqrt{3}\,\sqrt{5}$	roots of products as products of roots
$3\sqrt{3} - 3 \cdot 3\sqrt{2} - 6 \cdot 3\sqrt{5}$	definition of square root
$\mathbf{3\sqrt{3} - 9\sqrt{2} - 18\sqrt{5}}$	simplify

No further simplification is possible since none of the radical terms are like radical terms.

> **TEST YOUR UNDERSTANDING**
>
> (b) Add: $\sqrt{75} + \sqrt{48}$

17.3 equivalent equations

In Lesson 6 we defined equivalent equations to be equations that have the same solutions. Thus the solution sets for equivalent equations must be equal sets. The number 2 is a solution to both of the equations, (a) $x + 4 = 6$ and (b) $x^2 - 4 = 0$, as shown here.

(a) $x + 4 = 6$ (b) $x^2 - 4 = 0$

$(2) + 4 = 6$ $(2)^2 - 4 = 0$

$6 = 6$ True $4 - 4 = 0$ True

But these equations are not equivalent equations because -2 is a solution to equation (b) but is not a solution to equation (a).

(a) $x + 4 = 6$ (b) $x^2 - 4 = 0$

$(-2) + 4 = 6$ $(-2)^2 - 4 = 0$

$2 = 6$ False $4 - 4 = 0$ True

If every term of a particular equation is multiplied by the same non-zero quantity, the resulting equation is an equivalent equation to the original equation.† On the left below we write the equation $x + y = 6$. On the right we write the equation $2x + 2y = 12$, which is the original equation with every term having been multiplied by 2. The ordered pair (4, 2) is a solution to both equations.

$x + y = 6$ $2x + 2y = 12$

$4 + 2 = 6$ $2(4) + 2(2) = 12$

$6 = 6$ True $8 + 4 = 12$ True

Of course, there are an infinite number of ordered pairs of x and y that will satisfy either of these equations, but it can be shown that any ordered pair that satisfies either one of the equations will satisfy the other equation and thus we say that these equations are equivalent equations!

17.4 elimination

Thus far we have solved systems of two first-degree equations in two unknowns by solving one of the equations to find an expression for one of the variables that contains the other variable. This expression for the variable is then **SUBSTITUTED** for the variable in the second equation and then the resulting equation is solved for the value of the remaining variable. Here we will solve equation (a) for x and substitute the resulting expression for x in equation (b).

(a) $\begin{cases} x + 2y = 8 \longrightarrow x = 8 - 2y \\ 2x - 2y = -2 \longrightarrow 2(8 - 2y) - 2y = -2 \longrightarrow 16 - 4y - 2y = -2 \end{cases}$ (b)

$$\longrightarrow 16 - 6y = -2 \longrightarrow -6y = -18 \longrightarrow \mathbf{y = 3}$$

Now we will use equation (a) to find the corresponding value of x.

$$x + 2y = 8 \longrightarrow x + 2(3) = 8 \longrightarrow x + 6 = 8 \longrightarrow \mathbf{x = 2}$$

The solution is the ordered pair **(2, 3)**.

We can solve the same system of equations by another method that is called the **ELIMINATION METHOD** or possibly by the more meaningful name of the **ADDITION METHOD.**

$$\begin{array}{ll} \text{(a)} & \begin{cases} x + 2y = 8 \\ 2x - 2y = -2 \end{cases} \\ \text{(b)} & \end{array}$$

†The multiplicative property of equality can be used to prove this statement.

We assume that values of x and y exist that will make both of these equations true statements and that x and y in the equations represent these numbers.† Thus $x + 2y$ represents the number 8 and $2x - 2y$ represents the number -2. The additive property of equality permits the addition of equal quantities to both sides of an equation. Thus we can add $2x - 2y$ to the left side of the top equation and -2 to the right side of the top equation.

$$\begin{array}{lrcr} \text{(a)} & x+2y & = & 8 \\ \text{(b)} & 2x-2y & = & -2 \\ \hline & 3x \quad & = & 6 \end{array}$$

By doing this we have eliminated the variable y. Now we can solve the resulting equation $3x = 6$ for x, find that $\boldsymbol{x = 2}$, and use this value for x in *either* of the original equations to find that $\boldsymbol{y = 3}$.

IN EQUATION (a)	IN EQUATION (b)
$x + 2y = 8$	$2x - 2y = -2$
$(2) + 2y = 8$	$2(2) - 2y = -2$
$2y = 6$	$-2y = -6$
$\boldsymbol{y = 3}$	$\boldsymbol{y = 3}$

example 17.4.1 Solve by using the elimination method:

$$\begin{array}{l} \text{(a)} \\ \text{(b)} \end{array} \begin{cases} 2x - y = 13 \\ 3x + 4y = 3 \end{cases}$$

solution If we add the equations in their present form

$$\begin{array}{lrcr} \text{(a)} & 2x - y & = & 13 \\ \text{(b)} & 3x + 4y & = & 3 \\ \hline & 5x + 3y & = & 16 \end{array}$$

we find that we have accomplished nothing because we have not eliminated one of the variables. But by proper use of the multiplicative property of equality we can change the equations into equivalent equations that, when added, will result in the elimination of one of the variables. We choose to eliminate the variable y and thus we will multiply every term in equation (a) by 4 and every term in equation (b) by 1.‡ Now the equations are added and we find that we have eliminated the variable y.

$$\begin{array}{llllrcrl} \text{(a)} & 2x - y = 13 & \longrightarrow (4) \longrightarrow & & 8x - 4y & = & 52 & \text{multiplied by 4} \\ \text{(b)} & 3x + 4y = 3 & \longrightarrow (1) \longrightarrow & \S & 3x + 4y & = & 3 & \text{multiplied by 1} \\ \hline & & & & 11x \quad & = & 55 & \longrightarrow \boldsymbol{x = 5} \end{array}$$

---

†If values of x and y do not exist that will simultaneously make both equations true statements, the attempted solution will degenerate into an equation that contains only numbers such as $2 = 4$, $4 + 2 = 6$, $0 = 0$, or $0 = 5$. The reasons for results like this will be discussed in Lesson 34.

‡Which will leave equation (b) unchanged. We say that we multiply by 1 to establish a general procedure for this type of problem.

§The notations $\longrightarrow (4) \longrightarrow$ and $\longrightarrow (1) \longrightarrow$ are just bookkeeping notations and have no mathematical meaning. We find them convenient to help us remember the number that we have used as a multiplier.

The number 5 can now be used to replace x in either of the original equations or either of the equivalent equations to find the corresponding value of y. We will use the original equation (b).

$$3x + 4y = 3 \longrightarrow 3(5) + 4y = 3 \longrightarrow 15 + 4y = 3$$
$$\longrightarrow 4y = -12 \longrightarrow \mathbf{y = -3}$$

The solution is the ordered pair $(\mathbf{5}, \mathbf{-3})$

example 17.4.2 Solve by using the elimination method: $\begin{cases} 2x - 3y = 5 & \text{(a)} \\ 3x + 4y = -18 & \text{(b)} \end{cases}$

solution There are many ways that the multiplicative property of equality can be used to form equivalent equations that when added will result in one of the variables being eliminated. We will show one way here and then repeat the problem and show another way. Look at the x terms in both equations. If we multiply the x term in the top equation by -3, the product will be $-6x$. If we multiply the x term in the bottom equation by 2, the product will be $+6x$. Now if we add the equations we can eliminate x since the sum of $+6x$ and $-6x$ is zero.

$$\begin{array}{lllll} \text{(a)} & 2x - 3y = 5 & \longrightarrow (-3) \longrightarrow & -6x + 9y = -15 & \text{multiplied by } -3 \\ \text{(b)} & 3x + 4y = -18 & \longrightarrow (2) \longrightarrow & \underline{6x + 8y = -36} & \text{multiplied by } 2 \\ & & & 17y = -51 & \longrightarrow \mathbf{y = -3} \end{array}$$

Now we will use -3 for y in the original equation (a) to find the corresponding value for x.

$$\text{(a)} \quad 2x - 3y = 5 \longrightarrow 2x - 3(-3) = 5 \longrightarrow 2x + 9 = 5$$
$$\longrightarrow 2x = -4 \longrightarrow \mathbf{x = -2}$$

Thus the solution is the ordered pair $(\mathbf{-2}, \mathbf{-3})$.

example 17.4.3 Solve by using the elimination method but this time eliminate y.

$\begin{cases} 2x - 3y = 5 & \text{(a)} \\ 3x + 4y = -18 & \text{(b)} \end{cases}$

solution Look at the y terms in both equations. One of them already has a minus sign. If we multiply the y term in equation (a) by $+4$, the product will be $-12y$ and if we multiply the y term in equation (b) by $+3$, the product will be $12y$ and, of course, the sum of $-12y$ and $+12y$ is zero.

$$\begin{array}{lllll} \text{(a)} & 2x - 3y = 5 & \longrightarrow (4) \longrightarrow & 8x - 12y = 20 & \text{multiplied by } 4 \\ \text{(b)} & 3x + 4y = -18 & \longrightarrow (3) \longrightarrow & \underline{9x + 12y = -54} & \text{multiplied by } 3 \\ & & & 17x \qquad = -34 & \longrightarrow \mathbf{x = -2} \end{array}$$

Now we could use $x = -2$ in either of the original equations or either of the equivalent equations to find that the corresponding value of y is -3. Again we find that the solution is the ordered pair $(\mathbf{-2}, \mathbf{-3})$.

example 17.4.4 Use elimination to solve the system: $\begin{cases} 2x + 5y = -7 & \text{(a)} \\ 3x - 4y = 1. & \text{(b)} \end{cases}$

solution Since one of the y terms already has a minus sign, we can find satisfactory equivalent equations by multiplying both equations by positive numbers.

$$\begin{array}{llllll} \text{(a)} & 2x + 5y = -7 & \rightarrow (4) & \rightarrow & 8x + 20y = -28 & \text{multiplied by 4} \\ \text{(b)} & 3x - 4y = 1 & \rightarrow (5) & \rightarrow & 15x - 20y = 5 & \text{multiplied by 5} \\ \hline & & & & 23x \qquad = -23 & \rightarrow \mathbf{x = -1} \end{array}$$

Now we will use -1 for x in equation (a) and find the corresponding value of y.

$$\text{(a)} \quad 2x + 5y = -7 \rightarrow 2(-1) + 5y = -7 \rightarrow -2 + 5y = -7$$
$$\rightarrow 5y = -5 \rightarrow \mathbf{y = -1}$$

Thus the solution is the ordered pair $(\mathbf{-1}, \mathbf{-1})$.

TEST YOUR UNDERSTANDING

(c) Use elimination to solve the system: $\begin{cases} \text{(a)} \ 2x - 3y = -12 \\ \text{(b)} \ 3x + 4y = -1 \end{cases}$

17.5 complex fractions

In Lesson 15 we defined a **rational number** to be a number that **can** be expressed as a quotient (fraction) of integers. Thus the following are all rational numbers.

$$13, \quad \frac{4}{7}, \quad \frac{-5}{14}, \quad \frac{6}{-13}, \quad 2\frac{1}{3}$$

The number 13 is a rational number because the number 13 can be expressed as a quotient of integers by writing, say, $\frac{130}{10}$. Of course, the mixed number $2\frac{1}{3}$ can be expressed as a quotient of integers as $\frac{7}{3}$.

An algebraic expression containing variables that is written in fractional form is called a **RATIONAL EXPRESSION** because it has the same form as a rational number that is written as a quotient of integers. Thus all the following are **rational expressions.**

$$\frac{x + y}{4}, \quad \frac{a}{-b + c}, \quad 5, \quad \frac{-7}{x}, \quad \frac{a + b}{14 - x}$$

The number 5 is a rational number but is also considered to be a rational expression because rational expression is a general term that describes both rational numbers and rational expressions that include variables and/or numbers.

We have used the **denominator numerator same quantity theorem** to help us change denominators as required so that rational expressions may be added in three steps as we show here:

$$\frac{a}{b} + \frac{x}{y} \longrightarrow \quad (1) \quad \frac{}{by} + \frac{}{by}$$

$$(2) \quad \frac{ay}{by} + \frac{bx}{by}$$

$$(3) \quad \frac{ay + bx}{by}$$

We have also used this same theorem to help us simplify quotients of rational expressions (complex fractions).

$$\frac{\frac{x}{y}}{\frac{b}{c}} = \frac{\frac{x}{y} \cdot \frac{c}{b}}{\frac{b}{c} \cdot \frac{c}{b}} = \frac{\frac{xc}{yb}}{\frac{bc}{bc}} = \frac{\frac{xc}{yb}}{1} = \boldsymbol{\frac{xc}{yb}}$$

We will use both of these procedures when we simplify expressions that are quotients of sums of rational expressions. These expressions are also called **COMPLEX FRACTIONS.**

example 17.5.1 Simplify $\dfrac{\frac{x}{y} + \frac{1}{y}}{\frac{x}{y} - \frac{1}{y}}$.

solution The simplification is performed in two steps. The first step is to add the two expressions in the numerator and add the two expressions in the denominator.

$$\frac{\frac{x}{y} + \frac{1}{y}}{\frac{x}{y} - \frac{1}{y}} = \frac{\frac{x + 1}{y}}{\frac{x - 1}{y}}$$

Now use the **denominator numerator same quantity theorem** by multiplying both the numerator and denominator by $\dfrac{y}{x - 1}$ which is the reciprocal of the denominator $\dfrac{x - 1}{y}$.

$$\frac{\frac{x + 1}{\cancel{y}} \cdot \frac{\cancel{y}}{x - 1}}{\frac{\cancel{x - 1}}{\cancel{y}} \cdot \frac{\cancel{y}}{\cancel{x - 1}}} = \boldsymbol{\frac{x + 1}{x - 1}}$$

example 17.5.2 Simplify $\dfrac{1 + \frac{1}{x}}{7}$.

solution First we add in the numerator.

$$\frac{1 + \frac{1}{x}}{7} = \frac{\frac{x + 1}{x}}{7}$$

Now we finish by multiplying the denominator and the numerator by the reciprocal of 7, which is $\frac{1}{7}$.

$$= \frac{\frac{x + 1}{x} \cdot \frac{1}{7}}{\frac{\cancel{7}}{1} \cdot \frac{1}{\cancel{7}}} = \boldsymbol{\frac{x + 1}{7x}}$$

example 17.5.3

Simplify $\dfrac{\dfrac{1}{x}}{1-\dfrac{1}{x}}$.

solution First we add in the denominator.

$$\frac{\dfrac{1}{x}}{1-\dfrac{1}{x}} = \frac{\dfrac{1}{x}}{\dfrac{x-1}{x}}$$

Now we finish by multiplying the denominator and the numerator by $\dfrac{x}{x-1}$, which is the reciprocal of $\dfrac{x-1}{x}$.

$$= \frac{\dfrac{1}{x}\,\dfrac{x}{x-1}}{\dfrac{\cancel{x-1}}{\cancel{x}}\,\dfrac{\cancel{x}}{\cancel{x-1}}} = \frac{1}{x-1}$$

example 17.5.4

Simplify $\dfrac{\dfrac{a}{b}+1}{\dfrac{x}{b}+4}$.

solution First we add in the numerator and add in the denominator.

$$\frac{\dfrac{a}{b}+1}{\dfrac{x}{b}+4} = \frac{\dfrac{a+b}{b}}{\dfrac{x+4b}{b}}$$

Now we finish by multiplying both the denominator and the numerator by the reciprocal of the denominator.

$$= \frac{\dfrac{a+b}{\cancel{b}}\,\dfrac{\cancel{b}}{x+4b}}{\dfrac{\cancel{x+4b}}{\cancel{b}}\,\dfrac{\cancel{b}}{\cancel{x+4b}}} = \frac{\mathbf{a+b}}{\mathbf{x+4b}}$$

TEST YOUR UNDERSTANDING

(d) Simplify: $\dfrac{\dfrac{m}{x}+1}{\dfrac{a}{x}-4}$

lesson checklist 17

- ☐ Like radical expressions
- ☐ Addition of like radical expressions
- ☐ Simplification of radical expressions
- ☐ Elimination method of solving systems of equations
- ☐ Definition of rational expression
- ☐ Simplification of rational expressions

ANSWERS TO TEST YOUR UNDERSTANDING

(a) $15\sqrt{3}$ (b) $9\sqrt{3}$ (c) $(-3, 2)$ (d) $\dfrac{m + x}{a - 4x}$

problem set 17

1. Richard and Howard multiplied the number by 2 and found that this product is the same number as that obtained by adding 15 to 3 times the number. What was their number?
2. John and Evelyn added 54 to the product of 3 and a certain number. They found that they could get the same result by subtracting 14 from 4 times the certain number. What was the number?
3. Wayne multiplied a number by 3. Bonnie said she got the same result by decreasing the opposite of the number by 32. What was the number?
4. Mickey multiplied the sum of a number and 4 by 2 and told Sherry that the result was 38. What was the number?
5. Sheri and Laurie both think of the same number. They subtract 4 from the number and multiply the difference by 2 and find that the result is 48. What was their number?
6. Earl built the new driveway 42 meters long. This was only 5 percent as long as Jan wanted it to be. How long did Jan want the driveway to be?
7. Drennan had 400 watermelons. He gave 8 percent of them away. How many watermelons did he give away?
8. Forty-two is what percent of 105?

Simplify:

***9.** $\sqrt{8} - \sqrt{50} + \sqrt{98}$ ***10.** $\sqrt{18} + \sqrt{8}$

***11.** $\sqrt{27} - 3\sqrt{18} - 6\sqrt{45}$ **12.** $\sqrt{18} + \sqrt{50} - \sqrt{32}$

Use elimination:

***13.** $\begin{cases} x + 2y = 8 \\ 2x - 2y = -2 \end{cases}$ ***14.** $\begin{cases} 2x - y = 13 \\ 3x + 4y = 3 \end{cases}$ ***15.** $\begin{cases} 2x - 3y = 5 \\ 3x + 4y = -18 \end{cases}$

Use substitution:

16. $\begin{cases} 3x + y = 9 \\ 2x - 3y = -5 \end{cases}$ **17.** $\begin{cases} -3x + 4y = 3 \\ 2x - y = 3 \end{cases}$

---

*The problems marked with an asterisk were worked out in this lesson.

Simplify:

18. $\dfrac{\frac{x}{y}}{\frac{a}{c}}$ **19.** $\dfrac{a}{\frac{b}{c}}$ **20.** $\dfrac{\frac{a}{m}}{x}$

***21.** $\dfrac{\frac{x}{y}+\frac{1}{y}}{\frac{x}{y}-\frac{1}{y}}$ ***22.** $\dfrac{1+\frac{1}{x}}{7}$ ***23.** $\dfrac{\frac{1}{x}}{1-\frac{1}{x}}$ ***24.** $\dfrac{\frac{a}{b}+1}{\frac{x}{b}+4}$

Graph on a number line:

25. $x - 3 \leq 0$, $D = \{\text{Reals}\}$ **26.** $x + 2 > 0$, $D = \{\text{Integers}\}$

Graph on a rectangular coordinate system:

27. $y - x - 3 = 0$ **28.** $7y - 5x + 7 = 0$

Add:

29. $\dfrac{x}{y} + \dfrac{m}{y^2b}$ **30.** $\dfrac{a}{b^3} + c + \dfrac{d}{b^2(a + b)}$

Add like terms:

31. $\dfrac{3k^{-3}p^9m^2}{k} + \dfrac{2k^{-2}}{k^2p^0m^{-2}} - \dfrac{5k^0m^2}{k^4}$ **32.** $\dfrac{6a^2x^0p^{-2}y}{y^5} - \dfrac{2a^0}{a^{-2}p^2y^4} + \dfrac{y^0y^3}{a^{-2}p^2}$

Multiply:

33. $\dfrac{3x^0m^2p^{-2}}{y}\left(\dfrac{4y^2p^0}{m} + \dfrac{2x^{-3}m}{yp^4}\right)$ **34.** $\dfrac{x^2p}{x^{-1}}\left(x^3p^{-1} - \dfrac{5}{x^3}\right)$

Simplify:

35. $|-4 + 5|$ **36.** $(x^2 - 4x) - (x^3 - 2x^2 - 3x)$

37. $-5 - (-6)(-2) + 7$ **38.** $-4[-3(-2 + 4)]$

Factor:

39. $ka^2m - 3k^2ap$

Evaluate:

40. $-a(b - 3a)(-b)$ if $a = -2$, $b = 2$

41. $x - (-y)[x - y(x + y)]$ if $x = 2$, $y = -3$

Solve:

42. $x - 3(-x + 5) = 4 - (-8)(x + 3)$

43. $2x - 5(x + 6) = 2(x - 8)$

Simplify:

44. $(-3)^{-3}$ **45.** -3^3 **46.** 3^{-3} **47.** $\dfrac{3}{(-3)^{-3}}$

48. True or false: $\{\text{Integers}\} \subset \{\text{Rationals}\}$

49. $-3 \in \{\text{what sets of numbers?}\}$

Simplify:

50. $-4[-2(7 - 11)(-3)](-2)$

lesson eighteen

Factoring trinomials
Difference of two squares
Introduction to scientific notation

18.1 factoring trinomials

To begin a quick review of the nomenclature of polynomials in one variable, we say that a monomial is a single expression of the form ax^n where **a is any real number and n is any whole number.** Thus the following expressions are monomials.

$$4, \qquad 6x^2, \qquad -2x^{15}, \qquad 4.163x^4$$

The number 4 can be classified as a monomial because it can be thought of as $4x^0$ and if x is any non-zero real number, then x^0 equals 1, so $4x^0 = 4 \cdot 1 = 4$.

A binomial is the indicated algebraic sum of two monomials and a trinomial is the indicated algebraic sum of three monomials. We use the word polynomial as the general descriptive term to describe monomials, binomials, trinomials, and algebraic expressions that are the indicated sum of four or more monomials.

We are familiar with the vertical format for multiplying binomials as shown here.

(a)
$$\begin{array}{r} x - 6 \\ x + 3 \\ \hline x^2 - 6x \qquad \\ +3x - 18 \\ \hline x^2 - 3x - 18 \end{array}$$

(b)
$$\begin{array}{r} x - 6 \\ x - 3 \\ \hline x^2 - 6x \qquad \\ -3x + 18 \\ \hline x^2 - 9x + 18 \end{array}$$

(c)
$$\begin{array}{r} x + 6 \\ x + 3 \\ \hline x^2 + 6x \qquad \\ +3x + 18 \\ \hline x^2 + 9x + 18 \end{array}$$

In each of these three examples the product is a trinomial. We call these trinomials QUADRATIC TRINOMIALS IN x or more simply QUADRATIC TRINOMIALS. The word QUADRATIC tells us that the highest power of the variable is 2.

To reverse the process and factor the trinomials into a product of binomials, we must observe the pattern that developed when we did the multiplications. Note that:

1. The first term of the trinomial is the product of the first terms of the binomials.
2. The last term of the trinomial is the product of the last terms of the binomials.

3. The *coefficient* of the middle term of the trinomial is the *sum* of the last two terms of the binomials.
4. If all signs in the trinomial are positive, all signs in both binomials are positive. If a negative sign appears in the trinomial, at least one of the terms of the binomials is negative.

We use these observations to help us factor trinomials. To factor the trinomial

$$x^2 - 3x - 18$$

we first write down two sets of parentheses to form an indicated product.

$$(\quad)(\quad)$$

Since the first term in the trinomial is the product of the first terms of the binomials, we enter x as the first term of each binomial.

$$(x\quad)(x\quad)$$

Now the product of the last terms of the binomials must equal -18, their sum must equal -3 and at least one of them must be negative. There are six pairs of integral† factors of -18.

$$(-18)(1) = -18 \qquad (2)(-9) = -18 \qquad (3)(-6) = -18$$

$$(18)(-1) = -18 \qquad (-2)(9) = -18 \qquad (-3)(6) = -18$$

Their sums are

$$(-18) + (1) = -17 \qquad (2) + (-9) = -7 \qquad (3) + (-6) = -3$$

$$(18) + (-1) = 17 \qquad (-2) + (9) = 7 \qquad (-3) + (6) = 3$$

Note that while all six pairs have a product of -18, only one pair, 3, -6, sums to -3. Therefore, the last terms of the binomials are 3 and -6, and so $(x + 3)(x - 6)$ are the factors of $x^2 - 3x - 18$ because

$$(x + 3)(x - 6) = x^2 - 3x - 18$$

Thus the general approach to factoring a quadratic trinomial that has a leading coefficient of 1 is to determine the pairs of integral factors of the last term of the trinomial that sum to the coefficient of the middle term. To factor $x^2 - 8x + 16$, we list the factors of $+16$ and see which pair, if any, sum to -8. If no pair of integral factors has a sum of -8, the trinomial cannot be factored over the integers.

PRODUCT	SUM
$(16)(1) = 16$	$(16) + (1) = 17$
$(-16)(-1) = 16$	$(-16) + (-1) = -17$
$(2)(8) = 16$	$(2) + (8) = 10$
$(-2)(-8) = 16$	$(-2) + (-8) = -10$
$(4)(4) = 16$	$(4) + (4) = 8$
$(-4)(-4) = 16$	$(-4) + (-4) = -8$

Thus we find that the factors of $x^2 - 8x + 16$ are $(x - 4)$ and $(x - 4)$ because the product of the first terms is x^2, the product of the last terms is $+16$ and the sum of the last terms is -8. This may seem to be a complicated procedure, but there is no short cut until the reader becomes sufficiently familiar with the process to perform some of the calculations mentally. We will check our solution by multiplying the factors.

†Since $(2\sqrt{3})(3\sqrt{3}) = 18$, both $2\sqrt{3}$ and $3\sqrt{3}$ are factors of 18. We will not consider non-integral factors such as these and will concentrate on factors that are integers. The process of factoring a polynomial into expressions all of whose coefficients are integers is defined as factoring over the set of integers.

$$\begin{array}{l} x \quad -4 \\ x \quad -4 \\ \hline x^2 - 4x \\ \quad\;\; -4x + 16 \\ \hline x^2 - 8x + 16 \end{array} \qquad \text{Check}$$

example 18.1.1 Factor $x^2 - 14x - 15$.

solution The last term of the trinomial is -15, so the products of the constant terms in the binomial must be -15. Four pairs of integral factors have a product of -15,

$$(3)(-5) = -15, \quad (-3)(5) = -15, \quad (-15)(1) = -15, \quad (15)(-1) = -15$$

but only one pair sums to -14,

$$(3) + (-5) = -2, \quad (-3) + (5) = 2, \quad \underline{(\mathbf{-15}) + (\mathbf{1}) = \mathbf{-14}}, \quad (15) + (-1) = 14$$

Thus the constant terms of the binomials are -15 and 1 because these are the only two integers whose product is -15 and whose sum is -14. So $x^2 - 14x - 15$ in factored form is $(x - 15)(x + 1)$. We will check by multiplying the two factors:

$$\begin{array}{l} x \quad -15 \\ x \quad +\;1 \\ \hline x^2 - 15x \\ \quad\; + \quad x - 15 \\ \hline x^2 - 14x - 15 \end{array} \qquad \text{Check}$$

We note that the trinomial in this problem was written as $x^2 - 14x - 15$ with the powers of the variable x in descending order. **If the trinomial is not in this form, the first step in factoring is to write the trinomial in descending powers of the variable.**

example 18.1.2 Factor $-5x + x^2 + 6$

solution We begin by writing the trinomial in descending powers of the variable as

$$x^2 - 5x + 6$$

The minus sign in the middle term indicates that at least one of the constant terms is a negative number. The last term, $+6$, is a positive number and is a product of the constant terms so both of the constants must be negative for their product is positive. Two pairs of negative integers have a product of $+6$,

$$(-3)(-2) = 6 \quad \text{and} \quad (-1)(-6) = 6$$

but only the first pair sums to -5

$$(-3) + (-2) = -5$$

Thus $$x^2 - 5x + 6 = (x - 3)(x - 2)$$

example 18.1.3 Factor $x^2 + 5 + 6x$.

solution We begin by writing the trinomial in descending powers of the variable

$$x^2 + 6x + 5$$

There are no minus signs in the trinomial so all constants in the binomial factors will be positive. The constants therefore are positive integers whose product is $+5$ and whose sum is $+6$. The constants are $+5$ and $+1$ because

$$(5)(1) = 5 \quad \text{and} \quad 5 + 1 = 6$$

Thus $$x^2 + 6x + 5 = (x + 5)(x + 1)$$

As we see in the next example, we always begin the process of factoring by factoring the greatest common factor if it exists. Then we factor the resulting factors if possible.

example 18.1.4 Factor $x^3 + 6x^2 + 5x$.

solution If we first factor the greatest common factor x, we find

$$x^3 + 6x^2 + 5x = x(x^2 + 6x + 5)$$

and now the trinomial can be factored as in the last problem to get

$$\mathbf{x(x + 5)(x + 1)}$$

example 18.1.5 Factor $4bx^3 - 4bx^2 - 80bx$.

solution Here we see that the greatest common factor of all three terms is $4bx$, and if we factor $4bx$, we find

$$4bx^3 - 4bx^2 - 80bx = 4bx(x^2 - x - 20)$$

and now the trinomial can be factored and the final result is

$$4bx(x - 5)(x + 4)$$

example 18.1.6 Factor $-x^2 + x + 20$.

solution **To factor trinomials in which the coefficient of the second-degree term is negative, it is helpful first to factor a negative quantity.** Here we will factor (-1) and get

$$-x^2 + x + 20 = (-1)(x^2 - x - 20)$$

and now we factor the trinomial to get

$$(-1)(x - 5)(x + 4)$$

Finally, the (-1) can be multiplied by either of the binomials to yield two possible final results,

$$(-x + 5)(x + 4) \quad \text{or} \quad (x - 5)(-x - 4)$$

either of which is correct.

example 18.1.7 Factor $-3x^3 - 6x^2 + 72x$.

solution First we factor the greatest common factor $-3x$, and then we factor the resulting trinomial.

$$-3x(x^2 + 2x - 24) = \mathbf{-3x(x + 6)(x - 4)}$$

Thus we again find that the original trinomial has 3 factors.

In the next example we show that sometimes the greatest common factor is in the form of a sum or is a product in which a sum is a factor.

example 18.1.8 Factor: $m(x - 1)x^2 + 7mx(x - 1) + 10m(x - 1)$

solution The greatest common factor of each term of the trinomial is $m(x - 1)$. If we begin by factoring this term, we get

$$m(x - 1)(x^2 + 7x + 10)$$

Now we complete the solution by factoring the trinomial $x^2 + 7x + 10$ and the result is

$$\mathbf{m(x - 1)(x + 2)(x + 5)}$$

In this book we have restricted our investigation of the factoring of trinomials to trinomials such as

$$x^2 - 2x + 1$$

where the coefficient of the second-degree term is 1 or to trinomials such as the one in the last example that can be reduced to this form by factoring a common factor. Some trinomials in which the coefficient of the second-degree term is not 1 such as

$$2x^2 - 7x - 15$$

can also be factored. Factoring of this type and other advanced types of factoring are discussed in the appendix on factoring.

> **TEST YOUR UNDERSTANDING**
>
> Factor:
>
> (a) $x^2 - 2x - 24$ (b) $a^2x^2 - 2a^2x - 24a^2$

18.2 difference of two squares

Since each of the terms in the binomial is a perfect square, binomials such as

$$x^2 - y^2 \qquad 4p^2 - 25 \qquad m^2 - 16$$

are sometimes called the **difference of two squares** and can be generated by multiplying the sum and difference of two monomials.

$$\begin{array}{lll} x + y & 2p + 5 & m + 4 \\ x - y & 2p - 5 & m - 4 \\ \hline x^2 + xy & 4p^2 + 10p & m^2 + 4m \\ \quad - xy - y^2 & \quad - 10p - 25 & \quad - 4m - 16 \\ \hline x^2 \quad - y^2 & 4p^2 \quad - 25 & m^2 \quad - 16 \end{array}$$

We note in each case that the middle term is eliminated because the addends that are to form the middle term have the same absolute value but are opposite in sign.

If we are asked to factor a binomial that is the difference of two squares such as

$$9m^2 - 49$$

the problem is a problem in recognition. There is no procedure to follow. We **recognize** that each term of the binomial is a perfect square and that the binomial can be written as

$$(3m)^2 - (7)^2$$

Now from the pattern developed above, we can write

$$9m^2 - 49 = (3m + 7)(3m - 7)$$

example 18.2.1 Factor: $-4 + x^2$

solution **We recognize that both of the terms are perfect squares.** We begin by reversing the order of the terms and designating the squared terms as

$$x^2 - 4 = (x)^2 - (2)^2$$

and now we can write the factored form as

$$(x + 2)(x - 2)$$

example 18.2.2 Factor: $49m^2 - a^2$

solution **We recognize that each of the terms is a perfect square** and that the binomial can be written as

$$(7m)^2 - (a)^2$$

and the factored form of this binomial is

$$(7m + a)(7m - a)$$

example 18.2.3 Factor: $-36a^2 + 25y^2$

solution **We recognize that both of the terms are perfect squares.** We begin by reversing the order of the terms and designating the squared terms as

$$25y^2 - 36a^2 = (5y)^2 - (6a)^2$$

Now we write the factored form as

$$(5y + 6a)(5y - 6a)$$

TEST YOUR UNDERSTANDING

Factor:

(c) $x^2 - 16$

(d) $4p^2 - 9a^2$

18.3 scientific notation

In science courses, it is sometimes necessary to use extremely large numbers and extremely small numbers. For example, to calculate the number of molecules in 1000 liters of a gas, it would be necessary to multiply 1000 times 1000 times a very large number such as 26,890,000,000,000,000,000, which represents the number of molecules in a gas in a cubic centimeter. Besides requiring large pieces of paper, multiplying these numbers in their present form is cumbersome and often leads to errors as it is easy to miscount the number of zeroes. If we use a type of mathematical shorthand called SCIENTIFIC NOTATION, however, computations such as the above can be performed easily and accurately.

To write a number in scientific notation, the numerator and the denominator are multiplied by the required power of 10 that will place the decimal point immediately to the right of the first non-zero digit in the number **(denominator numerator same quantity theorem)**. For example, if we wish to write the number

$$.0000416$$

in scientific notation, we would like to place the decimal point between the 4 and the 1.

$$4.16$$

To accomplish this we multiply the number by 10^5, and we must also divide by 10^5 to keep from changing the value of the expression.

$$.0000416 = \frac{.0000416}{1}\,\frac{10^5}{10^5} = \frac{4.16}{10^5}$$

Now, if we remember that $\frac{1}{10^5}$ can be written as 10^{-5}, we can write

$$.0000416 = \frac{4.16}{10^5} = 4.16 \times 10^{-5}$$

We have described the algebraically correct procedure, but since scientific notation is used so often, we prefer to use another thought process. This thought process is much easier to use, but it is not quite so rigorous.

When we look at numbers written in scientific notation such as

$$4.16 \times 10^{+b} \quad \text{and}^{\dagger} \quad 4.16 \times 10^{-b}$$

we think of 10^{+b} as a decimal point indicator that tells us that the **true location** of the decimal point **is really** b places **to the right** of where it is written and 10^{-b} as a decimal point indicator that tells us that the **true location** of the decimal point **is really** b places **to the left** of where it is written. If we use this thought process, the 10^{-7} in the notation

$$4.165 \times 10^{-7}$$

tells us that the decimal point **is really** seven places **to the left** of where it is written, giving

$$.0000004165$$

as the number being designated. In a like manner, the exponential 10^7 in the notation

$$4.165 \times 10^{7}$$

tells us that the decimal point **is really** seven places **to the right** of where it is written, giving

$$41{,}650{,}000$$

which is the number being designated.

It is helpful to use a two-step procedure to write a number in scientific notation. The first step is to place the decimal point immediately to the right of the first non-zero digit in the number. Then we follow this notation with the power of 10 that designates the true location of the decimal point. If we use this procedure to write

$$714{,}600{,}000$$

in scientific notation, we begin by placing the decimal point immediately to the right of the first non-zero digit (which is 7) and dropping the terminal zeroes.

$$7.146$$

Now we follow this with $\times 10^8$ to indicate that the decimal point **is really eight places to the right** of where we have written it.

$$7.146 \times 10^{8}$$

example 18.3.1 Write .000316 in scientific notation.

solution **We always begin by writing the decimal point immediately after the first non-zero digit.**

$$3.16 \times 10^{?}$$

Now we must choose an exponent for 10 that tells us where the decimal point **really is.** Since it **really is** four places to the left of where we have written it, the proper exponent is -4. Thus

$$.000316 \quad \text{equals} \quad \mathbf{3.16 \times 10^{-4}}$$

example 18.3.2 Write $.000316 \times 10^{-7}$ in scientific notation.

solution We note that **the true location of the decimal point** is seven places to the left of where it is

†The replacements for b are restricted to positive integers.

written. Begin by placing the decimal point immediately after the first non-zero digit

$$3.16 \times 10^?$$

Now to find the proper exponent. The location of the decimal point in the original notation was seven places to the left of where it was written. We have moved it four places to the right of that point, so we have written it 11 places from its true location. Thus

$$.000316 \times 10^{-7} \quad \text{equals} \quad \mathbf{3.16 \times 10^{-11}}$$

example 18.3.3 Write $.000316 \times 10^7$ in scientific notation.

solution We note that the true location of the decimal point is seven places to the right of where it is shown. Now we begin by writing

$$3.16 \times 10^?$$

and since the true location of the decimal point is three places to the right of where we show it, the proper exponent is +3. Thus

$$.000316 \times 10^7 \quad \text{equals} \quad \mathbf{3.16 \times 10^3}$$

example 18.3.4 In scientific notation, write

(a) 47,800 (b) $47{,}800 \times 10^{-7}$ (c) $47{,}800 \times 10^7$

solution

(a)	47,800	equals	$\mathbf{4.78 \times 10^4}$
(b)	$47{,}800 \times 10^{-7}$	equals	$\mathbf{4.78 \times 10^{-3}}$
(c)	$47{,}800 \times 10^7$	equals	$\mathbf{4.78 \times 10^{11}}$

A careful study of the six examples above is recommended. Often the student searches for a simple rule or short cut that he can use to solve the problem. No such short cut exists for writing a number in scientific notation, as the examples above will verify.

Test Your Understanding

Write the following numbers in scientific notation:

(e) 4732 (f) .0004732 (g) 4732×10^{-5}

lesson checklist 18

- ☐ Factoring trinomials
- ☐ Difference of two squares
- ☐ Scientific notation

Answers to Test Your Understanding

(a) $(x - 6)(x + 4)$ (b) $a^2(x - 6)(x + 4)$ (c) $(x + 4)(x - 4)$
(d) $(2p - 3a)(2p + 3a)$ (e) 4.732×10^3 (f) 4.732×10^{-4}
(g) 4.732×10^{-2}

problem set 18

1. If 8 is added to a number and the result multiplied by 2, the product equals 36. What is the number?

2. If the product of a number and 4 is increased by 14, the result is 30. What is the number?

3. Six times the sum of a number and 4 equals 42. What is the number?

4. An alloy of silver and copper is 15 percent silver. If I have 80 pounds of the alloy, how many pounds of silver do I have?

5. Jack had 120 raisins before Betty took 6 of them. What percent of Jack's raisins did Betty take?

6. Dan took 15 percent of Jaci's marbles and hid them. If he hid 6 marbles, how many marbles did Jaci have originally?

Simplify:

7. $3\sqrt{27} - \sqrt{18} + 2\sqrt{45}$

8. $-\sqrt{18} + 2\sqrt{50} - 3\sqrt{32}$

Use elimination:

9. $\begin{cases} 4x - 3y = 10 \\ 3x + 4y = -5 \end{cases}$

10. $\begin{cases} 7x + 3y = -13 \\ -2x - 2y = -2 \end{cases}$

Use substitution:

11. $\begin{cases} 4x - 3y = 17 \\ x + 2y = -4 \end{cases}$

Simplify:

12. $\dfrac{\dfrac{4}{a}}{x}$

13. $\dfrac{\dfrac{a}{b} - 1}{4 - \dfrac{a}{b}}$

14. $\dfrac{-2^{-2}}{2^2}$

15. $\dfrac{4}{(-3)^{-2}}$

16. $\dfrac{4}{-3^{-2}}$

Graph on a number line:

17. $x - 2 \nless 0$, $D = \{\text{Reals}\}$

18. $4\sqrt{2} \in$ {what sets of numbers?}

Write the following numbers in scientific notation:

*19. .00031

*20. $.000316 \times 10^{-7}$

*21. $.000316 \times 10^{7}$

*22. 47,800

*23. $47{,}800 \times 10^{-7}$

*24. $47{,}800 \times 10^{7}$

Factor the following polynomials. Always begin by checking to see if the terms have a common factor. If they do, factor the greatest common factor as the first step.

*25. $x^2 - 14x - 15$

*26. $x^2 - y^2$

*27. $x^3 + 6x^2 + 5x$

*28. $-5x + x^2 + 6$

*29. $-36a^2 + 25y^2$

*30. $4bx^3 - 4bx^2 - 80bx$

*31. $x^2 + 5 + 6x$

*32. $49m^2 - a^2$

*33. $-x^2 + x + 20$

34. $x^2 - 5x - 24$

35. $-25a^2 + 16x^2$

*36. $-3x^3 - 6x^2 + 72x$

37. $-14 + 5x + x^2$

38. $5a^2 - 5x^2$

*39. $m(x - 1)x^2 + 7mx(x - 1) + 10m(x - 1)$

---

*The problems marked with an asterisk were worked out in this lesson.

40. $-11x + x^2 + 24$

41. $-9 + a^2$

42. $b^2x^2 + 2b^2x - 15b^2$

43. $5 - 6x + x^2$

44. $18x^2 - 8a^2$

45. $(x + 3)x^2 + (x + 3)x - 12(x + 3)$

46. Graph on a rectangular coordinate system: $x = -y + 4$

47. Add: $\frac{m}{a + b} + \frac{x + y}{b}$

48. Multiply: $3x^{-2}\left(x^2y^0 - \frac{2y^2}{x^{-2}}\right)$

49. Simplify: $\frac{x^2(y^{-2})^0z^3}{(z^3)^{-3}x^{-2}}$

lesson nineteen

Closure
Integer word problems
Solution of rational equations

19.1 closure

The concept of **CLOSURE** of a given set under a given operation is interesting, important, and is a simple concept. In fact, it is so simple that the simplicity itself often causes students difficulty, for they believe that surely there must be more to it than there is. The word **closure** will appear in almost all higher-level algebra books and will not cause apprehension if we devote sufficient time to understanding what it means now. We will try to be as straightforward as possible with the explanation.

Suppose we want to use a particular operation and only work with a designated set of numbers. If we restrict ourselves to this set of numbers, it seems reasonable to say that we also restrict our answers to members of this set of numbers. If we don't do this, then we are not restricting ourselves to working with the designated set of numbers and we are not doing what we said we would do at the outset because we are accepting numbers for answers that are outside the set that we said we would work with exclusively.

We will draw a circle and consider it to be an **ENCLOSURE**. If we restrict ourselves to working with a particular set, say, the positive integers, we will mentally place these numbers inside our enclosure.

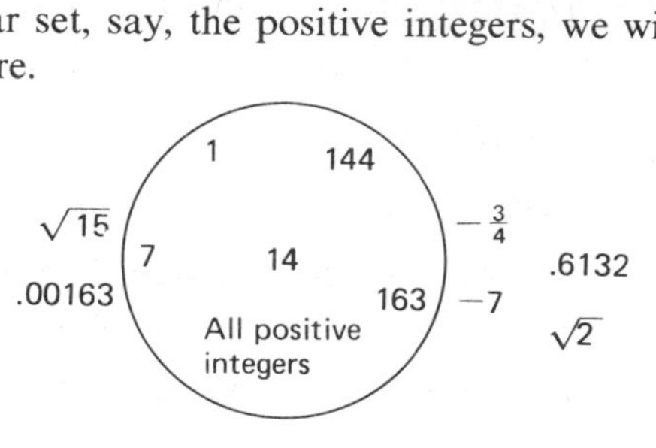

If we use these numbers in a particular operation and if **every possible result or answer is a number that is a member of the set of positive integers and thus is inside the enclosure that we have drawn, we say that the set is closed under this particular operation.**

If any of the members of the set, when used in the particular operation, result in an answer that is not inside the enclosure, we say that the set is **not closed** under the particular operation. A set cannot be partially closed and partially not closed. **If any result is a number that is not inside the enclosure, the operation is not closed regardless of the number of results that are inside the enclosure.** Note that we do not use the word **open** to designate this result but use the words **not closed.**

Let's see if the set of positive integers is closed under the operation of addition. Now

$$2 + 2 = 4, \qquad 51 + 75 = 126, \qquad 7 + 13 = 20$$

In fact, the sum of any two positive integers is also a positive integer and lies inside our enclosure so we say that the set of positive integers is closed under the operation of addition.

Let's try subtraction.

$$20 - 4 = 16, \qquad 14 - 2 = 12, \qquad 8 - 12 = \mathbf{-4}$$

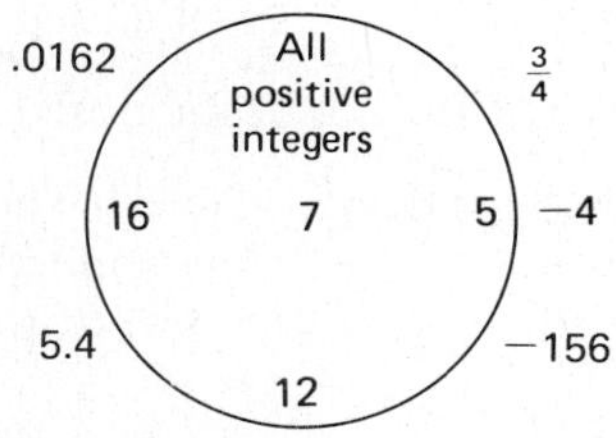

The first two answers are positive integers and thus lie inside our enclosure, but the third answer does not lie inside the enclosure because -4 is not a member of the set of positive integers. **This one example is all that is needed to state that the set of positive integers is not closed under the operation of subtraction.**

Let's try multiplication.

$$4 \cdot 2 = 8, \qquad 71 \cdot 3 = 213, \qquad 48 \cdot 2 = 96$$

All these products are positive integers. It can be proved (a topic for a more advanced course) that all products of positive integers are also positive integers. Thus we say that the set of positive integers is closed under the operation of multiplication.

What about division? One example of a quotient of two integers that is not an integer is $\frac{1}{2}$. Thus we say that the set of positive integers is not closed under the operation of division.

example 19.1.1 For what operations is the set $\{0, 1\}$ closed?

solution We will begin by writing down A, S, M, D to represent the operations of addition, subtraction, multiplication, and division. When we investigate a particular operation and find that the set is not closed, we will draw a slash line through the letter that represents the particular operation. We use a check mark to designate closure.

ADDITION:	Not closed. $1 + 1 = 2$	~~A~~SMD†
SUBTRACTION:	Not closed. $0 - (1) = -1$	~~A~~~~S~~MD
MULTIPLICATION:	Closed. The product of any two numbers of this set is also a member of the set.	~~A~~~~S~~M✓D
DIVISION:	Not closed. $\frac{1}{0}$ is not defined.	~~A~~~~S~~M✓~~D~~

Thus the set $\{0, 1\}$ is closed for the operation of multiplication and is not closed for the operations of addition, subtraction, and division.

†Note that if a given number is a member of a set, we may use the number more than once, in fact, as often as we desire.

example 19.1.2 For what operations is the set of negative integers closed?

solution

ADDITION:	Closed. The sum of any two negative integers is a negative integer.	✓ ASMD
SUBTRACTION:	Not closed. $-4-(-8) = +4$	✓ A~~S~~MD
MULTIPLICATION:	Not closed. $(-4)(-2) = +8$	✓ A~~S~~~~M~~D
DIVISION:	Not closed. $\frac{-4}{-2} = +2$	✓ A~~S~~~~M~~~~D~~

Thus the set of negative integers is closed for the operation of addition and is not closed for subtraction, multiplication, or division.

example 19.1.3 For what operations is the set $\{-1, 0, 1\}$ closed?

solution

ADDITION:	Not closed. $1 + 1 = 2$	~~A~~ S M D
SUBTRACTION:	Not closed. $1-(-1) = 2$	~~A~~ ~~S~~ M D
MULTIPLICATION:	Closed. The product of any combination of the numbers. $-1, 0, 1$ is one of these numbers.	~~A~~ ~~S~~ ✓M D
DIVISION:	Not closed, $\frac{-1}{0}$ is not defined.	~~A~~ ~~S~~ ✓M ~~D~~

Thus the set $\{-1, 0, 1\}$ is closed for the operation of multiplication but is not closed for addition, subtraction, or division.

example 19.1.4 For what operations is the set of integers closed?

solution

ADDITION:	Closed. The sum of any two integers is an integer.	✓ ASMD
SUBTRACTION:	Closed. The difference of any two integers is an integer.	✓✓ ASMD
MULTIPLICATION:	Closed. The product of any two integers is an integer.	✓✓✓ ASMD
DIVISION:	Not closed. $\frac{7}{13}$ is not an integer.	✓✓✓ ASM~~D~~

Thus the set of integers is closed for the operations of addition, subtraction, and multiplication but is not closed for division.

example 19.1.5 For what operations is the set of real numbers closed?

solution

ADDITION:	Closed. The sum of any two real numbers is a real number.	✓ ASMD
SUBTRACTION:	Closed. The difference of any two real numbers is a real number.	✓✓ ASMD
MULTIPLICATION:	Closed. The product of any two real numbers is a real number.	✓✓✓ ASMD
DIVISION:	Not closed. $\frac{23}{0}$ is not defined.	✓✓✓ ASM~~D~~

Thus we see that the set of real numbers is closed for the operations of addition, subtraction, and multiplication but is not closed for division because division of a non-zero real number by zero is not defined.

TEST YOUR UNDERSTANDING

(a) For what operations is the set $\{1, 2, 3\}$ closed?

19.2 integer word problems

If we use the letter N to designate an unspecified integer and then look at the number line

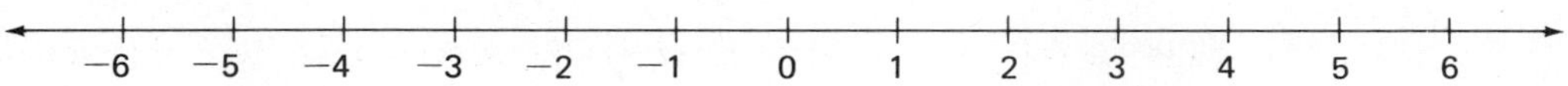

we see that the integer to the right of any given integer is one greater than the given integer. Thus we may use

$$N + 1$$

to designate the next greater integer and $N + 2$ to designate the next greater integer, etc.

example 19.2.1 Find three consecutive integers such that the sum of the first and third is 146.

solution We will designate the consecutive integers as

$$N \qquad N + 1 \qquad N + 2$$

The sum of the first integer and the third integer is 146 so we write

$$\underbrace{N} + \underbrace{N + 2} = 146$$

and now we solve for N.

$$2N + 2 = 146$$

$$2N = 144$$

$$N = 72$$

Thus $N + 1 = 73$, and $N + 2 = 74$. The desired integers are **72, 73,** and **74.**

CHECK

$$72 + 74 = 146$$

$$146 = 146 \quad \text{Check}$$

even and odd integers **If N is the letter we use to designate any unspecified integer, then certainly $2N$ will be an even integer** because the product of 2 and any integer is an even integer. If we use $2N$ to designate any even integer and look at the number line, we see that there are two units between any two even integers. For example, -2 lies two units to the right of -4, and 6 lies two units to the right of 4. Thus if we are asked to designate several consecutive even integers, we designate the first even integer as $2N$, the next as $2N + 2$, the next as $2N + 4$, etc.

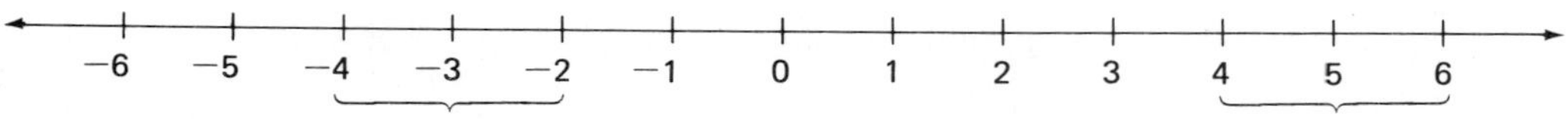

example 19.2.2 Find three consecutive even integers such that the sum of the first and the second equals the sum of the third and -10.

solution We will designate the three even integers as

$$2N \qquad 2N + 2 \qquad 2N + 4$$

The sum of the first and the second equals the sum of the third and -10.

$$\underbrace{2N} + \underbrace{2N + 2} = \underbrace{2N + 4} + \underbrace{(-10)}$$

Now solve

$$\begin{aligned} 2N + 2N + 2 &= 2N + 4 - 10 \\ 4N + 2 &= 2N - 6 \\ 2N &= -8 \longrightarrow N = -4 \end{aligned}$$

Thus

$$2N = \mathbf{-8}, \qquad 2N + 2 = \mathbf{-6} \qquad \text{and} \qquad 2N + 4 = \mathbf{-4}.$$

CHECK

$$\begin{aligned} -8 + -6 &= -4 + (-10) \\ -14 &= -14 \quad \text{Check} \end{aligned}$$

example 19.2.3 Find three consecutive odd integers such that the sum of the first and the second equals the third decreased by 15.

solution **Now if $2N$ is an even integer, then $2N + 1$ will certainly be an odd integer.** Since on the number line odd integers are also 2 units apart, the next odd integer greater than $2N + 1$ will be $2N + 3$; the next will be $2N + 5$, etc.

We will designate the three consecutive odd integers as

$$2N + 1 \qquad 2N + 3 \qquad 2N + 5$$

The sum of the first and second equals the third decreased by 15.

$$\underbrace{2N + 1} + \underbrace{2N + 3} = \underbrace{2N + 5} - \underbrace{15}$$

Now solve

$$\begin{aligned} 2N + 1 + 2N + 3 &= 2N - 10 \\ 4N + 4 &= 2N - 10 \\ 2N &= -14 \\ N &= -7 \end{aligned}$$

So

$$2N + 1 = \mathbf{-13}, \qquad 2N + 3 = \mathbf{-11}, \qquad 2N + 5 = \mathbf{-9}$$

CHECK

$$\begin{aligned} -13 + (-11) &= -9 - 15 \\ -24 &= -24 \quad \text{Check} \end{aligned}$$

Test Your Understanding

(b) Find three consecutive even integers such that five times the first plus 4 equals two times the sum of the second and the third.

19.3 solution of rational equations

Often we need to solve equations in which some of the terms of the equation have denominators that contain variables or the denominators are numbers other than the number 1. The equations

$$\frac{y}{2} + \frac{1}{4} = \frac{y}{6} \qquad \frac{3y}{2} + \frac{8 - 4y}{7} = 3$$

are equations of this type. Because the terms of the equation are all rational expressions, we can classify these equations by calling them **RATIONAL EQUATIONS**. There are many ways that these equations can be solved, **but the most straightforward method of attack is to eliminate the denominators first by a judicious application of the multiplicative property of equality.** Since Lesson 10 we have been adding rational expressions by using the least common multiple of the original denominators as the new denominator. To eliminate the denominators in these equations, we will again use the least common multiple, but we will use it in a different way. As permitted by the multiplicative property of equality, we will **multiply the numerator of every term in the equation by the least common multiple of all the denominators of the terms of the equation.** Since every denominator is guaranteed to be a factor of the least common multiple, we are able to eliminate the denominators in one step. The remainder of the solution is straightforward.

example 19.3.1 Solve $\frac{y}{2} + \frac{1}{4} = \frac{y}{6}$.

solution The least common multiple of the denominators is 12. We will multiply the numerator of every term by 12 and cancel the denominators.

$$\frac{12 \cdot y}{2} + \frac{12 \cdot 1}{4} = \frac{12 \cdot y}{6}$$

Now divide and find

$$6y + 3 = 2y$$

Now solve for y

$$4y = -3$$
$$y = -\frac{3}{4}$$

example 19.3.2 Solve $\frac{3y}{2} + \frac{8 - 4y}{7} = 3$.

solution We will work this problem twice. This time we will rewrite the second term so that it is expressed as the algebraic sum of two terms.

$$\frac{3y}{2} + \frac{8}{7} - \frac{4y}{7} = 3$$

Now we multiply every numerator by 14, which is the least common multiple of all of the denominators.

$$\frac{14 \cdot 3y}{2} + \frac{14 \cdot 8}{7} - \frac{14 \cdot 4y}{7} = 14 \cdot 3$$

Now divide and simplify

$$21y + 16 - 8y = 42$$

Now solve

$$13y = 26$$
$$y = \mathbf{2}$$

example 19.3.3 Solve $\frac{3y}{2} + \frac{8 - 4y}{7} = 3$.

solution The beginner often makes mistakes when he tries to eliminate denominators when one or more of the terms has a binomial expression as the numerator. A simple ploy that will prevent the most common mistake is to enclose the binomials in the numerators in parentheses first.

$$\frac{3y}{2} + \frac{(8 - 4y)}{7} = 3$$

Now multiply every numerator by the least common multiple of the denominators

$$\frac{14 \cdot 3y}{2} + \frac{14 \cdot (8 - 4y)}{7} = 14 \cdot 3$$

Now divide, simplify, and complete the solution.

$$21y + 2(8 - 4y) = 42 \longrightarrow 21y + 16 - 8y = 42$$
$$\longrightarrow 13y = 26 \longrightarrow y = \mathbf{2}$$

example 19.3.4 Solve $\frac{x + 1}{4} - \frac{3}{2} = \frac{2x - 9}{10}$.

solution First we enclose the binomials in the numerators in parentheses.

$$\frac{(x + 1)}{4} - \frac{3}{2} = \frac{(2x - 9)}{10}$$

Now multiply every numerator by the least common multiple of the denominators.

$$\frac{20(x + 1)}{4} - \frac{20 \cdot 3}{2} = \frac{20(2x - 9)}{10}$$

Now, divide, simplify, and complete the solution.

$$5(x + 1) - 30 = 2(2x - 9) \longrightarrow 5x + 5 - 30 = 4x - 18$$
$$\longrightarrow 5x - 25 = 4x - 18 \longrightarrow x = 7$$

example 19.3.5 Solve $\frac{4x + 3}{15} - \frac{2x - 3}{9} = \frac{3x + 2}{3} - x$.

solution First we enclose the numerators in parentheses.

$$\frac{(4x + 3)}{15} - \frac{(2x - 3)}{9} = \frac{(3x + 2)}{3} - x$$

Now we multiply every numerator by the least common multiple of the denominators.

$$\frac{45(4x + 3)}{15} - \frac{45(2x - 3)}{9} = \frac{45(3x + 2)}{3} - 45x$$

Now divide, simplify, and complete the solution.

$$\longrightarrow 3(4x + 3) - 5(2x - 3) = 15(3x + 2) - 45x$$
$$\longrightarrow 12x + 9 - 10x + 15 = 45x + 30 - 45x$$
$$\longrightarrow 2x + 24 = 30 \longrightarrow 2x = 6 \longrightarrow x = 3$$

TEST YOUR UNDERSTANDING

(c) Solve: $\frac{4x - 8}{5} + \frac{2x - 4}{2} = 9$

---

lesson checklist 19

- ☐ Definition of closure
- ☐ Consecutive integers
- ☐ Consecutive even integers
- ☐ Consecutive odd integers
- ☐ Solution of rational equations

ANSWERS TO TEST YOUR UNDERSTANDING

(a) This set is not closed for any of the four basic operations.

(b) 8, 10, 12

(c) $x = 7$

problem set 19

*1. Find three consecutive integers such that the sum of the first and third is 146.

*2. Find three consecutive even integers such that the sum of the first and second equals the sum of the third and negative 10.

*3. Find three consecutive odd integers such that the sum of the first and second equal the third decreased by 15.

4. If the sum of three times a number and 7 is multiplied by 4, the result is -20. What is the number?

5. James and Myra sold their old car for 18 percent of the original price. If they got $549 in the sale, what was the original price of the car?

6. Ham has $6 and Elizabeth has $48. Ham's money is what percent of Elizabeth's money?

For what operations are the following sets closed?

*7. $\{0, 1\}$

*8. {Positive integers}

*9. {Negative integers}

*10. {Integers}

*11. {Real numbers}

Solve:

*12. $\frac{y}{2} + \frac{1}{4} = \frac{y}{6}$

*13. $\frac{3y}{2} + \frac{8 - 4y}{7} = 3$

*14. $\frac{x + 1}{4} - \frac{3}{2} = \frac{2x - 9}{10}$

Write the following numbers in scientific notation:

15. $.000456 \times 10^{-11}$

16. $.000456$

17. $.000456 \times 10^{11}$

18. $456{,}000{,}000$

Factor:

19. $x^3 - 6x^2 + 9x$

20. $t^2 - 25$

21. $4t^2 - 25$

22. $x^2 + x - 12$

23. $x^3 - x^2 - 12x$

24. $m^2px - 4pxm + p^2xm^2$

Solve:

25. $p - 3(-p + 5) = 4 + (-4) - 4p$

26. $2x - 4(x + 6) = 2(x - 8)$

Evaluate:

27. $a(a - b) + (-b)$ if $a = 2, b = -4$

28. $-x(-y)[x - (-y)]$ if $x = -2, y = -3$

Simplify:

29. $-4[-3(-3 + 4)]$

30. $|-11 - 5|$

31. $3\sqrt{50} - 2\sqrt{72} + \sqrt{98}$

32. $-2\sqrt{27} + 4\sqrt{75} - 2\sqrt{108}$

33. $4\sqrt{48} - 2\sqrt{300}$

Use elimination:

34. $\begin{cases} 4x - 6y = -4 \\ 3x + 5y = 16 \end{cases}$

35. $\begin{cases} 7x + y = 25 \\ 2x - 5y = -14 \end{cases}$

36. $\begin{cases} 2x - 3y = 16 \\ 3x + 4y = -10 \end{cases}$

*The problems marked with an asterisk were worked out in this lesson.

Use substitution:

37. $\begin{cases} 7x + y = 25 \\ 2x - 5y = -14 \end{cases}$

Simplify:

38. $\dfrac{\dfrac{a}{b}}{\dfrac{1}{c}}$

39. $\dfrac{a}{\dfrac{b}{c}}$

40. $\dfrac{t + \dfrac{1}{t}}{t - \dfrac{1}{t}}$

41. $\dfrac{b - \dfrac{a}{b}}{b + \dfrac{a}{b}}$

Graph on a number line:

42. $x - 3 \not< 2$, $D = \{\text{Integers}\}$

Graph on a rectangular coordinate system:

43. $4y = -3x + 8$

Add:

44. $\dfrac{1}{x^2} + \dfrac{b + c}{x} + \dfrac{d}{x^3}$

45. $3x^{-3}y^2z - \dfrac{2y^2z}{x^3} - \dfrac{4y^2z^{-1}}{x^3}$

Multiply:

46. $4x^2y^{-2}\left(\dfrac{x^{-2}}{y^{-2}} + \dfrac{3x^0}{y^2}\right)$

47. $\dfrac{4}{-3^{-3}}$

48. True or false: $\{\text{Irrational}\} \subset \{\text{Rational}\}$

49. $\dfrac{47}{5} \in$ {what sets of numbers?}

lesson twenty

Systems of equations
Four equations in four unknowns
More on scientific notation

20.1 systems of equations

In Lesson 30 we will introduce uniform motion word problems. These problems will require the use of variables that represent rate or velocity and variables that represent time. Rather than use the usual variables x, y, and z, we will use variables that are easy to associate with the word or words in the problem. If we need a variable to represent the rate of Mike, we will use R_M, which can be read as *the rate of Mike* or *R sub M*. If we need a variable to represent Joanie's rate, we will use R_J, which can be read as *the rate of Joanie* ***or*** *R sub J*. In the same way, Bud's time and Sadie's time will be represented by the variables T_B and T_S, which can be read as *the time of Bud* and *the time of Sadie* or as *T sub B* and *T sub S*. The first problem we will solve involves the rates and times of Anne and Pat and will use the variables R_A, T_A and R_P, T_P.

example 20.1.1 Solve the following system of equations for R_A.

$$R_AT_A + R_PT_P = 320, \qquad R_P = 50, \qquad T_P = 4, \qquad T_A = 3$$

solution In the left equation we will substitute 50, 4, and 3 for R_P, T_P, and T_A, respectively.

$$R_A(3) + (50)(4) = 320$$

and solve

$$3R_A + 200 = 320 \longrightarrow 3R_A = 120 \longrightarrow \boldsymbol{R_A = 40}$$

In the next problem are the equations that come from a story problem about a turtle and a rabbit. Thus R_R and T_R stand for the rate of the rabbit and the time of the rabbit and R_T and T_T stand for the rate of the turtle and the time of the turtle.

example 20.1.2 Solve the following system of equations for T_R and T_T.

$$R_TT_T + 120 = R_RT_R, \qquad R_T = 2, \qquad R_R = 10, \qquad T_T = T_R$$

solution We begin by substituting 2 and 10 for R_T and R_R in the first equation.

$$\text{(a)}\quad 2T_T + 120 = 10T_R$$

We have used the first three given equations thus far. The remaining given equation is $T_T = T_R$. We can use this equation to change T_T to T_R in equation (a) above or to change T_R to T_T. We choose to change T_T to T_R. We do this and then complete the solution.

$$\begin{array}{rrl}
2T_R + 120 = & 10T_R & \text{substituted } T_R \text{ for } T_T \\
-2T_R \qquad & -2T_R & \text{add } -2T_R \text{ to both sides} \\
\hline
120 = & 8T_R & \\
\dfrac{120}{8} = & \dfrac{8T_R}{8} & \text{divide both sides by 8} \\
\mathbf{15} = & \mathbf{T_R} & \text{and since } T_T = T_R, \quad \mathbf{T_T = 15}
\end{array}$$

The next problem is about a motorist and a highway patrolman, hence the variables R_M, T_M, R_P, and T_P.

example 20.1.3 Solve the following system of equations.

$$R_M T_M + 4 = R_P T_P, \qquad R_M = 55, \qquad R_P = 63, \qquad T_P = T_M$$

solution

$$\begin{array}{rrl}
55T_M + 4 = & 63T_M & \text{substituted for } R_M, R_P, \text{ and } T_P \\
-55T_M \qquad & -55T_M & \text{add } -55T_M \text{ to both sides} \\
\hline
4 = & 8T_M & \\
\dfrac{1}{2} = & T_M & \text{divided both sides by 8}
\end{array}$$

Now we will work a problem in which Little Brother and Sis take a trip. We will use R_L and T_L for the rate and time of Little Brother and R_S and T_S for the rate and time of Sis.

example 20.1.4 Solve the following system of equations for T_L and T_S.

$$R_L T_L = R_S T_S, \qquad R_L = 40, \qquad R_S = 80, \qquad T_S = T_L - 5$$

solution We begin with the first equation by replacing R_L with 40 and R_S with 80.

$$40T_L = 80T_S$$

Now we will replace T_S with $T_L - 5$ and multiply out using the distributive property.

$$40T_L = 80(T_L - 5) \longrightarrow 40T_L = 80T_L - 400$$

Now we complete the solution.

$$\begin{array}{rrl}
40T_L = & 80T_L - 400 & \\
-40T_L \quad & -40T_L \qquad & \text{add } -40T_L \text{ to both sides} \\
\hline
0 = & 40T_L - 400 & \\
+400 \quad & +400 & \text{add 400 to both sides} \\
\hline
400 = & 40T_L & \\
\dfrac{400}{40} = & \dfrac{40T_L}{40} & \text{divide both sides by 40}
\end{array}$$

so $\mathbf{T_L = 10}$ and since $T_S = T_L - 5$, $T_S = 10 - 5 \longrightarrow \mathbf{T_S = 5}$.

The following equations are from a problem about a freight train and an express train. Thus the rate and time of the freight are symbolized by R_F and T_F and those for the express by R_E and T_E.

example 20.1.5 Solve the following system of equations for R_F and R_E.

$$R_FT_F = R_ET_E, \qquad T_F = 16, \qquad T_E = 12, \qquad R_E = R_F + 15$$

solution In the left equation we will substitute 16, 12, and $R_F + 15$ for T_F, T_E, and R_E, respectively.

$$R_FT_F = R_ET_E \longrightarrow R_F(16) = (R_F + 15)(12) \longrightarrow 16R_F = 12R_F + 180$$
$$\longrightarrow 4R_F = 180 \longrightarrow \mathbf{R_F = 45}$$

and since $R_E = R_F + 15$, $R_E = 45 + 15 \longrightarrow \mathbf{R_E = 60}$.

The equations in the next problem come from a word problem in which one man walks and another man rides a motorcycle. Thus we use the variables R_W, T_W, R_M, and T_M.

example 20.1.6 Solve the following system of equations.

$$R_MT_M = R_WT_W, \qquad R_M = 8, \qquad R_W = 2, \qquad T_M = 5 - T_W$$

solution In the left equation we substitute 8 and 2 for R_M and R_W, respectively.

$$8T_M = 2T_W$$

Now we substitute $5 - T_W$ for T_M and solve the resulting equation for T_W.

$$8(5 - T_W) = 2T_W$$
$$40 - 8T_W = 2T_W$$
$$40 = 10T_W$$
$$\mathbf{4 = T_W}$$

Since $T_M + T_W = 5$ and $T_W = 4$,

$$\mathbf{T_M = 1}$$

Now since $T_M = 5 - T_W$, we substitute 4 for T_W and find that $T_M = 1$.

$$T_M = 5 - (4) = \mathbf{1}$$

TEST YOUR UNDERSTANDING

(a) Solve the following system of equations:

$$R_MT_M + R_XT_X = 270, \qquad R_M = 40, \qquad R_X = 30, \qquad T_X = T_M + 2$$

20.2 scientific notation

In section 18.3 we introduced the topic of scientific notation and discussed the method of writing a number in scientific notation. Scientific notation is particularly useful in problems that require the multiplication and division of very large or very small numbers. We will discuss multiplication first.

multiplication We begin by multiplying the numbers 4,000,000 and 20,000,000 by using scientific notation. As the first step we write both numbers in scientific notation.

$$4{,}000{,}000 = 4 \times 10^6 \qquad 20{,}000{,}000 = 2 \times 10^7$$

Then we note that the numbers are to be multiplied by writing

$$(4 \times 10^6)(2 \times 10^7)$$

Since the order of multiplication of real numbers does not affect the value of the product, we may rearrange the order of the multiplication and place the exponentials last.

$$(4 \cdot 2)(10^6 \cdot 10^7)$$

Now we multiply 4 by 2 and get 8 and multiply the exponentials by using the product theorem for exponents to get 10^{13}. Thus our answer is the number

$$\mathbf{8 \times 10^{13}}$$

example 20.2.1 Write the numbers $.003 \times 10^{-4}$ and 2×10^{20} in scientific notation and then multiply.

solution First we write the numbers in scientific notation.

$$(.003 \times 10^{-4})(2 \times 10^{20}) = (3 \times 10^{-7})(2 \times 10^{20})$$

Next we rearrange the order of the factors and then we multiply:

$$(3 \cdot 2)(10^{-7} \cdot 10^{20}) = \mathbf{6 \times 10^{13}}$$

example 20.2.2 Multiply $(.00004 \times 10^{-5})(700{,}000)$.

solution We write the numbers in scientific notation, rearrange the order of factors, and then multiply.

$$(4 \times 10^{-10})(7 \times 10^5) = (4 \cdot 7)(10^{-10} \cdot 10^5) = 28 \times 10^{-5}$$
$$= \mathbf{2.8 \times 10^{-4}}$$

division A similar procedure is used to divide numbers written in scientific notation. The exponentials are handled separately from the other numbers. To divide 20,000,000 by 4,000,000, we first write both numbers in scientific notation.

$$\frac{20{,}000{,}000}{4{,}000{,}000} = \frac{2 \times 10^7}{4 \times 10^6}$$

which we think of as a product of quotients. We then simplify both expressions.

$$\left(\frac{2}{4}\right) \cdot \left(\frac{10^7}{10^6}\right) = .5 \times 10^1 = \mathbf{5}$$

example 20.2.3 Divide .0016 by 400,000.

solution We write both numbers in scientific notation as the numerator and denominator of a fraction,

$$\frac{1.6 \times 10^{-3}}{4 \times 10^5}$$

which we think of as a product of quotients (fractions). Then we simplify both expressions.

$$\left(\frac{1.6}{4}\right)\left(\frac{10^{-3}}{10^5}\right) = .4 \times 10^{-8}$$

Now to finish we write $.4 \times 10^{-8}$ in scientific notation as

$$= \mathbf{4 \times 10^{-9}}$$

multiplication and division The procedure for simplifying a problem such as

$$\frac{(.06 \times 10^5)(300{,}000)}{(1000)(.00009)}$$

is first to simplify both the numerator and denominator by using scientific notation. Next we rearrange the expression into a product of fractions and then simplify each fraction.

$$\frac{(6 \times 10^3)(3 \times 10^5)}{(1 \times 10^3)(9 \times 10^{-5})} = \underbrace{\frac{6 \cdot 3}{1 \cdot 9}} \times \underbrace{\frac{10^3 \cdot 10^5}{10^3 \cdot 10^{-5}}} = \underbrace{\frac{18}{9}} \times \underbrace{\frac{10^8}{10^{-2}}} = 2 \times 10^{10}$$

example 20.2.4 Simplify $\dfrac{(.0007 \times 10^{-23})(4000 \times 10^6)}{(.00004)(7{,}000{,}000)}$.

solution We will begin by writing every number in scientific notation. Next we rearrange the expression into a product of two fractions and then simplify each fraction.

$$\frac{(7 \times 10^{-27})(4 \times 10^9)}{(4 \times 10^{-5})(7 \times 10^6)} = \underbrace{\frac{4 \cdot 7}{4 \cdot 7}} \times \underbrace{\frac{10^{-27} \cdot 10^9}{10^{-5} \cdot 10^6}} = \underbrace{\frac{28}{28}} \times \underbrace{\frac{10^{-18}}{10^1}} = 1 \times 10^{-19}$$

example 20.2.5 Simplify $\dfrac{(20 \times 10^{-45})(400 \times 10^{20})}{(100{,}000)(.0008 \times 10^{-15})}$.

solution First we write all numbers in scientific notation.

$$\frac{(2 \times 10^{-44})(4 \times 10^{22})}{(1 \times 10^5)(8 \times 10^{-19})}$$

Now we group the exponentials and the other numbers and simplify.

$$\underbrace{\frac{2 \cdot 4}{1 \cdot 8}} \times \underbrace{\frac{10^{-44} \cdot 10^{22}}{10^5 \cdot 10^{-19}}} = \underbrace{\frac{8}{8}} \times \underbrace{\frac{10^{-22}}{10^{-14}}} = \mathbf{1 \times 10^{-8}}$$

TEST YOUR UNDERSTANDING

(b) Simplify.

$$\frac{(4000)(270 \times 10^4)}{(.009)(3000 \times 10^{-5})}$$

lesson checklist 20

- ☐ Use of substitution to solve four equations in four unknowns
- ☐ Multiplication with scientific notation
- ☐ Division with scientific notation

ANSWERS TO TEST YOUR UNDERSTANDING

(a) $T_M = 3, T_X = 5$ (b) 4×10^{13}

problem set 20

1. Find four consecutive integers such that the sum of the first and the fourth is 37.
2. Find three consecutive even integers such that the sum of the first and three times the third equals 44.
3. Find three consecutive odd integers such that the sum of the second and two times the third equals 19.
4. If a number is decreased by 7 and this difference is multiplied by 5, the result is 2 less than twice the number. What is the number?
5. Harold has 16% as many chickens as Anne does. If Harold has 180 chickens, how many does Anne have?
6. Betsy has 320 seashells while Louise has 800. The number of Betsy's shells represents what percent of the number of Louise's shells?

*7. $R_A T_A + R_P T_P = 320, \quad R_P = 50, \quad T_P = 4, \quad T_A = 3;$ find R_A.

*8. $R_F T_F = R_E T_E, \quad T_F = 16, \quad T_E = 12, \quad R_E = R_F + 15;$ find R_E, R_F.

*9. $R_M T_M = R_W T_W, \quad R_M = 8, \quad R_W = 2, \quad T_M = 5 - T_W;$ find T_M, T_W.

Simplify:

*10. $(.003 \times 10^{-4})(2 \times 10^{20})$

*11. $\dfrac{.0016}{400{,}000}$

*12. $\dfrac{(.06 \times 10^5)(300{,}000)}{(1000)(.00009)}$

*13. $\dfrac{(.0007 \times 10^{-23})(4000 \times 10^6)}{(.00004)(7{,}000{,}000)}$

For what operations are the following sets closed?

14. $\{-1, 0, 1\}$
15. {Negative real numbers}

Solve:

16. $\dfrac{m}{5} + 6 = \dfrac{m}{2}$

17. $\dfrac{2x}{3} + \dfrac{6 - 3x}{4} = 7$

18. $\dfrac{n + 3}{6} - \dfrac{1}{3} = \dfrac{2n - 2}{5}$

Factor:

19. $x^3 - x^2 - 20x$
20. $x^2 - 12x + 35$
21. $x^3 + 2x^2 - 35x$
22. $x^2 - 6x + 8$
23. $x^2 - 3x + 2$
24. $x^2 - 36$
25. $25x^2 - 36$
26. $bpx^2 - bpx - 20bp$

---

*The problems marked with an asterisk were worked out in this lesson.

Simplify:

27. $\dfrac{4(x^{-2}y^2)^3p^0z}{x^{-2}y^5}$

28. $\left(\dfrac{x^2}{z^{-4}}\right)^2\left(\dfrac{x^3y}{z^2}\right)^{-1}$

Multiply:

29. $4x^{-2}y\left(\dfrac{3x^2}{y} - 4xy^2\right)$

30. $p^0x^2\left(\dfrac{4x^{-2}}{y} + \dfrac{m}{x^2p^0}\right)$

Solve:

31. $4(-x + 2) + 3x - 4 = 7(x - 11) + 4$

Evaluate:

32. $-a[-b + (-a)] + 6(a - b)$ if $a = -2, b = 5$

33. $-b - (-a)(a - b)$ if $a = 3, b = 4$

Graph on a number line:

34. $x + 3 > 2, D = \{\text{Reals}\}$

35. $x - 3 > -4, D = \{\text{Integers}\}$

Graph on a rectangular coordinate system:

36. $x + y + 2 = 0$

37. $y = -2x$

Add:

38. $\dfrac{a + b}{x} + c + \dfrac{1}{x^2}$

39. $\dfrac{a}{b} + \dfrac{c}{b + x}$

Simplify:

40. $\dfrac{a + \dfrac{1}{b}}{m - \dfrac{1}{b}}$

41. $\dfrac{a - \dfrac{1}{x}}{\dfrac{1}{x} - a}$

Use elimination:

42. $\begin{cases} 3x + 2y = -2 \\ 2x - 5y = 24 \end{cases}$

Use substitution:

43. $\begin{cases} 7x + 2y = 9 \\ 2x - y = 3 \end{cases}$

Simplify:

44. $\dfrac{\dfrac{a}{b}}{a + b}$

45. $\dfrac{\dfrac{x}{y + m}}{\dfrac{1}{c}}$

46. $3\sqrt{8} - \sqrt{50} + 2\sqrt{98}$

47. $2\sqrt{27} - 3\sqrt{18} + \sqrt{45}$

48. $|-23|$

49. $-5(-6 + 4) - 7(4) - (-6)$

Add like terms:

50. $x^2y^5m^{-2} + \dfrac{3y^6m^{-4}}{ym^{-2}x^{-2}} + 5x^2m^{-5}y$

lesson twenty-one

Finding the equation of a given line
Multiplication of radical expressions

21.1 the equation of a line

In Lesson 12 we learned to graph a linear equation. In this lesson we will learn the process of writing the equation of the line if we are given the graph of the line. We remember that the equations of vertical and horizontal lines are of the form $x = \pm k$ or $y = \pm k$ such as

$$x = 4 \qquad x = -3 \qquad y = 4 \qquad y = -3$$

If the given graph is a vertical or horizontal line, we write the equation of the line by inspection.

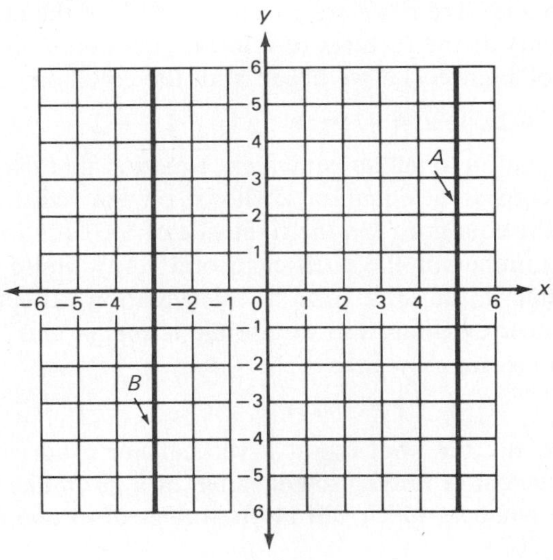

Figure 21.1

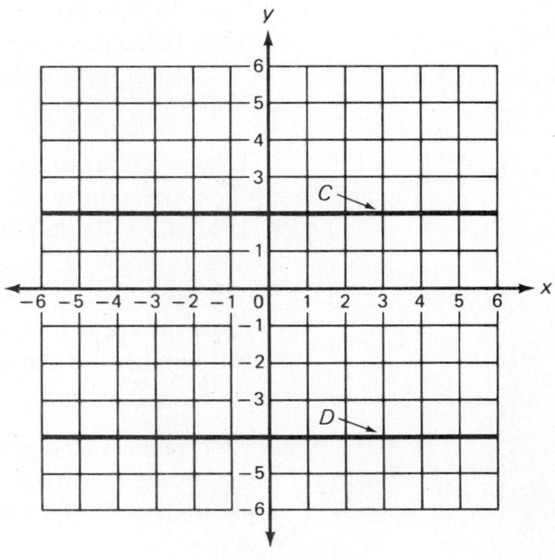

Figure 21.2

In Figure 21.1 the x-coordinate of every point on line A is $+5$ and thus the equation of this line is $x = 5$. The equation of line B is $x = -3$ because the x-coordinate of every point on the line is -3. Since the y-coordinate of every point on line C in Figure 21.2 is $+2$, we say that the equation of this line is $y = 2$. The equation of line D is $y = -4$ because the y-coordinate of every point on this line is -4.

In Lesson 12 we also discussed the method of graphing lines that are not vertical lines or horizontal lines. The equations of these lines can be written in the form $ax + by + c = 0$ where a and b are non-zero real numbers and c is any real number. To graph an equation of this type, we (a) solve the equation for y, (b) make a table and insert values of x, (c) use the equation to find the corresponding values of y, and (d) put the values of y in the table and then graph the points designated by the ordered pairs of x and y in the table.

To graph $y + 2x - 2 = 0$,

(a) $y = -2x + 2$

(b)

x	0	2	-1
y			

(c)

x	0	2	-1
y	2	-2	4

(d)

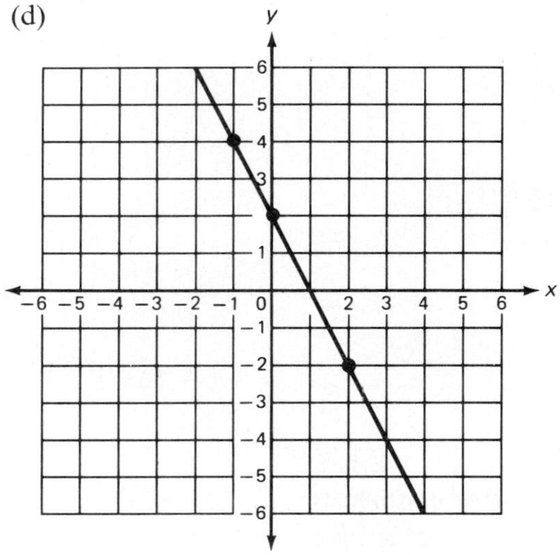

Figure 21.3

Now we will investigate the procedure for reversing the process. We will learn to find the equation of a particular line if we are given the graph of the line. To do this we must first take a detailed look at the features of a linear equation.

In graphing the line of Figure 21.3 we began with the equation on the left.

$$y + 2x - 2 = 0 \longrightarrow y = -2x + 2$$

Then we rearranged this equation to get an equivalent equation that expressed y in terms of x and constants. This equivalent equation is shown on the right. Note that $+y$ is isolated on the left side of the equation. On the right side of the equation, we have x and two numbers. One of the numbers is the coefficient of x. Any first-degree equation in two unknowns can be written in the same form as this equation. **This form is called the slope-intercept form of a linear equation, and we use the letters m and b to represent the two numbers in the equation as we show here.**

$$y = mx + b$$

Since the equation of any line that is not a vertical nor a horizontal line can be written in this form, **the problem of finding the equation of a given line is reduced to the problem of finding the two numbers which will be the values of m and b in the equation above.**

$$y = -6x + 2 \qquad y = \tfrac{2}{3}x - 5 \qquad y = .0007x + 3$$

Here we show the equations of three different lines. Note that each equation has both a y and an x. The only difference in the equations is that the numbers are different!

21.2 intercept

In the slope-intercept form of the equation, $y = mx + b$, we call the constant b the INTERCEPT of the equation because b represents the value of y when x has a value of zero. **Thus b is the y-coordinate of the line at the point where the line intercepts the y-axis.** In Figure 21.4 we show the graphs of two lines. The y-coordinate of the upper line at the point where this line intercepts the y-axis is $+4$, so the intercept b in the equation of this line has a value of 4. The y-coordinate of the lower line at the point where this line intercepts the y-axis is -3, so the intercept b in the equation of this line has a value of -3.

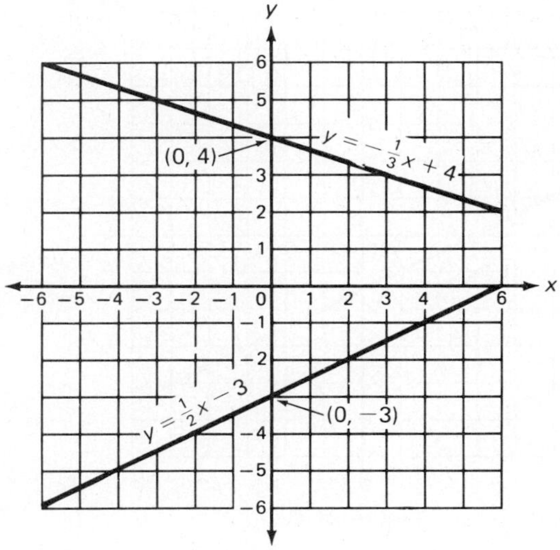

Figure 21.4

21.3 slope

In the slope-intercept form, $y = mx + b$, we call the constant m the slope of the line. Thus in the equation $y = -2x + 6$, we say that the slope of this line is -2 because the coefficient of x is -2. **We note that the slope has both a sign and a magnitude.**† A line represented by a line segment that points toward the upper right part of the coordinate plane has a positive slope. A line represented by a line segment that points toward the lower right part of the coordinate plane has a negative slope.

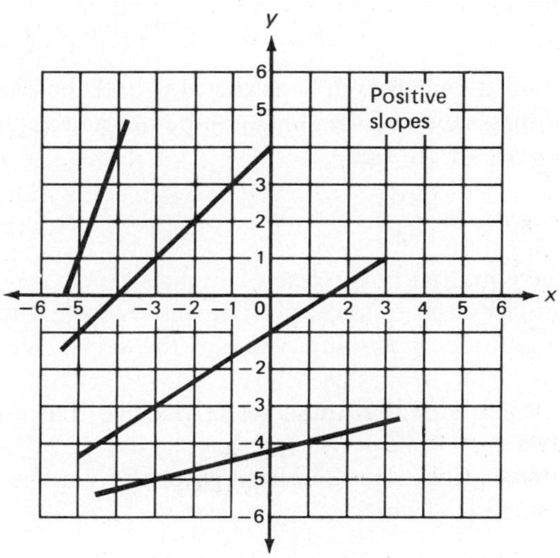

Figure 21.5

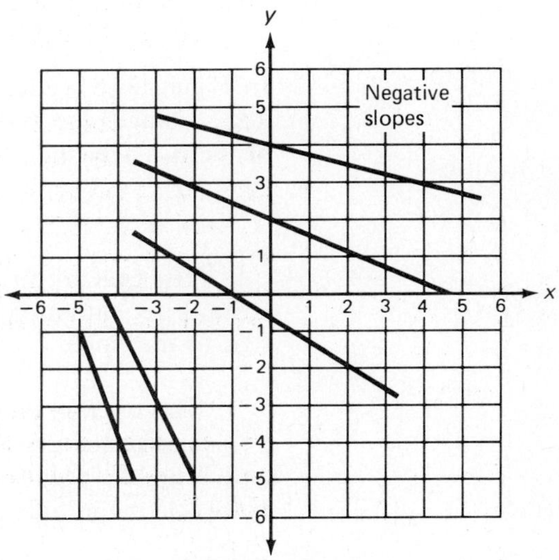

Figure 21.6

†Absolute value.

All the lines† in Figure 21.5 have positive slopes and all the lines in Figure 21.6 have negative slopes.

The magnitude or absolute value of the slope is defined to be the ratio of the absolute value of the change in the *y*-coordinate to the absolute value of the change in the *x*-coordinate as we move from one point on the line to another point on the line.

$$|m| = \frac{|\text{change in } y|}{|\text{change in } x|}$$

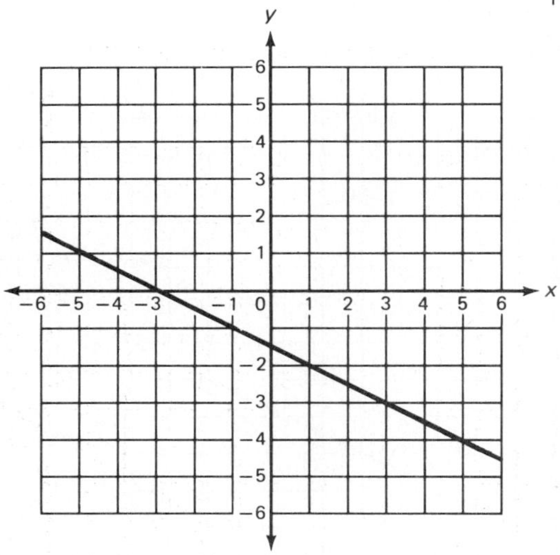

Figure 21.7

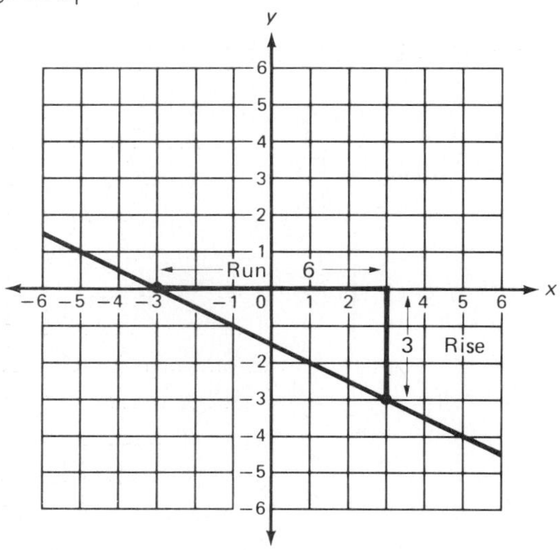

Figure 21.8

In Figure 21.7 we show the graph of a line. To find the magnitude of the slope of this line, we arbitrarily choose two points on the line and connect the two points with lines drawn parallel to the coordinate axes. The line in Figure 21.8 is the same line and we have arbitrarily selected two points and connected them with the required perpendicular lines. The length of the horizontal line is 6 and is the difference of the x-coordinate of the two points. The length of the vertical line is 3 and this is the difference of the y-coordinates of the two points. Since the magnitude of the slope is the ratio of the absolute value of the change in the y-coordinate to the absolute value of the change in the x-coordinate, we see that the magnitude or absolute value of the slope of this line is one half.

$$|m| = \frac{|\text{change in } y|}{|\text{change in } x|} \longrightarrow |m| = \frac{3}{6} \longrightarrow |m| = \frac{1}{2}$$

In Figure 21.8 we have labeled the change in x as the RUN and the change in y as the RISE. Using these words, the magnitude of the slope can be defined as the **absolute value of the rise over the absolute value of the run.**

$$|\text{slope}| = \frac{|\text{rise}|}{|\text{run}|} \quad \text{or} \quad |m| = \frac{|\text{rise}|}{|\text{run}|}$$

The general form of the equation of the line of Figure 21.7 is $y = mx + b$ and to write the equation of this line, we need to know (1) The value of the intercept b, (2) The sign of the slope, (3) The magnitude or absolute value of the slope. We see that

1. The y value of the coordinate of the point where the line intercepts the y-axis is approximately -1.4 so $b = -1.4$.
2. The line points to the lower right and thus the *sign* of the slope is negative.
3. The magnitude of the slope is $\frac{3}{6}$ which is equivalent to $\frac{1}{2}$.

---

†We always use line segments to indicate the position of mathematical lines and although these segments are shorter than normal they indicate adequately the position of the lines under consideration.

So the equation of this line is

$$y = -\tfrac{1}{2}x - 1.4$$

Not everyone will read the intercept on the graph as -1.4. Some will say -1.5 or some other number close to -1.4. Also not everyone will compute the magnitude of the slope to be exactly $\frac{1}{2}$. However, the values found for these numbers should be close to -1.4 and $-\frac{1}{2}$.

example 21.3.1 Find the equations of the lines graphed in the accompanying figures.

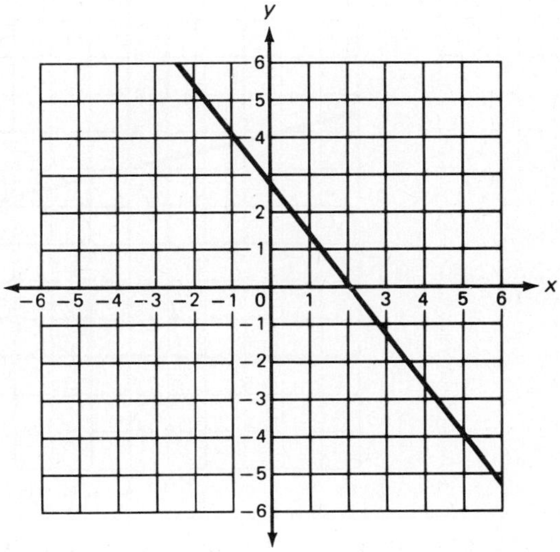

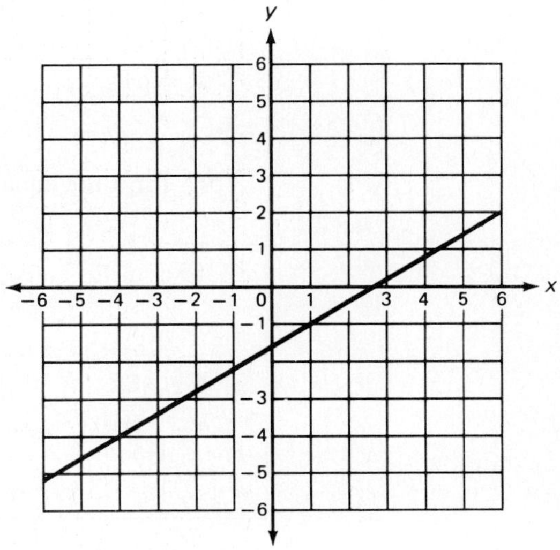

solution The desired equation is $y = mx + b$ and we need to find m and b.

By inspection, $b = +2.8$
By inspection, the sign of m is $-$

The desired equation is $y = mx + b$ and we need to find m and b.

By inspection, $b = -1.7$
By inspection, the sign of m is $+$

Now we need to find the magnitudes or absolute values of the slopes. We will arbitrarily choose two points on each of the lines, draw the triangles, and compute the slopes.

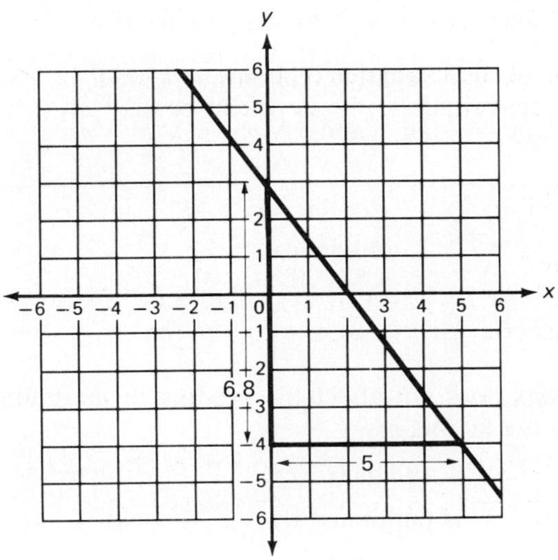

$$|m| = \frac{6.8}{5} = 1.36$$

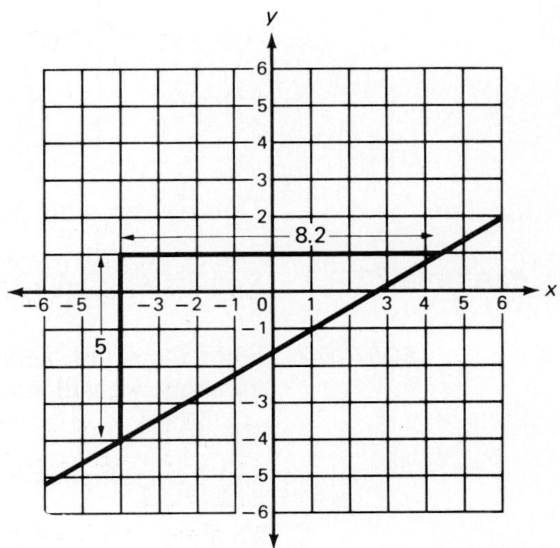

$$|m| = \frac{5}{8.2} = .61$$

So $b = +2.8$ and $\boldsymbol{m} = -\mathbf{1.36}$
Using these values in $y = mx + b$ yields

$$y = -1.36x + 2.8$$

So $b = -1.7$ and $\boldsymbol{m} = +\mathbf{.61}$
Using these values in $y = mx + b$ yields

$$y = .61x - 1.7$$

TEST YOUR UNDERSTANDING

(a) Find the equation of line a.

(b) Find the equation of line b.

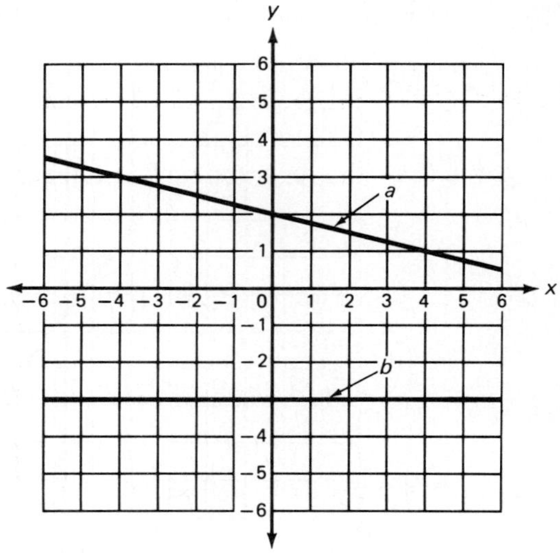

21.4 multiplication of radical expressions

The product theorem for square roots is restated.

THE PRODUCT OF SQUARE ROOTS THEOREM

If m and n are non-negative real numbers, then

$$\sqrt{n}\,\sqrt{m} = \sqrt{mn} \quad \text{and} \quad \sqrt{mn} = \sqrt{m}\,\sqrt{n}$$

Thus we can write $\sqrt{3 \cdot 2}$ as $\sqrt{3}\,\sqrt{2}$ and we can write $\sqrt{3}\,\sqrt{2}$ as $\sqrt{3 \cdot 2}$ or as $\sqrt{6}$.

example 21.4.1 Simplify $4\sqrt{3} \cdot 3\sqrt{2}$.

solution Since the order of the factors does not affect the product in the multiplication of real numbers, we will rearrange the factors as

$$4 \cdot 3\sqrt{3}\,\sqrt{2}$$

and 4 is multiplied by 3 and $\sqrt{3}$ is multiplied by $\sqrt{2}$ to yield

$$\mathbf{12\sqrt{6}}$$

example 21.4.2 Simplify $4\sqrt{3} \cdot 6\sqrt{6}$.

solution Again we will rearrange the order of the indicated multiplication to get

$$4 \cdot 6\sqrt{3}\,\sqrt{6}$$

and multiply 4 by 6 and $\sqrt{3}$ by $\sqrt{6}$ to get

$$24\sqrt{18}$$

and since $\sqrt{18}$ can be written as $3\sqrt{2}$, we can write

$$24\sqrt{18} = 24(3\sqrt{2}\,) = \mathbf{72\sqrt{2}}$$

example 21.4.3 Simplify $4\sqrt{3}\,(5\sqrt{2} + 6\sqrt{3}\,)$.

solution The notation indicates that $4\sqrt{3}$ is to be multiplied by both terms within the parentheses:

$$4\sqrt{3}\,(5\sqrt{2} + 6\sqrt{3}\,) = 4\sqrt{3} \cdot 5\sqrt{2} + 4\sqrt{3} \cdot 6\sqrt{3}$$

Now we rearrange the order of the factors in each term to get

$$4 \cdot 5\sqrt{3}\,\sqrt{2} + 4 \cdot 6\sqrt{3}\,\sqrt{3}$$

and lastly we perform the indicated multiplications.

$$\mathbf{20\sqrt{6} + 72}$$

example 21.4.4 Simplify $4\sqrt{2}\,(3\sqrt{2} + 5)$.

solution First we multiply as indicated

$$4\sqrt{2} \cdot 3\sqrt{2} + 4\sqrt{2} \cdot 5$$

and now simplify

$$\mathbf{24 + 20\sqrt{2}}$$

example 21.4.5 Simplify $5\sqrt{2}\,(2\sqrt{3} + \sqrt{6}\,)$.

solution

$$5\sqrt{2} \cdot 2\sqrt{3} + 5\sqrt{2}\,\sqrt{6}$$

which leads to

$$10\sqrt{6} + 5\sqrt{12}$$

and we write $\sqrt{12}$ as $2\sqrt{3}$ and get

$$10\sqrt{6} + 5(2\sqrt{3}\,) = \mathbf{10\sqrt{6} + 10\sqrt{3}}$$

TEST YOUR UNDERSTANDING

Multiply and simplify:

(c) $3\sqrt{2}\,(7\sqrt{2} - \sqrt{6}\,)$

lesson checklist 21

- ☐ Equations of vertical and horizontal lines
- ☐ Slope intercept form of a linear equation
- ☐ Intercept
- ☐ Sign of the slope of a line
- ☐ Magnitude of the slope of a line
- ☐ Rise over run
- ☐ Multiplication of radical expressions

ANSWERS TO TEST YOUR UNDERSTANDING

(a) $y = -\frac{1}{4}x + 2$ (b) $y = -3$ (c) $42 - 6\sqrt{3}$

problem set 21

1. Find three consecutive odd integers such that five times the first equals the sum of the third and -32.
2. Find four consecutive integers such that twice the sum of the first two equals the sum of the fourth and 32.
3. Find three consecutive even integers such that three times the first is 8 less than five times the third.
4. If twice a number is increased by 5, and this sum multiplied by 3, the result equals 57. Find the number.
5. On the way to Oklahoma, Keith loses 65 cents. If his total worth at that time was 13 dollars, what percent of his total worth did he lose?
6. Twenty percent of Stillson's total bill was 8 dollars. What was the total bill?

Simplify:

*7. $4\sqrt{3}(5\sqrt{2} + 6\sqrt{3})$ 8. $4\sqrt{3} \cdot 6\sqrt{12}$ *9. $4\sqrt{2}(3\sqrt{2} + 5)$

10. $5\sqrt{3}(4 - 2\sqrt{5})$ 11. $5\sqrt{5}(3 - 4\sqrt{5})$

12. $3\sqrt{2}(4\sqrt{6} - 5)$ 13. $-\sqrt{2}(6\sqrt{2} - 15)$ 14. $2\sqrt{50} - 3\sqrt{32}$

15. $5\sqrt{18} + \sqrt{72}$ 16. $2\sqrt{75} + 3\sqrt{48}$ 17. $3\sqrt{20} - 2\sqrt{80}$

18. $R_HT_H + R_ST_S = 180$, $R_H = 70$, $R_S = 20$, $T_H = T_S$; find T_H and T_S.

19. $R_FT_F = R_ST_S$, $T_S = 6$, $T_F = 5$, $R_F - 10 = R_S$; find R_S and R_F.

20. $R_MT_M = R_RT_R$, $R_M = 8$, $R_R = 2$, $5 - T_M = T_R$; find T_M and T_R.

Simplify:

21. $$\frac{(.00003)(.006 \times 10^{-6})(200 \times 10^{4})}{(1800 \times 10^{15})(100{,}000)}$$

*The problems marked with an asterisk were worked out in this lesson.

22. Find the equation of the line.

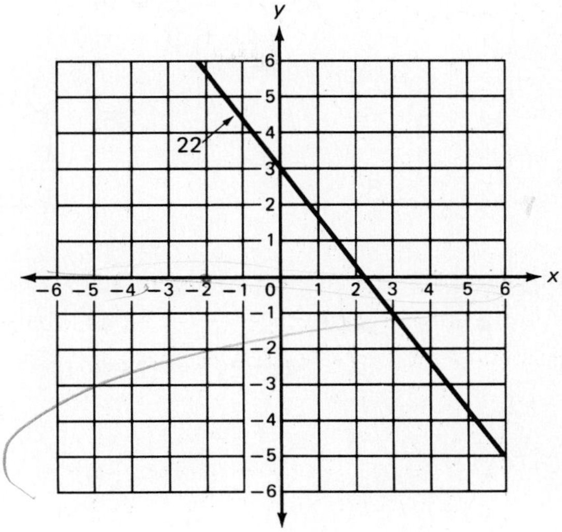

23. Find the equation of the line.

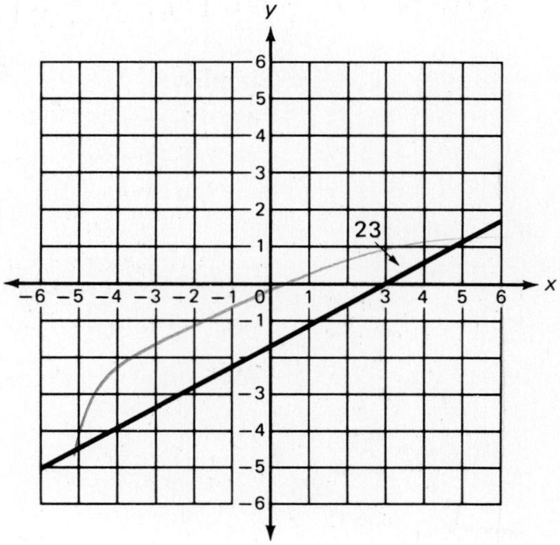

24. The set of positive even integers is closed for what operations?

Solve:

25. $\dfrac{4 + x}{2} - \dfrac{1}{3} = 6$

26. $\dfrac{4 - x}{3} + \dfrac{1}{6} = 5$

The equations $y = -2$, $x = -2$, and $y = -2x$ look very similar, but the graphs of these equations are very different. Graph the following equations on a rectangular coordinate system.

27. $x = -2$

28. $y = -2x$

Add:

29. $\dfrac{a + b}{x + y} + \dfrac{a}{x}$

30. $\dfrac{a}{x^2(a + b)} + \dfrac{c}{x}$

Add like terms:

31. $\dfrac{3a^{-2}y}{b} + 7b^{-1}ya^{-2} - \dfrac{5b^{-1}}{a^2y^{-1}} - \dfrac{2b^{-1}a^{-2}}{y}$

Multiply:

32. $\dfrac{4x^{-2}}{b}\left(x^{-3}b^2 - \dfrac{4b}{x^2}\right)$

33. $\dfrac{2a^4}{b^{-2}}\left(\dfrac{b^2}{a^4} - 3b^{-2}a^4\right)$

Simplify:

34. $\dfrac{(p^2m^2)^0k^4}{k^{-3}m^{-2}}$

35. $\dfrac{a^2my^4}{m^{-3}y^2a^{-2}}$

36. $|-15 + 2|$

Solve:

37. $2(a - 3) + 6a = 5(-a + 2)$

38. $3x + 2(-x) - 2x = 5[4 - (-3)]$

Evaluate:

39. $x(-a + x) - (-a)$ if $a = 2$, $x = -3$

40. $k[-a(k) + c]$ if $a = -2$, $c = 3$, $k = 4$

Use elimination:

41. $\begin{cases} 5x - 3y = 22 \\ 2x + 4y = -12 \end{cases}$

42. $\begin{cases} 3x - 2y = 19 \\ 4x + 3y = -3 \end{cases}$

Simplify:

43. $\dfrac{\dfrac{a}{b}}{a + b}$

44. $\dfrac{\dfrac{a}{b} - 1}{1 - \dfrac{a}{b}}$

45. $\dfrac{\dfrac{x}{y} - \dfrac{1}{y}}{a + \dfrac{1}{y}}$

Factor. Remember to begin by factoring the greatest common factor if one exists.

46. $x^2 - 9y^2$

47. $7x^2 - 63y^2$

48. $ap^2x^2 - 4ap^2y^2$

49. $mbx^2 + mbx - 20mb$

50. $2x^2 + 12x + 18$

lesson twenty-two

Division of polynomials
Zero factor theorem
Solution of quadratic equations by factoring

22.1 division of polynomials

In the problem sets since Lesson 11 we have been finding the product of two polynomials. Now we investigate the inverse process, the division of polynomials.

The simplest type of polynomial division is the division of a polynomial by a monomial. The desired result can be obtained by dividing each term of the polynomial by the monomial as shown in the following examples.

example 22.1.1 Divide $3x^3 + 7x^2 - x$ by x.

solution $$\frac{3x^3}{x} + \frac{7x^2}{x} - \frac{x}{x} = \mathbf{3x^2 + 7x - 1}$$

example 22.1.2 Divide $12x^{12} - 8x^8 + 4x^6$ by $4x^4$.

solution $$\frac{12x^{12}}{4x^4} - \frac{8x^8}{4x^4} + \frac{4x^6}{4x^4} = \mathbf{3x^8 - 2x^4 + x^2}$$

Before demonstrating the method of dividing a polynomial by a binomial we will complete a long division problem involving the natural numbers, an algorithm with which we are familiar. The method of dividing a polynomial by a binomial is very similar.

$$\begin{array}{r|l} & 4 \\ \hline 12 & 49 \\ & \underline{48} \\ & 1 \end{array} \qquad \text{so } 49 \div 12 = 4\tfrac{1}{12}$$

We multiplied 4 by 12 and recorded the product of 48 below the 49. Next we **mentally changed the sign** of the $+48$ to -48 and then added algebraically to find the remainder of 1.

example 22.1.3 Divide $-2x^2 + 3x^3 + 5x + 50$ by $-3 + x$.

solution The first step is to write both polynomials in descending powers of the variable and use the same format for division as we used above.

$$x - 3\,\overline{)\,3x^3 - 2x^2 + 5x + 50}$$

To determine the first term of the quotient, divide the first term of the divisor into the first term of the dividend; in this case, divide x into $3x^3$ and get $3x^2$. Record as indicated.

$$\begin{array}{r l}
 & 3x^2 \\
x - 3 & \overline{)\,3x^3 - 2x^2 + 5x + 50}
\end{array}$$

Now multiply the term $3x^2$ by $x - 3$ and record as shown below:

$$\begin{array}{r l}
 & 3x^2 \\
x - 3 & \overline{)\,3x^3 - 2x^2 + 5x + 50} \\
 & \underline{3x^3 - 9x^2}
\end{array}$$

Now **mentally change the sign** of both $3x^3$ and $-9x^2$ and add algebraically:

$$\begin{array}{r l}
 & 3x^2 \\
x - 3 & \overline{)\,3x^3 - 2x^2 + 5x + 50} \\
 & \underline{3x^3 - 9x^2} \\
 & \quad +7x^2
\end{array}$$

Now bring down the $+5x$.

$$\begin{array}{r l}
 & 3x^2 \\
x - 3 & \overline{)\,3x^3 - 2x^2 + 5x + 50} \\
 & \underline{3x^3 - 9x^2} \\
 & \quad +7x^2 + 5x
\end{array}$$

Now divide the x of $x - 3$ into $7x^2$, get $7x$, and record as shown.

$$\begin{array}{r l}
 & 3x^2 + 7x \\
x - 3 & \overline{)\,3x^3 - 2x^2 + 5x + 50} \\
 & \underline{3x^3 - 9x^2} \\
 & \qquad 7x^2 + 5x
\end{array}$$

Multiply $7x$ by $x - 3$ and record

$$\begin{array}{r l}
 & 3x^2 + 7x \\
x - 3 & \overline{)\,3x^3 - 2x^2 + 5x + 50} \\
 & \underline{3x^3 - 9x^2} \\
 & \qquad 7x^2 + 5x \\
 & \qquad \underline{7x^2 - 21x}
\end{array}$$

Now **mentally change the sign** of both $7x^2$ and $-21x$ and add algebraically. Repeat the procedure until the remainder of 128 is obtained.

$$\begin{array}{r}
3x^2 + 7x + 26 \\
x - 3 \overline{\smash{)}\, 3x^3 - 2x^2 + 5x + 50} \\
\underline{3x^3 - 9x^2 } \\
7x^2 + 5x \\
\underline{7x^2 - 21x } \\
26x + 50 \\
\underline{26x - 78} \\
128
\end{array}$$

Thus we find that

$$\frac{3x^3 - 2x^2 + 5x + 50}{x - 3} = \mathbf{3x^2 + 7x + 26 + \frac{128}{x - 3}}$$

example 22.1.4 Divide $-2x + 5 + 3x^3$ by $-3 + x$.

solution Again we begin by writing both polynomials in descending order of the variables and using the format for long division.

$$x - 3 \overline{\smash{)}\, 3x^3 - 2x + 5}$$

We note that the dividend has an x^3 term and an x term but no x^2 term. A good ploy to avoid confusion is to insert an x^2 term with zero as its coefficient as shown below. Of course, zero multiplied by x^2 yields zero so the polynomial is really unchanged.

$$\begin{array}{r}
3x^2 + 9x + 25 \\
x - 3 \overline{\smash{)}\, 3x^3 + 0x^2 - 2x + 5} \\
\underline{3x^3 - 9x^2 } \\
9x^2 - 2x \\
\underline{9x^2 - 27x } \\
25x + 5 \\
\underline{25x - 75} \\
80
\end{array}$$

Thus $\dfrac{3x^3 - 2x + 5}{x - 3} = \mathbf{3x^2 + 9x + 25 + \dfrac{80}{x - 3}}$

TEST YOUR UNDERSTANDING

(a) Divide: $3x^3 - x - 7$ by $x - 5$

22.2 quadratic equations

Quadratic equations are second-degree polynomial equations that can be written in one of two forms shown here.

$$ax^2 + bx + c = 0 \qquad ax^2 + bx + c = y$$

In this book we will restrict our investigation of quadratic equations to equations in the form of the equation shown on the left. This equation is a second degree polynomial equation in one unknown. If the equation is written in descending powers of the variable with all non-zero members on the left side of the equation, we say that the equation is written in **STANDARD FORM**. Thus the equation

$$x^2 - 3x - 10 = 0$$

is a **quadratic equation in standard form.** To designate a general quadratic equation, we use the letter a to represent the coefficient of x^2, the letter b to represent the coefficient of x, and the letter c to represent the constant term. Using these letters to represent the constants in the equation, we can write **a general quadratic equation in standard form** as

$$ax^2 + bx + c = 0$$

Thus if $a = 1$, $b = -3$, and $c = -10$, we have the equation

$$x^2 - 3x - 10 = 0$$

If we substitute either 5 or -2 for the variable x in the quadratic equation $x^2 - 3x - 10 = 0$, the equation will be transformed into a true statement as shown here,

if $x = 5$	if $x = -2$
$(5)^2 - 3(5) - 10 = 0$	$(-2)^2 - 3(-2) - 10 = 0$
$25 - 15 - 10 = 0$	$4 + 6 - 10 = 0$
$0 = 0$	$0 = 0$

> If the replacement of the variable by a number makes the equation a true statement, we say that the number **satisfies** the equation and we call the number a **root** of the equation or a **solution** of the equation. We say that the set of all numbers that satisfy the equation constitutes the **solution set** of the equation.

The numbers 5 and -2 are the only numbers that will satisfy the equation above. Every quadratic equation has at most† two numbers that will make the equation a true statement. For that matter, every third-degree polynomial equation in one variable has at most three numbers that will satisfy the equation; every fourth-degree polynomial equation in one variable has at most four numbers that will satisfy the equation. To generalize, we can say that every nth-degree polynomial equation in one variable (n is a natural number) has at most n roots.

†We say *at most* because some quadratic equations have only one root. For instance, the only root of the equation $x^2 - 4x + 4 = 0$ is the number $+2$.

22.3 solution of quadratic equations by factoring

> ZERO FACTOR THEOREM
>
> If p and q are any real numbers and if $p \cdot q = 0$, then either $p = 0$ or $q = 0$, or both.

This says that **if the product of two real numbers is zero, one or both of the numbers is zero.** Thus if we indicate the product of 4 and an unspecified number by writing

$$4(\quad) = 0$$

the only number that we can place in the parentheses that will make the equation a true statement is the number zero.

In the same way, if we indicate the product of two unspecified numbers by writing

$$(\quad)(\quad) = 0$$

the quantity in the first parentheses or the quantity in the second parentheses must equal zero or the product will not equal zero.

Now let's look at the equation

$$(x - 3)(x + 5) = 0$$

Here we have two quantities multiplied and the product is equal to zero. From the **zero factor theorem**, we know that at least one of the quantities must equal zero if the product is to equal zero, so either

$$x - 3 = 0 \qquad \text{or} \qquad x + 5 = 0$$

But if $x - 3 = 0$, $\boldsymbol{x = 3}$ and if $x + 5 = 0$, $\boldsymbol{x = -5}$

Thus the two values of x that satisfy the condition stated are 3 and -5.

We can use the zero factor theorem to help us solve quadratic equations that can be factored. We do this by first writing the equation in standard form and factoring the polynomial. Then we set each of the factors equal to zero and solve for the values of the variable.

example 22.3.1 Find the roots of $x^2 - 18 = 3x$ by the factor method.

solution First we write the equation in standard form and then factor.

$$x^2 - 3x - 18 = 0 \longrightarrow (x + 3)(x - 6) = 0$$

Since the product $(x + 3)(x - 6)$ equals zero, by the zero factor theorem, we know that one of these factors must equal zero.

If $x + 3 = 0$, $\boldsymbol{x = -3}$. If $x - 6 = 0$, $\boldsymbol{x = +6}$

To check, we will use -3 and $+6$ as values for x in the original equation

if $x = -3$	if $x = 6$
$(-3)^2 - 3(-3) - 18 = 0$	$(6)^2 - 3(6) - 18 = 0$
$9 + 9 - 18 = 0$	$36 - 18 - 18 = 0$
$0 = 0$ Check	$0 = 0$ Check

example 22.3.2 Find the roots of $-25 = -4x^2$.

solution First we write the equation in standard form, $4x^2 - 25 = 0$, and then we factor to get $(2x - 5)(2x + 5) = 0$. For this to be true, either $2x - 5$ equals zero or $2x + 5$ equals zero.

$$\text{If } 2x - 5 = 0 \qquad\qquad \text{If } 2x + 5 = 0$$
$$2x = 5 \qquad\qquad 2x = -5$$
$$\boldsymbol{x = \frac{5}{2}} \qquad\qquad \boldsymbol{x = -\frac{5}{2}}$$

To check, we will use $\frac{5}{2}$ and $-\frac{5}{2}$ as values for x in the original equation

$$\text{if } x = \frac{5}{2} \qquad\qquad \text{if } x = -\frac{5}{2}$$
$$4\left(\frac{5}{2}\right)^2 - 25 = 0 \qquad\qquad 4\left(-\frac{5}{2}\right)^2 - 25 = 0$$
$$4\left(\frac{25}{4}\right) - 25 = 0 \qquad\qquad 4\left(\frac{25}{4}\right) - 25 = 0$$
$$25 - 25 = 0 \qquad\qquad 25 - 25 = 0$$
$$0 = 0 \quad \text{Check} \qquad\qquad 0 = 0 \quad \text{Check}$$

example 22.3.3 Find the values of x that satisfy $x - 56 = -x^2$.

solution First we rewrite the equation in standard form.

$$x^2 + x - 56 = 0$$

Now we factor

$$(x + 8)(x - 7) = 0$$

For this to be true, either $x + 8$ equals zero or $x - 7$ equals zero.

$$\text{If } x + 8 = 0, \boldsymbol{x = -8} \qquad\qquad \text{If } x - 7 = 0, \boldsymbol{x = 7}$$

To check, we will use -8 and $+7$ as replacements for x in the original equation.

$$\text{If } x = -8 \qquad\qquad \text{If } x = 7$$
$$(-8)^2 + (-8) - 56 = 0 \qquad\qquad (7)^2 + (7) - 56 = 0$$
$$64 - 8 - 56 = 0 \qquad\qquad 49 + 7 - 56 = 0$$
$$0 = 0 \quad \text{Check} \qquad\qquad 0 = 0 \quad \text{Check}$$

example 22.3.4 Solve $x^2 - x = 42$.

solution First we write the equation in standard form

$$x^2 - x - 42 = 0$$

and now we factor

$$(x - 7)(x + 6) = 0$$

The zero factor theorem says that if a product of two factors equals zero, either one factor or the other factor must equal zero.

$$\text{If } x - 7 = 0, \boldsymbol{x = 7} \qquad\qquad \text{If } x + 6 = 0, \boldsymbol{x = -6}$$

and check.

$$(7)^2 - (7) - 42 = 0 \qquad (-6)^2 - (-6) - 42 = 0$$

$$49 - 7 - 42 = 0 \qquad 36 + 6 - 42 = 0$$

$$0 = 0 \quad \text{Check} \qquad 0 = 0 \quad \text{Check}$$

Test Your Understanding

Solve by factoring and then using the zero factor theorem:

(b) $-x^2 = -3x - 40$

lesson checklist 22

- ☐ Division of a polynomial by a polynomial
- ☐ Descending powers of the variable
- ☐ Division of a polynomial by a binomial
- ☐ Standard form of a quadratic equation
- ☐ Zero factor theorem
- ☐ Solution of a quadratic equation by factoring

Answers to Test Your Understanding

(a) $3x^2 + 15x + 74 + \dfrac{363}{x - 5}$ (b) $8, -5$

problem set 22

1. Ward had some pennies. If he added 5 more pennies and multiplied the total by 13, he would have 208 pennies. How many pennies did Ward begin with?
2. Betsy misplaced 15 percent of her marbles. If she misplaced 33 marbles, how many did she have to begin with?
3. Rue began with D dollars. She then got 4 more dollars and multiplied all she had by 12. Now she has 136 more dollars than she had at the beginning. How many dollars did she begin with?
4. Zwaan lost 55 grains of rice out of his bowl. If his bowl originally had 1100 grains of rice in it, what percent of his rice did he lose?
5. Jeff added 5 apples to the ones he already had. Then he magically multiplied them by 8 and found that now he had 54 more apples than he had at the beginning. How many did he have at the beginning?
6. Julie had 1500 roses. She gave 18 percent of these to Kristen. How many roses did Julie give to Kristen?

Divide:

*7. $12x^{12} - 8x^8 + 4x^6$ by $4x^4$

*8. $-2x^2 + 3x^3 + 5x + 50$ by $-3 + x$

*9. $-2x + 5 + 3x^3$ by $-3 + x$

10. $2x^3 - x - 13$ by $x + 5$

Solve:

*11. $x^2 - 18 = 3x$

*12. $4x^2 - 25 = 0$

*13. $x - 56 = -x^2$

14. $\frac{6 + x}{5} + \frac{1}{3} = 8$

15. $\frac{4 - x}{6} - \frac{1}{2} = 11$

Add:

16. $\frac{m^2 + b}{a + b} + \frac{m}{b}$

17. $\frac{a}{x + y} + \frac{b + c}{x^2}$

Add like terms:

18. $\frac{7a^{-2}y}{x^2} - \frac{4x^{-2}y}{a^2} + 2a^{-2}yx^2$

19. $7a^2x^{-4} + 3x^{-4}a^2 - \frac{2x^{-4}}{a^{-2}}$

20. Find the equation of the line shown.

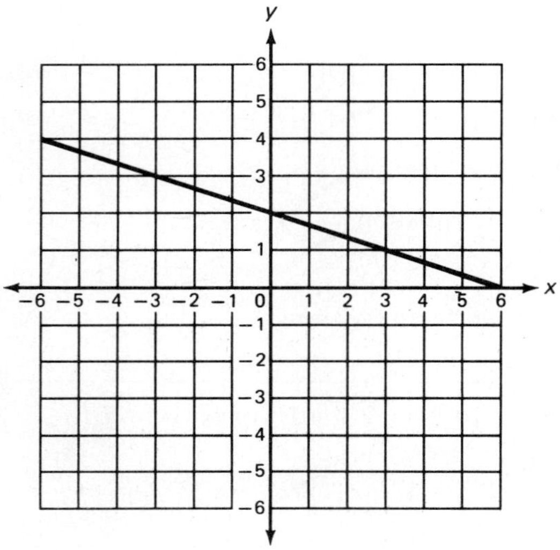

21. Find the equation of the line shown.

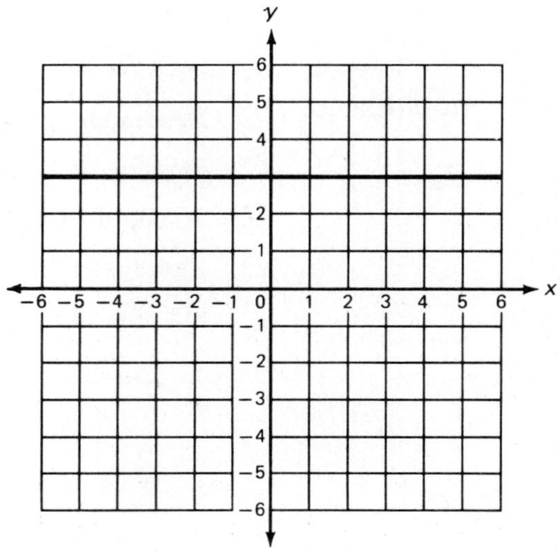

Graph on a rectangular coordinate system:

22. $y = -3x + 2$

23. $y = -3$

24. $R_G T_G + R_B T_B = 100$, $R_G = 4$, $R_B = 10$, $T_B = T_G + 3$; find T_G and T_B.

25. $R_K T_K = R_N T_N$, $R_K = 6$, $R_N = 3$, $T_K = T_N - 8$; find T_K and T_N.

Simplify:

26. $\frac{(.00004 \times 10^{-13})(700 \times 10^{-2})}{(14 \times 10^{25})(100{,}000)}$

27. $\frac{500 \times 10^{-23}}{(25{,}000)(.00002)}$

28. $4\sqrt{3}\,(5\sqrt{3} - 2)$

29. $2\sqrt{12}\,(4 - 3\sqrt{2})$

30. $2\sqrt{75} - 6\sqrt{48}$

31. $-5\sqrt{18} + 2\sqrt{50}$

Multiply:

32. $\frac{z^5y^{-2}}{p^4}\left(\frac{3z^4}{y^{-2}} + \frac{2y^2p^{-7}}{z^5}\right)$

Simplify:

33. $\dfrac{(x^{-4}y^0)k^3p^{-5}}{p^2y^5x^4k^0}$

34. $\dfrac{(a^2xy^0)y^2x^4}{x^{-4}y^{15}}$

35. $\dfrac{(-4a^0b)^0(x^{-3}y)^{-3}}{x^{-3}y^2}$

36. $|-5 - 14(-2)|$

Solve:

37. $3(x + 2) - 5x = 7(x - 3) + 5$

38. $4x + (-3)(-x - 6) = -x - 4 + (-2)$

Evaluate:

39. $(a - c^2)(c - 4)$ if $a = 3$, $c = -2$

40. $x + (-y - x)(-x)$ if $x = -4$, $y = -2$

41. $(-5)^{-2}$

42. -5^{-2}

43. $-\dfrac{1}{5^{-2}}$

44. True or false: {Rational} $\in$ {Reals}

45. $4\frac{1}{3} \in$ {what sets of numbers?}

Use elimination:

46. $\begin{cases} 3x + y = 3 \\ 2x - 3y = 13 \end{cases}$

Simplify:

47. $\dfrac{\dfrac{x}{y}}{1 - \dfrac{1}{y}}$

48. $\dfrac{\dfrac{x}{y} + 1}{x + \dfrac{a}{y}}$

lesson twenty-three

Value word problems
General word problems in two unknowns
Properties of inequality

23.1 value word problems

In this lesson we will present word problems whose solutions require the solution of systems of two or more simultaneous equations. **To find the necessary equations, we read the problem and try to find statements about things that are equal.** Each time we find such a statement, we will write the statement as an algebraic equation using variables and numbers as required. **When we have the same number of equations as we have variables, we will use the substitution and/or the elimination methods to find the values of the variables that simultaneously satisfy all equations.**

example 23.1.1 Jack and Betty have 28 coins that are nickels and dimes. If the value of the coins is \$1.95, how many coins of each type do they have?

solution This problem is a typical value word problem. Hidden in the wording of this problem are two statements about things that are equal. One statement is that the **number of nickels** plus the **number of dimes** equals 28. We will use N_N for the number of nickels and N_D for the number of dimes.

$$N_N + N_D = 28$$

The second statement is that the **value of the nickels** plus the **value of the dimes** equals the **total value** of \$1.95.

IF	THEIR VALUE WOULD BE
we have 2 nickels	$5(2) = 10¢$
we have 6 nickels	$5(6) = 30¢$
we have N_N nickels	$\mathbf{5N_N}$ **cents**

IF	THEIR VALUE WOULD BE
we have 2 dimes	$10(2) = 20¢$
we have 6 dimes	$10(6) = 60¢$
we have N_D dimes	$\mathbf{10N_D}$ **cents**

Thus the algebraic statement that the value of the nickels plus the value of the dimes equals the total value can be written as

$$5N_N + 10N_D = 195$$

Note that we are expressing our values in cents. If we wished to express our values in dollars, this equation would be written

$$.05N_N + .10N_D = 1.95$$

and we would have to contend with decimals. Either way is correct, but most students prefer to work with coefficients that are integers. **It is necessary, however, to use the same units throughout a problem. If one value is expressed in pennies, then all values must be expressed in pennies! If one value is expressed in dollars, then all values must be expressed in dollars!**

Now we have the two unknowns N_N and N_D and we also have two equations, **the same number of equations that we have unknowns!**

$$N_N + N_D = 28$$
$$5N_N + 10N_D = 195$$

The values of N_N and N_D that will simultaneously satisfy both equations may be found using either the substitution method or the elimination method.

SUBSTITUTION

$$5(28 - N_D) + 10N_D = 195$$
$$140 - 5N_D + 10N_D = 195$$
$$140 + 5N_D = 195$$
$$5N_D = 55$$
$$\mathbf{N_D = 11}$$

and since $N_N + N_D = 28$

$\mathbf{N_N = 17}$ by inspection.

ELIMINATION

$$-5N_N - 5N_D = -140$$
$$\underline{5N_N + 10N_D = 195}$$
$$5N_D = 55$$
$$\mathbf{N_D = 11}$$

and since $N_N + N_D = 28$

$\mathbf{N_N = 17}$ by inspection.

example 23.1.2 Airline fares for a flight from Tifton to Adel are \$30 for first class and \$25 for tourist class. If a flight carried 52 passengers for a total fare of \$1360, how many first class passengers were on the flight?

solution In this problem the two statements about things that are equal are similar to the two statements from the previous problem. The **number** of first class tickets plus the **number** of tourist class tickets equals a total **number** of 52 tickets. Also the **value** of the first class tickets plus the **value** of the tourist class tickets equals a total **value** of \$1360. We can use dollars as our unit of currency in this problem without having to use decimal numbers.

(a) $$N_F + N_T = 52$$

(b) $$30N_F + 25N_T = 1360$$

We will substitute the expression for N_F from equation (a), $(N_F = 52 - N_T)$ for N_F in equation (b).

$$30(52 - N_T) + 25N_T = 1360 \longrightarrow 1560 - 30N_T + 25N_T = 1360$$
$$\longrightarrow 1560 - 5N_T = 1360 \longrightarrow -5N_T = -200 \longrightarrow \mathbf{N_T = 40}$$

And since $N_F + N_T = 52$, then $\mathbf{N_F = 12}$, by inspection.

example 23.1.3 Toy has \$4.45 in quarters and dimes. She has 8 more quarters than dimes. How many coins of each type does she have?

solution Again we have two statements about quantities that are equal:
The **value** of the quarters plus the **value** of the dimes equals a total **value** of 445 pennies.

(a) $$25N_Q + 10N_D = 445$$

The number of quarters is 8 greater than the number of dimes. This equation is a little tricky. Let's write

(b) $$N_Q = N_D \quad \text{False}$$

which we know is UNTRUE. We wish to make this a true statement of equality by adding 8 to one side or the other. **The problem says that we have 8 more quarters than dimes so we must add 8 to the number of dimes to have equal quantities on both sides of the equation.**

$$N_Q = N_D + 8$$

A careless approach will result in 8 being added to N_Q, which, of course, is incorrect as N_Q is already greater than N_D. To solve we will substitute the expression for N_Q in equation (b) for N_Q in equation (a).

$$25(N_D + 8) + 10N_D = 445$$
$$25N_D + 200 + 10N_D = 445$$
$$35N_D + 200 = 445$$
$$35N_D = 245$$
$$N_D = 7$$

and since $N_Q = N_D + 8$, we know that $N_Q = (7) + 8$ or $N_Q = 15$.

CHECK

$25N_Q + 10N_D = 445$	$N_Q = N_D + 8$
$25(15) + 10(7) = 445$	$(15) = (7) + 8$
$375 + 70 = 445$	$15 = 15$ Check
$445 = 445$ Check	

example 23.1.4 Timpson's dress shop had a big sale. They sold the cheaper dresses for \$20 each and sold the more expensive ones for \$45 each and took in \$1375 on the sale. How many of each kind did they sell if they sold 20 more of the cheaper dresses than the expensive dresses?

solution The equations for this problem are very similar to the equations used in the last problem. The first equation says that the value of the cheaper dresses plus the value of the expensive dresses equals total receipts from the sale.

(a) $$20N_C + 45N_E = 1375$$

The other equation relates the numbers of dresses sold. We begin by writing

$$N_C = N_E \quad \text{False}$$

which we know to be untrue because 20 more of the cheaper dresses were sold. **Thus we**

must add 20 to the number of expensive dresses so that the sides of the equation will be equal.

(b) $$N_C = N_E + 20$$

We will solve these equations by replacing N_C in equation (a) with its equivalence from equation (b).

$$20(N_E + 20) + 45N_E = 1375$$
$$20N_E + 400 + 45N_E = 1375$$
$$65N_E = 975$$
$$N_E = \mathbf{15}$$

and since there were 20 more cheap dresses sold, we find that the number of cheaper dresses sold equals 35.

$$N_C = N_E + 20 \rightarrow N_C = 15 + 20 = \mathbf{35}$$

TEST YOUR UNDERSTANDING

(a) Wesley and Jackie saved nickels and dimes. One day Jackie broke open the piggy bank and found $1.45 and 5 more nickels than dimes. How many dimes did she find?

23.2 general word problems in two unknowns

Word problems of the type just discussed are called **value problems**. As we have seen, they can be worked rather easily if we recognize the statements made in the problem concerning things that are equal and write algebraic equations that correspond to these verbal statements of equality. Many types of word problems can be worked using this approach. We will present several here. We will not work the problems all the way out but will note the verbal statement of equality and write the equations implied by these statements.

example 23.2.1 Together Charles and Nelle picked 92 quarts of berries. If Charles picked 6 more quarts than Nelle picked, how many quarts did he pick?

solution (a) Quarts Charles picked + quarts Nelle picked equals 92

(b) Quarts Nelle picked plus 6 equals quarts Charles picked

(a) $N_C + N_N = 92$

(b) $N_N + 6 = N_C$

example 23.2.2 The number of boys in Sarah's class exceeded the number of girls by 7. If there were a total of 29 pupils in the class, how many were boys and how many were girls?

solution (a) The number of boys equals the number of girls plus 7.

(b) The number of boys plus the number of girls equals 29.

$$\text{(a)}\quad N_B = N_G + 7$$

$$\text{(b)}\quad N_B + N_G = 29$$

example 23.2.3 Phillip cut a 38-meter rope into two pieces. The long piece was 9 meters longer than the short piece. What were the two lengths?

solution (a) Length of long piece plus length of short piece equals 38 meters.

(b) The long piece was 9 meters longer than the short piece.

$$\text{(a)}\quad L + S = 38$$

$$\text{(b)}\quad S + 9 = L$$

example 23.2.4 The sum of two numbers is 72. The difference of the numbers is 26. What are the numbers?

solution (a) The large number plus the small number equals 72.

(b) The large number minus the small number equals 26.

$$\text{(a)}\quad L + S = 72$$

$$\text{(b)}\quad L - S = 26$$

As we see, the types of problems that can be solved using two or more variables are myriad.

23.3 properties of inequality

With one glaring exception the properties of inequality are the same as the properties of equality.

> ADDITIVE PROPERTY
>
> The same quantity may be added to both sides of a statement of equality or a statement of inequality without altering the truth or falsity of the statement of equality or the statement of inequality.

> MULTIPLICATIVE PROPERTY FOR POSITIVE REAL NUMBERS
>
> Every term of a statement of equality or inequality may be multiplied or divided by the same positive real number without altering the truth or falsity of the statement of equality or the statement of inequality.

The glaring exception occurs when we multiply by a negative real number! The truth of a statement of equality is not altered by multiplying by a negative number.

$$5 = 2 + 3 \quad \text{True}$$

Now multiply every term by -2 and get

$$5(-2) = 2(-2) + 3(-2) \longrightarrow -10 = -4 - 6 \quad \text{Still true!}$$

But the truth of a statement of inequality is altered!

$$8 > 5 \quad \text{True}$$

Now we multiply every term by -2 and get

$$-16 > -10 \quad \text{Now false!}$$

Thus when every term on both sides of an inequality is multiplied by a negative real number, the sense of the inequality symbol must be reversed so that the truth or falsity of the statement of inequality will not be altered.†

$$8 > 5 \quad \text{True!}$$

Now we multiply every term by -2 and reverse the sense of the inequality symbol.

$$-16 < -10 \quad \text{Still true!}$$

example 23.3.1 Graph the solution of $-x \geq 2$, $D = \{\text{Reals}\}$.

solution We solve the given inequality for $+x$ by multiplying both sides by (-1) and **reversing the sense of the inequality symbol.**

$$-x \geq 2 \longrightarrow (-1)(-x) \leq (-1)(2) \longrightarrow x \leq -2$$

Thus we want to graph the solution of $x \leq -2$.

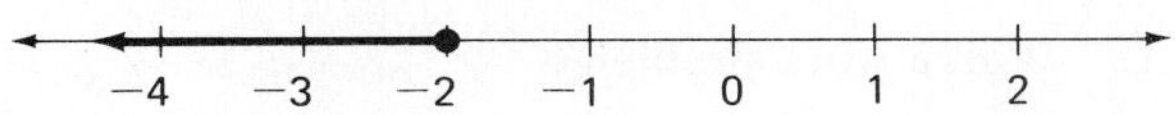

The graph indicates that the number -2 and all real numbers less than -2 satisfy the stated inequality.

example 23.3.2 Graph the solution of $4 - x \leq 6$, $D = \{\text{Integers}\}$.

solution We first isolate $-x$ by adding -4 to both sides. **Note that we do not reverse the symbol when adding a negative quantity.**

$$\begin{array}{r} 4 - x \leq 6 \\ -4 \qquad -4 \\ \hline -x \leq 2 \end{array}$$

Now we multiply both sides by -1 and **reverse the sense of the inequality symbol** to get

$$x \geq -2$$

And if we graph $x \geq -2$ over the integers, we find

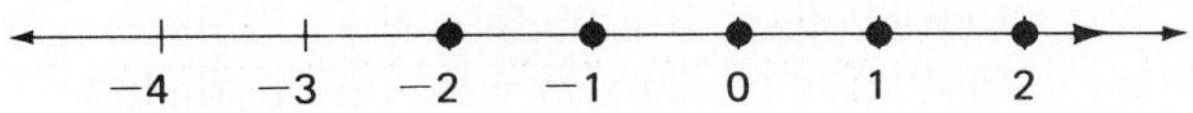

†The sense of a *greater than* symbol is reversed when the symbol is reversed so that it reads *less than* or when a *less than* symbol is reversed so that it reads *greater than.*

example 23.3.3 Graph the solution of $-3x + 4 \le 13$, $D = \{\text{Reals}\}$.

solution We add -4 to both sides to get $-3x \le 9$ and then divide both sides by -3 (or *multiply* both sides by $-\frac{1}{3}$) **and reverse the sense of the inequality symbol** to get

$$x \ge -3$$

and we graph the solution to the inequality to indicate all real numbers that are greater than or equal to -3.

TEST YOUR UNDERSTANDING

(b) Graph the solution of $-2x + 5 > 13$, $D = \{\text{Integers}\}$.

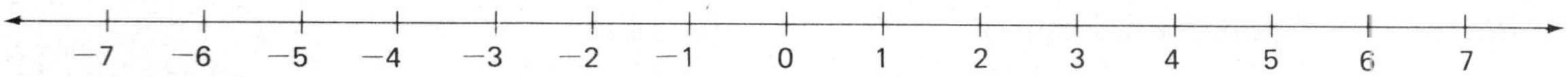

lesson checklist 23

- ☐ Number of equations equals number of variables
- ☐ Value word problems
- ☐ Properties of inequality
- ☐ Reversing the sense of the inequality symbol

ANSWERS TO TEST YOUR UNDERSTANDING

(a) $N_D = 8$, $N_N = 13$

(b)

problem set 23

*1. Jack and Betty have 28 coins that are nickels and dimes. If the value of the coins is $1.95, how many coins of each type do they have?

*2. Airline fares for a flight from Tifton to Adel are $30 for first class and $25 for tourist class. If a flight carried 52 passengers for a total fare of $1360, how many first class passengers were on the flight?

*3. Timpson's dress shop had a big sale. They sold the cheaper dresses for $20 each and sold the more expensive ones for $45 each and took in $1375 on the sale. How many of each kind did they sell if they sold 20 more of the cheaper dresses than the expensive dresses?

*The problems marked with an asterisk were worked out in this lesson.

*4. Toy has $4.45 in quarters and dimes. She has 8 more quarters than dimes. How many coins of each type does she have?

*5. Together Charles and Nelle picked 92 quarts of berries. If Charles picked 6 more quarts than Nelle picked, how many quarts did he pick?

*6. The number of boys in Sarah's class exceeded the number of girls by 7. If there were a total of 29 pupils in the class, how many were boys and how many were girls?

7. Find three consecutive odd integers such that three times the first exceeds the third by 2.

8. On his way to Oxford, Bruce got his pocket picked in the Paris Metro. If he lost $300 and this represented 96 percent of his total worth at that time, what was he worth?

Graph on a number line:

*9. $4 - x \le 6$, $D = \{\text{Integers}\}$

*10. $-3x + 4 \le 13$, $D = \{\text{Reals}\}$

Simplify:

11. $2\sqrt{2}\,(3\sqrt{2} - 4\sqrt{3})$

12. $3\sqrt{3}\,(2\sqrt{3} + 3\sqrt{2})$

13. $5\sqrt{7}\,(2\sqrt{7} - 3\sqrt{2})$

14. $3\sqrt{5}\,(2\sqrt{3} - 3\sqrt{5})$

15. $3\sqrt{50} - \sqrt{72}$

16. $2\sqrt{54} - 2\sqrt{125}$

Divide

17. $x^3 - 2x^2 + x - 5$ by $x + 4$

18. $x^3 - 5$ by $x - 4$

Solve:

19. $-x^2 = -36$

20. $x^2 + 7x + 12 = 0$

21. $-x^2 = 4x - 12$

22. $x^2 + 21 = -10x$

23. $R_T T_T = R_B T_B$, $R_T = 4$, $R_B = 12$, $T_T = T_B + 4$, find T_B and T_T.

Simplify:

24. $$\frac{(.0035 \times 10^{-4})(200 \times 10^{6})}{(700 \times 10^{5})(.00005)}$$

25. Find the equation of the line.

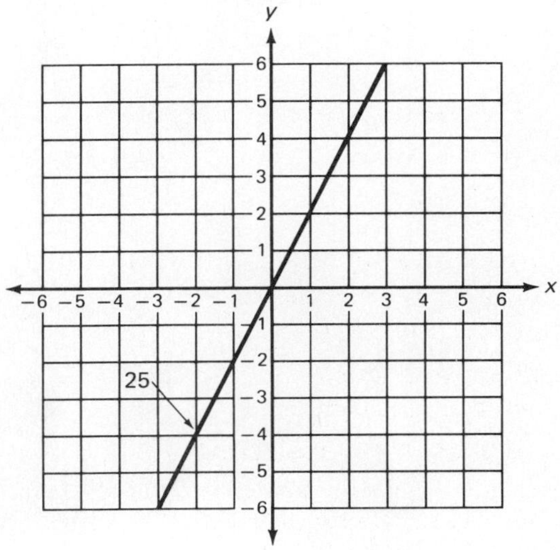

26. Find the equation of the line

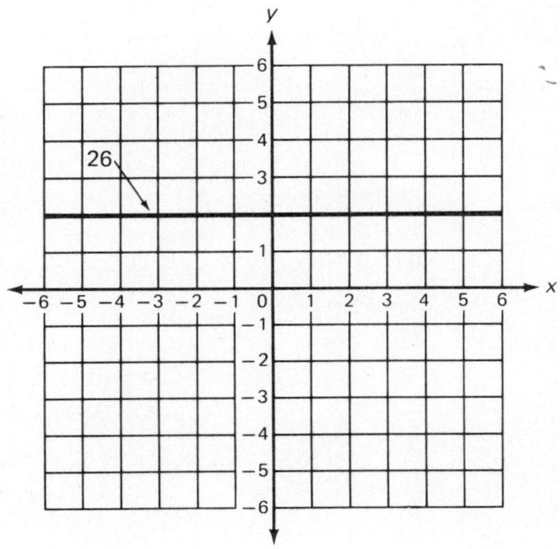

27. The set of whole numbers is closed for what operations?

Solve:

28. $\dfrac{3x - 2}{3} + 4 = \dfrac{1}{6}$

Graph on a rectangular coordinate system:

29. $y - x - 3 = 0$

Add:

30. $\dfrac{y + a}{x^2(m + c)} + \dfrac{b}{m + c}$

31. $\dfrac{a^2xy}{m} - \dfrac{3m^{-1}ay}{a^{-1}} + 2xm^{-1}ya^2$

Multiply:

32. $\dfrac{a^{-4}b}{x}\left(\dfrac{x}{a^{-4}b} + \dfrac{a^3b^{-1}}{x^2}\right)$

33. $ab^2\left(a^{-1}b^2 + \dfrac{4a}{b^2}\right)$

Simplify:

34. $\dfrac{(k^0L^{-2})^{-3}m^2}{(k^2)^{-3}L^6}$

35. $\dfrac{(a^2b^{-1})^2c^{-2}}{a^{-3}c^5b^2}$

36. $|-23 - 5|$

Solve:

37. $3(2x - 6) = 2[2(3 - x) - 5]$

38. $4m + 2(-4 - m) = 3m(-2) + (-4)$

Evaluate:

39. $-x(a - c)$ if $a = 2$, $x = -3$, $c = -4$

40. $x + a(-m)$ if $a = 3$, $x = 5$, $m = -4$

Simplify:

41. $(-2)^{-2}$

42. $\dfrac{1}{2^{-2}}$

43. $\dfrac{1}{(-2)^{-2}}$

44. True or false: {Whole} $\in$ {Integers}

45. $2\sqrt{3} \in$ {what sets of numbers?}

lesson twenty-four

Properties of real numbers
Properties of a field
Products of rational expressions
Quotients of rational expressions

24.1 the properties of real numbers

In this book we restrict our investigation of numbers to the set of **REAL NUMBERS**. We multiply, divide, add, and subtract real numbers. We take the roots of real numbers and raise real numbers to powers. We use real numbers as coefficients of variables in expressions, equations, and inequalities. We find the real numbers that make particular equations and inequalities true statements and say that these real numbers satisfy the equations and inequalities and thus are said to be solutions of or roots of the equations or inequalities.

In advanced mathematics courses, we will be introduced to and work with numbers that are not called real numbers. We will work with numbers that are called imaginary numbers and we will work with numbers that are called complex numbers. We will also learn to work with arrays of numbers that are called matrices. We will find that while matrices consist of more than one number, they can often be treated as if they were a single number in that they can be added, subtracted, and multiplied.

When we add, subtract, multiply, and divide matrices, real numbers, imaginary numbers, and complex numbers, we find that the same rules do not necessarily apply to all. We find that each set has its own peculiarities or properties when its members are used as elements in a particular operation. One of the tasks of the mathematician is to discover the peculiarities or properties of the various sets of numbers and to name, define, and classify these peculiarities or properties. In this lesson we will review the properties of the set of real numbers that we have already discussed and also introduce several new properties. We will begin by discussing the word operation, which of course is not a property.

24.2 binary operations

The four basic operations of algebra are addition, subtraction, multiplication, and division. It is interesting to note that each of these operations is a **BINARY OPERATION**† in that **only two numbers can be used in the operation at one time**. For instance,

(a) $10 + 3 + 2$ (b) $10 \cdot 3 \cdot 2$

Neither of the expressions above can be simplified in **ONE STEP**. In (a) we can compute the sum of the three numbers by adding 10 and 3 to get 13 and then adding 2 to get the final result of 15. This calculation could not be performed in one step because addition is a binary operation. The notation in (b) can be used to demonstrate that multiplication is also a binary operation. We can compute the product by multiplying 10 by 3 to get 30 and then multiplying 30 by 2 to get 60. This calculation could not be performed in one step because multiplication is a binary operation. Subtraction and division are also binary operations.

24.3 closure

If we exclude division by zero, the set of real numbers is closed for the operations of addition, subtraction, multiplication, and division. The emphasis that is given to closure may be bewildering to some, for it seems that we are making a mountain out of a mole hill. Maybe we are, but it is necessary to realize that the result of using two real numbers in any of the four basic operations (division by zero excluded) will be a real number.

24.4 properties of zero and one

Now we begin our discussion of properties. We note that the numbers one and zero have some special properties that no other numbers have. Because no other numbers have these properties, we can use the word **unique** to describe them and speak of the **unique properties of the number one and the unique properties of the number zero.**

24.5 multiplicative and divisive properties of one

The product of a particular real number and the number one is the particular real number.

$$4 \cdot 1 = 4$$

The quotient of any particular real number divided by the number one is the particular real number.

$$\frac{4}{1} = 4$$

These properties are not properties of any other real number and because they are important, we give these properties special names.

> MULTIPLICATIVE PROPERTY OF ONE
>
> The product of any real number a and the number 1 is the real number itself.
>
> $$a \cdot 1 = 1 \cdot a = a$$

†Bi from the Latin word *bis* meaning two as in bicycle.

DIVISIVE PROPERTY OF ONE

The quotient of any real number a divided by the number 1 is the real number itself

$$\frac{a}{1} = a$$

24.6 multiplicative and additive properties of zero

The product of any particular real number and zero is zero.

$$4 \cdot 0 = 0$$

The sum of any particular real number and zero is the particular real number.

$$4 + 0 = 4$$

We also give special names to these properties of the number zero.

MULTIPLICATIVE PROPERTY OF ZERO

The product of any real number a and the number zero is zero.

$$a \cdot 0 = 0 \cdot a = 0$$

ADDITIVE PROPERTY OF ZERO

The sum of any real number a and the number zero is the real number itself.

$$a + 0 = 0 + a = a$$

TEST YOUR UNDERSTANDING

Here we demonstrate four properties of the numbers 1 and 0. To the right of each example, name the property being demonstrated.

(a) $0 + 232 = 232$

(b) $1 \times 416 = 416$

(c) $0 \times 17 = 0$

(d) $\frac{416}{1} = 416$

24.7 properties of a field

The set of real numbers has nine special properties that apply when real numbers are used in the operations of multiplication and addition, and we say that any set that has these properties under the closed operations of multiplication and addition constitutes a **FIELD**. There is no special significance attached to the word field. We have to use some name and field is possibly as good a name as any other. It is the word that we use to describe a set that, when used in the operations of addition and multiplication, has these nine properties. The chart below gives the names of the properties and both a numerical and an abstract example of each property.

PROPERTIES OF A FIELD

Addition	*Name*	*Multiplication*
$4 + 0 = 4$	IDENTITY	$4 \cdot 1 = 4$
$a + 0 = a$		$a \cdot 1 = a$
$4 + (-4) = 0$	INVERSE	$4 \cdot \frac{1}{4} = 1$
$a + (-a) = 0$		$a \cdot \frac{1}{a} = 1$
$4 + (3 + 2) = (4 + 3) + 2$	ASSOCIATIVE PROPERTY	$(4 \cdot 3) \cdot 2 = 4 \cdot (3 \cdot 2)$
$a + (b + c) = (a + b) + c$		$(a \cdot b) \cdot c = a \cdot (b \cdot c)$
$4 + 2 = 2 + 4$	COMMUTATIVE PROPERTY	$4 \cdot 2 = 2 \cdot 4$
$a + b = b + a$		$ab = ba$
	DISTRIBUTIVE PROPERTY	
	$4(3 + 2) = 4 \cdot 3 + 4 \cdot 2$	
	$a(b + c) = ab + ac$	

We note that there are four properties listed under addition and that the same four properties are listed under multiplication. These are the **IDENTITY**, the **INVERSE**, the **ASSOCIATIVE PROPERTY**, and the **COMMUTATIVE PROPERTY**. At the bottom of the chart we show the **DISTRIBUTIVE PROPERTY** in the middle of the page as the **distributive property includes both addition and multiplication.**

24.8 identity

In the chart above the first entry in the middle column is the word identity. To the left and right of this word are examples under the column labeled addition and under the column labeled multiplication.

To have a mathematical system all that is necessary is a set of numbers to work with and an operation to use under which the particular set is closed. But to have a useful system, it is necessary to be able to perform an operation on a particular number and have the result equal the particular number. We begin with a number, then perform the operation with a special number and have the result **IDENTICALLY** equal the number with which we began. We can add the number zero to any particular real number and the sum is **identically** the particular real number. We can multiply any particular real number by the number 1 and the product is identically the particular real number. **Thus**

we call the numbers zero and one the identities for the operations of addition and multiplication, respectively.

IDENTITY

The number zero is the additive identity because:	The number one is the multiplicative identity because:

For any real number a

$$a + 0 = a \qquad\qquad a \cdot 1 = a$$

We note that the numbers 0 and 1 are the identities for all real numbers under the operations of addition and multiplication, respectively.

24.9 inverse

In Figure 24.1 the second entry in the middle column is the word **inverse**. To the left and right of this word are examples under the column labeled addition and under the column labeled multiplication.

Another feature of a useful system is the ability to perform an operation and have the result equal the identity for that operation. We give the name **inverse** to a number that when used with a particular number in an operation yields a result that is the **identity** for that operation. We note that if we add to a particular number the *opposite of* the number, the sum is zero, the identity for addition.

$$4 + (-4) = 0 \qquad -3 + (3) = 0 \qquad -\tfrac{1}{5} + \left(\tfrac{1}{5}\right) = 0$$

Thus every number has its own unique additive inverse under the operation of addition. **Further, we see that another name for the opposite of a number is the additive inverse of the number.**

We note that the product of a particular number and the **reciprocal** of the number is the number one, the **identity** for multiplication.

$$-5 \cdot \left(\frac{1}{-5}\right) = 1 \qquad 13 \cdot \left(\frac{1}{13}\right) = 1 \qquad -\frac{1}{14}(-14) = 1$$

Thus every number has its own unique **MULTIPLICATIVE INVERSE** under the operation of multiplication, and we see that another name for the **reciprocal** of a number is the **multiplicative inverse** of the number.

INVERSE

For any real number a the additive inverse is $-a$	For any non-zero real number a, the multiplicative inverse is $\frac{1}{a}$

because

$$a + (-a) = 0 \qquad\qquad a \cdot \left(\frac{1}{a}\right) = 1$$

It is important to remember that the numbers zero and one are the additive and multiplicative identities, respectively, for all real numbers but that every real number has its own unique additive and multiplicative inverse.

24.10 associative property

The third entry in Figure 24.1 deals with the **ASSOCIATIVE PROPERTY**. If we are asked to compute the sum of

$$4 + 3 + 2$$

we get the same result if we **ASSOCIATE** the first two numbers together for addition

$$(4 + 3) + 2$$

$$7 + 2$$

$$9$$

or if we **ASSOCIATE** the second two numbers together.

$$4 + (3 + 2)$$

$$4 + (5)$$

$$9$$

This property of real numbers is called the associative property of real numbers under addition.

Real numbers can be **associated** in the same manner when computing a product.

$4 \cdot 3 \cdot 2$	$4 \cdot 3 \cdot 2$
$(4 \cdot 3) \cdot 2$	$4 \cdot (3 \cdot 2)$
$12 \cdot 2$	$4 \cdot 6$
24	24

We obtain the same result whether we **associate** the first two numbers as shown on the left or whether we **associate** the second two numbers as shown on the right. This property of real numbers is called the associative property of real numbers under multiplication.

> ASSOCIATIVE PROPERTY
>
> For any three real numbers a, b, and c,
>
> $$a + (b + c) = (a + b) + c \quad \text{and} \quad a(bc) = (ab)c$$

24.11 commutative property

An operation performed with members of a particular set is said to be **COMMUTATIVE** if the order in which the operation is performed does not affect the result. Since the order of addition or multiplication of two real numbers does not affect the result, we can say that addition and multiplication of real numbers are **commutative** operations.

$$4 + 2 = 6 \qquad 2 + 4 = 6 \qquad 4 \cdot 2 = 8 \qquad 2 \cdot 4 = 8$$

> COMMUTATIVE PROPERTY
>
> For any two real numbers a and b,
>
> $$a + b = b + a \quad \text{and} \quad ab = ba$$

24.12 distributive property

Lastly we define the **DISTRIBUTIVE PROPERTY** of real numbers. This property of real numbers permits multiplication to be distributed over addition as shown here.

$$4(3 + 2) = 4 \cdot 3 + 4 \cdot 2$$

> DISTRIBUTIVE PROPERTY
>
> For any three real numbers a, b, and c,
>
> $$a(b + c) = ab + ac$$

We note that while there are statements of the identity, inverse, associative property, and commutative property for both multiplication and addition, there is only one statement of the distributive property and that this single statement includes both multiplication and addition.

> TEST YOUR UNDERSTANDING
>
> Here we give examples of four of the nine properties of a field. To the right of each example, write the name of the field property demonstrated.
>
> (e) $4 + (3 + 5) = (4 + 3) + 5$
>
> (f) $4 \cdot 7 = 7 \cdot 4$
>
> (g) $7(3 + 2) = 7 \cdot 3 + 7 \cdot 2$
>
> (h) $5 + (-5) = 0$

24.13 products of rational expressions

Fractions are multiplied by multiplying the numerators to form the numerator of the product and multiplying the denominators to find the denominator of the product. Thus to multiply

$$\frac{4x}{5} \cdot \frac{3xa}{y}$$

we multiply the numerators to get $12x^2a$ and multiply the denominators to get $5y$.

$$\frac{4x}{5} \cdot \frac{3xa}{y} = \frac{12x^2a}{5y}$$

Sometimes[†] we encounter indicated products of rational expressions whose simplification is facilitated if the terms are first factored and such canceling as is possible performed before any multiplication is done.

[†]Seldom in real life but often in algebra textbooks. These problems are excellent problems to give the reader practice in factoring and canceling.

example 24.13.1 Simplify $\dfrac{x^2 - 25}{x^2 - 7x} \cdot \dfrac{x^2 + 3x}{x^2 - 2x - 15}$.

solution If we multiply the expressions in their present form, we will get a very complicated expression for a product, and it would be very difficult to simplify this product. If we factor and cancel before we multiply, however, the simplified expression can be obtained quickly and easily.

$$\frac{\cancel{(x-5)}(x+5)}{\cancel{x}(x-7)} \cdot \frac{\cancel{x}\cancel{(x+3)}}{\cancel{(x+3)}\cancel{(x-5)}} = \mathbf{\frac{x+5}{x-7}}$$

example 24.13.2 Simplify $\dfrac{x^2 + x - 6}{x^2 - 4x - 21} \cdot \dfrac{x^2 - 8x + 7}{x^2 - x - 2}$.

solution We factor and cancel factors where possible.

$$\frac{\cancel{(x+3)}\cancel{(x-2)}}{\cancel{(x-7)}\cancel{(x+3)}} \cdot \frac{\cancel{(x-7)}(x-1)}{\cancel{(x-2)}(x+1)} = \mathbf{\frac{x-1}{x+1}}$$

Test Your Understanding

(i) Simplify: $\dfrac{x^2 + 5x + 6}{x^2 - 4} \cdot \dfrac{x^2 - 2x}{x^2 + 5x}$

24.14 quotients of rational expressions

In Lesson 14 we learned to simplify expressions such as

$$\frac{\frac{a}{b}}{\frac{c}{d}}$$

by using the **denominator numerator same quantity theorem** to justify multiplying both the denominator and the numerator by $\dfrac{d}{c}$, which is the reciprocal of the denominator.

$$\frac{\frac{a}{b}}{\frac{c}{d}} = \frac{\frac{a}{b} \cdot \frac{d}{c}}{\frac{c}{d} \cdot \frac{d}{c}} = \frac{\frac{ad}{bc}}{1} = \mathbf{\frac{ad}{bc}}$$

If the same division problem had been stated by writing

$$\frac{a}{b} \div \frac{c}{d}$$

we see that the same result can be obtained by inverting the divisor and multiplying

$$\frac{a}{b} \div \frac{c}{d} = \frac{a}{b} \cdot \frac{d}{c} = \mathbf{\frac{ad}{bc}}$$

We can use the same procedure to simplify quotients of more complicated rational expressions.

example 24.14.1 Simplify $\dfrac{x^2 - 2x}{x^2 + 2x - 8} \div \dfrac{x^2 + 5x}{x^2 + 7x + 12}$

solution As the first step we invert the divisor and change the division symbol to a dot that indicates multiplication. Then we factor and cancel as in the two previous examples.

$$\frac{x^2 - 2x}{x^2 + 2x - 8} \cdot \frac{x^2 + 7x + 12}{x^2 + 5x} = \frac{\cancel{x}\cancel{(x - 2)}}{\cancel{(x + 4)}\cancel{(x - 2)}} \cdot \frac{\cancel{(x + 4)}(x + 3)}{\cancel{x}(x + 5)} = \mathbf{\frac{x + 3}{x + 5}}$$

example 24.14.2 Simplify $\dfrac{x^2 - x - 6}{x^2 - 3x - 10} \div \dfrac{x^2 + 5x + 4}{x^2 - x - 20}$.

solution Again as the first step we invert the divisor and indicate multiplication rather than division. Then we factor and cancel.

$$\frac{x^2 - x - 6}{x^2 - 3x - 10} \cdot \frac{x^2 - x - 20}{x^2 + 5x + 4} = \frac{(x - 3)\cancel{(x + 2)}}{\cancel{(x - 5)}\cancel{(x + 2)}} \cdot \frac{\cancel{(x - 5)}\cancel{(x + 4)}}{\cancel{(x + 4)}(x + 1)} = \mathbf{\frac{x - 3}{x + 1}}$$

TEST YOUR UNDERSTANDING

(j) Simplify: $\dfrac{x^2 - 4}{x^2 + 7x + 10} \div \dfrac{x - 2}{x^2 - 25}$

24.15 difference of two squares

In Lesson 16 we introduced the topic of square roots and noted that every positive real number has a positive square root and a negative square root. Because 2 times 2 equals 4 and -2 times -2 also equals 4,

$$(2)(2) = 4 \quad \text{and} \quad (-2)(-2) = 4$$

we say that the two square roots of 4 are $+2$ and -2. **But when we use the radical to indicate the square root of a positive number as**

$$\sqrt{4}$$

we are indicating only the principal or positive square root. If we wish to indicate the negative square root, we must use a minus sign in front of the expression as

$$-\sqrt{4} = -2$$

If we are asked to find the numbers that satisfy the equation

$$x^2 = 4$$

we know that the numbers are $+2$ and -2 because both of these numbers satisfy the equation as we show here.

$$(+2)^2 = 4 \qquad \text{and also} \qquad (-2)^2 = 4$$

The general form of the equation $x^2 = 4$ is

$$p^2 = q^2$$

If we add $-q^2$ to both sides, we can factor

$$p^2 = q^2 \longrightarrow p^2 - q^2 = 0 \longrightarrow (p + q)(p - q) = 0$$

to solve

$$\begin{array}{rrr} \text{if} & p+q= & 0 \\ & -q & -q \\ \hline & p & =-q \end{array} \qquad\qquad \begin{array}{rrr} \text{if} & p-q= & 0 \\ & +q & +q \\ \hline & p & = q \end{array}$$

and use the zero factor theorem to solve. If $p + q = 0$, then $p = -q$ and if $p - q = 0$, then $p = q$.

The theorem that describes the solution to an equation of the form $p^2 = q^2$ is called the **difference of two squares theorem.**

> DIFFERENCE OF TWO SQUARES THEOREM
>
> If p and q are real numbers and if $p^2 = q^2$, then
>
> $$p = q \quad \text{or} \quad p = -q$$

example 24.15.1 Solve $h^2 = 16$.

solution We know that the general equation $p^2 = q^2$ has two solutions, which are $p = q$ and $p = -q$. In the same way the given equation has two solutions, which are

$$\mathbf{h = +4} \quad \text{and} \quad \mathbf{h = -4}$$

example 24.15.2 Solve $p^2 = 41$.

solution To use the form $p^2 = q^2$, it is helpful to write the given equation as

$$p^2 = (\sqrt{41}\,)^2$$

If we do this, we can write the answer as

$$\mathbf{p = \sqrt{41}} \quad \text{or} \quad \mathbf{p = -\sqrt{41}}$$

example 24.15.3 Solve $k^2 = 13$.

solution We omit the intermediate step and write the solution by inspection.

$$\mathbf{k = \sqrt{13}} \quad \text{or} \quad \mathbf{k = -\sqrt{13}}$$

Test Your Understanding

(k) Solve: $p^2 = 39$

lesson checklist 24

- ☐ Binary operations
- ☐ Closure
- ☐ Properties of the numbers zero and one
- ☐ The nine properties of a field
- ☐ Products and quotients of rational expressions
- ☐ Difference of two squares

Answers to Test Your Understanding

(a) Additive property of zero (b) Multiplicative property of one

(c) Multiplicative property of zero

(d) Divisive property of one (e) Associative property of addition

(f) Commutative property of multiplication (g) Distributive property

(h) Additive inverse (i) $\frac{x + 3}{x + 5}$ (j) $x - 5$ (k) $p = \pm\sqrt{39}$

problem set 24

1. Obie has \$3.20 in quarters and dimes. If he has 3 more quarters than he has dimes, how many dimes does he have?
2. The Radio Shop sold some transistor radios for \$7 each and some for \$9 each. If a total of 14 were sold for a total price of \$110, how many of each type were sold?
3. A board 20 meters long is cut into two uneven lengths. If the long piece is 8 meters longer than the short piece, how long are the pieces?
4. Use the letters a, b, and c as required to
   (a) State the associative property for addition.
   (b) State the commutative property for multiplication.
   (c) State the distributive property.
5. Find three consecutive even integers so that twice the sum of the first two exceeds the third by 48.
6. (a) What is the identity for addition?
   (b) What is the identity for multiplication?
   (c) What is the multiplicative inverse of a?
7. Eight percent of a number is 762. What is the number?
8. Find the number that is 14 percent of 1200.

Graph on a number line:

9. $3 - x \nless 0$, $D = \{\text{Integers}\}$

Simplify:

10. $3\sqrt{2}\,(2\sqrt{2} + 3\sqrt{5}\,)$

11. $3\sqrt{2}\,(4\sqrt{3} + 2\sqrt{6}\,)$

12. $3\sqrt{2}\,(\sqrt{6} - 2\sqrt{18}\,)$

13. $3\sqrt{128} - \sqrt{8} + 2\sqrt{18}$

14. $3\sqrt{5} \cdot 2\sqrt{6} \cdot 4\sqrt{2}$

Divide:

15. $x^3 - x^2 + 10$ by $x + 4$

16. $x^4 - 3x + 5$ by $x - 1$

Solve:

17. $25x^2 - 36 = 0$

18. $x^2 = 5x - 6$

19. $x^2 - 11x + 28 = 0$

20. $x^2 - 5x = 50$

21. $R_L T_L = R_S T_S$, $R_L = 120$, $R_S = 280$, $20 - T_L = T_S$; find T_L and T_S

Simplify:

22. $\dfrac{(.0007 \times 10^{-23})(4000 \times 10^6)}{(.00004)(7{,}000{,}000)}$

23. Find the equation of the line.

24. Find the equation of the line.

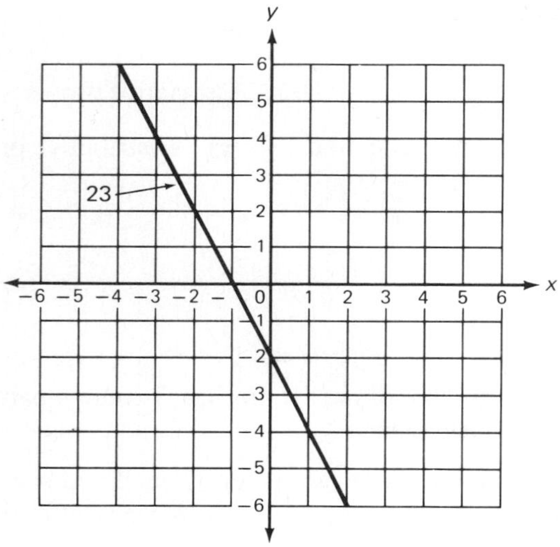

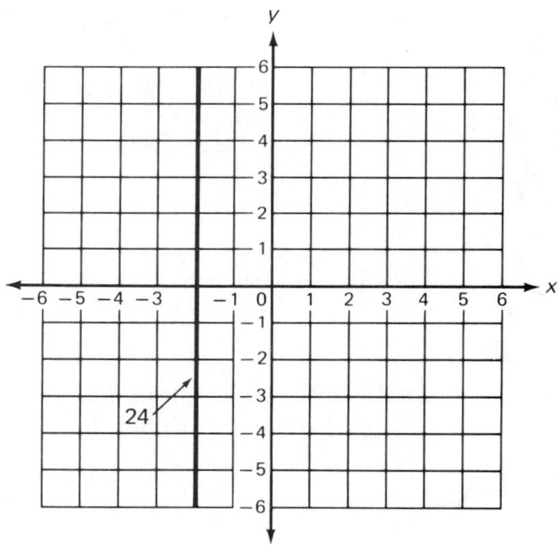

25. The set $\{0, 1\}$ is closed for what operations?

Solve:

26. $\dfrac{2x}{3} + \dfrac{x - 3}{5} = 1$

Graph on a rectangular coordinate system:

27. $y = -3x$

28. $y = -3x + 2$

Add:

29. $\dfrac{a}{x^2} + \dfrac{b}{x^3c} + \dfrac{d}{x^2c^2}$

30. $\dfrac{a}{k + m} + \dfrac{1}{m}$

31. $-4a^{-2}b + \dfrac{3ab}{a^3} - \dfrac{2b}{a^2}$

Multiply:

32. $\dfrac{a^{-2}b}{x}\left(\dfrac{x^2b^{-3}}{a^4x} + \dfrac{2ab}{a^3}\right)$ **33.** $a^4b\left(a^{-4}b^2 + \dfrac{2b^{-1}}{a^{-4}}\right)$

Simplify:

34. $\dfrac{(m^2)^0a^{-2}xp}{(p^{-2})^{-1}m^{-2}}$ **35.** $\dfrac{(a^2)^{-3}x^0}{a^{-4}x^{-2}}$ **36.** $|-5 + 12|$

Solve:

37. $x - 2 + (-4)(x - 2) = 3x + 5$ **38.** $y - 3(-3 - y) = -y[2 - (-3)]$

Evaluate:

39. $c[a - (c)]$ if $c = -3, a = 5$

40. $-[-a(-b) - c]$ if $a = 2, b = -3, c = 4$.

Simplify:

***41.** $\dfrac{x^2 - 25}{x^2 - 7x} \cdot \dfrac{x^2 + 3x}{x^2 - 2x - 15}$ ***42.** $\dfrac{x^2 - x - 6}{x^2 - 3x - 10} \div \dfrac{x^2 + 5x + 4}{x^2 - x - 20}$

43. $\dfrac{1}{-4^{-2}}$

44. True or false: {Whole} $\subset$ {Integers}

45. $\dfrac{53}{3} \in$ {what sets of numbers?}

Solve:

***46.** $p^2 = 41$ ***47.** $k^2 = 13$ **48.** $p^2 = 4$

---

*The problems marked with an asterisk were worked out in this lesson.

lesson twenty-five

Angles, triangles, and Pythagoras
Solution of right triangles
Definitions, axioms, and proofs
More on systems of equations

25.1 angles and triangles

There is no czar of mathematics who is empowered to enforce conformity on authors of mathematics textbooks. Each author is more or less free to define terms and concepts as he deems necessary and appropriate. Thus, while most definitions are constant from author to author, not all are. The definition of an **ANGLE** is a case in point. All authorities agree that four angles are formed by the intersection of two non-parallel straight lines in the same plane.

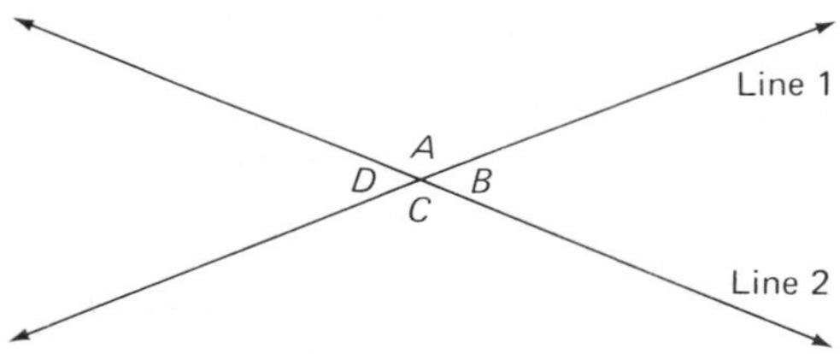

Figure 25.1

In Figure 25.1 we show two intersecting lines and have designated the four angles formed by the letters A, B, C, and D. If we focus our attention on just one of the angles, say angle B, we see that this angle is formed by the parts of line 1 and line 2 that lie on or to the right of the point of intersection of the lines (see Fig. 25.2). Since the mathematical lines are infinite in length and we have divided each line into two parts, we will call the two parts **HALF LINES** or **RAYS**.

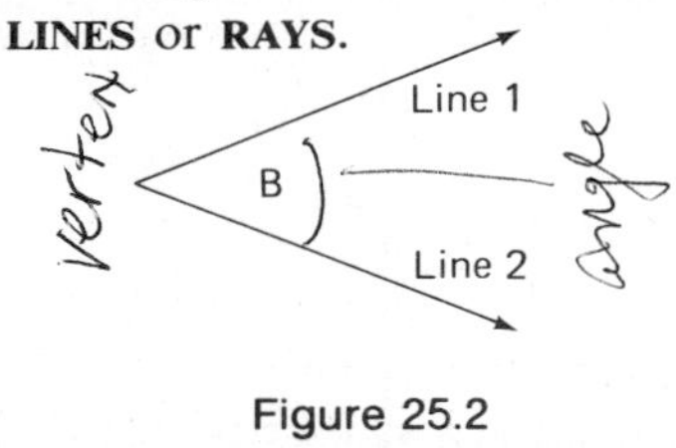

Figure 25.2

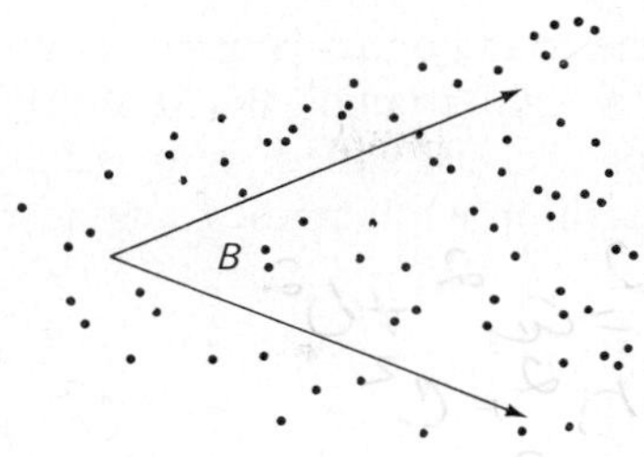

Figure 25.3

The point of contention is whether the angle is the opening between the rays or whether the angle is the rays themselves. We discuss this question in the language of mathematics by speaking of the points in the coordinate plane. If we consider all the points in the plane as in Figure 25.3, we see that we have the points that lie on the two rays, the points that lie *inside* the rays and the points that lie *outside* the rays.

European authors generally define the angle to be the opening between the rays so they say that the angle is the set of points that is bounded by the rays. Many American authors define the angle to be the rays themselves so they say the angle is the set of points that make up the rays. Some say that the two half lines or rays are the sides of the angle but don't say what the angle is. Others don't speak of the opening at all but define an angle to be a rotation of a ray about its end point. No one of these definitions is necessarily more desirable than another. Since a precise definition is not required in this book, we will sidestep the issue and say that

An angle is formed by two half lines or rays that are in the same plane and that have a common end point.

If two straight lines intersect and are perpendicular one to the other, we define the measure of each of the four angles created to be 90 degrees. Instead of writing the word degrees, it is customary to place a small elevated circle after the number that designates the number of degrees. Thus 90 degrees can be written as 90°, and 47 degrees and 135 degrees can be written as 47° and 135°. We show here two intersecting perpendicular lines with the resulting 90° angles (Fig. 25.4).

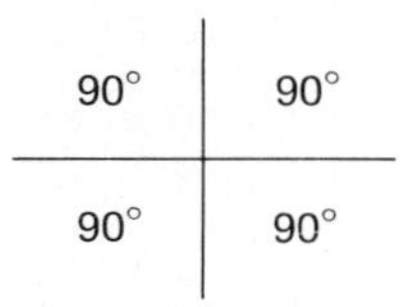

Figure 25.4

With this definition of a 90° angle and two axioms, it can be proved, by using geometry, that **the sum of the interior angles of any triangle is 180°**. We show three triangles and note that the sum of the three interior angles in each triangle is 180° (Fig. 25.5).

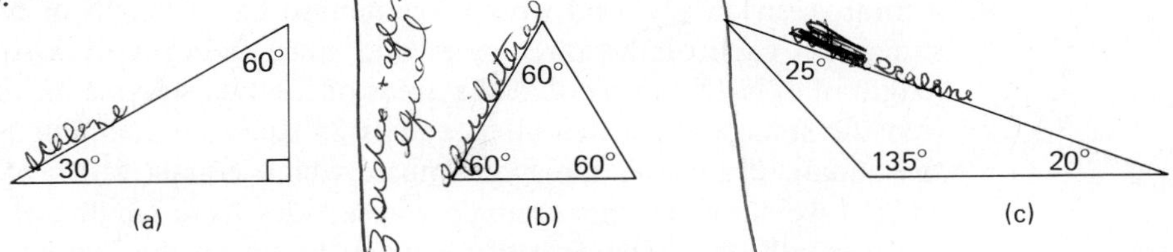

Figure 25.5

The triangle on the left has one angle that has a measure of 90°. It is customary to designate a 90° angle by a small square as shown in the figure. Since another name for a 90° angle is a **RIGHT ANGLE** and this triangle contains a right angle, we call this triangle a **RIGHT TRIANGLE**. Any triangle that contains a right angle is called a right triangle and the side of the triangle that is opposite the right angle is always the longest side. We call this side of a right triangle the **HYPOTENUSE** (see Fig. 25.6). The other two sides are called legs or are just called sides. Right triangles have a special property that makes them very useful in mathematics, engineering, and physics.

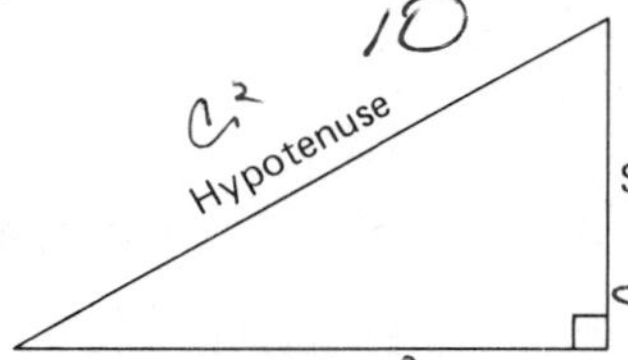

Figure 25.6

25.2 pythagorean theorem

It can be shown that **the square drawn on the hypotenuse of a right triangle has the same area as the sum of the areas of the squares drawn on the other two sides.** While this theorem was known to the Egyptians as early as the Middle Kingdom,† the geometric proof of the theorem is normally attributed to a Greek philosopher and mathematician named **PYTHAGORAS** who was born on the Aegean island of Samos and who was later associated with a school or brotherhood in the town of Crotona on the Italian peninsula in the sixth century B.C. We call the theorem for which he supposedly developed the proof, the **PYTHAGOREAN THEOREM.**

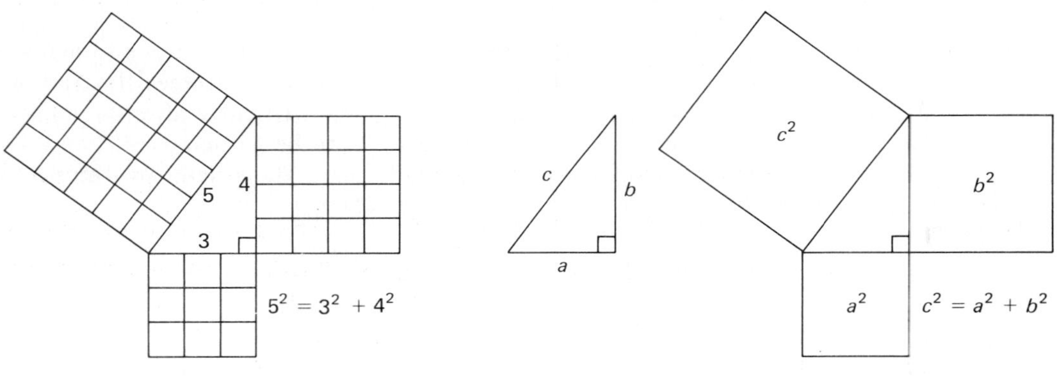

Figure 25.7

In Figure 25.7(a) we have drawn a right triangle whose sides have lengths of 3 and 4 units, respectively, and whose hypotenuse has a length of 5 units. We have drawn a square on each of the three sides and since the area of a square whose sides have a length of L is L^2, we see that the areas of the two squares on the sides are 4^2 and 3^2 and that the sum of these two areas equals 25 square units. This is the same as the area of the square drawn on the hypotenuse, which equals 5^2 or 25 square units. In Figure 25.7(b) we show another triangle whose sides have lengths of a and b and hypotenuse has a length of c. The area of a square drawn on the hypotenuse would be c^2, and the areas of the squares drawn on sides a and b would be a^2 and b^2, respectively, as shown by Figure 25.7(c). We normally label the hypotenuse as c and the other two sides as a and b. Thus the general algebraic expression of the **Pythagorean theorem** is

$$a^2 + b^2 = c^2$$

†Circa 2000 B.C.

where c **is the length of the hypotenuse and** a **and** b **represent the lengths of the other two sides.**

This theorem can be used to find the length of a side of a right triangle if the lengths of the other two sides are known.

example 25.2.1 Given the triangle with the lengths of the sides as shown, use the Pythagorean theorem to find the length of side a.

solution

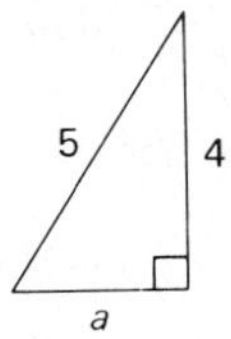

The square on the hypotenuse equals the sum of the squares on the other two sides. Thus,

$$5^2 = 4^2 + a^2 \longrightarrow 25 = 16 + a^2 \longrightarrow 9 = a^2$$

We use the difference of two squares theorem to finish the solution.

$$a^2 = 9 \qquad \text{which leads to} \qquad a = +3 \quad \text{or} \quad a = -3$$

While -3 is a solution to the equation $a^2 = 9$, it is not a solution to the problem at hand because physical lengths are designated by positive numbers. Thus we reject this solution and say that

$$\boldsymbol{a = +3}$$

example 25.2.2 Find the side p in the triangle shown.

solution

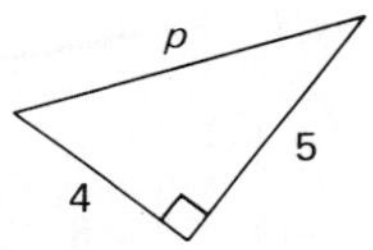

We apply the Pythagorean theorem to this triangle to write

$$p^2 = 5^2 + 4^2$$

Now we simplify and use the difference of two squares theorem.

$$p^2 = 41 \longrightarrow p^2 = (\sqrt{41})^2 \longrightarrow \boldsymbol{p = \sqrt{41}} \quad \text{or} \quad \boldsymbol{p = -\sqrt{41}}$$

But sides of triangles do not have negative lengths so we discard the negative result and say

$$\boldsymbol{p = \sqrt{41}}$$

example 25.2.3 Find side k.

solution

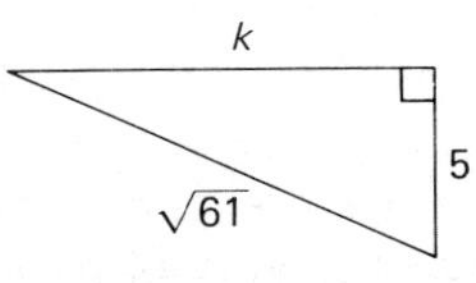

We use the Pythagorean theorem to write

$$(\sqrt{61})^2 = k^2 + 5^2$$

and now we simplify and solve

$$61 = k^2 + 25 \longrightarrow k^2 = 36 \longrightarrow \boldsymbol{k = 6}$$

Now we finish the solution by using the difference of two squares theorem.

$$k^2 = 36 \longrightarrow k^2 = (6)^2$$

so

$$\boldsymbol{k = 6} \quad \text{or} \quad \boldsymbol{k = -6}$$

but since -6 has no meaning as the length of a side of a triangle, we say

$$\boldsymbol{k = 6}$$

example 25.2.4 Find side m.[†]

solution

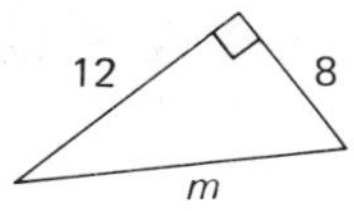

By using the Pythagorean theorem we can write

$$m^2 = 12^2 + 8^2$$

and now we simplify and solve.

$$\longrightarrow \quad m^2 = 208 \quad \longrightarrow \quad m^2 = (\sqrt{208})^2$$
$$\longrightarrow \quad m = \sqrt{208} \quad \longrightarrow \quad m = 4\sqrt{13}$$

example 25.2.5 Find side f.

solution

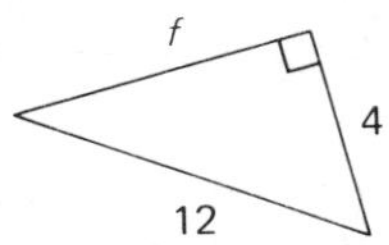

First use the Pythagorean theorem to write

$$12^2 = f^2 + 4^2$$

and now we simplify and solve.

$$\longrightarrow \quad 144 = f^2 + 16 \quad \longrightarrow \quad f^2 = 128$$
$$\longrightarrow \quad f^2 = (\sqrt{128})^2 \quad \longrightarrow \quad f = \sqrt{128}$$
$$\longrightarrow \quad \mathbf{f = 8\sqrt{2}}$$

Test Your Understanding

(a) Find side k:

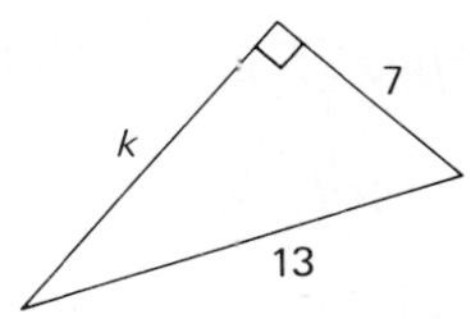

The Pythagorean theorem can also be used to find the distance between two points. We demonstrate this procedure in the following problem.

[†]The Greeks must have drawn some of their right triangles as this one is drawn because the word hypotenuse comes from the Greek words *hupo* meaning *under* and *teinein* meaning *to stretch*, so hypotenuse means *stretched under*.

example 25.2.6 Find the distance between the points whose coordinates are (4, 2) and (−3, −2).

solution The first step is to graph the points as we have done in the figure at the left.

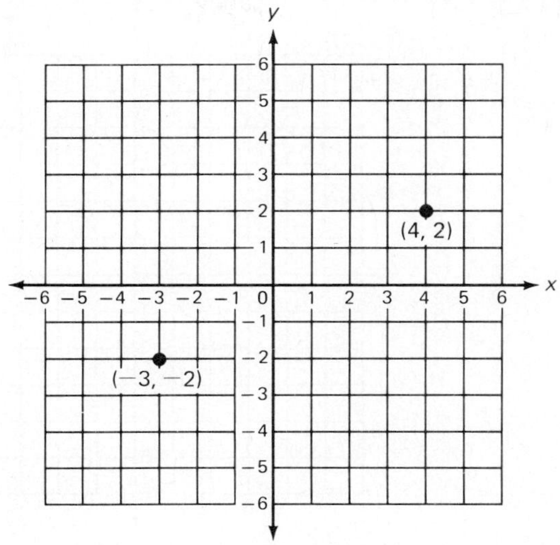

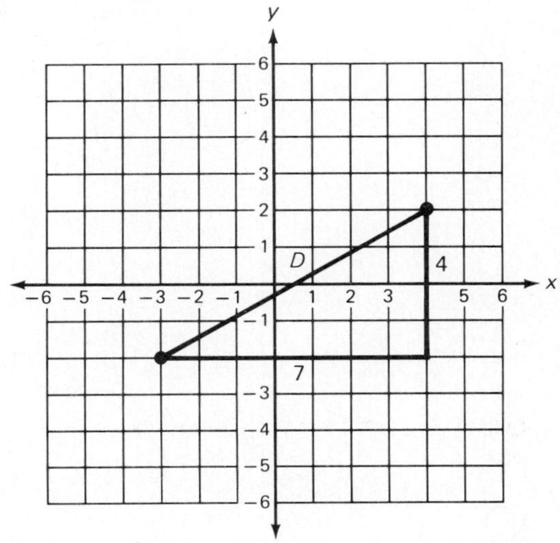

Then we draw a right triangle whose hypotenuse terminates on the points and whose sides are parallel to the coordinate axes as shown in the figure at the right. Since the sides are parallel to the coordinate axes, they are perpendicular to each other and meet in a right angle. The exact lengths of the sides of the triangle are determined by using the given x- and y-coordinates of the points and in this triangle the two sides have lengths of 4 and 7. Now we can use the theorem of Pythagoras to find the length of the hypotenuse, which is the distance between the two points.

$$D^2 = 7^2 + 4^2 \longrightarrow D^2 = 65 \longrightarrow D = \sqrt{65}$$

example 25.2.7 Find the distance between the points (3, −4) and (−5, 2).

solution We graph the points and draw the required triangle as shown in the accompanying figure. The distance between the points is found by using the theorem of Pythagoras as

$$D^2 = 8^2 + 6^2 \longrightarrow D^2 = 100$$

$$\longrightarrow D = \sqrt{100} \longrightarrow \mathbf{D = 10}$$

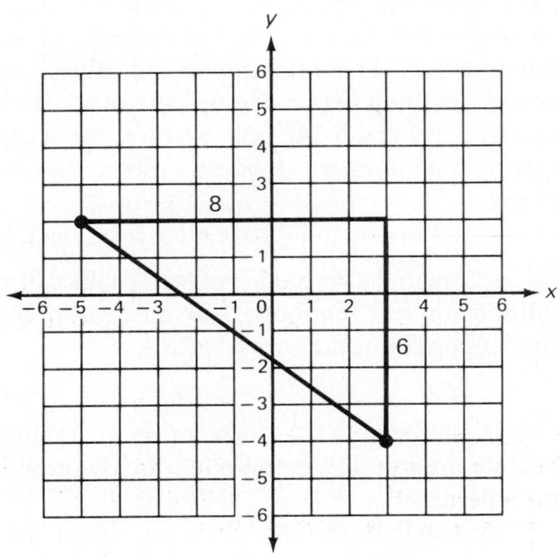

TEST YOUR UNDERSTANDING

(b) Find the distance between the points (4, −4) and (−4, −1)

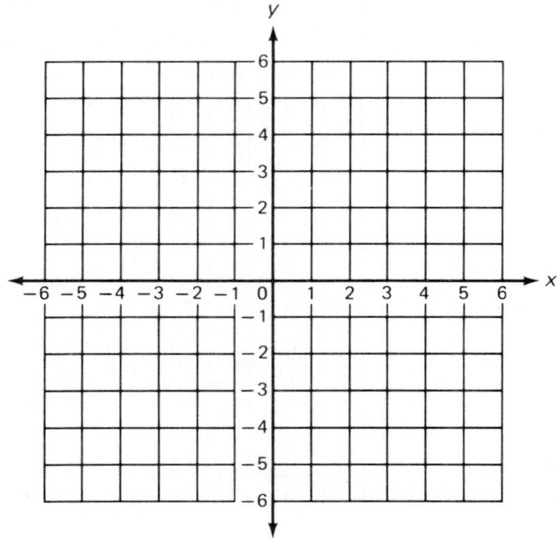

25.3 definitions, axioms, and proofs

The development of algebra rests on a foundation of definitions and axioms. We must first define what we mean when we write certain symbols. In Lesson 5 we defined the notation

$$x^4$$

to mean $x \cdot x \cdot x \cdot x$. It was not necessary to choose the notation x^4. We could just as easily have chosen the notation

$$4_x$$

—but we didn't. Later on we defined the notation

$$xc$$

to indicate the product of x and c. We could have defined this notation to indicate the quotient of x divided by c—but we didn't. Thus we began our development by carefully defining all notations and symbols.

Unfortunately, we cannot build a mathematical system using definitions alone. To the definitions we must add statements of observations that everyone accepts as true without proof. These statements are called **AXIOMS** or **POSTULATES**. The statements that concern real numbers make up part of what we call the properties of real numbers as we discussed in Lesson 24. For instance, we accept without proof that the order of addition of two real numbers does not affect the resultant sum. We state this abstractly by writing

$$a + b = b + a$$

and in Lesson 24 we said that we would call this property the commutative property of addition for real numbers. Also in Lesson 24 we stated the associative property of real numbers under addition by writing

$$(a + b) + c = a + (b + c)$$

In mathematics we take pride in holding our list of definitions and axioms to an absolute minimum. Then we use the definitions and axioms to prove all other statements and assertions.

You will note that when we stated the commutative property we restricted our statement of this property to a statement about only two numbers and we restricted our statement of the associative property to a statement about only three numbers.

In Lesson 1 we said that the order of addition of any number of signed numbers did not affect the resultant sum. This statement cannot be justified by the commutative property alone because the commutative property is applicable only to the order of addition of two numbers. In the following examples we will take these two restricted statements and use them to prove relationships that involve three or more terms or three or more factors.

example 25.3.1 If a, b, and c represent unspecified real numbers, use the associative and commutative properties to show that

$$a + b + c = c + b + a$$

solution We will begin with the expression $a + b + c$ as it appears on the left and use the associative and commutative properties to rearrange the terms so that the resulting expression is $c + b + a$, the same as the right side. We begin by associating b and c by using a set of parentheses.

$$a + (b + c)$$

Next we change the order inside the parentheses as permitted by the commutative property and then remove the parentheses.

$$a + (b + c) = a + (c + b) = a + c + b$$

Now c is the middle term instead of the last term. One more step should do it. We enclose a and c in parentheses, change the order of a and c, and remove the parentheses.

$$(a + c) + b = (c + a) + b = c + a + b$$

Now we have c as the first term but since our goal is $c + b + a$, we must change the order of a and b so we need one more step. As before, we use parentheses, change the order inside the parentheses, and then remove the parentheses.

$$c + (a + b) = c + (b + a) = \mathbf{c + b + a}$$

example 25.3.2 If x, y, and m represent unspecified real numbers, use the commutative and associative properties to show that

$$x + y + m = m + x + y$$

Skip Sec 25.3

solution We will work with the left side and transform it into the same form as the right side.

$x + y + m$	given
$x + (y + m)$	associative property
$x + (m + y)$	commutative property
$(x + m) + y$	associative property
$(m + x) + y$	commutative property
$m + x + y$	removed parentheses

example 25.3.3 If a, b, c, and d represent unspecified real numbers, show that

$$a + b + c + d = b + a + d + c$$

solution The expression on the right is the same as that on the left except that the order of the first two symbols and the last two symbols is reversed. We need only four steps.

$a + b + c + d$	given
$(a + b) + (c + d)$	associative property
$(b + a) + (d + c)$	commutative property (twice)
$b + a + d + c$	removed parentheses

In Lesson 2 we stated that the order in which real numbers are multiplied does not affect the value of the product. In the following examples we will use the associative and commutative properties of multiplication to prove this assertion for some expressions that have a finite number of factors.

example 25.3.4 If a, b, c, and d represent unspecified real numbers, show that

$$abcd = bdac$$

solution This one looks more difficult. Let's go about it step-by-step and feel our way along. We will transform the expression on the left into the same form as the expression that we have on the right. On the right the first factor is b, so

$abcd$	given
$(ab)cd$	associative property
$(ba)cd$	commutative property
$bacd$	removed parentheses

Now we have moved b to the front; on the right above we note that c is the last factor, so

$bacd$	from above
$ba(cd)$	associative property
$ba(dc)$	commutative property
$badc$	removed parentheses

We wanted to end up with $bdac$. Thus far, we have moved b to the front and c to the end, and all that is amiss is the order of the two middle factors.

$badc$	from above
$b(ad)c$	associative property
$b(da)c$	commutative property
$bdac$	removed parentheses

As you see, these examples are almost the same as those involving addition. In these we use the properties of associativity and commutativity for multiplication instead of the same properties for addition.

example 25.3.5 If P, Q, and L are unspecified real numbers, show that $PQL = LPQ$.

solution We want to prove that $PQL = LPQ$. We will begin with PQL and change this to LPQ

PQL	given
$P(QL)$	associative
$P(LQ)$	commutative
$(PL)Q$	associative
$(LP)Q$	commutative
LPQ	removed parentheses

Test Your Understanding

(c) Use the associative and commutative properties of addition to show that $x + y + m = x + m + y$.

25.4 more on systems of equations

When we solve systems of equations using the substitution method, sometimes it is necessary to rearrange one of the equations before the necessary substitutions can be accomplished. In the example problem shown here, we will have to rearrange the equation that relates the time variables before we can substitute.

example 25.4.1 $R_E T_E = R_W T_W$, $R_E = 200$, $R_W = 250$, $T_E + T_W = 9$; find T_E and T_W.

solution We begin by replacing R_E with 200 and R_W with 250 to find

$$200T_E = 250T_W$$

Now we must use the equation $T_E + T_W = 9$ to substitute for T_E or for T_W. The equation cannot be used in its present form and must be solved for T_E by adding $-T_W$ to both sides or be solved for T_W by adding $-T_E$ to both sides. We work the problem twice and use both procedures to show that the final results will be the same.

$$\begin{array}{rcl} T_E + T_W &=& 9 \\ -T_W && -T_W \\ \hline T_E &=& 9 - T_W \end{array}$$

Now we will substitute $9 - T_W$ for T_E and solve the resulting equation for T_W:

$$\begin{array}{rcl} 200(9 - T_W) &=& 250T_W \\ 1800 - 200T_W &=& 250T_W \\ +200T_W && +200T_W \\ \hline 1800 &=& 450T_W \end{array}$$

so $T_W = \dfrac{1800}{450}$ or $\boldsymbol{T_W = 4}$.

$$\begin{array}{rcl} T_E + T_W &=& 9 \\ -T_E && -T_E \\ \hline T_W &=& 9 - T_E \end{array}$$

Now we will substitute $9 - T_E$ for T_W and solve the resulting equation for T_E:

$$\begin{array}{rcl} 200T_E &=& 250(9 - T_E) \\ 200T_E &=& 2250 - 250T_E \\ 250T_E && +250T_E \\ \hline 450T_E &=& 2250 \end{array}$$

so $T_E = \dfrac{2250}{450}$ or $\boldsymbol{T_E = 5}$.

Since $T_E + T_W = 9$, we can solve for T_E by replacing T_W with 4:

$$\begin{array}{rcr} T_E + 4 & = & 9 \\ -4 & & -4 \\ \hline T_E & = & \mathbf{5} \end{array}$$

Since $T_E + T_W = 9$, we can solve for T_W by replacing T_E with 5.

$$\begin{array}{rcr} 5 + T_W & = & 9 \\ -5 & & -5 \\ \hline T_W & = & \mathbf{4} \end{array}$$

Thus we see that while one procedure leads to our solving for T_E first and the other leads to our solving for T_W first, both procedures finally yield the same answers.

lesson checklist 25

- ☐ Half lines or rays
- ☐ Definition of an angle
- ☐ Definition of a right angle
- ☐ Hypotenuse
- ☐ Pythagorean theorem
- ☐ Distance between two points
- ☐ Use of associative and commutative properties in proofs

ANSWERS TO TEST YOUR UNDERSTANDING

(a) $2\sqrt{30}$ (b) $\sqrt{73}$ (c)

$x + y + m$	given
$x + (y + m)$	associative property
$x + (m + y)$	commutative property
$x + m + y$	removed parentheses

problem set 25

1. The Girl's Club collects 50 dimes and quarters that have a total value of \$11.00. How many quarters did they collect?
2. Frank and Mary cut the rope into two pieces. If the rope was originally 535 meters long and the short piece is 15 meters shorter than the other piece, what are the lengths of the two pieces?
3. Football tickets cost 1 dollar for adults and 50 cents for children. Thirteen hundred people paid 1000 dollars for tickets. How many adults bought tickets?
4. Laura and Lois bought 40 stamps. Some of the stamps cost 13 cents each and the others cost 20 cents each. If the total cost of the stamps was \$6.60, how many 20-cent stamps did they buy?
5. Seventeen percent of the cost of the house was \$6800. What was the full cost of the house?

6. James had to make a down payment of 20 percent on his new car. If the cost of the car was \$6400, what was the amount of his down payment?

7. Find three consecutive odd integers such that 10 times the first exceeds 4 times the third by 2.

8. Find three consecutive integers such that three times the first exceeds twice the third by 17.

Use the Pythagorean theorem to find the lengths of the unknown sides of the triangle.

9.
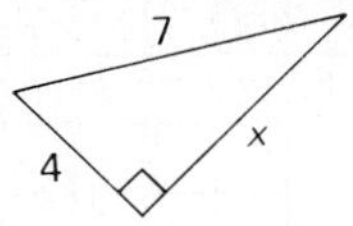

10.
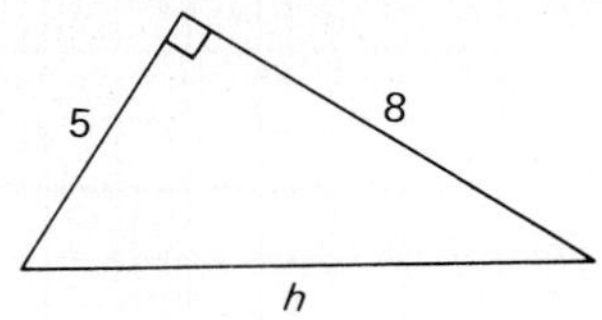

*11. Find the distance between (4, 2) and (−3, −2).

12. Use the associative and commutative properties to show that $xyz = yxz$.

13. Use the associative and commutative properties to show that $x + y + z = z + y + x$.

14. Identify the special properties of the numbers zero and one.

15. List the nine properties of a field and use letters or numbers to demonstrate each one.

16. Which field property pertains to both multiplication and addition?

17. What is another name for the multiplicative inverse of a number?

18. What is another name for the additive inverse of a number?

19. Use the letters a and b to demonstrate the commutative property of real numbers under addition.

20. Use the numerals 4, 7, and 8 to demonstrate the associative property of real numbers under multiplication.

Simplify:

21. $3\sqrt{5}(2\sqrt{5} - 4)$ 22. $3\sqrt{21}\sqrt{7}\sqrt{8}$ 23. $3\sqrt{180} + \sqrt{135}$

Divide:

24. $x^4 - x^2 + 2x + 3$ by $x - 2$

Solve:

25. $9x^2 - 49 = 0$ 26. $x^2 - 5x = -4$ 27. $x^2 - 35 = -2x$

*28. $R_E T_E = R_W T_W$, $R_E = 200$, $R_W = 250$, $T_E + T_W = 9$; find T_E, T_W

Simplify:

29. $$\frac{(.00077 \times 10^{-3})(40 \times 10^{6})}{(.00011 \times 10^{5})(140{,}000)}$$

Use elimination:

30. $$\begin{cases} 2x + 5y = -10 \\ x - 3y = 17 \end{cases}$$

---

*The problems marked with an asterisk were worked out in this lesson.

31. Find the equation of the line:

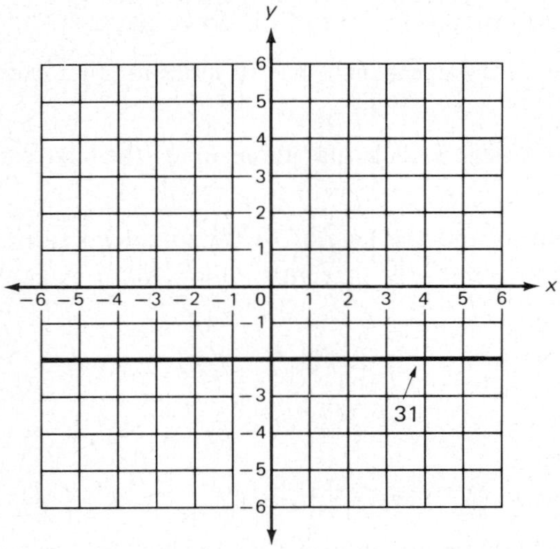

32. Find the equation of the line:

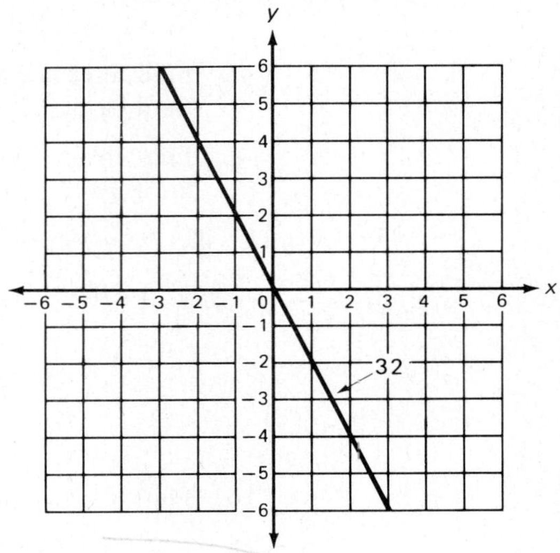

33. The set of negative real numbers is closed for what operations?

Solve:

34. $\dfrac{5x - 2}{6} = \dfrac{1}{5}$

Graph on a rectangular coordinate system:

35. $x = -3$

36. $y + 2x - 5 = 0$

Simplify:

37. $\dfrac{\dfrac{x}{a + b}}{\dfrac{1}{b}}$

38. $\dfrac{a + \dfrac{a}{b}}{x - \dfrac{1}{b}}$

Add:

39. $\dfrac{m}{a^2} + \dfrac{b}{a(c + d)}$

40. $\dfrac{k}{b^4} + \dfrac{L}{b + c}$

Multiply:

41. $2x^2(x^{-4}a + a^{-2}x)$

Simplify:

42. $\dfrac{x^2 - 2x}{x^2 + 2x - 8} \div \dfrac{x^2 + 5x}{x^2 + 7x + 12}$

43. $\dfrac{(p^{-4}y)^{-2}y^4}{(p^0y)^2p^4}$

44. $\dfrac{mx^2ym^{-4}(y^0)^2}{m^{-3}(x^2)^0y^{-2}}$

Solve:

45. $4(x - 2) + 2x - 4 = x + 6(-x + 3)$

Evaluate:

46. $-a[a - (-b)] - b$ if $a = -2, b = 4$

47. $\dfrac{-1}{-3^{-2}}$

48. True or false: {Rationals} $\in$ {Reals}

49. $0 \in$ {what sets of numbers?}

50. Graph on a number line $4 - x \not> 5$, $D =$ {Integers}

lesson twenty-six

Roots of squares of prime numbers
Significant digits • rounding off
Square root tables

26.1 simplifying square roots

We have learned to simplify square roots of positive integers by first writing the integer as a product of prime numbers and then using the product of square roots theorem as we show here.

$$\sqrt{50} = \sqrt{5 \cdot 5 \cdot 2} = \sqrt{5}\,\sqrt{5}\,\sqrt{2} = \mathbf{5\sqrt{2}}$$

A similar but slightly different thought process makes the simplification of some of these expressions somewhat easier. Instead of expressing the integer as a product of prime numbers, we express it as a product of a prime number and the squares of prime numbers. For example, in the problem above, as the first step we would write 50 as the product of 2 and 25.

$$\sqrt{50} = \sqrt{25 \cdot 2}$$

and then use the product of square roots theorem to complete the simplification.

$$\sqrt{25 \cdot 2} = \sqrt{25}\,\sqrt{2} = \mathbf{5\sqrt{2}}$$

This thought process is especially helpful when the integer is a very large number and a large perfect square factor of this number is easily identifiable. The square root of 50,000 can be quickly simplified when we realize that 50,000 can be written as the product of 5 and 10,000 and, of course, 10,000 is equal to the square of 100.

$$\sqrt{50{,}000} = \sqrt{10{,}000 \cdot 5} = \sqrt{10{,}000}\,\sqrt{5} = \mathbf{100\sqrt{5}}$$

example 26.1.1 Simplify $\sqrt{360{,}000}$.

solution 360,000 has two perfect square factors, 36 and 10,000.

$$\sqrt{360{,}000} = \sqrt{36 \cdot 10{,}000} = \sqrt{36}\,\sqrt{10{,}000} = 6 \cdot 100 = \mathbf{600}$$

example 26.1.2 Simplify $\sqrt{7500}$.

solution 7500 has two perfect square factors, 25 and 100.

$$\sqrt{7500} = \sqrt{3 \cdot 25 \cdot 100} = \sqrt{3}\,\sqrt{25}\,\sqrt{100} = \sqrt{3} \cdot 5 \cdot 10 = \mathbf{50\sqrt{3}}$$

example 26.1.3 Simplify $\sqrt{16{,}200}$.

solution The perfect square factors of 16,200 are 81 and 100.

$$\sqrt{16{,}200} = \sqrt{2 \cdot 81 \cdot 100} = \sqrt{2}\,\sqrt{81}\,\sqrt{100} = \sqrt{2} \cdot 9 \cdot 10 = \mathbf{90\sqrt{2}}$$

TEST YOUR UNDERSTANDING

Look for perfect squares in the radicand to simplify the following:

(a) $\sqrt{300}$ (b) $\sqrt{2000}$

26.2 significant digits

A digit is any one of the 10 Arabic numerals 0, 1, 2, 3, 4, 5, 6, 7, 8, 9. When we represent a number, we do it by using one or more of these digits.

In physics and engineering problems the data that we work with are obtained by measurement or observation. The accuracy of the data depend to a large extent on the mechanical limitations of the instrument used to take the measurement and are, in reality, approximations.

If we use a ruler to measure a particular length and we say that the length is

1.42 inches

we have given an estimation of the length that is correct to 3 **significant digits**. We have stated that the length is somewhere between 1.425 and 1.415 inches, which is the same thing as saying that it is closer to 1.42 than it is to 1.41 or to 1.43 as shown in Figure 26.1.

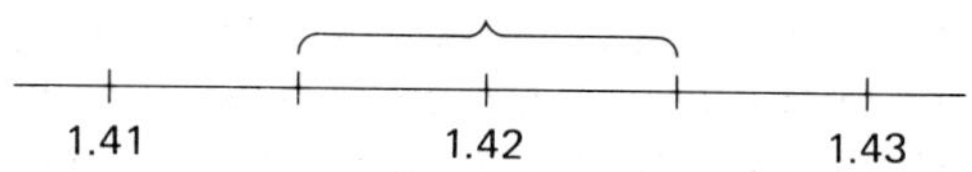

Figure 26.1

If we use a more accurate means of measuring the same length, we could arrive at an estimate that is correct to more significant digits. If the new measurement is

1.4187 inches

we are now saying that the length is closer to 1.4187 than it is to 1.4186 or to 1.4188 as shown in Figure 26.2.

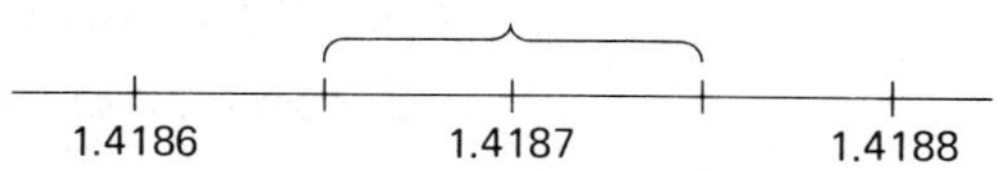

Figure 26.2

This new estimation of the length of 1.4187 inches is correct to 5 significant digits and is more exact than our original estimate, 1.42 inches, which is correct to only 3 significant digits.

Zeros that appear after the decimal and before the first non-zero digit are not considered significant digits.

.0132	3 significant digits
.00132	3 significant digits

These zeros are significant if preceded by a non-zero digit.

4.0132	5 significant digits
400.0132	7 significant digits
4060.0132	8 significant digits

And zeros that appear after the decimal and after the first non-zero digit are considered to be significant digits.

.132	3 significant digits
.1320	4 significant digits
.13200	5 significant digits
10.13200	7 significant digits

In numbers that have terminal zeros before the decimal, such as 180,000, the number of significant digits cannot be determined without additional information that states the accuracy of the measurement.

26.3 rounding off

It is often desirable to express a number with fewer digits than were originally used. This process is called **rounding off** the number and these are the rules that are often used.

1. If the part to be dropped amounts to less than 5 units in the first discarded place, the last retained digit will not be changed. Thus 41,438 rounded off to 3 significant digits is 41,400.
2. If the part to be dropped amounts to more than 5 units in the first discarded place, increase the last retained digit by one. Thus 86,273 rounded off to 3 significant digits is 86,300.
3. If the digit to be dropped is a terminal 5 or 5 followed by zeros and the digit just to the left of the 5 is even, make no change in the digit. Thus 4165 rounded off to 3 significant digits is 4160 and 416,500 rounded off to 3 significant digits is 416,000.
4. If the digit to be dropped is a terminal 5 or 5 followed by zeros and the digit just to the left of the 5 is odd, write this digit as the next higher even digit. Thus 4175 rounded off to 3 significant digits is 4180, and 41,750 rounded off to 3 significant digits is 41,800.

example 26.3.1 Round off 4.17654748 to 8, 7, 6, 5, 4, and 3 significant digits.

solution

8 significant digits	4.1765475
7 significant digits	4.176547
6 significant digits	4.17655
5 significant digits	4.1765

4 significant digits	4.177
3 significant digits	4.18

TEST YOUR UNDERSTANDING

Round off the following numbers as indicated:

(c) 7.4750 to 3 significant digits

(d) 7.4850 to 3 significant digits

(e) 7.48752 to 4 significant digits

26.4 square root tables

In Lesson 16 we learned how to approximate the square root of a number by a process of cut-and-try. Of course, the easiest way to find the square root of a particular positive real number is to insert the number into our handy pocket calculator and then hit the square root key.

Tables of square roots can be used to find an approximation of the square root of any positive number by anyone who knows how to write a number in scientific notation and who knows how and when to use the product of square roots theorem. Practice in solving problems such as the following should increase our feel for square roots and allow us to practice scientific notation. Tables of square roots customarily give the square root of numbers between 1 and 10. Thus our procedure will be

1. Write the number in scientific notation and round off as necessary.
2. If the power of 10 is not an even power of 10, express it as 10 times an even power of 10.

example 26.4.1 Use the square root tables to approximate $\sqrt{416.23}$.

solution The square root table in this book has only 3 digits so we round off to 3 significant digits and write the number in scientific notation.

$$\sqrt{4.16 \times 10^2}$$

Now use the product of square roots theorem

$$\sqrt{4.16}\ \sqrt{10^2}$$

and look up $\sqrt{4.16}$ in the table and write $\sqrt{10^2}$ as 10.

n	n^2	$\sqrt{n}$
4.10	16.8100	2.02485
4.11	16.8921	2.02731
4.12	16.9744	2.02978
4.13	17.0569	2.03224
4.14	17.1396	2.03470
4.15	17.2225	2.03715
4.16	17.3056	2.03961
4.17	17.3889	2.04206
4.18	17.4724	2.04450
4.19	17.5561	2.04695

$$\sqrt{4.16}\sqrt{10^2} = \mathbf{2.03961 \times 10^1}$$

example 26.4.2 Use the square root tables to approximate $\sqrt{4156}$.

solution As the first step we round off and write the number in scientific notation.

$$\sqrt{4160} = \sqrt{4.16 \times 10^3}$$

We do not have 10 to an even power as a factor, and 10 to an odd power is never a perfect square. We use a subterfuge and write 10^3 as $10 \cdot 10^2$.

$$\sqrt{4.16 \times 10^3} = \sqrt{4.16 \cdot 10 \cdot 10^2}$$

Now we use the product of square roots theorem and the table of square roots to complete the solution.

n	n^2	$\sqrt{n}$
4.10	16.8100	2.02485
4.11	16.8921	2.02731
4.12	16.9744	2.02978
4.13	17.0569	2.03224
4.14	17.1396	2.03470
4.15	17.2225	2.03715
4.16	17.3056	2.03961
4.17	17.3889	2.04206
4.18	17.4724	2.04450
4.19	17.5561	2.04695

n	n^2	$\sqrt{n}$
9.91	98.2081	3.14802
9.92	98.4064	3.14960
9.93	98.6049	3.15119
9.94	98.8036	3.15278
9.95	99.0025	3.15436
9.96	99.2016	3.15595
9.97	99.4009	3.15753
9.98	99.6004	3.15911
9.99	99.8001	3.16070
10.00	100.0000	3.16228

$$\sqrt{4.16 \cdot 10 \cdot 10^2} = \sqrt{4.16}\sqrt{10}\sqrt{10^2} = \mathbf{(2.03961)(3.16228) \times 10^1}$$

Leave the answers in the form of an indicated multiplication. In science classes you will complete the multiplication and find a numerical answer, but at this time we are interested in the process—not the answer.

example 26.4.3 Use the square root tables to approximate $\sqrt{.0005273}$.

solution

$$\sqrt{.0005273} = \sqrt{5.27 \times 10^{-4}}$$

The power of 10 is even so we have no problem.

n	n^2	$\sqrt{n}$
5.20	27.0400	2.28035
5.21	27.1441	2.28254
5.22	27.2484	2.28473
5.23	27.3529	2.28692
5.24	27.4576	2.28910
5.25	27.5625	2.29129
5.26	27.6676	2.29347
5.27	27.7729	2.29565
5.28	27.8784	2.29783
5.29	27.9841	2.30000

$$\sqrt{5.27 \times 10^{-4}} = \sqrt{5.27}\sqrt{10^{-4}} = \mathbf{2.29565 \times 10^{-2}}$$

example 26.4.4 Use the square root tables to approximate $\sqrt{.00005273}$.

solution

$$\sqrt{.00005273} = \sqrt{5.27 \times 10^{-5}}$$

Here the power of 10 is again odd so we must again use a subterfuge, but be careful because

$$10^{-5} \neq 10 \cdot 10^{-4} \quad \text{but instead} \quad 10^{-5} = 10 \cdot 10^{-6}$$

$$\sqrt{5.27 \times 10^{-5}} = \sqrt{5.27 \times 10 \times 10^{-6}} = \sqrt{5.27}\sqrt{10}\sqrt{10^{-6}}$$

n	n^2	$\sqrt{n}$
5.20	27.0400	2.28035
5.21	27.1441	2.28254
5.22	27.2484	2.28473
5.23	27.3529	2.28692
5.24	27.4576	2.28910
5.25	27.5625	2.29129
5.26	27.6676	2.29347
5.27	27.7729	2.29565
5.28	27.8784	2.29783
5.29	27.9841	2.30000

n	n^2	$\sqrt{n}$
9.91	98.2081	3.14802
9.92	98.4064	3.14960
9.93	98.6049	3.15119
9.94	98.8036	3.15278
9.95	99.0025	3.15436
9.96	99.2016	3.15595
9.97	99.4009	3.15753
9.98	99.6004	3.15911
9.99	99.8001	3.16070
10.00	100.0000	3.16228

$$\mathbf{(2.29565)(3.16228)} \times \mathbf{10^{-3}}$$

example 26.4.5 Use the square root tables to approximate $\sqrt{462 \times 10^{-15}}$

solution

$$\sqrt{462 \times 10^{-15}} = \sqrt{4.62 \times 10^{-13}} = \sqrt{4.62 \times 10 \times 10^{-14}}$$

n	n^2	$\sqrt{n}$
4.60	21.1600	2.14476
4.61	21.2521	2.14709
4.62	21.3444	2.14942
4.63	21.4369	2.15174
4.64	21.5296	2.15407
4.65	21.6225	2.15639
4.66	21.7156	2.15870
4.67	21.8089	2.16102
4.68	21.9024	2.16333
4.69	21.9961	2.16564

n	n^2	$\sqrt{n}$
9.91	98.2081	3.14802
9.92	98.4064	3.14960
9.93	98.6049	3.15119
9.94	98.8036	3.15278
9.95	99.0025	3.15436
9.96	99.2016	3.15595
9.97	99.4009	3.15753
9.98	99.6004	3.15911
9.99	99.8001	3.16070
10.00	100.000	3.16228

$$= \sqrt{4.62} \times \sqrt{10} \times \sqrt{10^{-14}}$$
$$= \mathbf{(2.14942)(3.16228)} \times \mathbf{10^{-7}}$$

TEST YOUR UNDERSTANDING

Use the square root table on pages 432–434 to help approximate:

(f) $\sqrt{364{,}000}$ (g) $\sqrt{.0000364}$

26.5 addition of expressions with factorable denominators

Algebra books tend to emphasize the factoring of trinomials and binomials because the ability to factor is important and also because these exercises give the student experience in manipulating expressions that contain variables. Thus all algebra books contain problems such as this one:

$$\text{Simplify:} \quad \frac{x^2 + x - 20}{x^2 - x - 12} \div \frac{x^2 + 7x + 10}{x^2 + 9x + 14}$$

For the same reason, algebra books present problems in addition of rational expressions whose denominators are factorable. These problems are designed so that the addition is facilitated if one or more of the denominators are factored before the addition is attempted. The key to these problems is recognizing that they are contrived problems designed to give the student practice in factoring.

example 26.5.1 Add $\dfrac{6x}{x^2 - x - 12} - \dfrac{p}{x - 4}$.

solution We recognize this problem as a problem designed to give the student practice in factoring. We begin by factoring the first denominator.

$$\frac{6x}{(x-4)(x+3)} - \frac{p}{x-4}$$

and now we can add with a new denominator of $(x - 4)(x + 3)$.

$$\frac{\quad}{(x-4)(x+3)} - \frac{\quad}{(x-4)(x+3)} \longrightarrow \frac{6x}{(x-4)(x+3)} - \frac{p(x+3)}{(x-4)(x+3)}$$

$$= \mathbf{\frac{6x - px - 3p}{x^2 - x - 12}}$$

example 26.5.2 Add $\dfrac{7}{x^2 - 5x - 6} - \dfrac{5}{x^2 - 6x}$.

solution As the first step we factor both denominators:

$$\frac{7}{(x-6)(x+1)} - \frac{5}{x(x-6)}$$

and we see that the least common multiple of the denominators is $x(x - 6)(x + 1)$. We use this as our new denominator

$$\frac{\quad}{x(x-6)(x+1)} - \frac{\quad}{x(x-6)(x+1)} \longrightarrow \frac{7x}{x(x-6)(x+1)} - \frac{5(x+1)}{x(x-6)(x+1)}$$

$$= \mathbf{\frac{2x - 5}{x(x-6)(x+1)}}$$

TEST YOUR UNDERSTANDING

(h) Add: $\dfrac{p}{x^2 - 9} + \dfrac{2k}{x^2 - 3x}$

lesson checklist 26

- ☐ Roots of products of squares
- ☐ Significant digits

- ☐ Rounding off
- ☐ Square root tables
- ☐ Addition with factorable denominators

ANSWERS TO TEST YOUR UNDERSTANDING

(a) $10\sqrt{3}$ (b) $20\sqrt{5}$ (c) 7.48 (d) 7.48 (e) 7.488

(f) $(1.90788)(3.16228) \times 10^2$ (g) $(1.90788)(3.16228) \times 10^{-3}$

(h) $\dfrac{px + 2k(x + 3)}{x(x - 3)(x + 3)}$

problem set 26

1. Cakes sell for \$1.50 each and cookies sell for 15 cents each. Gerald bought 40 cakes and cookies for \$46.50. How many cakes did he buy?
2. A 98-foot rope is cut into two pieces. One piece is 10 feet shorter than twice the length of the other piece. What is the length of each piece?
3. Jars of pickled peaches cost \$3.00 each and jars of peanut butter cost \$1.50 each. Harry bought twice as many jars of peaches as peanut butter and spent \$30.00. How many jars of peaches did he buy?
4. Mickey has \$27.50 in quarters and nickels. If he has twice as many quarters as he has nickels, how many quarters does he have?
5. Mary Esther made a down payment of 6 percent. If her check for the down payment was for \$10.80, what was the total value of her purchase?
6. Esther was a putative tomato farmer. She planted 1500 seeds but got only 90 tomato plants. What percent of her seeds sprouted?
7. Find three consecutive integers such that five times the first exceeds three times the third by 10.
8. Find three consecutive odd integers such that eight times the first exceeds three times the third by 3.

Simplify:

*9. $\sqrt{50{,}000}$ *10. $\sqrt{360{,}000}$

Use the square root tables to approximate:

*11. $\sqrt{416.23}$ *12. $\sqrt{.0005273}$

*13. $\sqrt{.00005273}$ *14. $\sqrt{462 \times 10^{-15}}$

15. Find *m*.

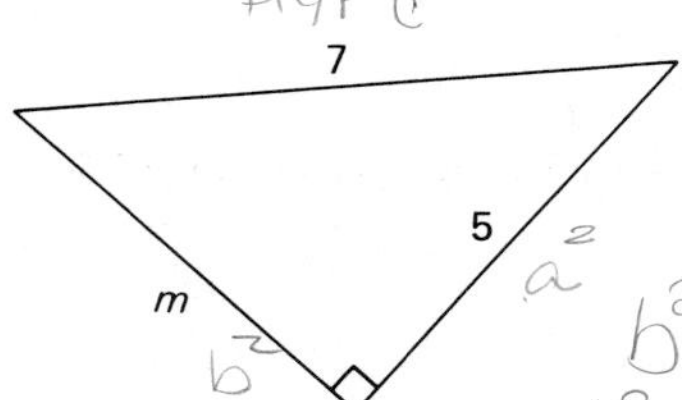

*The problems marked with an asterisk were worked out in this lesson.

16. Find the distance between (6, 3) and (−2, 4)

17. Use the associative and commutative properties to show that $xyzk = xkyz$.

18. Use two real numbers to demonstrate the commutative property of real numbers under multiplication.

19. List the nine properties of a field and use numbers or letters to give an example of each property.

20. What is the multiplicative inverse of $-2\frac{1}{2}$?

Simplify:

21. $2\sqrt{14} \cdot \sqrt{3} \cdot 2\sqrt{7}$

22. $4\sqrt{15} \cdot \sqrt{5} \cdot 2\sqrt{6}$

23. $2\sqrt{50} - 3\sqrt{200}$

Divide:

24. $4x^2 - 2x + 5$ by $x + 2$

Solve:

25. $t^2 - 25 = 0$

26. $p^2 - 42 = -p$

27. $k^2 - 4k = 45$

28. $R_LT_L = R_ST_S$, $R_L = 120$, $R_S = 280$, $T_S = 20 - T_L$; find T_S, T_L

Simplify:

29. $\dfrac{(3000 \times 10^{-14})(.00008)}{(.0002 \times 10^5)(200{,}000)}$

Use elimination:

30. $3x + 2y = 2$
$5x + 3y = 4$

31. Find the equation of the line:

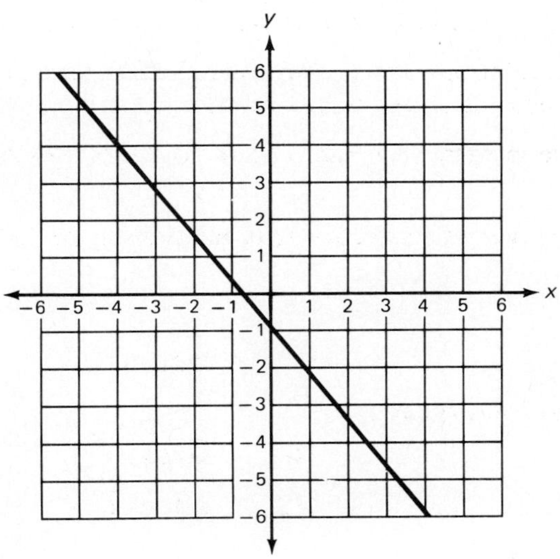

32. Find the equation of the line:

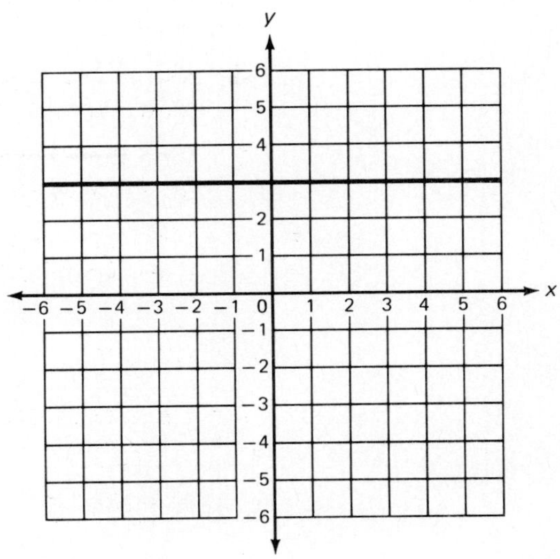

33. The set of positive real numbers is closed for what operations?

Solve:

34. $\dfrac{3x}{5} + \dfrac{1}{2} = x$

Graph on a rectangular coordinate system:

35. $y = -3x$

36. $2x + 3y - 2 = 0$

Simplify:

37. $\dfrac{\dfrac{1}{c}}{a + b}$

38. $\dfrac{m - \dfrac{4}{c}}{\dfrac{k}{c} + m}$

Add:

39. $\dfrac{m}{k + b} + \dfrac{1}{b^2}$

40. $\dfrac{x}{a} + \dfrac{m}{a^2} + \dfrac{k}{a + b}$

Multiply:

41. $\dfrac{4a^2}{a^{-2}}\left(\dfrac{x^{-2}}{a^2} - \dfrac{a^2}{x^{-2}}\right)$

Simplify:

42. $\dfrac{x^2 + x - 20}{x^2 - x - 12} \div \dfrac{x^2 + 7x + 10}{x^2 + 9x + 14}$

43. $\dfrac{x^2m^{-4}x^0m^2}{x^{-4}m^0}$

44. $\dfrac{(x^2y^{-2})^2y^{-2}}{(x^0)^2(y^{-3}x)^2}$

Solve:

45. $-3(p + 6) - 4 = -2(p - 3) - 8$

Evaluate:

46. $c - (-d)(c)$ if $c = -2, d = 4$

Simplify:

47. 3^{-2}

48. True or false: $5 \in \{\text{Irrationals}\}$

49. $-7 \in \{\text{what sets of numbers?}\}$

Add:

***50.** $\dfrac{6x}{x^2 - x - 12} - \dfrac{p}{x - 4}$

***51.** $\dfrac{7}{x^2 - 5x - 6} - \dfrac{5}{x^2 - 6x}$

lesson twenty-seven

Absolute value inequalities
Graphical solution of systems of linear equations

27.1 absolute value inequalities

We review the concept of absolute value by saying that every real number except zero has two parts. One of the parts is the plus or minus sign and the other part is the numerical part. We can think of the numerical part as designating the quality of "bigness" of the number, and we call this quality the absolute value of the number. Since we customarily use positive numbers to designate degrees of bigness, we will use a positive number to designate the absolute value of all real numbers except zero. Thus we say that the two numbers

$$3 \quad \text{and} \quad -3$$

both have an absolute value of 3 although one of them is a positive number and one of them is a negative number.

We designate the absolute value of a number by enclosing the number within vertical lines. Thus we designate the absolute value of 3 by writing $|3|$, and we designate the absolute value of -3 by writing $|-3|$. Of course, the absolute value of both of these numbers is 3.

$$|3| = 3 \qquad |-3| = 3$$

It is difficult to describe the absolute value of a number by using words. Most authors do not like to speak of the "bigness" of a number as we have done, for they feel that bigness can be confused with the concept of *greater than* that is used to compare numbers. For this reason, many prefer the formal definition of absolute value used in more advanced courses. This definition does not use words as it is stated using only symbols. The definition is in three parts and we have avoided using the definition thus far because the third part is confusing to some people.

(a) if $x > 0$, $\quad |x| = x$

(b) if $x = 0$, $\quad |x| = 0$

(c) if $x < 0$, $\quad |x| = -x$

Part (a) speaks of the absolute value of positive numbers, which are numbers that are greater than zero. Part (b) describes the absolute value of zero. Part (c) can be confusing because of the minus sign. Part (c) describes the absolute value of negative numbers, which are numbers that are less than zero. We know that the absolute value of a negative number is the opposite of the number, as

$$|-3| = -(-3) = 3$$

Thus when we write

$$\text{if } x < 0, |x| = -x$$

we are not saying that the absolute value is a negative number but that the absolute value of a negative number is the opposite of the negative number, which is a positive number.

> The absolute value of a non-zero real number is a positive number.† The absolute value of zero is zero.

$$|3| = +3 \qquad |-3| = +3 \qquad |0| = 0$$

In an attempt to describe the absolute value of a number by using words, some authors define the absolute value of a number to be the number that describes the distance on the number line from the origin to the graph of the number being considered. If we use this definition, we will find that $+3$ and -3 have the same absolute value, for they are both 3 units from the origin, and thus both numbers have an absolute value of 3 (see Fig. 27.1).

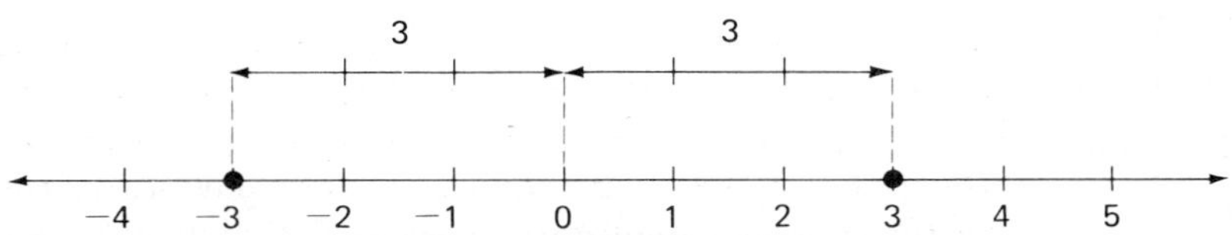

Figure 27.1

Every real number except zero has an opposite and both members of a pair of opposites have the same absolute value. **For this reason, if we ask what non-zero numbers have a specified absolute value, we will receive two answers to our question.** For example,

$$|\text{What numbers?}| = 4$$

Here we ask what numbers have an absolute value of 4. Of course, the answers are $+4$ and -4 because both of these numbers have an absolute value of 4. It is customary to use a single letter variable as the unknown so we will restate the question by writing

$$|x| = 4$$

and as we have said, the two values of x that satisfy this condition are $+4$ and -4.

Often it is desirable to display the solution of an absolute value equation or inequality in graphical form. The graph of the solution to the equation $|x| = 4$ is the graph of the numbers $+4$ and -4 as we show in Fig. 27.2.

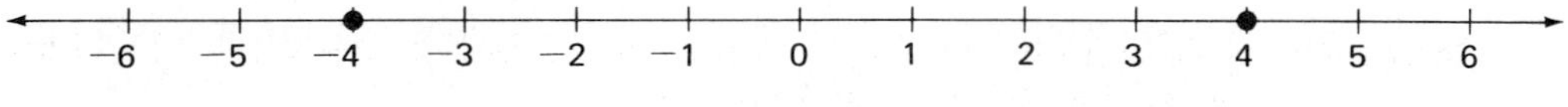

Figure 27.2

†Some might like to say that the absolute value of a number is the positive value of the number, but this would be incorrect. A number is either a positive number or a negative number. Numbers do not have two values.

example 27.1.1 Graph $|x| > 2$, $D = \{\text{Reals}\}$.

solution We are asked to indicate on the number line every real number whose absolute value is greater than 2.

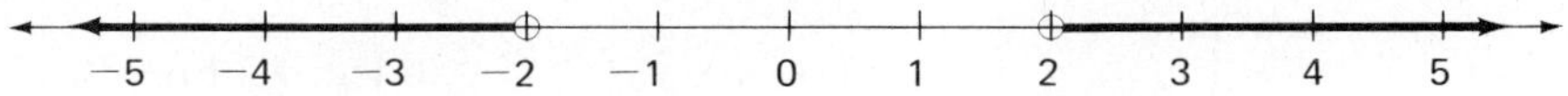

We note that part of the solution set to this inequality consists of negative numbers. This is correct, for if we think carefully, we will realize that while these negative numbers are less than 2, their absolute values are greater than 2.

example 27.1.2 Graph $|x| \leq 3$, $D = \{\text{Integers}\}$.

solution We are asked to indicate on the number line the location of all integers whose absolute value is equal to or less than 3.

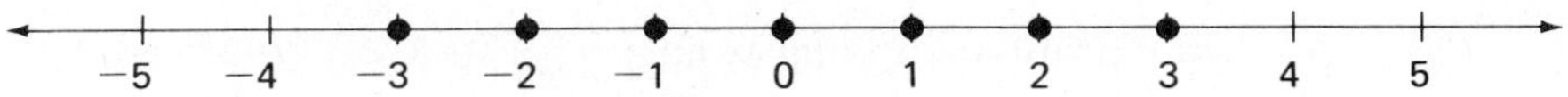

We note that all integers that are greater than or equal to -3 and that are also less than or equal to 3 have an absolute value that is equal to or less than 3.

example 27.1.3 Graph $|x| < -4$, $D = \{\text{Reals}\}$.

solution

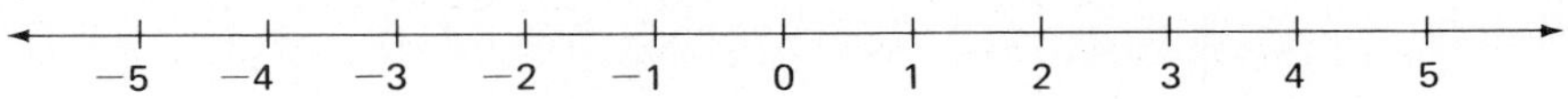

We have not graphed the solution for the given condition because there are no real numbers that satisfy the given condition. The statement $|x| < -4$ asks for the real number replacements for x whose absolute values are less than -4. There are no real numbers that satisfy this condition since the absolute value of any non-zero real number is greater than zero. If we use the formal language of sets, instead of saying that we have not graphed the solution, we say that the solution is the empty set $\{\ \}$, or the null set $\varnothing$, and we say that the bare number line shown is the graph of this set.

example 27.1.4 Graph $|x| > -4$, $D = \{\text{Reals}\}$.

solution

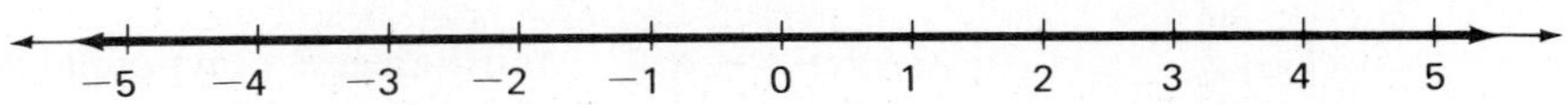

This one is also tricky. The absolute value of every real number is zero or a number greater than zero. Certainly, then, if the absolute value of every real number is equal to or greater than zero, the absolute value of every real number is also greater than -4. Thus the solution to the stated condition is the set of real numbers, which is graphed by indicating the entire number line.

example 27.1.5 Graph $-|x| \geq -2$, $D = \{\text{Reals}\}$.

solution We begin by multiplying both sides of the inequality by -1 and also **reversing the sense of the inequality symbol** and we find

$$|x| \leq 2$$

which is graphed below.

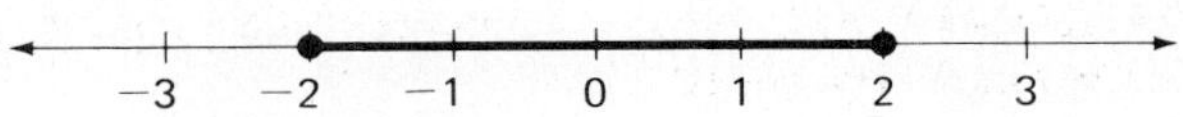

example 27.1.6 Graph $-|x| - 2 > -5$, $D = \{\text{Integers}\}$.

solution First we must solve the inequality for $+|x|$ by isolating $|x|$ on one side of the inequality.

$-\|x\| - 2 > -5$	given
$-\|x\| > -3$	added $+2$ to both sides
$\|x\| < 3$	multiplied both sides by -1 and **reversed the sense of the inequality symbol**

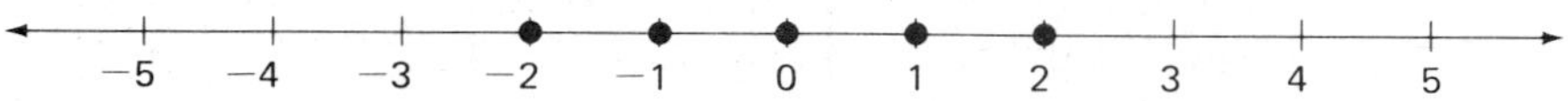

TEST YOUR UNDERSTANDING

(a) Graph on a number line: $-|x| + 3 < 1$, $D = \{\text{Reals}\}$

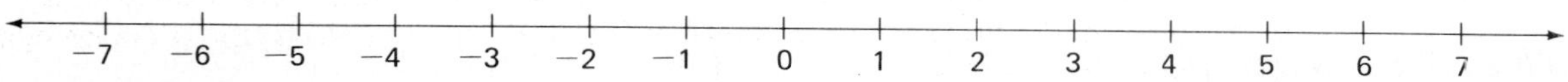

27.2 graphical solutions of systems of linear equations

As we remember from Lessons 11 and 12, the degree of a term of a polynomial is the sum of the exponents of the variables in the term. Thus $2xy^5$ is a sixth-degree term, $4x$ is a first-degree term, and xy is a second-degree term.

Also we remember that the degree of a polynomial is the same as the degree of its highest degree term and the degree of a polynomial equation is the same as the degree of the highest degree term in the equation. Thus

$x^4y + 4y$	is a fifth-degree polynomial
$x^4y + 4y = x$	is fifth-degree polynomial equation
$x + 2y = 4$	is a first-degree polynomial equation

We have learned how to solve second-degree polynomial equations (quadratic equations) by factoring, but thus far we have devoted most of our attention to first-degree polynomial equations. We have learned that the graph of a first-degree polynomial equation in two unknowns is a straight line and that we call this kind of equation a linear equation. We have learned to find the solution to a system of two linear equations by methods that we call the **substitution method** and the **elimination method.** Here we will show that we can find the solution to a system of two linear equations by **graphing** each of the equations and visually estimating the coordinates of the point on the graph where the two lines cross.

example 27.2.1 Find an approximate solution to the given system by graphing. $\begin{cases} y = x + 1 \\ y = -2x + 3 \end{cases}$

solution We have graphed the lines in the accompanying figure and we can see the point on the graph where they cross, but it is impossible to determine visually the exact coordinates of the crossing point. **The advantage to the graphical method is that we can see what we are doing** and this helps many people understand the problem better. As we can see in this figure, there are an infinite number of ordered pairs that satisfy the equation $y = x + 1$. Visually, this means there are an infinite number of points on the line. Also, there are an infinite number of points that lie on the line described by the equation $y = -2x + 3$.

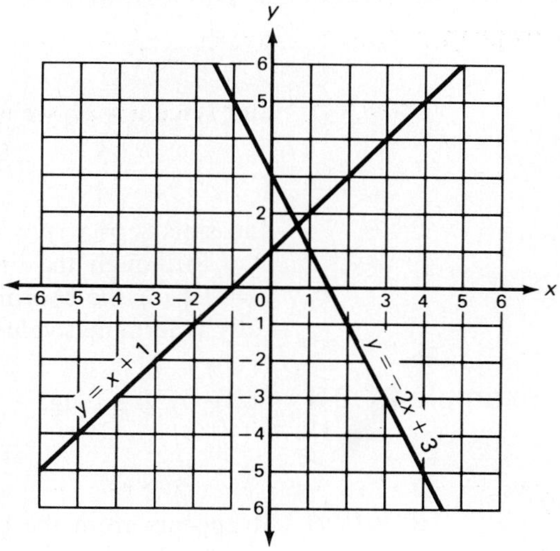

But from the figure we see that only one point lies on both of these lines and the coordinates of this point will therefore satisfy both of the given equations. We estimate that the coordinates of the point are **(.6, 1.5)**.

If we wish an exact answer, we must use either the substitution method or the elimination method. We will check our graphical solution by solving the problem again by the substitution method.

(a) $$(x + 1) = -2x + 3 \longrightarrow 3x = 2 \longrightarrow \boldsymbol{x = \frac{2}{3}}$$

(b) $$y = \frac{2}{3} + 1 \longrightarrow \boldsymbol{y = \frac{5}{3}}$$

Thus the exact coordinates are $(\frac{2}{3}, \frac{5}{3})$, which are approximately equal to the values of .6 and 1.5 that we found by visually estimating the coordinates of the point where the lines cross.

example 27.2.2 Solve by graphing.
$$\begin{cases} y = x \\ y = \frac{5}{4}x - 1 \end{cases}$$

solution We have graphed the lines in the accompanying figure. We see that the lines are almost parallel and this makes a visual determination of the coordinates of the crossing point very difficult. The graph gives us a visual representation of the problem but is of little use in finding an exact numerical answer. By looking at the graph we estimate that the coordinates of the crossing point are:

$$x = 3.5 \text{ and } y = 3.4.$$

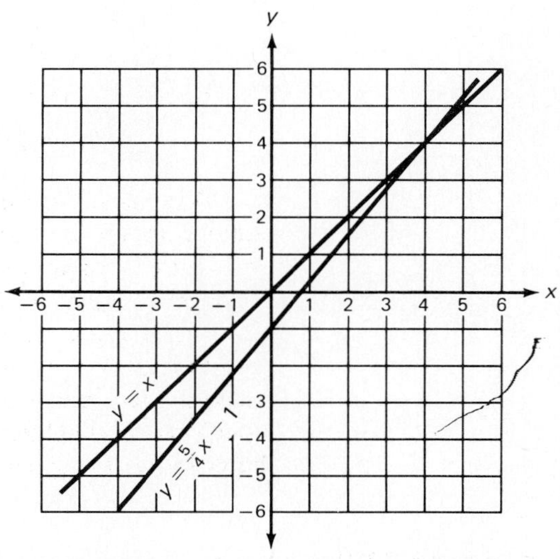

As in the last problem we can use either substitution or elimination if we wish to find the exact coordinates of this point. We will use substitution.

$$y = \frac{5}{4}x - 1$$

and since $y = x$, we substitute x for y and then solve for x.

$$x = \frac{5}{4}x - 1 \longrightarrow x - \frac{5}{4}x = -1 \longrightarrow -\frac{1}{4}x = -1 \longrightarrow \mathbf{x = 4}$$

and since $x = y$, $\mathbf{y = 4}$.

Although the lines were almost parallel, we see that our guesses of $x = 3.5$ and $y = 3.4$ for the coordinates of the crossing point were reasonably close to the algebraically determined values of $x = 4$ and $y = 4$.

example 27.2.3 Solve by graphing. $\begin{cases} y = x + 2 \\ y = x - 1 \end{cases}$

solution It appears from the graph that the lines are parallel and thus never intersect. If this is so, no point lies on both of the lines and no ordered pair will satisfy both equations. An attempt at solving this system by substitution or elimination will degenerate into a false numerical statement. We will look into this in detail in Lesson 34.

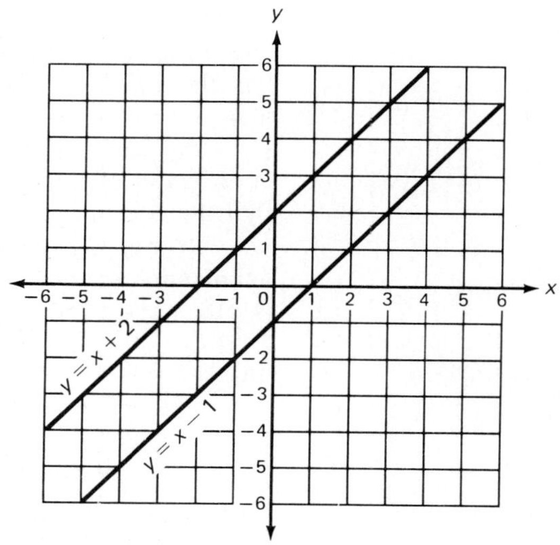

TEST YOUR UNDERSTANDING

(b) Solve the given system by graphing and check your results by using either substitution or elimination.

$\begin{cases} x + y = 2 \\ 2y - x = -8 \end{cases}$

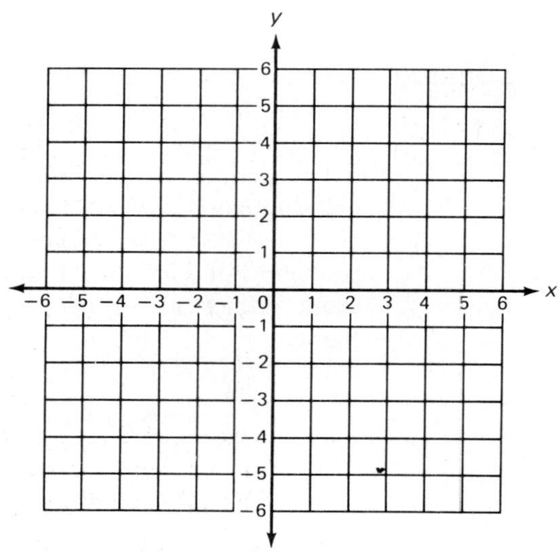

lesson checklist 27

- ☐ Definition of absolute value
- ☐ Absolute value inequalities
- ☐ Graphical solution of linear equations
- ☐ Estimating the coordinates of the crossing point

ANSWERS TO TEST YOUR UNDERSTANDING

(a)

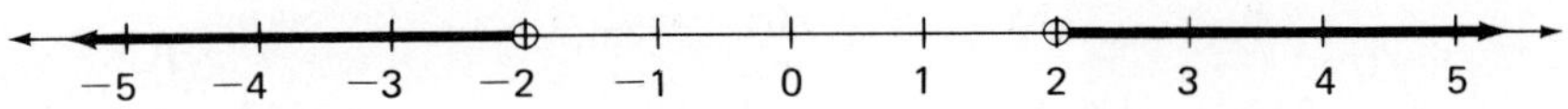

(b)

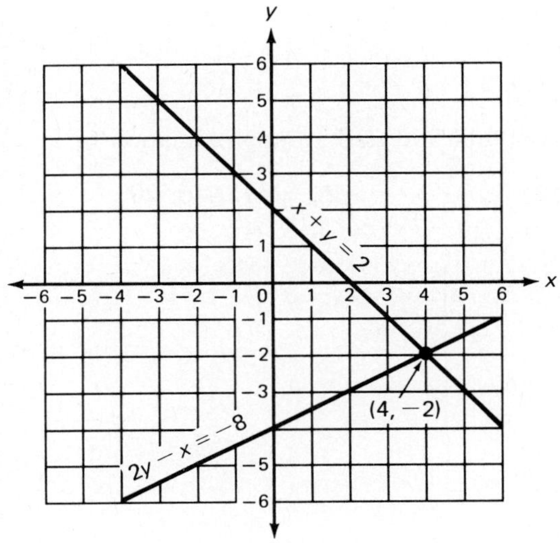

problem set 27

1. The Kwick shop sells sandwiches for 50 cents each and doughnuts for 20 cents each. During one 4-hour period 75 items were sold for 33 dollars. How many sandwiches were sold?

2. Brandy sold shoes for \$25 a pair and boots for \$35 a pair. One day the cash receipts totaled \$680 and she had sold twice as many pairs of shoes as pairs of boots. How many pair of each did she sell?

3. A tree 50 meters long was cut into two pieces. The long piece was 10 meters longer than three times the length of the short piece. How long was the long piece?

4. Bucky and Upshaw sold wildebeests for \$125 each and warthogs for \$75 each. One day they made \$3500 and sold 20 more warthogs than wildebeests. How many of each did they sell?

5. Homer Lee had 340 raisins and had only 20% as many raisins as Bodie had. How many raisins did Bodie have?

6. Ruby's peaches weighed only 15% as much as Franklin's peaches weighed. If Franklin's peaches weighed 800 kilograms, how much did Ruby's peaches weigh?

7. Find three consecutive odd integers such that 6 added to five times the first is 1 greater than four times the third.

8. Find two consecutive even integers such that four times the first is 16 larger than three times the second.

Graph on a number line:

9. $-x + 3 \leq -2$, $D = \{\text{Reals}\}$

*10. $|x| > -4$, $D = \{\text{Integers}\}$

*11. $-|x| \geq -2$, $D = \{\text{Reals}\}$

*12. $-|x| - 2 > -5$, $D = \{\text{Integers}\}$

Solve by graphing:

*13. $\begin{cases} y = x + 1 \\ y = -2x + 3 \end{cases}$

14. $\begin{cases} y = -x \\ y = x - 3 \end{cases}$

Use the square root tables to find an approximation for:

15. $\sqrt{7{,}141{,}652}$

16. $\sqrt{71{,}416{,}520}$

17. Find h.

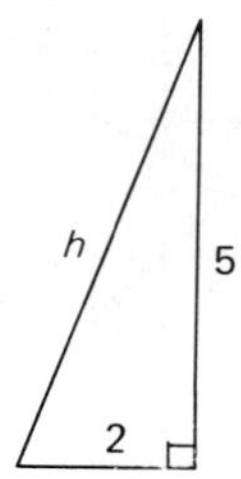

18. Find the distance between $(4, -2)$ and $(-3, 5)$

19. Use the commutative and associative properties to show that $a + b + c + d = c + b + a + d$.

20. List the properties of a field.

Simplify:

21. $2\sqrt{3}\,(3\sqrt{2} + 2\sqrt{6}\,)$

22. $2\sqrt{24} \cdot \sqrt{6} \cdot \sqrt{3}$

23. $5\sqrt{3}\,(3 - \sqrt{12}\,)$

Divide:

24. $x^3 - x^2 - 5x + 2$ by $x + 2$.

Solve:

25. $-x^2 = 7x - 30$

26. $x^2 + 49 = 14x$

27. $R_1T_1 + R_2T_2 = 190$, $T_1 = 2$, $T_2 = 3$, $R_2 - 2R_1 = 10$ find R_1.

Simplify:

28. $$\frac{(.00056 \times 10^4)(7 \times 10^3)}{(.00049 \times 10^{16})(.00002 \times 10^{-5})}$$

---

*The problems marked with an asterisk were worked out in this lesson.

Use elimination:

29. $\begin{cases} 7x + 2y = 11 \\ 2x - 3y = -4 \end{cases}$

30. Find the equation of the line:

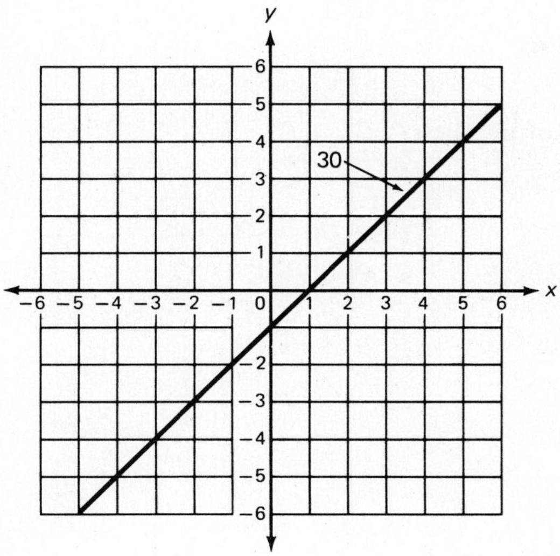

31. Find the equation of the line:

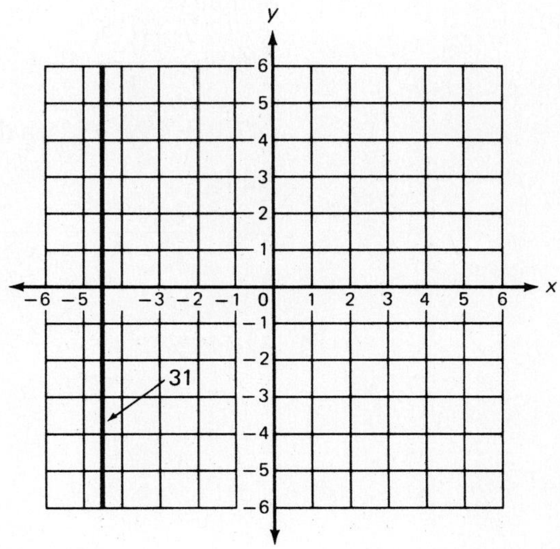

32. The set of positive integers is closed for what operations?

Solve:

33. $\dfrac{3x}{2} + \dfrac{x - 2}{5} = 1$

Graph on a rectangular coordinate system:

34. $y = 4x$

35. $y = -3x + 2$

Simplify:

36. $\dfrac{\dfrac{b}{a}}{\dfrac{1}{b + a}}$

37. $\dfrac{\dfrac{x - y}{a}}{\dfrac{x}{a} - 1}$

Add:

38. $\dfrac{a}{x} + \dfrac{d}{x^2c} + \dfrac{c}{x^3c^2}$

39. $\dfrac{k}{x + y} + \dfrac{1}{y}$

Multiply:

40. $\dfrac{ab^{-2}}{c}\left(\dfrac{c^{-3}b^2}{a^{-4}c} + \dfrac{3ba}{a^{-3}}\right)$

Add like terms:

41. $3a^{-2}b^{-1} + \dfrac{2a}{b^{-3}} + \dfrac{ab^2}{b}$

Simplify:

42. $\dfrac{(p^2)^0x^2a^{-1}m}{(m^{-1})^{-2}p^{-2}}$

43. $\dfrac{x^2 - 4x + 3}{x^2 + x - 12} \cdot \dfrac{x^2 + 12x + 32}{x^2 - 5x + 4}$

Solve:

44. $5x + 2 = (x - 2)(4) + (-2) + (-x)$

Evaluate:

45. $x[y - (-x)] + (y - x)$ if $x = 2, y = -3$

Simplify:

46. $\dfrac{(-3)^{-2}}{-3^2}$

47. $-4\sqrt{2} \in$ {what sets of numbers?}

Add:

48. $\dfrac{2x - 5}{x^2 - 3x} - \dfrac{2}{x - 3}$

49. $\dfrac{4xp}{x^2 - 5x - 6} - \dfrac{4p}{x - 6}$

lesson twenty-eight

Solution of rational equations
Dependent and independent variables
Functions

28.1 rational equations

In Lesson 19 we discussed the solution of rational equations in which all denominators are integers. In these problems we found that the recommended first step is to multiply the numerator of every term by the least common multiple of the denominators. Since every denominator is a factor of the least common multiple, this procedure permits us to cancel every denominator as we see in the following problem.

example 28.1.1 Solve $\frac{y}{2} + \frac{1}{4} = \frac{y}{6}$.

solution As the first step we will multiply every numerator by 12, the least common multiple of the denominators.

$$12\left(\frac{y}{2}\right) + 12\left(\frac{1}{4}\right) = 12\left(\frac{y}{6}\right) \longrightarrow 6y + 3 = 2y \longrightarrow 4y = -3 \longrightarrow y = -\frac{\mathbf{3}}{\mathbf{4}}$$

The denominators of the equation above were all real numbers. In this lesson we will discuss equations whose denominators contain variables. If there are variables in the denominator, the replacement values of the variable are restricted. For instance, if the given equation is

$$\frac{t-2}{t} = \frac{14}{3t} - \frac{1}{3}$$

The number zero would not be a permissible value for t, for if we substitute the number zero for t, we find

$$\frac{0-2}{0} = \frac{14}{3(0)} - \frac{1}{3} \longrightarrow \frac{\mathbf{-2}}{\mathbf{0}} = \frac{\mathbf{14}}{\mathbf{0}} - \frac{\mathbf{1}}{\mathbf{3}}$$

which is meaningless for division by zero is not defined.

If our equation is

$$\frac{n}{n+2} = \frac{3}{5}$$

we cannot accept -2 as a value for n for if we try to substitute -2 for n, we obtain

$$\frac{(-2)}{(-2)+2} = \frac{3}{5} \longrightarrow \frac{-2}{0} = \frac{3}{5}$$

an expression in which zero is the denominator of a fraction, and division of a non-zero real number by zero is not defined.

Thus as our first step in the solution of rational equations whose terms have variables in one or more denominators, we will list the unacceptable values of the variable which, of course, are those values of the variable that would cause any denominator to equal zero.

example 28.1.2 Solve $\frac{t-2}{t} = \frac{14}{3t} - \frac{1}{3}$.

solution ($t \neq 0$). As the next step we multiply every numerator by $3t$, the least common multiple of the denominators. This will allow us to cancel the denominators and solve the resulting equation.

$$3\cancel{t}\left(\frac{t-2}{\cancel{t}}\right) = \cancel{3t}\left(\frac{14}{\cancel{3t}}\right) - \cancel{3}t\left(\frac{1}{\cancel{3}}\right) \longrightarrow 3t - 6 = 14 - t \longrightarrow 4t = 20 \longrightarrow \mathbf{t = 5}$$

example 28.1.3 Solve $\frac{n}{n+2} = \frac{3}{5}$.

solution ($n \neq -2$). Now we multiply every term by $(5)(n+2)$, the least common multiple of the denominators, cancel the denominators, and solve.

$$(5)\cancel{(n+2)}\frac{(n)}{\cancel{(n+2)}} = \frac{3}{\cancel{5}}(\cancel{5})(n+2) \longrightarrow 5n = 3(n+2)$$

$$\longrightarrow 5n = 3n + 6 \longrightarrow 2n = 6 \longrightarrow \mathbf{n = 3}$$

example 28.1.4 Solve $\frac{2}{3n} = \frac{2}{n+4}$.

solution ($n \neq 0, -4$). Now we multiply every numerator by $(3n)(n+4)$, cancel the denominators, and solve.

$$\cancel{(3n)}(n+4)\frac{(2)}{\cancel{(3n)}} = \frac{(2)}{\cancel{(n+4)}}(3n)\cancel{(n+4)} \longrightarrow (n+4)2 = 6n \longrightarrow 2n + 8 = 6n$$

$$\longrightarrow 8 = 4n \longrightarrow \mathbf{n = 2}$$

example 28.1.5 Solve $\frac{4}{x} = \frac{7}{x-2}$.

solution ($x \neq 0, 2$). Now multiply each term by $x(x-2)$, cancel the denominators, and solve.

$$\cancel{x}(x-2)\frac{4}{\cancel{x}} = \frac{x\cancel{(x-2)}\,7}{\cancel{x-2}} \longrightarrow 4x - 8 = 7x$$

$$\longrightarrow -8 = 3x \longrightarrow \mathbf{x = -\frac{8}{3}}$$

example 28.1.6 Solve $\frac{4}{p} - \frac{3}{p-4} = 0$.

solution ($p \neq 0, 4$). We multiply each term by $p(p-4)$, which is the least common multiple of the denominators, cancel the denominators, and solve.

$$\not{p}(p-4)\frac{4}{\not{p}} - p\cancel{(p-4)}\frac{3}{\cancel{p-4}} = 0 \longrightarrow 4p - 16 - 3p = 0$$

$$\longrightarrow p - 16 = 0 \longrightarrow \mathbf{p = 16}$$

Test Your Understanding

(a) Solve: $\frac{5}{p} - \frac{3}{p+2} = 0$

28.2 dependent and independent variables (again)

If we consider the equation

$$y = x + 6$$

we see that this equation matches a value of y with any value of x. If we give x values of 4 or 23

If $x = 4$

$$y = (4) + 6 \longrightarrow \mathbf{y = 10}$$

If $x = 23$

$$y = (23) + 6 \longrightarrow \mathbf{y = 29}$$

we can use the equation to find the corresponding value of y. On the left we let $x = 4$ and find that $y = 10$, and on the right we let $x = 23$ and find that the matching value of y is 29. **We see that the value of y depends on the value that we assign to x**, so we say that y is the **DEPENDENT VARIABLE** and x is the **INDEPENDENT VARIABLE.** Of course, we could assign a value to y and use the equation to find the value of x, and this would reverse the names. **It is customary, however, to use the letter x to designate the independent variable and to use the letter y to designate the dependent variable. We will follow this convention in this book.**

28.3 functions

We use the word **FUNCTION** to describe a relationship between two sets in which

1. The first set is the domain and the domain is defined.
2. For each member of the domain there is exactly one answer in the second set.

If we consider the equation

$$y = 4x + 6 \qquad D = \{\text{Reals}\}$$

we see that this equation will give us **one answer** for y for any real number value of x. For example,

If we let $x = 2$

$$y = 4(2) + 6$$
$$\mathbf{y = 14}$$

If we let $x = 10$

$$y = 4(10) + 6$$
$$\mathbf{y = 46}$$

Thus the equation and the domain define a function because

1. The domain is specified.
2. For any value of the domain the equation will give us one and only one value (answer) for y.

Now let us consider the ordered pairs

$$(4, 7), \qquad (5, -2), \qquad (7, -4), \qquad (8, 3)$$

and ask if these ordered pairs designate a function. The answer is yes because

1. The domain is specified. The only x values we can use are the x values in the ordered pairs, which are the numbers $\{4, 5, 7, 8\}$.
2. For each value of the domain there is exactly one answer for y. If we let x be 4, then y is 7; if x is 5, then y is -2; if x is 7, then y is -4; if x is 8, then y is 3.

From this example we see that an equation is not necessary to designate† a function. We have fulfilled both of the requirements of a function by using the ordered pairs.

If we use proper nomenclature to describe a function, we must use the words **DOMAIN**, **IMAGE**, and **RANGE**. The domain, as we know, is the set of all permissible replacement values of x. We will use the word **image** instead of using the word **answer** to describe the value of y that is matched with each value of x by the function. We will use the word **range** to describe the set of all images.

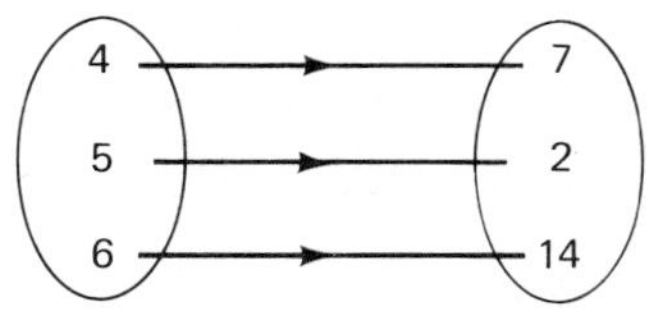

DOMAIN. The set of all permissible values of x.

RANGE. The set of all images (values of y or "answers")

> **DEFINITION**
>
> FUNCTION, DOMAIN, IMAGE, RANGE
>
> A **function** is a correspondence or mapping between two sets that associates with each element of the first set a unique element of the second set. The first set is called the **domain** of the function. For each element x of the domain, the corresponding element y of the second set is called the **image** of x under the function. The set of all images of the elements of the domain is called the **range** of the function.

There are two accepted definitions of a function. The definition above says that a function is the correspondence or mapping while the other definition says that a function is the set of ordered pairs.

---

†The equation itself cannot be called the function because the function is the mapping or pairing between the members of the domain and the answers for each member.

> A function is a set of ordered pairs such that no two ordered pairs have the same first element and different second elements.

Perhaps it will help if we show a few functions and a few non-functions. Remember, to be a function,

1. The domain must be specified.
2. A way must be provided to find the image for each value of x and there must be exactly one image for each value of x.

example 28.3.1 Does the diagram designate a function?

solution

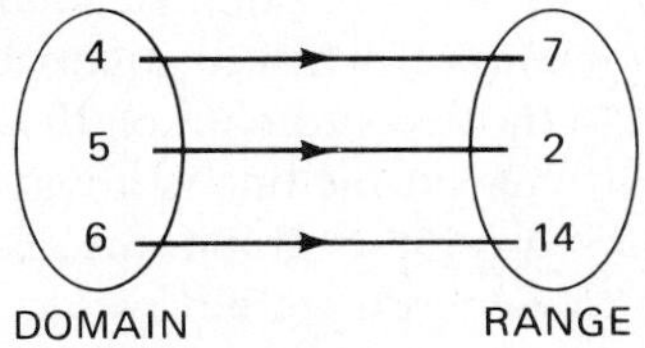

Yes. The elements of the domain are specified. There is one image and only one image for each member of the domain. We do not have an equation but can find the images by looking at the diagram.

example 28.3.2 Does the diagram designate a function?

solution

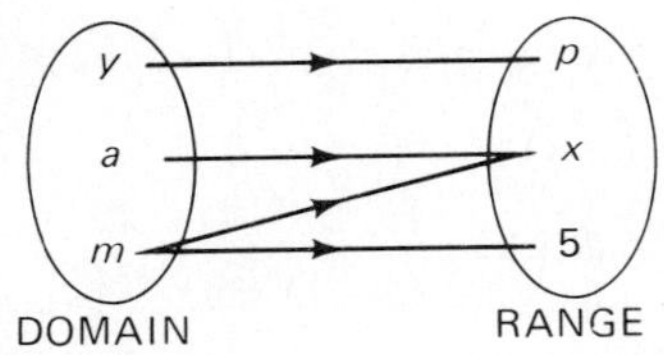

No. The members of the domain are specified, but m has two images and it is only allowed one.

example 28.3.3 Does the diagram designate a function?

solution

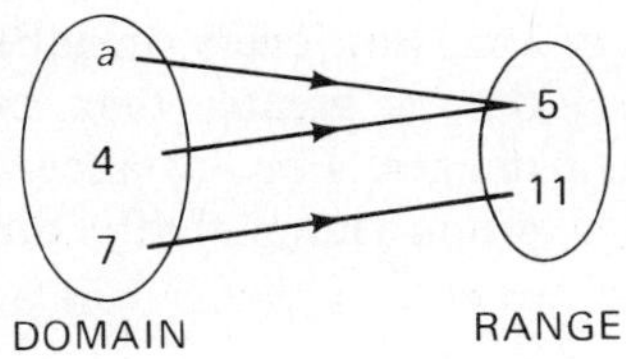

Yes. The domain is specified and each member of the domain has exactly one image. True, 5 is the image of both a and 4, but this is permissible.

example 28.3.4 Does the diagram designate a function?

solution

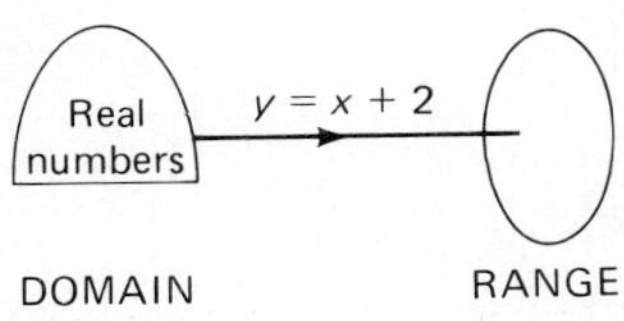

Yes. The diagram says that we can use any real number for x so the domain is specified. The members of the range are not specified, but the equation will allow us to find the value of y that is paired with any real number we use for x.

example 28.3.5 Does either of the sets of ordered pairs shown designate a function?

(a) $\{(4, 6), (7, 2), (4, 5)\}$ (b) $\{(4, 8), (15, 6), (11, 7)\}$

solution The set (a) does not designate a function, for 4 has two images. The set (b) does designate a function. The domain is specified and there is exactly one image for every member of the domain.

example 28.3.6 Does the diagram designate a function?

solution

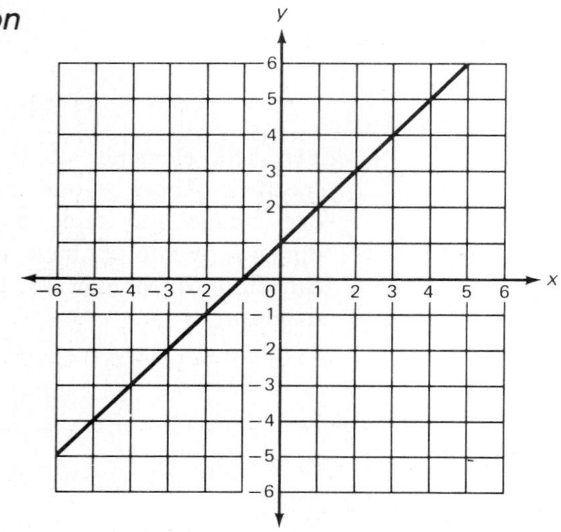

Yes, it certainly is a function. For every value of the x-coordinate on the line there is a corresponding y-coordinate. For instance, when x equals 2, it appears that the y-coordinate of the ordered pair that lies on the line is about 3.0. We can't tell from the graph the exact value of the y-coordinate but we do know that an exact value does exist.

example 28.3.7 Does the diagram designate a function? (The domain is the set of real numbers such that $x \geq 1$.)

solution

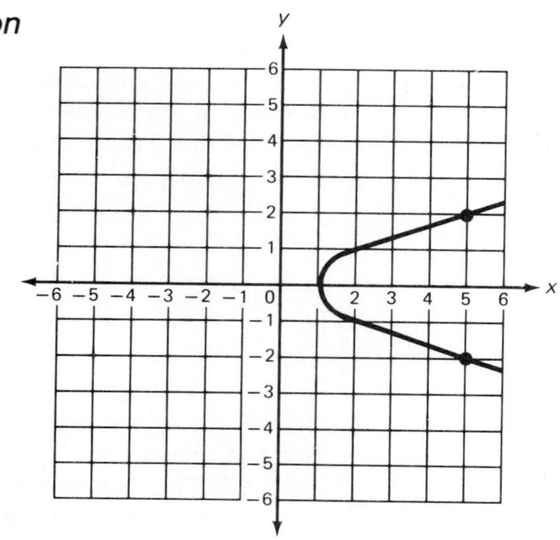

No. For every member of the domain that is greater than one, there are two images. For instance, when $x = 5$, the graph shows that y can be either $+2$ or -2.

28.4 relations

All functions can also be called relations because a relation is a pairing that matches each element of the domain with one or more images of the range.

example 28.4.1 Does the diagram designate a function?

solution

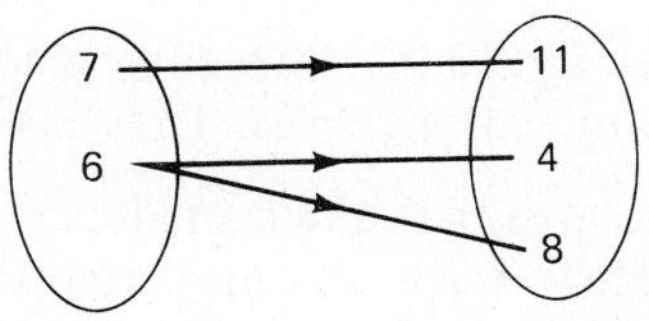

No. The number 6 has two images. The diagram designates a relation.

example 28.4.2 Does the diagram designate a function or a relation?

solution

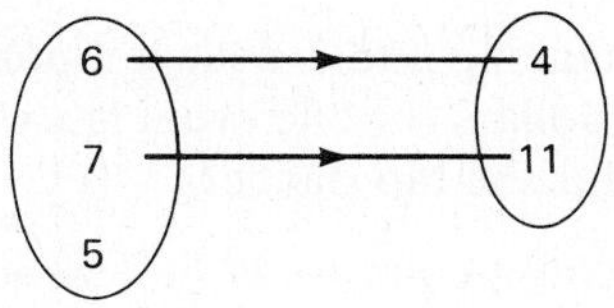

Neither. For a correspondence to be called a relation, there must be one or more images for each element of the domain. To be called a function there must be exactly one image for each element of the domain. In the diagram the number 5 does not have even one image, and thus the diagram does not designate either a function or a relation.

TEST YOUR UNDERSTANDING

Do the following designate functions?

(b) $\{(4, -2), (3, -5), (6, -3), (8, -5)\}$

(c)

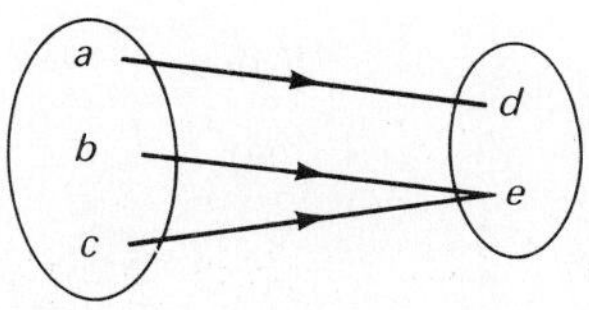

lesson checklist 28

- ☐ Unacceptable values of the variable
- ☐ Eliminate the denominators as the first step
- ☐ Dependent variable
- ☐ Independent variable
- ☐ Definition of a function
- ☐ Domain
- ☐ Image
- ☐ Range
- ☐ Exactly one image (answer)
- ☐ Relation

ANSWERS TO TEST YOUR UNDERSTANDING

(a) $p = -5$ (b) Yes (c) Yes

problem set 28

1. Skeet sold pies for \$2.00 and cakes for \$3.00. One day he took in \$200.00 and found that he had sold twice as many cakes as pies. How many pies did he sell?
2. Gus Brown divided the booty into a large part and a small part. The total value of the booty was 90 pieces of eight. How much was the large part if the large part was 18 greater than five times the small part?
3. Tommie and Anne were paid 10 cents to iron a shirt and 15 cents to block a hat. If they were paid \$6.25 for 55 pieces of work, how many hats did they block?
4. Zedie got 50 cents a minute for tap dancing and 3 dollars a minute for interpretative dancing. Miss Ruby paid her 95 dollars for one evening's work. If she danced for a total of 40 minutes, how long did she tap dance?
5. Pratt could count to 1230 and Iffus could count only 60% as high as Pratt could. How high could Iffus count?
6. Henson's golf score was 72 and was 80 percent of Hammack's score. What was Hammack's score?
7. Marguerite found four consecutive integers such that three times the sum of the first two was 1 larger than twice the sum of the last two. What were Marguerite's integers?
8. Find three consecutive even integers such that if 5 is added to the first integer and this sum multiplied by 6, the result is 6 larger than five times the third integer.

Solve:

*9. $\frac{t-2}{t} = \frac{14}{3t} - \frac{1}{3}$

10. $\frac{4x}{7} - \frac{1}{3} = x$

*11. $\frac{5x}{3} = \frac{4}{11}$

*12. $\frac{2}{3N} = \frac{2}{N+4}$

13. Which of the following do not represent functions. Why?

(a) (b) (c)

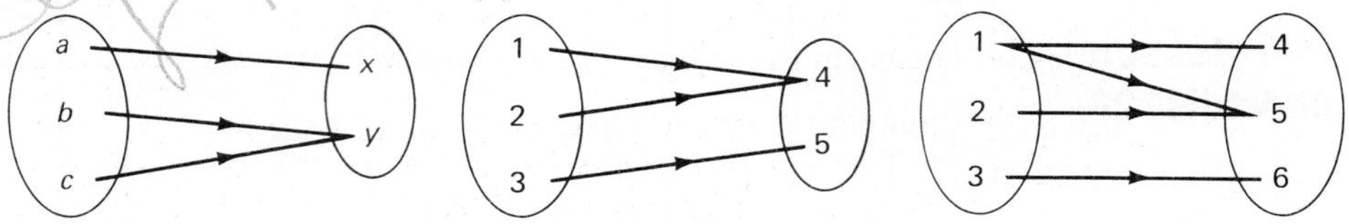

(d) (4, 7), (8, 5)
(7, 5), (9, 11)

Graph on a number line:

14. $4 - |x| \leq 2$, $D = \{\text{Reals}\}$

15. $-x + 4 < 1$, $D = \{\text{Integers}\}$

16. $|x| + 2 \leq 5$, $D = \{\text{Integers}\}$

Solve by graphing:

17. $\begin{cases} 3y - 2x = 6 \\ y = -x + 6 \end{cases}$

*The problems marked with an asterisk were worked out in this lesson.

Use square root tables to approximate:

18. $\sqrt{.01327 \times 10^{-6}}$ **19.** $\sqrt{.01327 \times 10^{-7}}$

20. Find m.

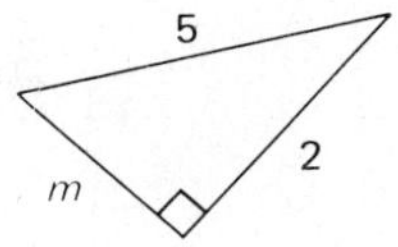

21. Find the distance between (4, 2) and (7, −5).

22. Use the commutative and associative properties to show that $x + y + m = x + m + y$.

Simplify:

23. $3\sqrt{2}\,(5\sqrt{8} + \sqrt{3})$ **24.** $3\sqrt{32} \cdot 2\sqrt{3} \cdot \sqrt{2}$ **25.** $3\sqrt{108} + 2\sqrt{52}$

26. Divide $2x^3 - 3x^2 - 5x - 3$ by $x - 2$.

Solve:

27. $x^2 = 7x - 10$

Simplify:

28. $\dfrac{2x + 14}{6x} \cdot \dfrac{9x^2}{x^2 + 7x}$

29. $R_1T_1 + R_2T_2 = 290$, $T_1 = 4$, $T_2 = 3$, $2R_1 + 20 = R_2$; find R_1

Simplify:

30. $\dfrac{(4{,}200{,}000)(.0001 \times 10^{-15})}{(3000 \times 10^{-13})(.000007)}$

Use elimination:

31. $\begin{cases} 9x + 4y = 6 \\ 3x - 3y = 15 \end{cases}$

32. Find the equation of the line: **33.** Find the equation of the line:

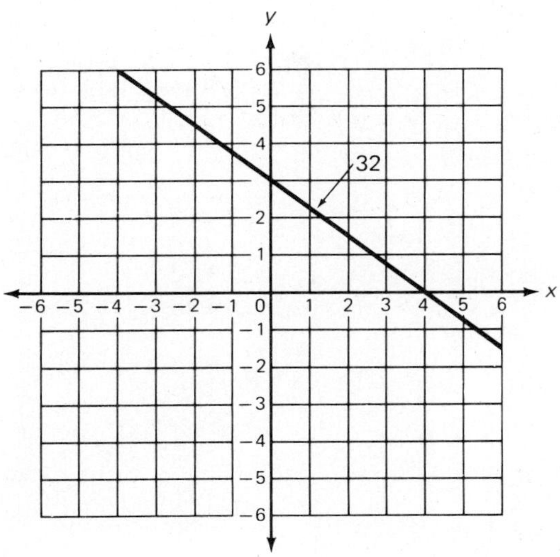

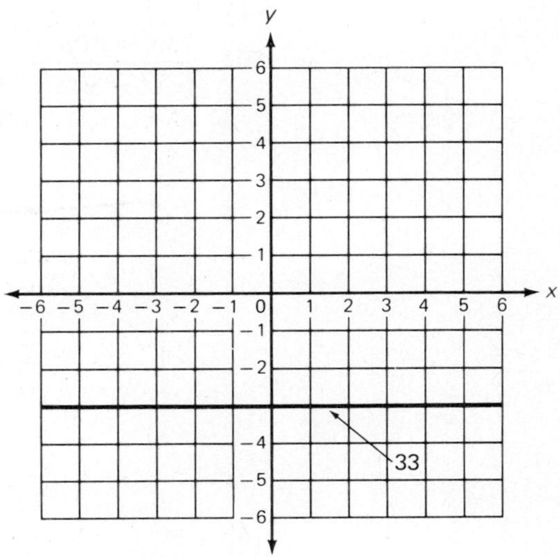

34. The set of negative integers is closed for what operations?

Solve:

35. $\dfrac{3x}{4} + \dfrac{x - 5}{7} = 10$

Graph on a rectangular coordinate system:

36. $y = -x$

37. $y = 2x - 3$

Simplify:

38. $\dfrac{\dfrac{a}{x}}{1 + \dfrac{a}{x}}$

39. $\dfrac{\dfrac{1}{b}}{a + b}$

Add:

40. $\dfrac{a}{b + c} - \dfrac{c}{b}$

41. $\dfrac{x^2}{y} + \dfrac{z}{y^2c} + \dfrac{1}{y^3c^2}$

Multiply:

42. $\dfrac{x^{-2}y}{b}\left(\dfrac{b^2y^{-3}}{x^4b} + \dfrac{2xy}{x^3}\right)$

Add like terms:

43. $-2x^{-4}y + \dfrac{3xy}{x^{-3}} - \dfrac{2y}{b^3}$

Simplify:

44. $\dfrac{(x^2y^4)^0a^{-2}y}{(y^{-2})^{-1}b^{-2}}$

45. $\dfrac{x^2 + 3x - 4}{x^2 - 4x + 3} \div \dfrac{x^2 + 2x - 8}{x^2 + 3x - 18}$

Solve:

46. $2x - (-2) + 3(4 + x) = 2(2x + 2)$

Evaluate:

47. $-[-b(a - c)]$ if $b = -2$, $a = 1$, $c = -3$

48. $\dfrac{-3^{-2}}{(-3)^2}$

49. True or false: {Rationals} $\subset$ {Naturals}

Add:

50. $\dfrac{2kx + 10p}{x^2 - 9} - \dfrac{2k}{x - 3}$

51. $\dfrac{-4x}{x^2 + 6x + 8} + \dfrac{4}{x + 4}$

lesson twenty-nine

Vertical line test for functions
Functional notation

29.1 functions

In Lesson 28 we discussed functions at some length but did not discuss why functions are important.

Algebra books published in the late nineteenth century and early twentieth century did not contain the word function. These books spoke of equations: linear equations, quadratic equations, and other types of equations. The calculus books of the 1930's and 1940's used the word function to describe some equations. The definition used, however, permitted more than one image for each member of the domain. If two images existed for some members of the domain, the function was called a **double-valued function.** If more than two images existed for some members of the domain, the function was called a **multivalued function.** The restriction on a function that it be single-valued is a rather recent development—say after 1950.

So we see that a function is the word we use now to describe what used to be called a single-valued equation. But because of the way the word is now defined, it covers any relationship that defines, for any value of the domain, one and only one image in the range. For instance, a set of ordered pairs such as (5, 4), (6, 2), and (3, −7) can be called a function even though there is no equation. The members of the domain are specified and each member of the domain is paired with exactly one image in the range. Many equations and other mathematical relationships are single-valued relationships, and it is convenient to have a word to describe them collectively.

example 29.1.1 Does the diagram designate a function?

solution

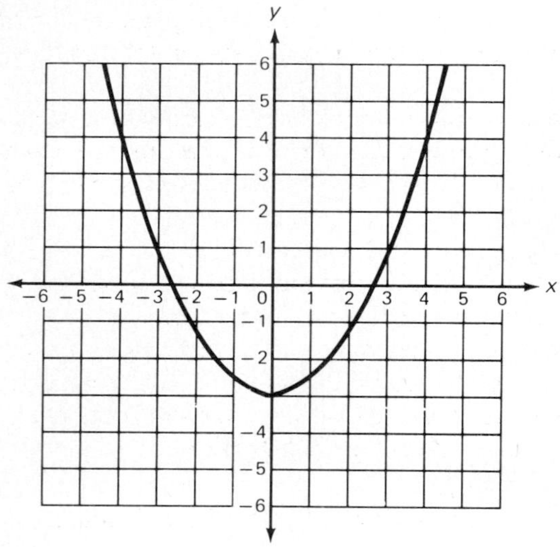

Yes—we assume that the figure is a portion of the entire graph and that the domain is all real numbers. If this is true, we see that the graph will match exactly one real value of y for each value of x.

example 29.1.2 Does the equation $x = y^2$ describe a function?

solution No. For example, if x is 4, then y could be either positive 2 or negative 2 because both

$$(2)^2 = 4 \quad \text{and} \quad (-2)^2 = 4$$

29.2 vertical line test

As we noted in section 28.2, it is customary to use the letter x to designate the independent variable in a two-variable equation, and it is customary to use the letter y to designate the dependent variable. There is no requirement that this convention be followed, but everyone does. It is also customary to graph the independent variable on the horizontal axis and the dependent variable on the vertical axis. If this convention is followed, the so-called **VERTICAL LINE TEST** can be used on the graph of a relation to see if the relation is also a function. Remember that a relation has one or more images for each element of the domain and a function is a relation that has exactly one image for every member of the domain.

In Figure 29.1, we have graphed three different relations. For each value of x, the graph designates at least one value of y. **If every possible vertical line that we can draw touches the graph of the relation at only one point, then the graph is the graph of a function.**

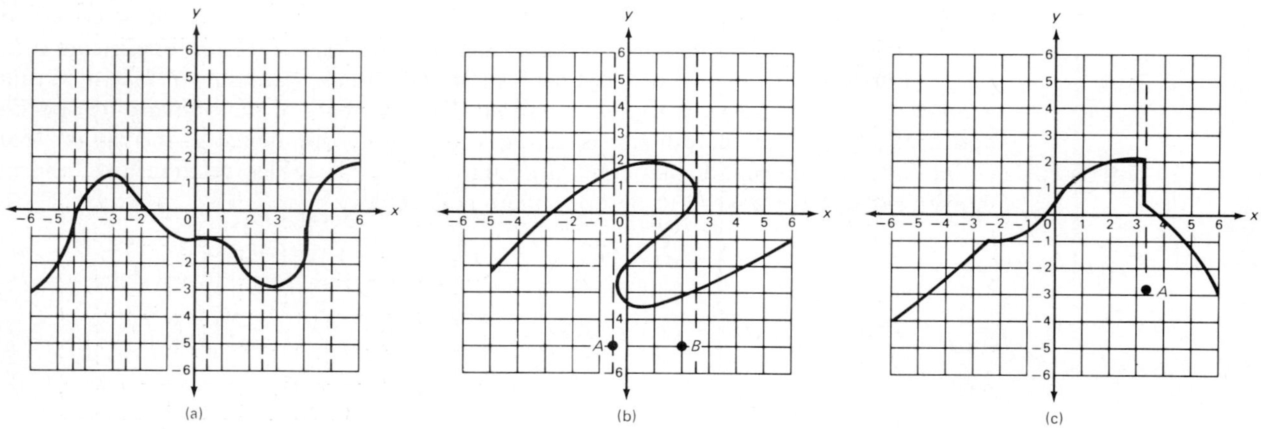

Figure 29.1

Look at Figure 29.1(a). We have drawn several vertical lines and each of these touches the graph at only one point. If we look carefully at the figure, we can tell that no matter where we draw the vertical line, it will touch the graph of the relation at only one point. Thus by using the vertical line test, we have shown that graph (a) is the graph of a function.

In Figure 29.1(b) we drew two vertical lines at values of x arbitrarily called A and B. We see that each of these lines or any vertical line drawn anywhere between them will touch the given graph in at least two points so the graph is not the graph of a function.

In Figure 29.1(c) the graph is vertical at one value of x that we have designated by the letter A. The vertical line drawn at this value of x touches the graph for its entire vertical length. Thus this graph is not the graph of a function.

The vertical line test can be used only if the convention of graphing the independent variable on the horizontal axis is followed. If the independent variable were graphed on the vertical axis, then we would have to use a horizontal line test. We will always follow the convention and thus will not dwell on other possibilities.

example 29.2.1 Is the relation graphed in the accompanying figure also a function?

solution

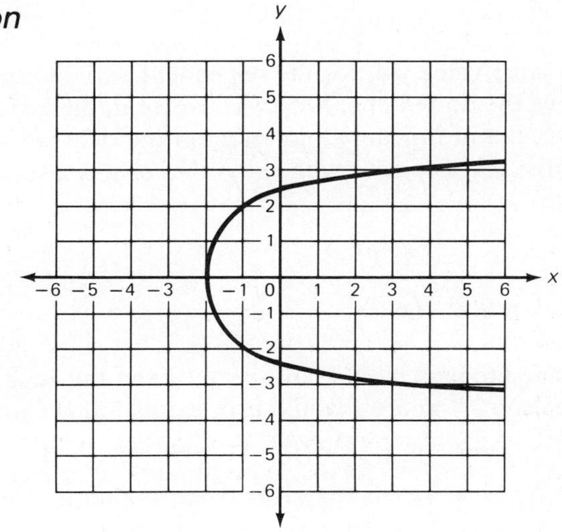

(The domain is assumed to be the values of x greater than -2.) *No.* Any vertical line drawn at a value of x such that $x > -2$ will touch the line in at least two points. This graph is the graph of a relation, not a function.

TEST YOUR UNDERSTANDING

(a) Use the vertical line test to determine if the graph at the right depicts a function.

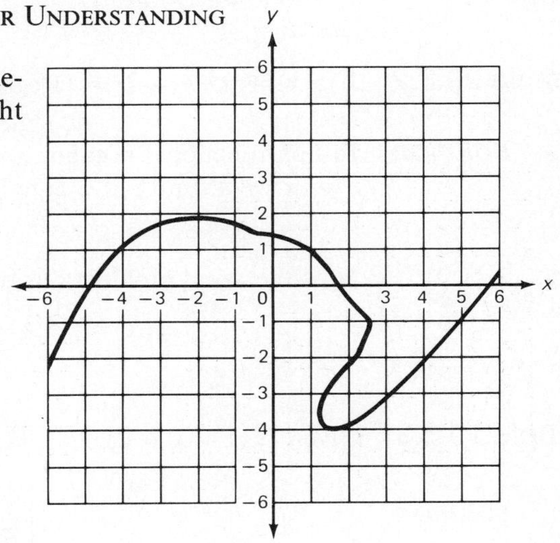

29.3 functional notation

In so-called functional notation we use symbols such as $f(x)$, $g(x)$, $h(x)$ to designate the dependent variable instead of using the letter y. Using functional notation the equation $y = 2x + 4$ could be written as

$$f(x) = 2x + 4 \quad \text{or} \quad g(x) = 2x + 4 \quad \text{or} \quad h(x) = 2x + 4$$

We do not restrict ourselves to the letters f, g, and h in designating functions, as any letter or symbol may be used.

One advantage of the functional notation is that the functional notation always indicates the element of the domain that is paired with a given image in the range. To demonstrate this on the left below we write the equation $y = 2x + 4$ and on the right we use functional notation for the same equation and write it as $f(x) = 2x + 4$. Then we replace x in each equation with the number 5 and solve to find the value of the dependent variable.

$y = 2x + 4$	$f(x) = 2x + 4$
$y = 2(5) + 4$	$f(5) = 2(5) + 4$
$y = 10 + 4$	$f(5) = 10 + 4$
- - - - - - - -	- - - - - - - -
$y = 14$	$f(5) = 14$

Of course, we get the same value, 14, for the dependent variable in both cases. If we cover up everything above the dotted line, however, we see in the last line on the left only the value for y is shown, but in functional notation on the right we see the value of y is shown, and the value of x that is paired with this value of y is also shown.

example 29.3.1 If $\phi(x) = 3x + 5$, $D = \{\text{Reals}\}$, find $\phi(-2)$.

solution In non-functional notation the problem would have read "given the equation $y = 3x + 5$, find y if x equals -2," and we would proceed as follows to find y.

$$y = 3x + 5 \longrightarrow y = 3(-2) + 5 \longrightarrow \mathbf{y = -1}$$

Now we will work the same problem again using functional notation as was used in the statement of the problem.

$$\phi(x) = 3x + 5 \longrightarrow \phi(-2) = 3(-2) + 5 \longrightarrow \boldsymbol{\phi(-2) = -1}$$

example 29.3.2 If $g(x) = x^2 + 4$, $D = \{\text{Reals}\}$, find $g(4)$.

solution In non-functional notation the problem would have read "given the equation $y = x^2 + 4$, find y if x equals 4."

$$y = x^2 + 4 \longrightarrow y = (4)^2 + 4 \longrightarrow \mathbf{y = 20}$$

Now the same problem but this time we will use functional notation,

$$g(x) = x^2 + 4 \longrightarrow g(4) = (4)^2 + 4 \longrightarrow \mathbf{g(4) = 20}$$

example 29.3.3 Given $f(x) = x + 2$, $D = \{\text{Reals}\}$, find $f(\frac{1}{2})$.

solution

$$f\left(\frac{1}{2}\right) = \left(\frac{1}{2}\right) + 2 \longrightarrow \mathbf{f\left(\frac{1}{2}\right) = \frac{5}{2}}$$

example 29.3.4 Given $f(x) = x + 2$, $D = \{\text{Integers}\}$, find $f(\frac{1}{2})$.

solution The domain of the function is the set of integers. The problem asks that we find the image of $\frac{1}{2}$. One-half does not have an image under this function because the replacement values of x are restricted to the set of integers. Thus we say that there is no real number that satisfies both conditions so the solution is the empty set { } or the null set $\varnothing$.

TEST YOUR UNDERSTANDING

If $f(x) = x^2 - 4$, $D = \{\text{Integers}\}$, and $g(x) = x + 2$, $D = \{\text{Negative integers}\}$

(b) Find $f(5)$

(c) Find $g(5)$

lesson checklist 29

- ☐ Double-valued function
- ☐ The new definition of function
- ☐ Vertical line test
- ☐ Functional notation

ANSWERS TO TEST YOUR UNDERSTANDING

(a) It does not depict a function (b) 21 (c) $\varnothing$

problem set 29

1. Pegasus collected horsefeathers. He paid \$3.00 for small horsefeathers and \$5.00 for large horsefeathers. If he had 153 horsefeathers and had paid \$559.00 for the lot, how many large horsefeathers did he have?
2. Demosthenes had 143 pebbles. If all the pebbles were either white or black and the number of black pebbles was 8 more than four times the number of white pebbles, how many pebbles were black?
3. Themistocles bought large ships and small ships. He paid 60 minas† for large ships and 20 minas for small ships. If he had 140 ships and paid 4000 minas for the lot, how many large ships did he have?
4. King Minos had 168 slave girls. If he paid 1 talent for smart girls and 3 talents for the other girls and had paid 258 talents for the bevy, how many other slave girls did he have?

†In Greek currency in the fifth century B.C., it took 60 minas to equal one talent.

5. Suela counted the votes and found that twenty percent of those present at the meeting voted aye. If 180 people voted aye, how many people were at the meeting?

6. There were 238 Arapahoes at the powwow. If a total of 1400 Indians attended the powwow, what percent of the total were Arapahoes?

7. Find three consecutive odd integers such that 5 times the sum of the first and 7 equals the opposite of 10 times the third.

8. Given that $f(x) = 4x + 2, D = \{\text{Reals}\}$; find $f(2)$

9. Given that $g(x) = x + 5,\ D = \{\text{Reals}\}$; find $g(4)$

*10. Given that $\phi(x) = 3x + 5,\ D = \{\text{Reals}\}$; find $\phi(-2)$

11. Which of the following represent functions? Use the vertical line as necessary.

(a)

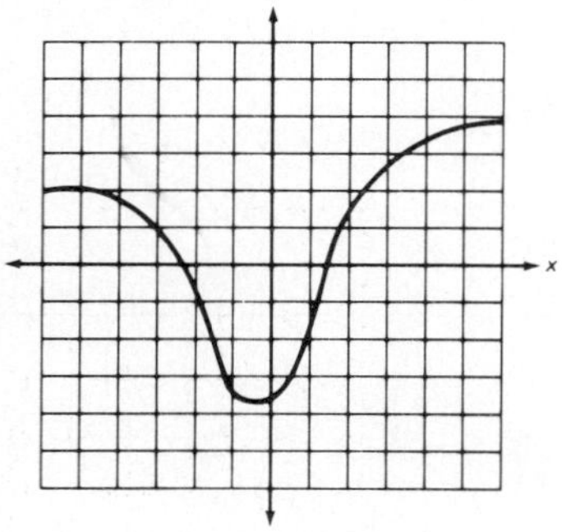

(b)

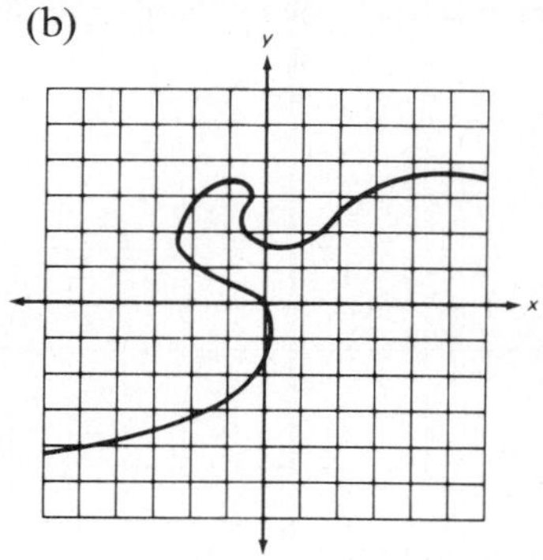

(c)

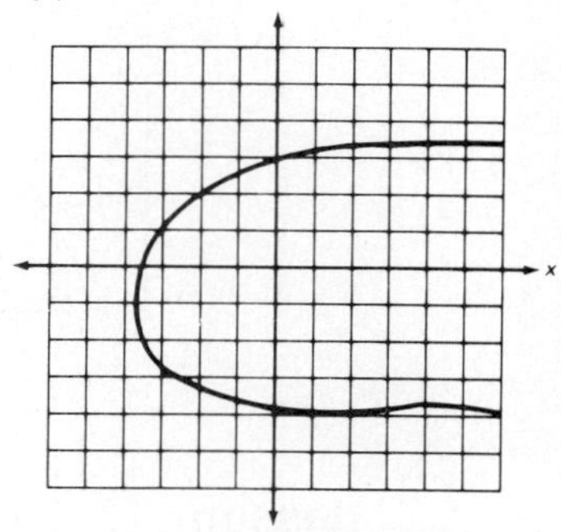

(d) (7, 6), (4, 2)
(5, 9), (7, 6)
(7, 6), (4, 3)

(e)

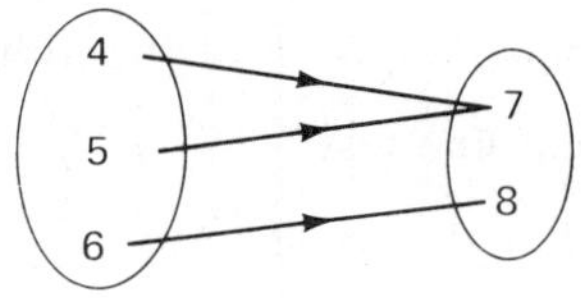

(f)

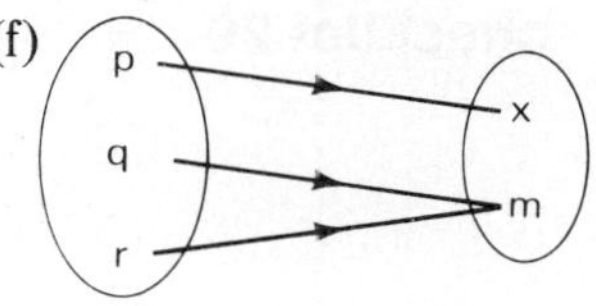

Solve:

12. $\dfrac{x+1}{x-1} = \dfrac{3}{2}$

13. $\dfrac{5}{x-3} = \dfrac{12}{x+1}$

Graph on a number line:

14. $-x - 2 \geq -4,\ D = \{\text{Integers}\}$

15. $|x| + 2 < 6,\ D = \{\text{Reals}\}$

16. $-|x| - 2 < -4,\ D = \{\text{Reals}\}$

Use square root tables to approximate:

17. $\sqrt{61{,}586 \times 10^5}$

18. $\sqrt{61{,}586 \times 10^4}$

19. Find h.

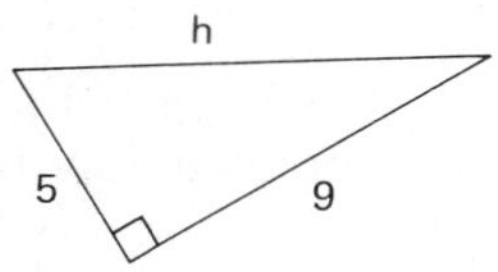

20. Use the commutative and associative properties to show that $m + x + y = x + m + y$.

Simplify:

21. $2\sqrt{2}\,(\sqrt{2} + 4)$

22. $4\sqrt{45} - 2\sqrt{180}$

23. $3\sqrt{2} \cdot 2\sqrt{6} \cdot \sqrt{12}$

*The problems marked with an asterisk were worked out in this lesson.

Divide:

24. $x^3 - x + 2$ by $x - 1$

Solve:

25. $x^2 + 8x = -12$

Simplify:

26. $\dfrac{x^2 - 2x + 1}{4x} \cdot \dfrac{8x^2}{x - 1}$

27. $\dfrac{(.00035 \times 10^{14})(7000 \times 10^{-21})}{50{,}000{,}000}$

Use elimination:

28. $4x + 2y = 12$
$3x - 3y = 18$

29. Find the equation of the line:

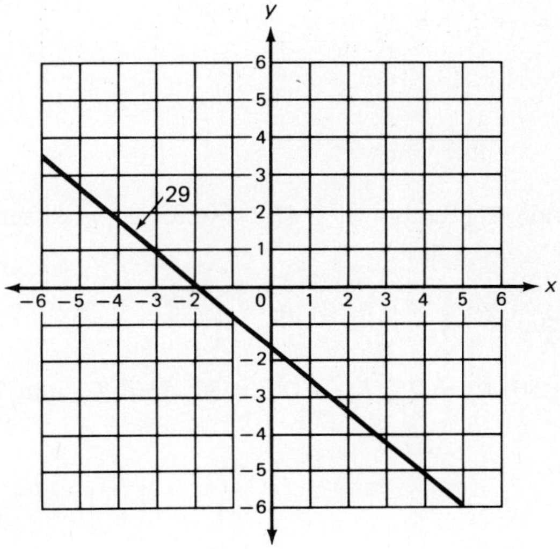

30. Find the equation of the line:

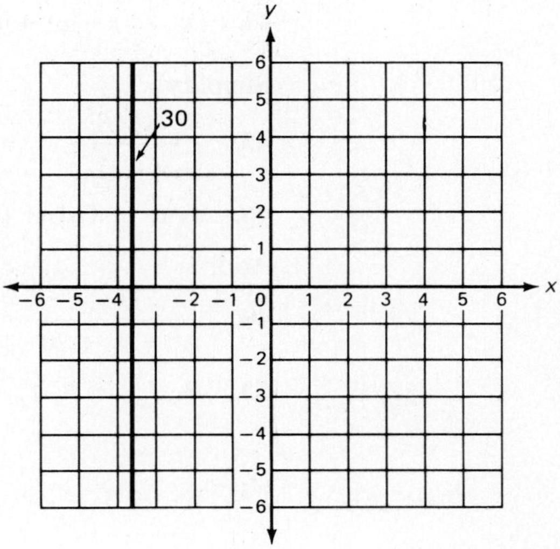

31. The set $\{0, 1, -1\}$ is closed for what operations?

Graph on a rectangular coordinate system:

32. $y = -4$

Solve by graphing:

33. $\begin{cases} y = -2x + 3 \\ y = -2 \end{cases}$

Simplify:

34. $\dfrac{a + \dfrac{b}{a}}{a^2 + b}$

35. $\dfrac{\dfrac{1}{a}}{\dfrac{a}{x}}$

Add:

36. $\dfrac{x}{a^2} + \dfrac{b}{a + c}$

37. $\dfrac{a}{yc} + \dfrac{b}{yc^2} + \dfrac{d}{y^2c}$

Add like terms:

38. $ax^2c + 2a^2cx - \dfrac{3ca}{x^{-2}} + \dfrac{4xa^2}{c^{-1}}$

Simplify:

39. $\dfrac{x^2 - 9}{x^2 - x - 6} \div \dfrac{x^2 + 10x + 21}{x^2 - 3x - 10}$

Multiply:

40. $ax^{-2}\left(\dfrac{x^2}{a^{-1}} - \dfrac{4x^2}{a}\right)$

Simplify:

41. $\dfrac{m^{-4}x^2(p^{-3})^2}{m^{-2}x^{-2}p^3y^0}$

42. $\dfrac{a^{-2}xax^{-2}}{py^0p^{-3}}$

Solve:

43. $4(x - 2) - (-4) = 2(x + 3)$

Evaluate:

44. $(x - y)(-x) + y$ if $x = -2, y = 5$

Simplify:

45. $\dfrac{-3^{-2}}{-(-3)^{-2}}$

46. True or False: {Natural} $\in$ {Reals}

47. $-6.35 \in$ {what sets of numbers?}

Add:

48. $\dfrac{3kx}{x^2 - x - 20} - \dfrac{3k}{x - 5}$

49. $\dfrac{3p}{x^2 - x} - \dfrac{2}{x}$

50. $R_M T_M = R_P T_P$, $R_M = 50$, $R_P = 25$, $T_M + T_P = 30$; find T_M *and* T_P.

lesson thirty

Uniform motion word problems

30.1 uniform motion word problems

Since Lesson 20 we have been using substitution to solve systems of four simultaneous equations involving four unknowns such as

$$R_F T_F = R_E T_E,\ T_F = 16,\ T_E = 12,\ R_E = R_F + 15$$

These equations are typical of the equations that we will learn to write in this lesson to help us solve uniform motion word problems. Uniform motion problems are so named because the statement of the problem tells about objects or things that move at a uniform velocity or at an average velocity.

The statements of equality made in uniform motion problems are statements of equality that concern **DISTANCE**, statements of equality that concern **RATE**, and statements of equality that concern **TIME**, and use the relationship that

$$\text{Distance} = \text{rate} \times \text{time}$$

or

$$D = RT$$

The statements about rate and time are not difficult to locate in the wording of the problem, but the beginner often has trouble identifying the statement that defines the relationship that concerns distances. **Since the distance equation is the troublesome one, we will consider this equation to be the key equation for this type of problem, and we will always write this equation first.** Then we will write the equations that concern time and the equations that concern rate. **When we have as many equations as we have variables, we will use the substitution method to solve for the variables.**

The statements of the distances discussed in the problems can be represented graphically by drawing diagrams in which arrows represent the distances. The problems in this book will usually discuss two distances and the diagrams will be one of three basic types.

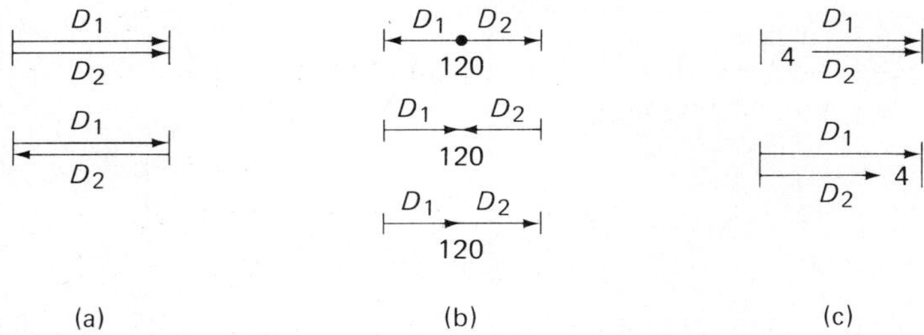

Figure 30.1

In Figure 30.1(a), we show the diagrams that will result when the problem states that the two distances are equal. In the upper diagram, the objects traveled in the same direction and in the lower diagram the objects traveled in opposite directions. Both cases lead to the same distance equation, which is

$$\text{Distance one} = \text{distance two} \quad \text{or} \quad D_1 = D_2$$

and since $D_1 = R_1t_1$ and $D_2 = R_2t_2$, we can write

$$R_1t_1 = R_2t_2$$

In Figure 30.1(b) we show the diagrams that result when the problem states that the sum of the distances traveled by two objects equals some given distance (here we use a distance of 120 units). In the upper diagram the objects began their movement from the same point and traveled in opposite directions until they were 120 units apart and in the middle diagram the objects began 120 units apart and traveled in opposite directions until they met. In the lower diagram they traveled in the same direction and the sum of the two distances equals 120. All three cases lead to the equation

$$\text{Distance one} + \text{distance two} = 120 \quad \text{or} \quad D_1 + D_2 = 120$$

and since $D_1 = R_1t_1$ and $D_2 = R_2t_2$, we can write

$$R_1t_1 + R_2t_2 = 120$$

Figure 30.1(c) shows the diagram that will result when one object travels farther by a specified distance (in this case 4 units) than the other object travels. From the diagrams we see that, in both cases, distance one is 4 units greater than distance two; so if we add 4 units to distance two, we get a statement of equality which is

$$\text{Distance one} = \text{distance two} + 4 \quad \text{or} \quad D_1 = D_2 + 4$$

and since $D_1 = R_1t_1$ and $D_2 = R_2t_2$, we can write

$$R_1t_1 = R_2t_2 + 4$$

example 30.1.1 On Tuesday the express train makes the trip in 12 hours. On Wednesday the freight train makes the same trip in 16 hours. Find the rate of each if the rate of the freight was 15 km/hr less than the rate of the express.

solution We read the problem and disregard statements about time and rate and look for the statement about distance. This information allows us to draw the diagram and from the diagram get the equation

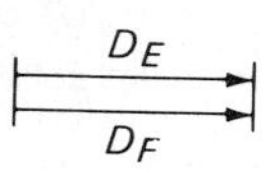

$$\text{Distance express} = \text{distance freight}$$

$$D_E = D_F$$

This is the distance equation which is the key equation for this uniform motion problem. Now the distance the express traveled equals the rate of the express times the time the express traveled, or

$$D_E = R_E T_E$$

and the distance the freight traveled equals the rate of the freight times the time the freight traveled, or

$$D_F = R_F T_F$$

Now if we substitute $R_E T_E$ for D_E and $R_F T_F$ for D_F in the equation $D_E = D_F$, we get

$$\mathbf{R_E T_E = R_F T_F}$$

which is the distance equation for this problem in final form.

The statement of the problem gives the time of the express as 12 hours and the time of the freight as 16 hours so

$$\mathbf{T_E = 12 \qquad T_F = 16}$$

Now, we have three equations, but have four unknowns, R_E, T_E, R_F, and T_F. **Thus we need one more equation so that the number of equations will equal the number of unknowns.** We get the final equation from the statement in the problem concerning rates which says that the rate of the express is 15 km/hr greater than the rate of the freight. **Writing this equation is tricky.** In an effort to avoid the common error of adding 15 to the wrong side, we first write

$$R_F = R_E \qquad \text{(false)}$$

which we know is incorrect for the rate of the express is greater than the rate of the freight. We add 15 to R_F so that the equation will have equal quantities on both sides.

$$\mathbf{R_E = R_F + 15}$$

We have found four equations in four unknowns,

$$\underbrace{R_E T_E = R_F T_F,} \qquad \underbrace{T_E = 12,} \qquad \underbrace{T_F = 16,} \qquad \underbrace{R_E = R_F + 15}$$

and we use the substitution method to solve for R_F and R_E.

$$R_E T_E = R_F T_F \longrightarrow (R_F + 15)12 = R_F(16) \longrightarrow 12R_F + 180 = 16R_F$$
$$\longrightarrow -4R_F = -180 \longrightarrow 4R_F = 180 \longrightarrow \mathbf{R_F = 45\ km/hr}$$

and since $R_E = R_F + 15$,

$$\mathbf{R_E = 60\ km/hr}$$

example 30.1.2 A robber has a 2-kilometer head start on a cop. How long will it take the cop to catch the robber if the robber travels at 70 km/hr and the cop travels at 100 km/hr?

solution The statement of the problem allows us to draw the following diagram that depicts the distances traveled.

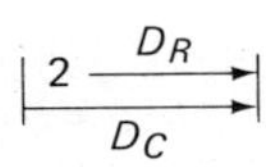

We see that the distance the cop traveled is 2 kilometers greater than the distance the robber traveled, which leads to

$$D_C = D_R + 2$$

and substituting $R_C T_C$ for D_C and $R_R T_R$ for D_R we find the final form of the distance equation.

$$\mathbf{R_C T_C = R_R T_R + 2}$$

We see that we have four unknowns and only one equation, so we need three more equations. The rate equations are

$$\mathbf{R_C = 100 \qquad R_R = 70}$$

and since the robber and the cop traveled for the same length of time, the time equation is

$$T_R = T_C$$

This gives us the last of our four equations.

$$\underbrace{R_C T_C = R_R T_R + 2,} \qquad \underbrace{R_C = 100,} \qquad \underbrace{R_R = 70,} \qquad \underbrace{T_R = T_C}$$

Now we use substitution to solve.

$$100T_C = 70T_C + 2 \longrightarrow 30T_C = 2 \longrightarrow T_C = \frac{1}{15} \textbf{ hr or 4 min}$$

example 30.1.3 The members of the Boys Club hiked to Chehaw State Park at 2 mph. Mr. Gillespie gave them a ride back to town in his delivery truck that traveled at 18 mph. If their total traveling time was 10 hours, how far was it to Chehaw State Park?

solution They hiked out and rode back for the same distance, which gives us the following distance diagram.

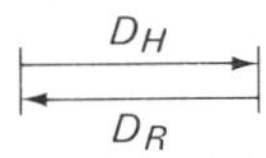

The distance equation is

$$D_H = D_R \quad \text{or} \quad \boldsymbol{R_H T_H = R_R T_R}$$

The rates were given in the statement of the problem

$$\boldsymbol{R_H = 2 \qquad R_R = 18}$$

The total time hiking and riding was 10 hours.

$$\boldsymbol{T_H + T_R = 10}$$

Now we have four equations in four unknowns.

$$\underbrace{R_H T_H = R_R T_R,} \qquad \underbrace{R_H = 2,} \qquad \underbrace{R_R = 18,} \qquad \underbrace{T_H + T_R = 10}$$

We use substitution to solve for T_R.

$$2(10 - T_R) = 18T_R \longrightarrow 20 - 2T_R = 18T_R \longrightarrow \boldsymbol{T_R = 1 \text{ hr}}$$

Since the rate riding was 18 mph and we have found the time riding to be 1 hour, we can find the distance by multiplying these two numbers.

distance = rate × time = 18 mph × 1 hour = **18 miles**

example 30.1.4 A southbound bus left Fort Walton Beach at 9 a.m. Two hours later a northbound bus left the same station. If the buses traveled at the same rate and were 352 kilometers apart at 2 p.m., find the rate of the buses.

solution The statement of the problem leads to this distance diagram.

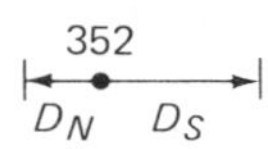

The distance equation is $D_N + D_S = 352$ or $\boldsymbol{R_N T_N + R_S T_S = 352}$. The southbound bus traveled for 5 hours and the northbound bus traveled for 3 hours.

$$\boldsymbol{T_N = 3 \qquad T_S = 5}$$

The rates were the same so $\boldsymbol{R_N = R_S}$.

Now we have four equations in four unknowns.

$$\underbrace{R_N T_N + R_S T_S = 352}, \quad \underbrace{T_N = 3}, \quad \underbrace{T_S = 5}, \quad \underbrace{R_N = R_S}$$

We use substitution to solve.

$$R_N(3) + R_N(5) = 352 \longrightarrow 8R_N = 352 \longrightarrow \mathbf{R_N = 44} \text{ km/hr}$$

and therefore

$$\mathbf{R_S = 44 \text{ km/hr}}$$

example 30.1.5 Harry and Jennet jog around a circular track that is 210 meters long. Jennet's rate is 230 meters per minute while Harry's rate is only 200 meters per minute. In how many minutes will Jennet be a full lap ahead?

solution This problem is simpler if we straighten it out and get the following distance diagram.

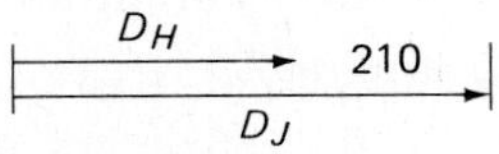

Thus the distance equation is $D_H = 210 + D_J$ or $\mathbf{R_H T_H + 210 = R_J T_J}$.

The time equation is $\mathbf{T_H = T_J}$.

The rate equations are $\mathbf{R_J = 230}$, $\mathbf{R_H = 200}$.

Thus our four equations are

$$\underbrace{R_H T_H + 210 = R_J T_J}, \quad \underbrace{T_H = T_J}, \quad \underbrace{R_J = 230}, \quad \underbrace{R_H = 200}$$

We use substitution to solve.

$$200T_H + 210 = 230T_H \longrightarrow 210 = 30T_H \longrightarrow T_H = 7 \text{ min}$$

Thus $T_J = \mathbf{7}$ **min** because $T_J = T_H$.

TEST YOUR UNDERSTANDING

(a) A southbound bus left Annville at 12 noon and headed south at 60 mph. One hour later a northbound bus traveling at 50 mph left the same terminal. At what time will the buses be 390 miles apart?

lesson checklist 30

- ☐ Distance equals rate times time
- ☐ Key equation
- ☐ Equating equal distances

ANSWERS TO TEST YOUR UNDERSTANDING

(a) $R_N T_N + R_S T_S = 390$, $R_S = 60$, $R_N = 50$, $T_S = T_N + 1$,

$T_S = 4$, $T_N = 3$ $\longrightarrow$ **4 p.m.**

problem set 30

*1. On Tuesday the express train makes the trip in 12 hours. On Wednesday the freight train makes the same trip in 16 hours. Find the rate of each if the rate of the freight was 15 km/hr less than the rate of the express.

*2. A robber has a 2-mile head start on a cop. How long will it take the cop to catch the robber if the robber travels at 70 mph and the cop travels at 100 mph?

*3. The members of the Boys Club hiked to Chehaw State Park at 2 km/hr. Mr. Gillespie gave them a ride back to town in his delivery truck that traveled at 18 km/hr. If their total traveling time was 10 hours, how far was it to Chehaw State Park?

*4. A southbound bus left Fort Walton Beach at 9 a.m. Two hours later a northbound bus left the same station. If the buses traveled at the same rate and were 352 kilometers apart at 2 p.m., find the rate of the buses.

5. Meme cut a 98-meter rope into two pieces so that the long piece was 10 meters shorter than twice the length of the short piece. How long were the pieces?

6. Bubber and Nancy sold 142 insurance policies. If Bubber sold 10 more than three times the number that Nancy sold, how many policies did Nancy sell?

7. Uly had to pay Wyatt 4 percent interest on his loan. If the loan was for 12,000 dollars, how much did Uly have to pay?

8. Find three consecutive odd integers such that 12 times the first is 10 greater than the product of 5 times the sum of the second and 3.

9. Given that $f(x) = x + 4$, $D = \{\text{Reals}\}$; find $f(-2)$

10. Given that $g(x) = x^2 - 1$, $D = \{\text{Reals}\}$; find $g(3)$

11. Which of the following represent functions? Use the vertical line test as necessary.

(a) (b) (c)

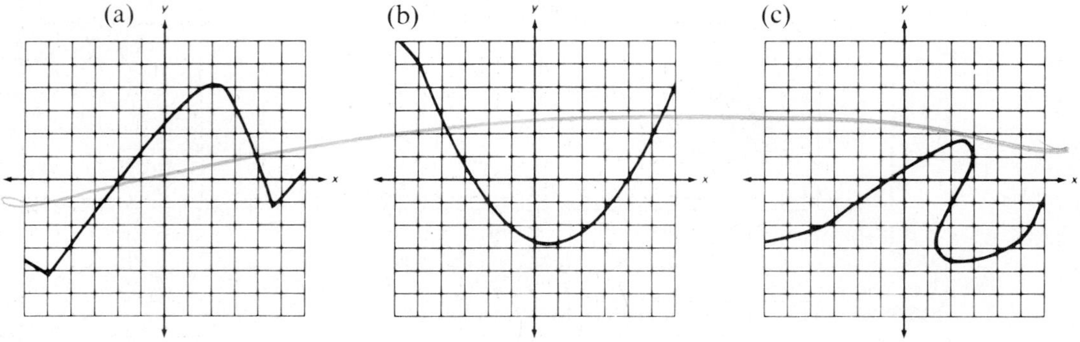

*The problems marked with an asterisk were worked out in this lesson.

(d) (7, 11), (14, 21), (15, 13), (−9, 11), (9, 18), (5, 21)

(e)

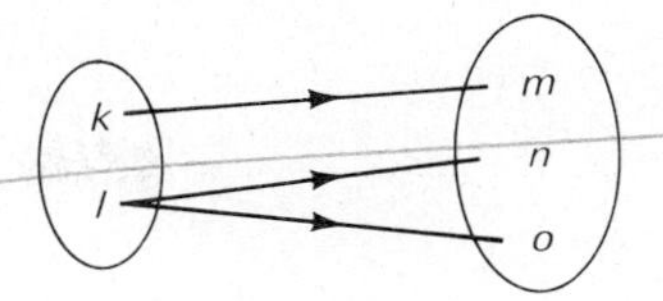

(f)

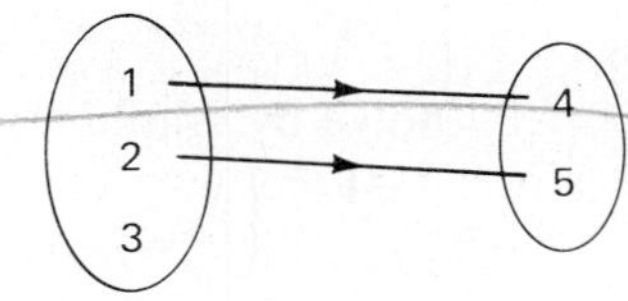

Solve:

12. $\dfrac{3x+4}{3} + \dfrac{4}{5} = 20$

13. $\dfrac{4}{x} = \dfrac{-1}{x+2}$

Graph on a number line:

14. $-x + 4 \geq 2$, $D = \{\text{Reals}\}$

15. $|x| - 2 > 1$, $D = \{\text{Integers}\}$

Use square root tables to approximate:

16. $\sqrt{.045086 \times 10^{-5}}$

17. Find the distance between $(-2, 4)$ and $(-5, -3)$.

18. Use the commutative and associative properties to show that $xyp = xpy$.

Simplify:

19. $4\sqrt{3}\,(\sqrt{12} - 5)$

20. $\sqrt{18} + \sqrt{50} - \sqrt{32}$

21. $3\sqrt{2} \cdot 3\sqrt{6} \cdot \sqrt{12}$

Divide:

22. $x^2 - 4$ by $x + 5$

Solve:

23. $x^2 = 12x - 36$

Simplify:

24. $\dfrac{6z^2 - 24}{3z + 6} \cdot \dfrac{2z + 3}{z - 2}$

25. $\dfrac{(4000 \times 10^{15})(.00072)}{(.03 \times 10^{15})(.0004 \times 10^{7})}$

Use elimination:

26. $\begin{cases} 7x + 2y = 15 \\ 3x - 4y = -13 \end{cases}$

27. Find the equation of the line:

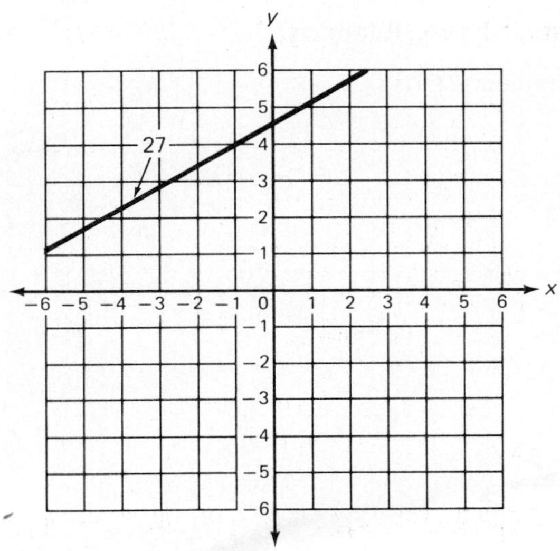

28. Find the equation of the line:

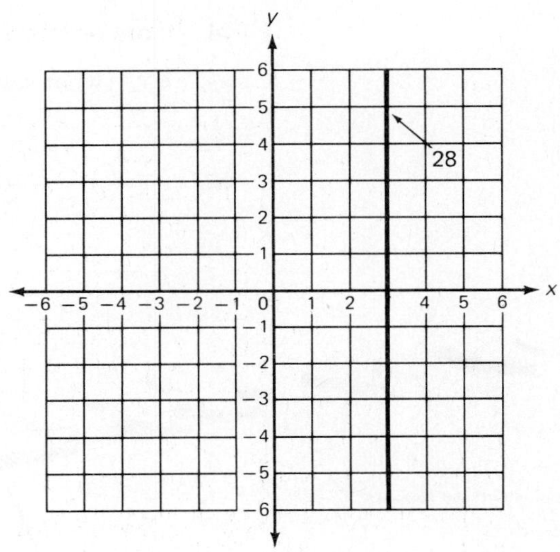

29. The set of real numbers is closed for what operations?

Graph on a rectangular coordinate system:

30. $x = -3$

Solve by graphing:

31. $\begin{cases} y - x = 4 \\ y = -x \end{cases}$

Simplify:

32. $\dfrac{x + y}{1 + \dfrac{x}{y}}$

33. $\dfrac{\dfrac{a}{1 + b}}{\dfrac{1}{b}}$

Add:

34. $\dfrac{a}{x^2} + \dfrac{c^2}{x + y}$

35. $\dfrac{m}{a^2} + \dfrac{x}{a^2b} + \dfrac{k}{b^2a}$

Add like terms:

36. $ax^2 - \dfrac{4a}{x^{-2}} + \dfrac{5a}{x^2}$

37. $x^2am^{-2} - \dfrac{4x^2}{m^2a^{-1}} + \dfrac{2m^{-2}}{x^{-2}a} + \dfrac{3x^2a^{-1}}{m^2}$

Simplify:

38. $\dfrac{x^2 - 36}{x^2 - 4x - 12} \div \dfrac{x^2 + 11x + 30}{x^2 - x - 6}$

39. $\dfrac{-8y^3x^2z^4}{2x^{-1}yz^5}$

40. $\dfrac{m^2xy^0x^{-2}m^4}{x^5y^{-3}x^{-3}m^2y}$

Solve:

41. $3(y - 5) - 6 + (-8) = 2(y - 2)$

Evaluate:

42. $-m[x - (-a)]$ if $m = 4$, $x = 2$, $a = -3$

Simplify:

43. $\dfrac{(-2)^{-2}}{-2^2}$

44. True or false: {Irrationals} $\subset$ {Rationals}

45. $4 \in$ {what sets of numbers?}

Add:

46. $\dfrac{5p}{x^2 - 25} + \dfrac{p}{x + 5}$

47. $\dfrac{6m}{x^2 + 2x - 8} + \dfrac{3m}{x + 4}$

lesson thirty-one

Equation of a line through two designated points
Equation of a line through a point and parallel to a given line
Equation of a line with a given slope and through a given point

31.1 linear equations

The graph of a first-degree equation in two unknowns is a straight line. This is the reason we call these equations linear equations. We also use the name linear equation to describe first-degree equations in more than two unknowns. The graph of a first-degree equation in three unknowns such as

$$4x + 2y - z = 5$$

is a plane and an equation in more than three unknowns cannot be graphed because graphs are restricted to the three dimensions available for graphing.

The so-called **standard form of the equation of a straight line** is

$$ax + by + c = 0$$

where a, b, and c are constants. The following are equations of straight lines in standard form

$$4x + y + 1 = 0 \qquad -2x - y - 11 = 0$$

If the equation of a line is written so that y is expressed as a function of x such as

$$y = mx + b$$

we say that we have written the equation in **slope-intercept form**. In the equation above, m represents the slope of the line and b represents the y-intercept of the line, which is the y-coordinate of the point where the line in question crosses the y-axis.

Thus far, we have learned how to draw the graph of a given linear equation and have learned how to find a good approximation of the equation of a given line. Both of these exercises have helped us to understand the relationship between the equation of a line and the graph of a line.

Algebra books usually contain three other types of straight line problems that are helpful in exploring this relationship. In the first type we are given the coordinates of two points and asked to find the equation of the line that passes through the two points.

31.2 equation of a line through two points

example 31.2.1 Find the equation of the line that goes through the points (4, 2) and (−5, −3).

solution The slope-intercept form of the desired equation is $y = mx + b$, and **we need to find the values of m and b.** First we graph the two points and draw the line (see figure at left). If we draw the slope triangle so that the sides of the triangle terminate on these points as we have done (in figure at right), we can determine the **exact value of the slope** to be $+\frac{5}{9}$. We say the exact value because we were given the exact coordinates of two points on the line. We did not estimate their value by looking at a graph.

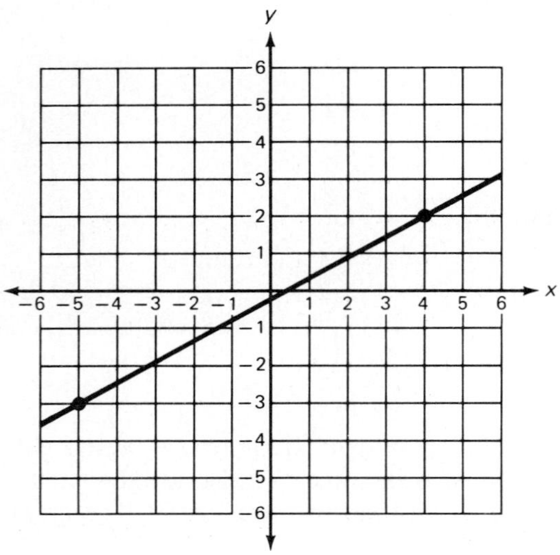

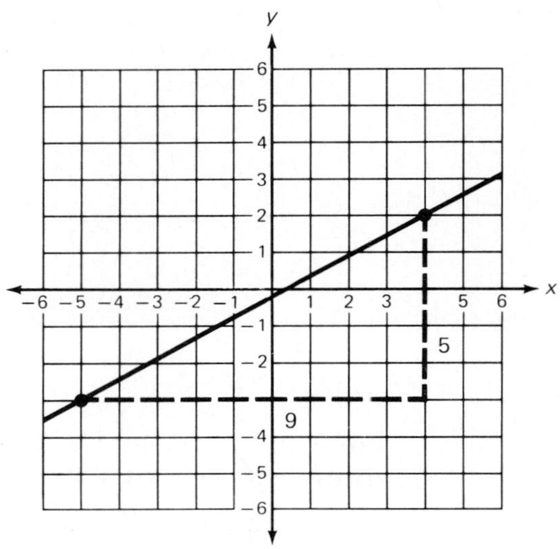

We can see from the graph that the value of the intercept is approximately −.3. **This estimated value of the intercept is not acceptable when the exact coordinates of two points on the graph are known, for we can find the exact value for the intercept.** We know the exact value of the slope so we can state the desired equation is

$$y = \frac{5}{9}x + b$$

We know the exact values of the coordinates of two points that lie on the line. We can use the coordinates of either of these points for x and y in the equation above and find the exact value of b algebraically.

USING (4, 2)	USING (−5, −3)
$y = \frac{5}{9}x + b$	$y = \frac{5}{9}x + b$
$(2) = \frac{5}{9}(4) + b$	$(-3) = \frac{5}{9}(-5) + b$
$\frac{18}{9} = \frac{20}{9} + b$	$-\frac{27}{9} = -\frac{25}{9} + b$
$-\frac{2}{9} = b$	$-\frac{2}{9} = b$

Now that we have the exact values of m and b, we can write the exact equation of the line that goes through the two points as

$$y = \frac{5}{9}x - \frac{2}{9}$$

example 31.2.2 Find the equation of the line that goes through the points (4, −2), (−3, 4).

solution

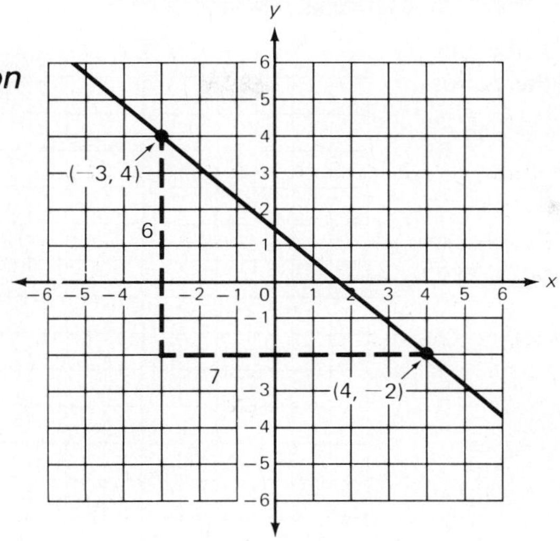

The general form of the desired equation is $y = mx + b$. We need to find the values of m and b. We graph the points, draw the line, and draw the triangle to find that the slope m is exactly $-\frac{6}{7}$.

Now we have

$$y = -\frac{6}{7}x + b$$

and we can find the exact value of b by substituting either (4, −2) or (−3, 4) for x and y and solving algebraically for b. We will use the point (4, −2).

$$-2 = -\frac{6}{7}(4) + b$$

$$-\frac{14}{7} = -\frac{24}{7} + b$$

$$\mathbf{\frac{10}{7} = b}$$

So the desired equation is $y = -\frac{6}{7}x + \frac{10}{7}$ and the values that we have found for the slope and the intercept are exact.

We see from these two problems that when we are given the coordinates of two points that lie on the line, the exact equation of the line can be determined. Estimated values of the slope and intercept are not acceptable for this type of problem.

example 31.2.3 Find the equation of the line through the points (4, 3), (4, −3).

solution

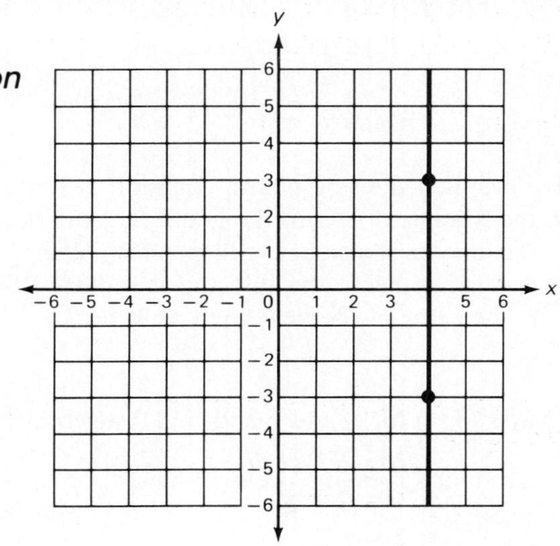

When we graph the points and draw the line, we find that the line is a vertical line. Vertical and horizontal lines can be thought of as special cases. By inspection the equation of this line is $x = 4$.

TEST YOUR UNDERSTANDING

(a) Find the equation of the line that passes through the points (5, −2) and (−3, 4).

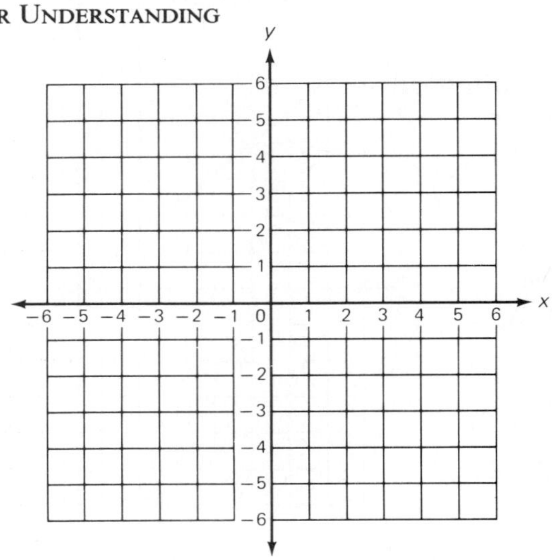

31.3 equation of a line through a point and parallel to a given line

To find the slope of a line through two specified points, we have to graph the points, draw the line, draw the slope triangle, and compute the slope. To find the slope of a line that is to be parallel to a given line, all that we need to do is to realize that two parallel lines have the same slope.

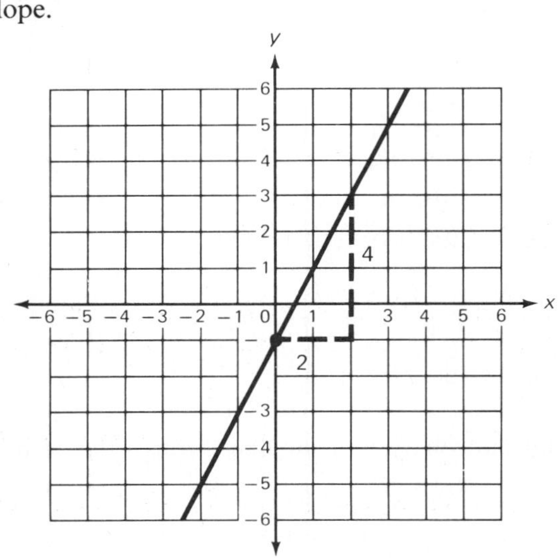

Figure 31.1

In Figure 31.1 we have graphed the line whose equation is $y = 2x - 1$. We see from the equation and from the triangle drawn in the graph that the slope of this line is 2. **Any line that is parallel to this line must also have a slope of 2.** Thus if we are asked to find the equation of the line that is parallel to this line and that passes through (3, 4), we are already halfway home, for the slope of the new line has to be 2.

$$y = 2x + b$$

Now we can use the coordinates (3, 4) for x and y and find b algebraically.

$$4 = 2(3) + b$$
$$4 = 6 + b$$
$$-2 = b$$

So the equation of the line is

$$y = 2x - 2$$

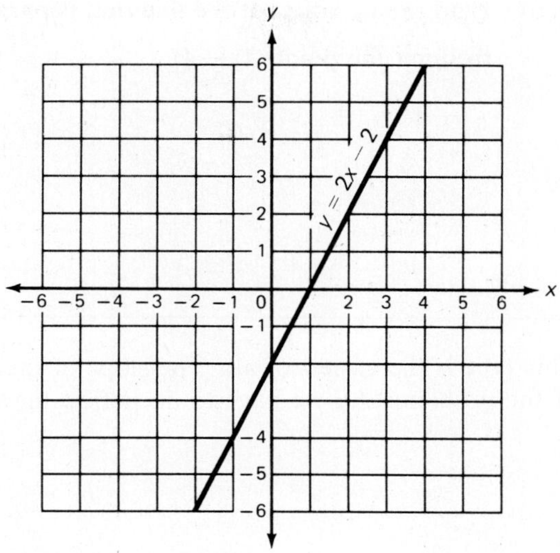

Figure 31.2

We have found the exact slope and the exact intercept (Figure 31.2). Estimated values for the slope and intercept will not be acceptable for this kind of problem.

example 31.3.1 Find the equation of the line through $(-1, -3)$ that is parallel to $4x + 3y = 7$.

solution The equation of the new line is

$$y = mx + b$$

and we have to find m and b. **If the new line is to be parallel to the line $4x + 3y = 7$, the new line must have the same slope as the line $4x + 3y = 7$.** If we write $4x + 3y = 7$ in slope intercept form, we find

$$y = -\frac{4}{3}x + \frac{7}{3}$$

and the slope of this line is $-\frac{4}{3}$. So the slope of the new line must be $-\frac{4}{3}$.

$$y = -\frac{4}{3}x + b$$

We find the value of b by using the coordinates $(-1, -3)$ for x and y and solving algebraically for b.

$$-3 = -\frac{4}{3}(-1) + b$$

$$-\frac{9}{3} = \frac{4}{3} + b$$

$$-\frac{13}{3} = b$$

Thus the desired equation is $\boldsymbol{y = -\frac{4}{3}x - \frac{13}{3}}$ and again we have found the exact value of the slope and the exact value of the intercept.

TEST YOUR UNDERSTANDING

(b) Find the equation of the line that is parallel to the line $y = \frac{1}{2}x + 2$ and passes through the point $(1, -4)$.

31.4 equation of a line with a given slope and through a given point

This type is the easiest of all. The slope of the desired equation is *given* in the statement of the problem. All we have to do is find the value of the intercept.

example 31.4.1 Find the equation of the line that goes through the point (3, 4) and has a slope of $-\frac{3}{4}$.

solution The equation of the line in question is $y = mx + b$ and we need to determine the proper values for m and b. The statement of the problem tells us that the slope is $-\frac{3}{4}$ and if we use this value for m in the equation, we find

$$y = -\frac{3}{4}x + b$$

If we use (3, 4) for x and y in this equation we can solve algebraically for b.

$$4 = -\frac{3}{4}(3) + b$$

$$4 = -\frac{9}{4} + b \longrightarrow \frac{16}{4} = -\frac{9}{4} + b \longrightarrow b = \frac{25}{4}$$

So the equation is

$$y = -\frac{3}{4}x + \frac{25}{4}$$

and the numbers $-\frac{3}{4}$ and $\frac{25}{4}$ are the exact values of the slope and the intercept.

example 31.4.2 Find the equation of the line that goes through the point $(-5, 11)$ and has a slope of $\frac{1}{7}$.

solution The statement of the problem gives us the slope, so we can write

$$y = \frac{1}{7}x + b$$

Now we use the values -5 and 11 for x and y and solve for b.

$$11 = \frac{1}{7}(-5) + b \longrightarrow \frac{77}{7} = -\frac{5}{7} + b \longrightarrow b = \frac{82}{7}$$

So the equation is

$$y = \frac{1}{7}x + \frac{82}{7}$$

and the numbers $\frac{1}{7}$ and $\frac{82}{7}$ are the exact values of the slope and the intercept.

TEST YOUR UNDERSTANDING

(c) Find the equation of the line whose slope is $-\frac{2}{3}$ and that passes through the point (2, 2).

lesson checklist 31

- ☐ Linear equation
- ☐ Standard form of a linear equation
- ☐ Slope-intercept form of a linear equation
- ☐ Equation of a line through two designated points
- ☐ Equation of a line parallel to a given line and passing through a designated point
- ☐ Equation of a line with a designated slope that passes through a designated point

ANSWERS TO TEST YOUR UNDERSTANDING

(a) $y = -\frac{3}{4}x + \frac{7}{4}$ (b) $y = \frac{1}{2}x - \frac{9}{2}$ (c) $y = -\frac{2}{3}x + \frac{10}{3}$

problem set 31

1. Harry and Jennet jog around a circular track that is 210 meters long. Jennet's rate is 230 meters per minute while Harry's rate is only 200 meters per minute. In how many minutes will Jennet be a full lap ahead?
2. Two ships are 120 nautical miles apart and are sailing toward each other on the great circle route from New York to England. The rate of one ship is 4 knots (nautical miles per hour) greater than the rate of the other ship. If the ships meet in 3 hours, find the rate of both ships.
3. A cargo plane left Fort Valley at 12 noon, flying west at 250 mph. At 2 p.m. a passenger plane left the same airport flying west at 350 mph. How long did it take the passenger plane to catch the cargo plane?
4. Rena sets out on a hike. After walking for a while at 5 km/hr, she gets a cramp in her leg. She catches a ride home in a truck whose speed is 20 km/hr. When she gets home, she finds she has been gone exactly one hour. How far had she walked?
5. Ellis and McWherter threw 752 rocks into Bobbins Mill. Ellis had a strong arm for he threw 144 more than 15 times the number that McWherter threw. How many did Ellis throw?

6. Zachary bought 23 pizzas for \$40.75. If cheese pizzas cost \$1.75 each and sausage pizzas cost \$2.25 each, how many cheese pizzas did he buy?

7. Tibbetts found that he had only 12 cents, which was 80 percent of the value of a large Varsity Orange. How much did a large Varsity Orange cost?

8. Find three consecutive integers such that five times the first is 6 greater than twice the sum of 4 and the third.

9. Find the equation of the line:

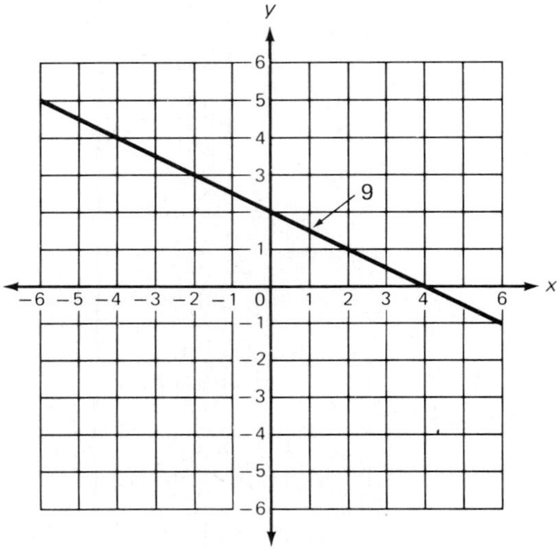

*10. Find the equation of the line through (4, 2) and $(-5, -3)$.

*11. Find the equation of the line that goes through $(-1, -3)$ and is parallel to $4x + 3y = 7$.

*12. Find the equation of the line through (3, 4) that has a slope of $-\frac{3}{4}$.

13. Given that $f(x) = x^2 + 4$, $D = \text{Reals}$; $g(x) = x^2$ $D = \text{Reals}$; $\phi(x) = x + 2$, $D = \text{Reals}$; find $\phi(0)$.

14. Which of the following represent functions? Use the vertical line test as necessary.

(a)

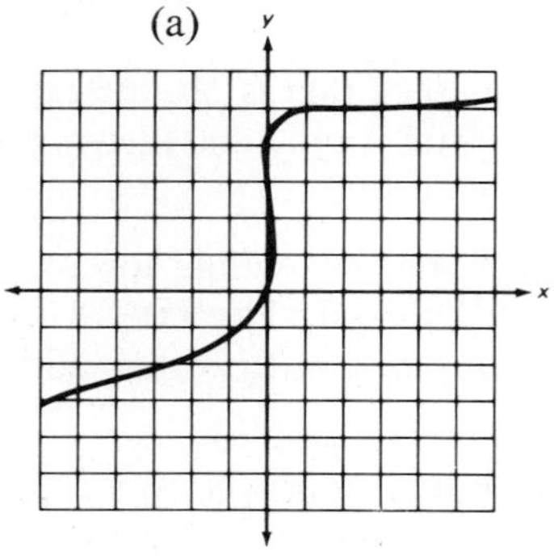

(b)

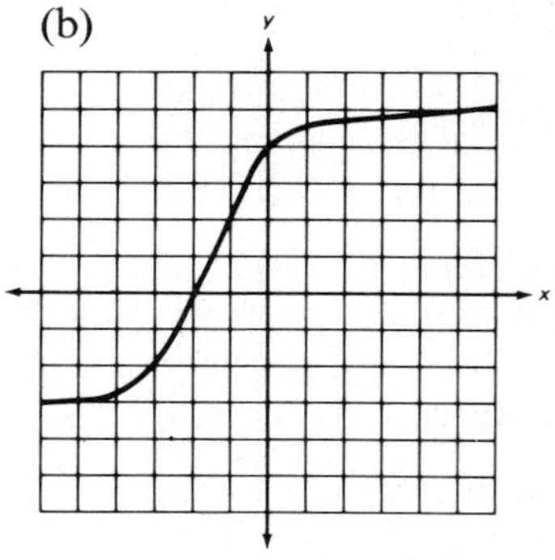

(c)

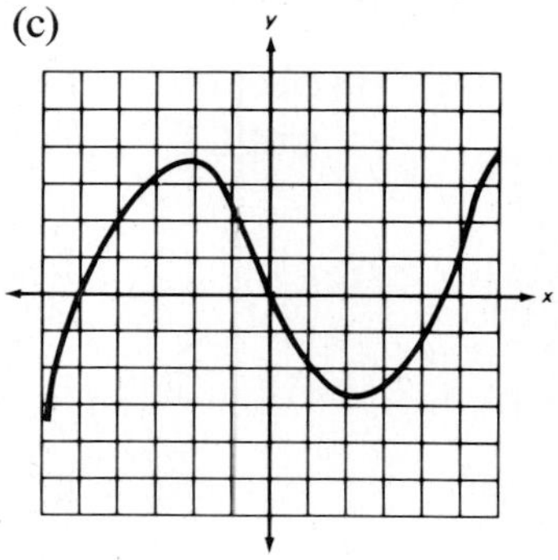

(d)
(5, 6), (7, 8), (9, 10)
(6, 7), (8, 9), (10, 11)

(e)

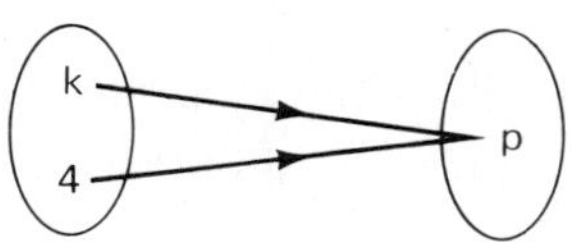

(f)

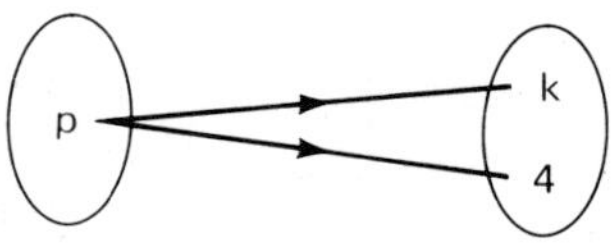

---

*The problems marked with an asterisk were worked out in this lesson.

Solve:

15. $\frac{4}{x-9} + \frac{5}{x} = 0$

16. $\frac{t}{4} + \frac{1}{5} = 14$

Graph on a number line:

17. $-4 - x > 2$, $D = \{\text{Integers}\}$

18. $|x| - 2 < 5$, $D = (\text{Reals})$

19. Find p.

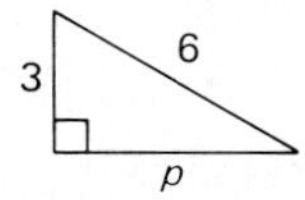

20. Use the commutative and associative properties to show that $xypm = xmyp$.

Simplify:

21. $\sqrt{12} - \sqrt{25} - \sqrt{27}$

22. $3\sqrt{2}\,(2\sqrt{2} - 3\sqrt{6}\,)$

23. $2\sqrt{7} \cdot \sqrt{2} \cdot \sqrt{14}$

24. Divide $x^3 + 3x^2 + 15x - 5$ by $x - 4$.

Solve:

25. $x^2 = 9x - 8$

Simplify:

26. $\frac{a^2 - b^2}{a^2 - 16} \cdot \frac{a + 4}{a + b}$

27. $\frac{(.005 \times 10^{-6})(400{,}000)}{.002 \times 10^{15}}$

Use elimination:

28. $4x - 7y = -16$
$3x + 2y = 17$

29. The set of negative integers is closed for what operations?

Graph on a rectangular coordinate system:

30. $x = 2$

Solve by graphing:

31. $\begin{cases} y - x + 5 = 0 \\ y = -x + 2 \end{cases}$

Simplify:

32. $\dfrac{x - \dfrac{1}{y}}{y - \dfrac{x}{y}}$

33. $\dfrac{c}{\dfrac{x}{y}}$

Add:

34. $\frac{1}{a} - \frac{x}{a + b}$

35. $\frac{a}{x^2b} + \frac{m}{x^2} - \frac{c}{b^2x}$

Add like terms:

36. $4abc^2 - \frac{3ac^2}{b^{-1}} + \frac{5ab}{c^2}$

37. $6x^2ym^2 - \frac{3x^2y}{m^{-2}} + \frac{2}{x^{-2}y^{-1}m^{-2}}$

Multiply:

38. $x^2ay\left(\frac{x^3}{a^{-3}y^2} + 5\right)$

Simplify:

39. $\frac{a^4x^3b^{-2}c}{c^{-3}x^2ba^0}$

40. $\frac{a^{-2}x^2p^{-4}k}{m^3xa^2p^{-2}}$

Solve:

41. $-3(x + 2) - (-5) = 5(x - 8)$

Evaluate:

42. $p - q(-p)(p + q)$ if $p = 2$, $q = -3$.

Simplify:

43. -2^{-2}

44. True or false: {Natural} $\subset$ {Whole}

45. $\sqrt{7} \in$ {what sets of numbers?}

Add:

46. $\dfrac{2p}{4 - x^2} - \dfrac{p}{2 + x}$

47. $\dfrac{-3mx}{x^2 + 5x + 4} + \dfrac{3m}{x + 4}$

lesson thirty-two

Review of square roots
Solution of radical equations

32.1 square roots revisited

Both -2 and $+2$ are square roots of 4 because

$$(-2)^2 = 4 \quad \text{and} \quad (+2)^2 = 4$$

But when we write $\sqrt{4}$, we are indicating the positive or principal square root of 4, which, of course, is the number 2.

> DEFINITION OF SQUARE ROOT
>
> If x is greater than or equal to zero, then $\sqrt{x}$ is the unique positive real number such that
>
> $$(\sqrt{x})^2 = x$$

We can state this definition in words by saying that the principal square root of a given positive number is that positive number which, multiplied by itself, yields the given number. Thus

$$\sqrt{2}\sqrt{2} = 2 \qquad \sqrt{7}\sqrt{7} = 7 \quad \text{and} \quad \sqrt{3.14}\sqrt{3.14} = 3.14$$

and also

$$(\sqrt{2})^2 = 2 \qquad (\sqrt{7})^2 = 7 \quad \text{and} \quad (\sqrt{3.14})^2 = 3.14$$

It is necessary to remember that algebraic expressions represent particular real numbers that are determined by the values assigned to the variables. Thus, if a particular algebraic expression represents a positive real number, the definition of the square root given in the box above applies to the expression. For example,

$$(\sqrt{x^2 + 4})^2 = x^2 + 4 \qquad \left(\sqrt{\frac{amx^2}{p}}\right)^2 = \frac{amx^2}{p} \qquad (\sqrt{x + 6})^2 = x + 6$$

$$(\sqrt{x^4 + 3x^2 + 5})^2 = x^4 + 3x^2 + 5$$

32.2 radical equations

There is only one number that will satisfy the equation $x = 2$, and that number is 2. If we replace x with 2 in this equation, we find

$$2 = 2$$

which is a true statement. If we square both sides of the original equation

$$(x)^2 = (2)^2 \longrightarrow x^2 = 4$$

the result is the equation $x^2 = 4$. While the equation $x = 2$ had only one solution, this equation has two numbers that satisfy it, the numbers $+2$ and -2.

Replacing x with $+2$	Replacing x with -2
$(+2)^2 = 4$	$(-2)^2 = 4$
$4 = 4$ True	$4 = 4$ True

We began with the equation $x = 2$, whose only solution is 2. We squared both sides and got the equation $x^2 = 4$ which also has the number 2 as a solution but has another solution, which is the number -2. **It can be shown that if both sides of an equation are squared, all of the solutions to the original equation (if any exist) are also solutions to the second equation, but the reverse is not true, for all of the solutions of the second equation are not necessarily solutions of the original equation.**

example 32.2.1 Solve $\sqrt{x - 2} + 2 = 0$.

solution We wish to **isolate the radical** on one side of the equation so that we may square both sides of the equation and use the fact that $(\sqrt{x})^2 = x$, so we add -2 to both sides of the equation and get

$$\sqrt{x - 2} = -2$$

Now square both sides and get

$$(\sqrt{x - 2})^2 = (-2)^2 \longrightarrow x - 2 = 4 \longrightarrow \mathbf{x = 6}$$

Now we must check our solution in the original equation.

$$\sqrt{6 - 2} + 2 = 0 \longrightarrow \sqrt{4} + 2 = 0 \longrightarrow 2 + 2 = 0 \longrightarrow 4 = 0 \quad \text{False}$$

We see that while 6 is a solution of the second equation, $x - 2 = 4$, it is not a solution to the original equation $\sqrt{x - 2} + 2 = 0$. Thus we see that there is no real number replacement for x that will satisfy the first equation, and we say that the solution set of this equation is the empty set.

example 32.2.2 Solve $\sqrt{x - 2} - 6 = 0$.

solution We first isolate the radical on one side by adding $+6$ to both sides of the equation.

$$\sqrt{x - 2} = 6$$

Now we square both sides to eliminate the radical and then solve the resulting equation.

$$(\sqrt{x - 2})^2 = (6)^2 \longrightarrow x - 2 = 36 \longrightarrow \mathbf{x = 38}$$

Now we will check this solution in the original equation.

$\sqrt{(38) - 2} - 6 = 0 \longrightarrow \sqrt{36} - 6 = 0 \longrightarrow 6 - 6 = 0 \longrightarrow 0 = 0$ Check

Thus **38** is a solution to the original equation.

example 32.2.3 Solve $\sqrt{x^2 + 9} - 5 = 0$.

solution First isolate the radical and get

$$\sqrt{x^2 + 9} = 5$$

Now eliminate the radical by squaring both sides of the equation

$$(\sqrt{x^2 + 9})^2 = (5)^2 \longrightarrow x^2 + 9 = 25$$

Now we will simplify, factor, and use the zero factor theorem to solve.

$$x^2 - 16 = 0 \longrightarrow (x + 4)(x - 4) = 0 \longrightarrow \mathbf{x = 4}, \quad \text{and} \quad \mathbf{x = -4}$$

Since neither of these solutions to the second equation is guaranteed to be a solution of the original equation, both solutions must be checked in the original equation.

Check $+4$	Check -4
$\sqrt{(4)^2 + 9} - 5 = 0.$	$\sqrt{(-4)^2 + 9} - 5 = 0$
$\sqrt{25} = 5$	$\sqrt{25} = 5$
$5 = 5$ Check	$5 = 5$ Check

Thus both $+4$ and -4 are solutions.

example 32.2.4 Solve $\sqrt{x - 1} - 3 + x = 0$.

solution We begin by adding $+3 - x$ to both sides of the equation to isolate the radical and get

$$\sqrt{x - 1} = 3 - x$$

Now we square both sides to eliminate the radical

$$(\sqrt{x - 1})^2 = (3 - x)^2 \longrightarrow x - 1 = 9 - 6x + x^2$$

Simplify, factor, and use the zero factor theorem to solve.

$$x^2 - 7x + 10 = 0 \longrightarrow (x - 2)(x - 5) = 0 \longrightarrow \mathbf{x = 2}, \quad \text{and} \quad \mathbf{x = 5}$$

Now we must check both 2 and 5 in the original equation.

Check $x = 2$	Check $x = 5$
$\sqrt{x - 1} = 3 - x$	$\sqrt{x - 1} = 3 - x$
$\sqrt{(2) - 1} = 3 - (2)$	$\sqrt{(5) - 1} = 3 - (5)$
$\sqrt{1} = 1$	$\sqrt{4} = -2$
$1 = 1$ Check. Thus $x = 2$ is a solution of the original equation.	$2 \neq -2$ Does not check. Thus $x = 5$ is not a solution of the original equation.

example 32.2.5 Solve $\sqrt{2x - 3} = \sqrt{x + 2}$.

solution One radical expression is isolated on each side of the equation so we begin by squaring both sides of the equation.

$$(\sqrt{2x - 3})^2 = (\sqrt{x + 2})^2 \longrightarrow 2x - 3 = x + 2 \longrightarrow x = 5$$

Now we will check $x = 5$ in the original equation.

$$\sqrt{2(5) - 3} = \sqrt{5 + 2} \longrightarrow \sqrt{7} = \sqrt{7} \quad \text{Check}$$

TEST YOUR UNDERSTANDING

(a) Solve: $\sqrt{x + 2} - 6 = 0$

lesson checklist 32

- ☐ Definition of square root
- ☐ Principal square root
- ☐ Solution of radical equations

ANSWERS TO TEST YOUR UNDERSTANDING

(a) $x = 34$

problem set 32

1. Charles went from Westminster to Oxford, a distance of 44 kilometers. He rode in a carriage at 5 km/hr for part of the trip and rode a horse at 8 km/hr for the rest of the trip. If the trip took a total of 7 hours, how long did he ride in the carriage?
2. The westbound stagecoach left Carson City 6 hours after the eastbound stagecoach left the same depot. Three hours after the westbound stagecoach left, the two coaches were 129 miles apart. If the speed of the westbound stagecoach was 10 mph, what was the speed of the eastbound stagecoach?
3. A train starts from Toledo at 11 a.m. and heads for Makinaw, 332 kilometers away. At the same time a train leaves Makinaw and heads for Toledo at 65 km/hr. If the trains meet at 1 p.m., what was the rate of the first train?
4. Henry can ride his horse at 4 mph and get to the battlefield on time. If he stops for 1 hour to make a speech to his troops, he finds he must ride at 5 mph to get to the battlefield on time. How far is it to the battlefield?
5. Jim and Jason collect coins. Their collection consists of pennies and quarters. They have 742 coins whose value is $17.02. How many pennies do they have?
6. Margaret wanted a new contemporary style house. Jay was agreeable but wanted to spend only 82 percent of what Margaret would spend. If Jay would agree to spending $93,480, how much did Margaret want to spend?

7. Sir Lancelot paid 20 farthings for his sword but only 4 farthings for his halberd. What percent of the sword's price was the halberd's price?

8. Find three consecutive even integers such that three times the sum of 4 and the first is 6 greater than twice the sum of 2 and the third.

*9. $\sqrt{x-2} - 6 = 0$ *10. $\sqrt{x^2+9} - 5 = 0$ *11. $\sqrt{x-1} - 3 + x = 0$

*12. $\sqrt{2x-3} = \sqrt{x+2}$

13. Find the equation of the line:

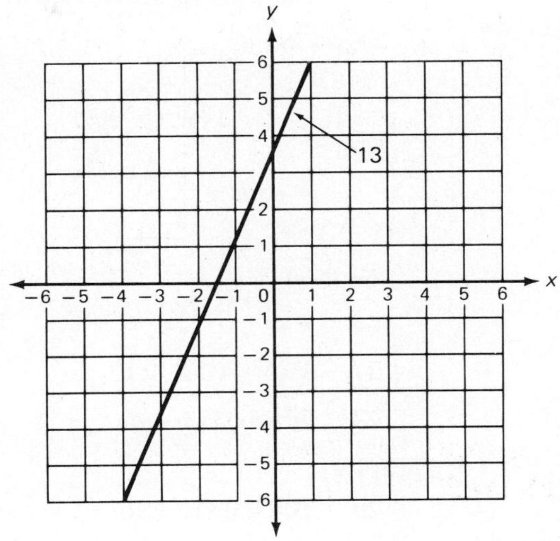

14. Find the equation of the line through $(3, -2)$ and $(-1, 1)$.

15. Find the equation of the line that goes through $(3, 1)$ and is parallel to $4x - 2y = 2$.

16. Find the equation of the line through $(-2, 3)$ that has a slope of $-\frac{4}{3}$.

17. Given that $f(x) = x^2 + 3$, $D = \{\text{Reals}\}$; find $g(2)$
$g(x) = x^2 - 4$, $D = \{\text{Reals}\}$;

18. Which of the following represent functions? Use the vertical line test as necessary.

(a)

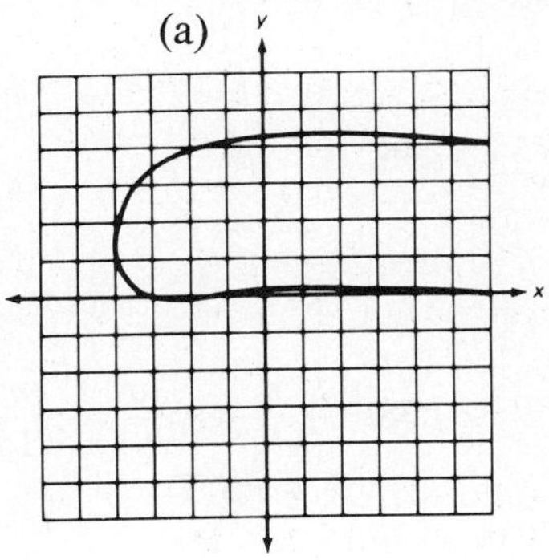

(b)

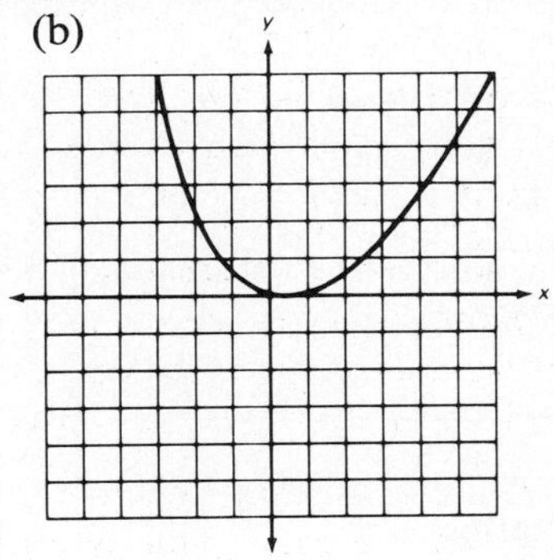

(c)

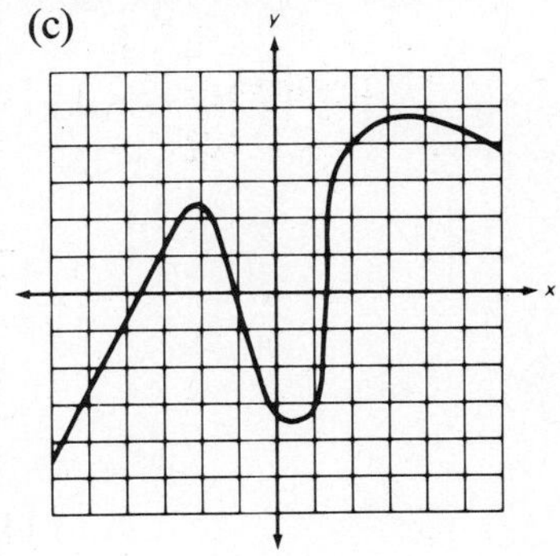

(d)
(1, 2), (3, 4), (5, 6)
(2, 3), (4, 5), (6, 7)

(e)

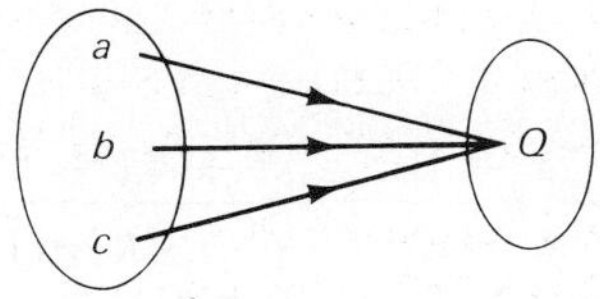

(f)

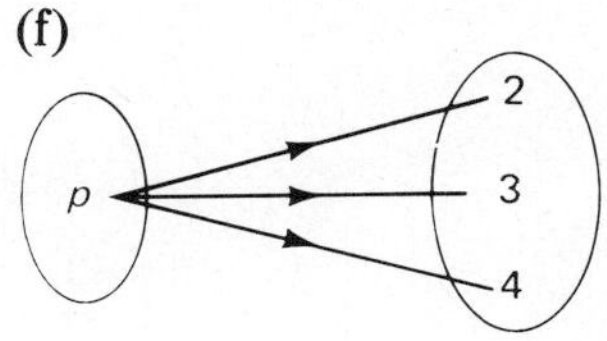

*The problems marked with an asterisk were worked out in this lesson.

Solve:

19. $\dfrac{y+2}{y} = \dfrac{4}{3}$

20. $\dfrac{x}{3} + \dfrac{2}{7} = 4$

Graph on a number line:

21. $-5 - x < 3$, $D = \{\text{Reals}\}$

22. $|y| + 6 < 9$, $D = \{\text{Reals}\}$

23. Use the commutative and associative properties to show that $abcy = ycba$.

24. Find p.

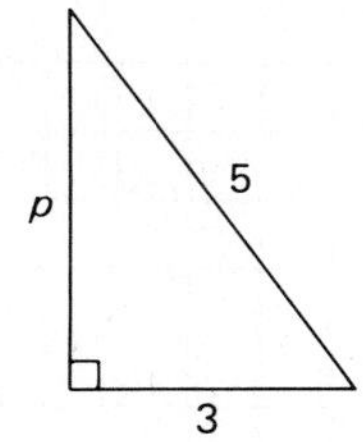

Simplify:

25. $\sqrt{25} - \sqrt{12} + \sqrt{32}$

26. $2\sqrt{3}\,(3\sqrt{3} - 2\sqrt{8}\,)$

27. $2\sqrt{5} \cdot \sqrt{5} \cdot \sqrt{10}$

28. Divide $3x^3 - 4x^2 - 18x + 9$ by $x - 3$.

Solve:

29. $2x^2 = 18x - 16$

Simplify:

30. $\dfrac{x-y}{x^2-25} \cdot \dfrac{x+5}{x^2-y^2}$

31. $\dfrac{(.0007 \times 10^{-7})(210{,}000)}{.0003 \times 10^{19}}$

32. Use elimination: $\begin{cases} 3x - 7y = 23 \\ 4x + 2y = 8 \end{cases}$

33. The set of $(-1, 0, 1)$ is closed for what operations?

Solve by graphing:

34. $\begin{cases} y - 2x + 3 = 0 \\ 4 = y \end{cases}$

Add:

35. $\dfrac{x + \dfrac{x}{y}}{y - \dfrac{1}{y}}$

36. $\dfrac{b}{\dfrac{x}{y} + 1}$

37. $\dfrac{2}{x} - \dfrac{c}{x^2 + y}$

38. $\dfrac{c}{b^2x^2} + \dfrac{a}{b^2} - \dfrac{m}{1}$

Add like terms:

39. $9x^2y^3z - \dfrac{4y^{-3}z}{x^{-2}} + \dfrac{2x^2}{y^3z^{-1}}$

Simplify:

40. $\dfrac{x^2 + 7x + 10}{x^2 + x - 12} \div \dfrac{x^2 + 4x - 5}{x^2 - 4x + 3}$

Multiply:

41. $ba^2y\left(\dfrac{a^3}{b^{-3}y^2} + \dfrac{7}{b}\right)$

Simplify:

42. $\dfrac{a^{-2}b^{3}x^{2}c}{c^{-3}x^{-2}ba^{0}}$

43. $\dfrac{y^{-3}m^{2}kp^{2}}{m^{-3}y^{4}k^{-3}p}$

Solve:

44. $-2(x + 3) - (-8) = -4(x - 1)$

Evaluate:

45. $-p + (-r - p)(r)$ if $p = 2$, $r = -1$.

Simplify:

46. $\dfrac{(-2)^{-2}}{-2^{-3}}$

47. True or false: {Whole} $\subset$ {Irrational}

48. $-6.5 \in$ {what sets of numbers?}

Add:

49. $\dfrac{kxy}{x^{2} - 36} - \dfrac{ky}{x - 6}$

lesson thirty-three

Slope formula
Ratio problems

33.1 slope formula

In Lesson 21 the slope of a straight line was defined to be the ratio of the change in the value of the y-coordinate to change in the value of the x-coordinate as we move from one point on the line to another point on the line. To demonstrate, we will find the slope of the line through the points (4, 3) and $(-2, -2)$ by first graphing the points and drawing the line as shown in Figure 33.1. The sign of the slope of this line is positive because the line segment graphed points toward the upper right. In Figure 33.2 we draw the triangle to determine the magnitude (absolute value) of the slope. From the triangle we see that the difference in the x-coordinates of the two points is 6 and the difference in the y-coordinates is 5. Thus the slope of this line is $\frac{5}{6}$.

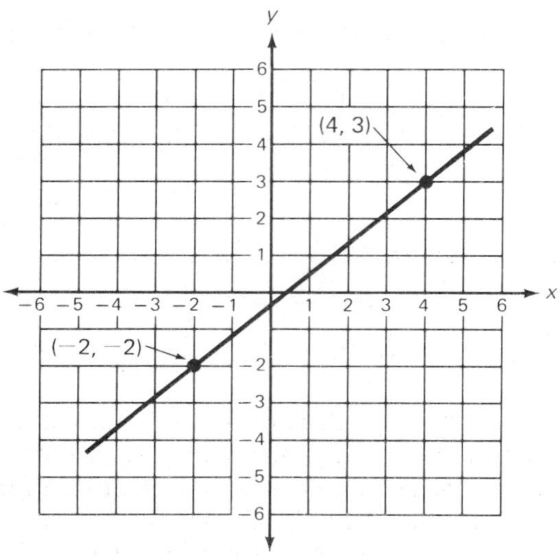

Figure 33.1

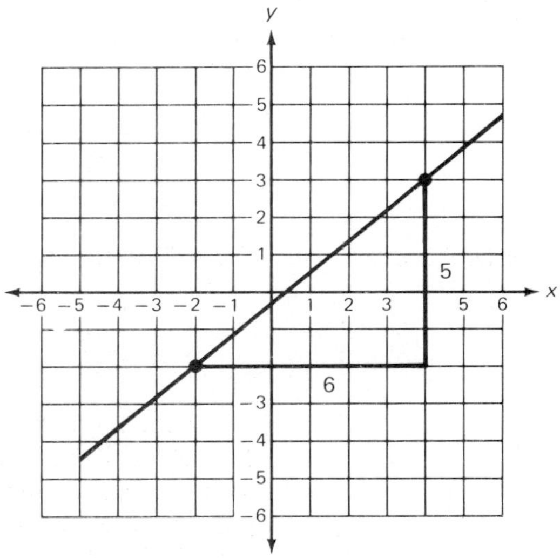

Figure 33.2

It is not necessary to graph the points to find the slope of the line. If we call the two points point 1 and point 2 and give them the coordinates (x_1, y_1) and (x_2, y_2), respectively, as in Figure 33.3, we can derive a relationship from which the slope for the line through any two points can be determined algebraically.

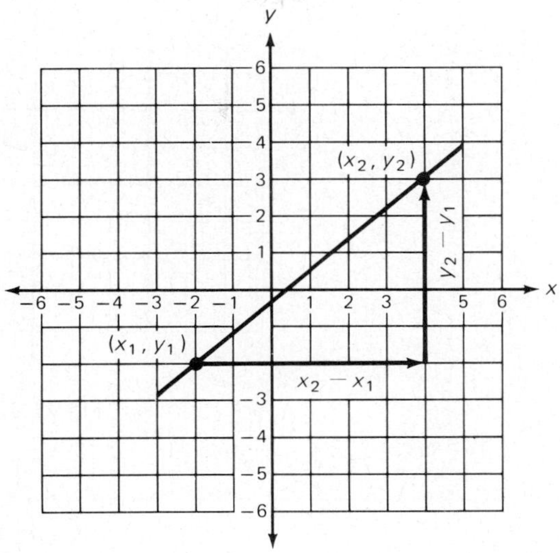

Figure 33.3

We begin at point 1 and move along the line to point 2. The change in the value of the x-coordinate is $x_2 - x_1$, and the change in the value of the y-coordinate is $y_2 - y_1$. The slope m is defined to be the ratio of the change in the y-coordinate to the change in the x-coordinate, so

$$m = \frac{y_2 - y_1}{x_2 - x_1}$$

We can use this relationship or formula to find the sign and the magnitude of the slope of the line through any two given points.

example 33.1.1 Find the slope of the line that passes through the points $(-3, 4)$ and $(5, -2)$.

solution Either point can be designated as point 1. We will use $(-3, 4)$ as point 1 and $(5, -2)$ as point 2.

$$m = \frac{y_2 - y_1}{x_2 - x_1} \longrightarrow m = \frac{-2 - (4)}{5 - (-3)} \longrightarrow m = \frac{-6}{8} \longrightarrow \mathbf{m = -\frac{3}{4}}$$

example 33.1.2 Work Example 33.1.1 again but this time use $(5, -2)$ as point 1 and $(-3, 4)$ as point 2.

solution

$$m = \frac{y_2 - y_1}{x_2 - x_1} \longrightarrow m = \frac{4 - (-2)}{-3 - (5)} \longrightarrow m = \frac{6}{-8} \longrightarrow \mathbf{m = -\frac{3}{4}}$$

The slope of the line was found to be $-\frac{3}{4}$ no matter which point was designated as point 1. We see from these two examples that the slope can be determined by using this formula and we also see that care must be exercised to prevent mistakes in handling the positive and negative signs of the numbers. In Figure 33.4 we use the old familiar graphical method. We see from this figure that the sign of the slope is negative and the magnitude is $\frac{6}{8}$, so the slope is $-\frac{6}{8}$ or $-\frac{3}{4}$, the same slope found by using the formula.

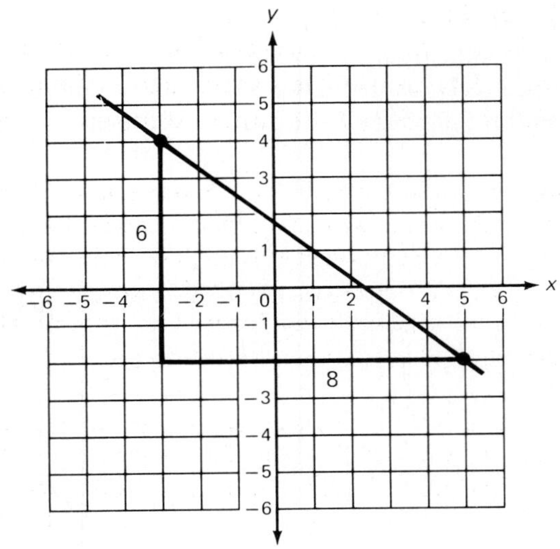

Figure 33.4

> **TEST YOUR UNDERSTANDING**
>
> (a) Use the slope formula to find the slope of the line that goes through the points $(-4, 2)$ and $(3, 5)$.

33.2 ratio

A RATIO is the name we use to describe a comparison of the relative magnitudes of two numbers when the numbers are written in the form of a fraction. Thus for the rectangle and the circle shown in Figure 33.5, we can compare the area of shaded portion of each figure to the area of the unshaded portion or compare either of these areas to the total area of the figure by making a statement about the ratios of the areas.

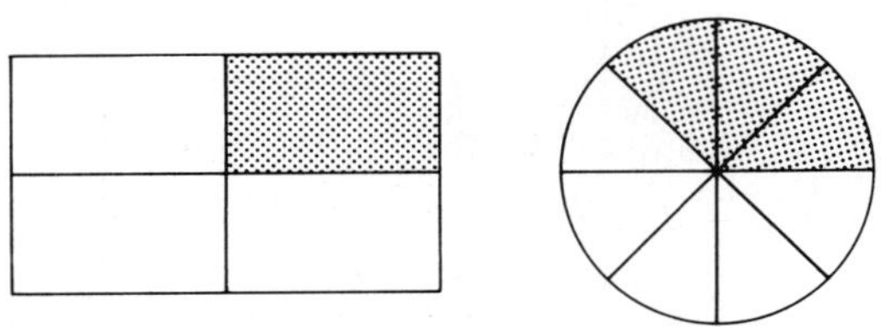

(a) (b)

Figure 33.5

(a) For the rectangle we can say
 (i) The ratio of the shaded area to the unshaded area is $\frac{1}{3}$.
 (ii) The ratio of the shaded area to the total area is $\frac{1}{4}$.
 (iii) The ratio of the unshaded area to the total area is $\frac{3}{4}$.

(b) For the circle we can say
  (i) The ratio of the shaded area to the unshaded area is $\frac{3}{5}$.
  (ii) The ratio of the shaded area to the total area is $\frac{3}{8}$.
  (iii) The ratio of the unshaded area to the total area is $\frac{5}{8}$.

Note the order in each of the statements above. The number associated with the first area mentioned was the numerator of the ratio and the number associated with the second area mentioned was the denominator of the ratio.

It is not necessary to use the word ratio in the word statement and sometimes the ratio is not written in the form of a fraction. The following three statements apply to the circle and all have the same meaning.

1. The ratio of the shaded area to the unshaded area is $\frac{3}{8}$.
2. The ratio of the shaded area to the unshaded area is 3 : 8.
3. The shaded area is to the unshaded area as 3 is to 8.

example 33.2.1 Farmer Bud plants wheat and barley in the ratio of 4 to 5. What other ratios are implied by this statement?

solution

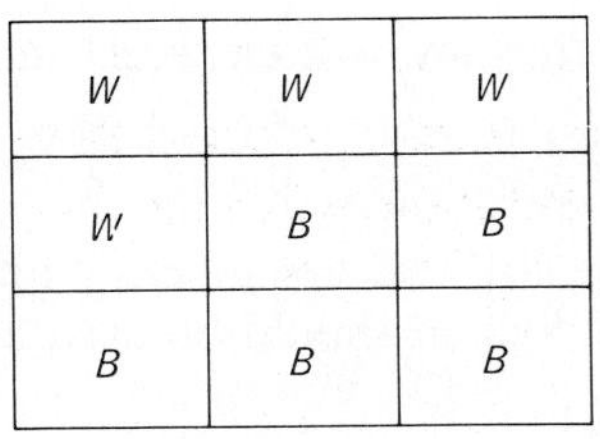

If we have 4 parts of wheat and 5 parts of barley we have a total of 9 parts as shown in the figure. Thus the relationship implied between wheat and total and barley and total can be stated as

Wheat is to total as 4 is to 9
Barley is to total as 5 is to 9

33.3 proportion

A proportion is a statement that two ratios are equal. We can say that the ratio of 1 to 4 equals the ratio of 3 to 12. This is a statement of a proportion and we can make the same statement using numerals by writing both of the ratios as fractions.

$$\frac{1}{4} = \frac{3}{12}$$

This fractional form of the statement of a proportion is useful in solving problems in which a constant ratio between parts of a whole is desired.

example 33.3.1 Farmer Bud wishes to plant 423 acres in wheat and barley in the ratio of 4 to 5. How many acres should he plant in wheat and how many acres should he plant in barley?

solution We can state the problem pictorially by drawing two rectangles.

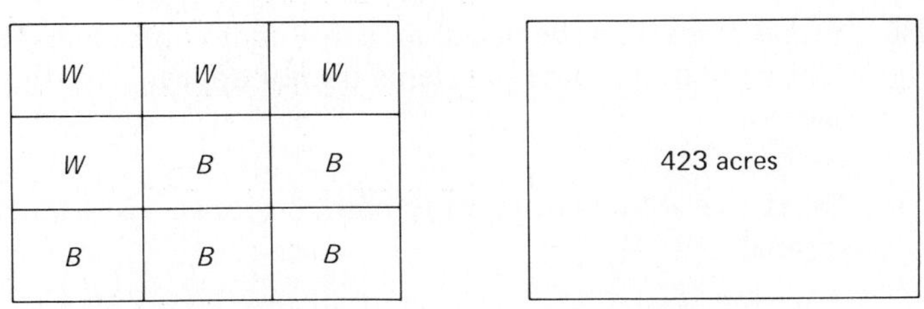

The rectangle on the left shows 4 parts wheat, 5 parts barley, and 9 parts total. From these relationships, we can get three statements of equal ratios.

$$\frac{\text{wheat}}{\text{barley}} = \frac{4}{5}, \qquad \frac{\text{wheat}}{\text{total}} = \frac{4}{9}, \qquad \frac{\text{barley}}{\text{total}} = \frac{5}{9}$$

If the 423 acres shown in the other rectangle is to be divided up as stated by these proportions, we can replace *total* with 423 in the second and third proportions and solve for wheat and barley.

$$\frac{W}{423} = \frac{4}{9} \longrightarrow \frac{W}{\cancel{423}}(\cancel{423}) = \frac{4}{9}(423) \longrightarrow W = \frac{4}{9}(423) \longrightarrow \mathbf{W = 188\ acres}$$

$$\frac{B}{423} = \frac{5}{9} \longrightarrow \frac{B}{\cancel{423}}(\cancel{423}) = \frac{5}{9}(423) \longrightarrow B = \frac{5}{9}(423) \longrightarrow \mathbf{B = 235\ acres}$$

example 33.3.2 The ratio of men to women in the Sylvester Tennis Club is 5 to 3. If the club has 184 members, how many are men?

solution Beginners often find it helpful to first write all the proportions stated directly or implied and then reread the problem to see which of these can be used to provide the desired answer. We were given parts of 5 and 3 which implies a total of 8.

Men = 5

Women = 3

Total = 8

We can use these relationships as they are written to get three different proportions.

$$\frac{M}{W} = \frac{5}{3}, \qquad \frac{M}{T} = \frac{5}{8}, \qquad \frac{W}{T} = \frac{3}{8}$$

If we reread the problem, we see that we are given a total of 184 and asked for the number of men so the second proportion should be used and 184 should be substituted for T in this proportion.

$$\frac{M}{T} = \frac{5}{8} \longrightarrow \frac{M}{184} = \frac{5}{8} \longrightarrow \cancel{184}\frac{M}{\cancel{184}} = \frac{5}{8} \cdot 184 \longrightarrow \mathbf{M = 115}$$

example 33.3.3 The ratio of Sally's weight on earth to her weight on the moon is 6 : 1. If she weighs 150 kg on earth, what would she weigh on the moon?

solution This problem states a particular ratio and there are no implied ratios for there is no total in this problem.

$$\frac{W_E}{W_M} = \frac{6}{1} \longrightarrow \frac{150}{W_M} = \frac{6}{1} \longrightarrow \cancel{W_M}\frac{150}{\cancel{W_M}} = \frac{6}{1} \cdot W_M$$

$$\longrightarrow 150 = 6W_M \longrightarrow \frac{150}{6} = \frac{\cancel{6}W_M}{\cancel{6}} \longrightarrow 25 \text{ kg} = W_M$$

33.4 implied ratios

Proportions can be used to solve many problems encountered in day-to-day living. Unfortunately, these problems do not appear with the words ratio or proportion printed thereon.

example 33.4.1 Detia uses 4 liters of paint to paint 3 rooms. How much paint will she require to paint 15 rooms?

solution The statement of the problem implies that the ratio of liters used to rooms painted is 4 to 3. There is no total.

$$\frac{L}{R} = \frac{4}{3} \longrightarrow \frac{L}{15} = \frac{4}{3} \longrightarrow \cancel{15}\frac{L}{\cancel{15}} = \frac{4}{3} \cdot 15 \longrightarrow \mathbf{L = 20\ liters}$$

example 33.4.2 In a factory, 15 defective parts were found in a batch of 9000 parts. How many non-defective parts would be in a batch of 180,000?

solution The statement of the problem implies that the ratio of defective parts is to total parts produced as 15 is to 9000. Thus there are 8985 non-defective parts.

15 = defective
8985 = non-defective
9000 = total

and these relationships are used to write the proportions

$$\frac{15}{8985} = \frac{D}{ND}$$
$$\frac{15}{9000} = \frac{D}{T}$$
$$\frac{8985}{9000} = \frac{ND}{T}$$

The problem gives us the total of 180,000 and asks for the number of non-defective parts, so we use the last proportion.

$$\frac{8985}{9000} = \frac{ND}{T} \longrightarrow \frac{8985}{9000} = \frac{ND}{180{,}000} \longrightarrow (180{,}000)\frac{8985}{9000} = \cancel{(180{,}000)}\frac{N_D}{\cancel{180{,}000}}$$
$$\longrightarrow \mathbf{ND = 179{,}700}$$

example 33.4.3 Betty and Jack use 10 gallons of gas in going 180 kilometers. How many gallons are required to go 306 kilometers?

solution The problem says that the ratio of gallons to distance is 10 to 180. There is no total in this problem.

$$\frac{G}{D} = \frac{10}{180} \longrightarrow \frac{G}{306} = \frac{10}{180} \longrightarrow \cancel{(306)}\frac{G}{\cancel{306}} = (306)\frac{10}{180} \longrightarrow \mathbf{G = 17\ gal}$$

example 33.4.4 A 6-foot man casts a 4-foot shadow. What is the length of the shadow of a 30-foot tree?

solution The problem implies that the ratio of the height of an object to length of its shadow is 6 to 4. There is no total.

$$\frac{H}{LS} = \frac{6}{4} \longrightarrow \frac{30}{LS} = \frac{6}{4} \longrightarrow (4)\cancel{(LS)}\frac{30}{\cancel{LS}} = \cancel{(4)}(LS)\frac{6}{\cancel{4}} \longrightarrow 120 = 6LS$$
$$\longrightarrow \frac{120}{6} = \frac{\cancel{6}LS}{\cancel{6}} \longrightarrow \mathbf{LS = 20\ ft}$$

TEST YOUR UNDERSTANDING

(b) Sam stays alive 5 days on six jars of peanut butter. How many jars would he need to stay alive 30 days?

lesson checklist 33

- ☐ Slope formula
- ☐ Ratio
- ☐ Proportion
- ☐ Implied proportion

ANSWERS TO TEST YOUR UNDERSTANDING

(a) $+\frac{3}{7}$ (b) 36 jars

problem set 33

*1. Farmer Bud wishes to plant 423 acres in wheat and barley in the ratio of 4 to 5. How many acres should he plant in wheat and how many acres should he plant in barley?

*2. The ratio of men to women in the Sylvester Tennis Club is 5 to 3. If the club has 184 members, how many are men?

*3. Detia and Rob use 4 liters of paint to paint 3 rooms. How many liters will they use to paint 15 rooms?

*4. A 6-foot man casts a 4-foot shadow. What is the length of the shadow of a 30-foot tree?

5. Hannibal lost 12 percent of his troops in the battle. If he lost 1800 troops, how many did he have when the battle began?

6. Two chariots left Rome at the same time headed for Brindisium. The rate of the fast chariot was 15 km/hr and the rate of the slower chariot was 8 km/hr. How long will it take before the fast chariot is 35 kilometers ahead of the slower chariot?

7. Sara and Susan have a collection of 123 nickels and dimes. They have 77 more dimes than nickels. How many dimes do they have?

8. Dr. Scott found three consecutive odd integers such that four times the first was 4 greater than twice the sum of the third and 5. What were Dr. Scott's integers?

9. Graph the points $(-3, 4)$ and $(5, -2)$ and use the graph to determine the slope of the line through these points.

*10. Use the slope formula to find the slope of the line that passes through the points given in Problem 9.

Solve:

11. $\sqrt{x^2 - 5x - 2} - 2 = 0$

12. $\sqrt{x - 1} - 6 = 0$

*The problems marked with an asterisk were worked out in this lesson.

13. Find the equation of the line:

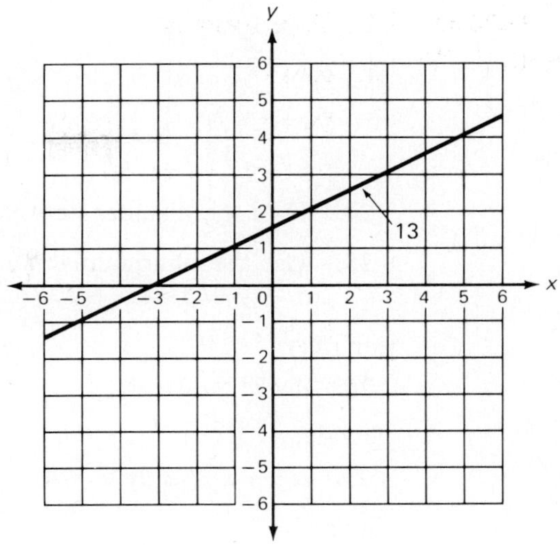

14. Find the equation of the line that goes through $(-4, -7)$ and $(5, -2)$.

15. Find the equation of the line through $(-5, 4)$ that is parallel to $2x + 2y - 7 = 0$.

16. Find the equation of the line through $(-5, -2)$ that has a slope of $\frac{3}{2}$.

17. (a) Given that $f(x) = 2x - 4$, $D = \{\text{Reals}\}$; find $f(-2)$.
(b) Given that $g(x) = x^2$, $D = \{\text{Integers}\}$; find $g(-2)$.
(c) Given that $h(x) = 3x + 10$, $D = \{\text{Neg Integers}\}$; find $h(-2)$.

18. Which of the following represent functions? Use the vertical line test as necessary.

(a)

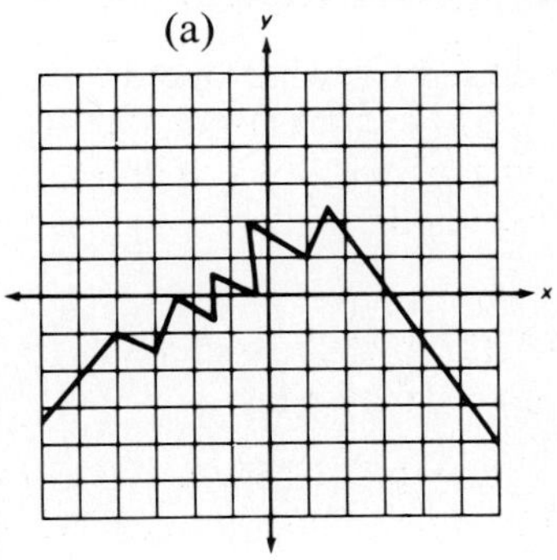

(b)

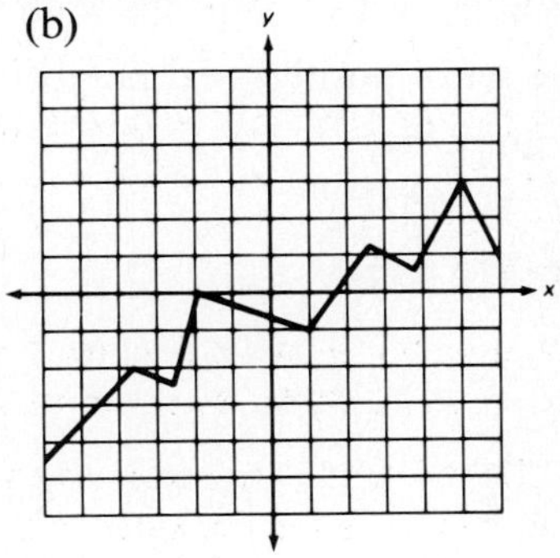

(c)

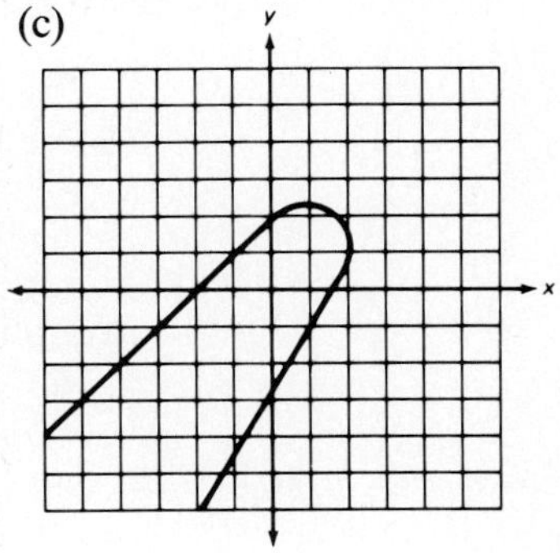

(d)

(4, 5), (7, 6), (8, 9)
(5, 6), (4, 7), (10, 8)

(e)

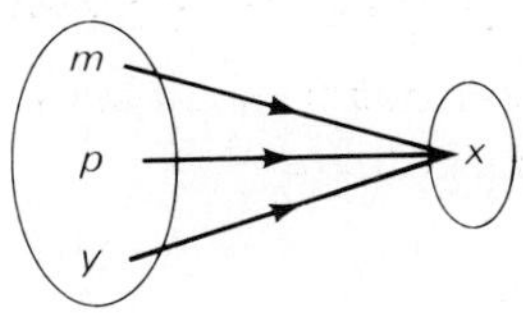

(f)

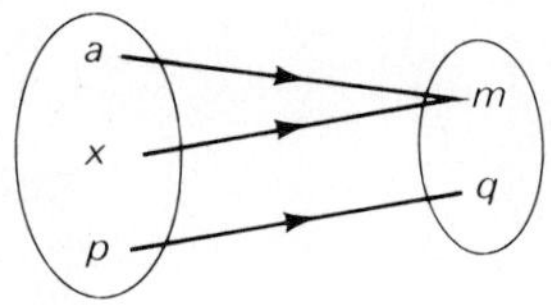

Solve:

19. $\dfrac{x - 4}{x} = \dfrac{7}{9}$

20. $\dfrac{x - 1}{x} = \dfrac{x + 1}{x + 3}$

Graph on a number line:

21. $-x + 3 > 4$, $D = \{\text{Integers}\}$

22. $|x| - 4 \le 1$, $D = \{\text{Reals}\}$

23. Find x.

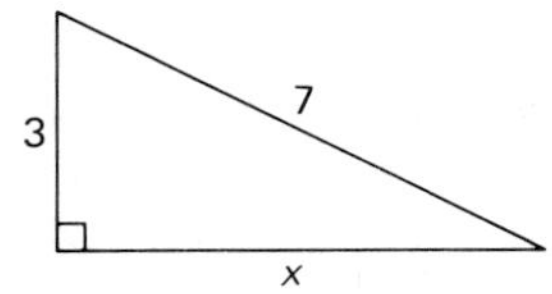

24. Find the distance between (4, 2) and (−3, 3).

25. Use the commutative and associative properties to show that $x + y + z = x + z + y$.

Simplify:

26. $4\sqrt{27} - 2\sqrt{300}$

27. $4\sqrt{3}\,(2\sqrt{3} - 2)$

28. $\sqrt{8} \cdot 2\sqrt{3}$

Divide:

29. $x^3 - x + 2$ by $x - 4$

Simplify:

30. $\dfrac{(.0032 \times 10^{-6})(4000 \times 10^{2})}{(8000 \times 10^{15})}$

Use elimination:

31. $4x - 2y = 16$
$3x - 3y = 18$

Solve:

32. $x^2 = 2x + 24$

Simplify:

33. $\dfrac{x^2 + x - 6}{x^2 + 8x + 15} \cdot \dfrac{x^2 + 4x - 5}{x^2 - 9x + 14}$

34. The set of positive real numbers is closed for what operations?

Solve by graphing:

35. $\begin{cases} y + x = 0 \\ y = x - 2 \end{cases}$

Simplify:

36. $\dfrac{a - \dfrac{b}{c}}{a + \dfrac{b}{c}}$

37. $\dfrac{\dfrac{1}{x}}{y}$

Add:

38. $\dfrac{1}{p^2} + \dfrac{1}{p + a}$

39. $\dfrac{a + b}{x^2} + \dfrac{c}{x} - \dfrac{d}{x^3}$

Add like terms:

40. $\dfrac{6ax^2}{m^2} - \dfrac{4am^{-2}}{x^{-2}} - \dfrac{3a^{-1}x^2}{m^2}$

41. $ac^2b - 3abc^2 + \dfrac{5b}{a^{-1}c^2}$

Simplify:

42. $\dfrac{x^2 - 10x + 25}{x^2 - 9} \div \dfrac{x^2 - 25}{x^2 + 2x - 15}$

43. $\dfrac{(a^{-4}b^0)^2x^{-2}y^4}{a^4b^{-2}x^2}$

44. $\dfrac{(m^{-2}y^3)^5x^{-4}}{m^{-3}y^4m^2y^{-2}}$

Solve:

45. $-2(p - 5) + (-p) = 3p + 8$

Evaluate:

46. $-a(-a - b)$ if $a = -2$, $b = 3$

Simplify:

47. $\dfrac{4}{(-2)^{-3}}$

48. True or false: {Real} $\subset$ {Integers}

49. $\dfrac{3}{2} \in$ {what sets of numbers?}

Add:

50. $\dfrac{3}{x + 4} - \dfrac{2}{x - 4} + \dfrac{5}{x^2 - 16}$

lesson thirty-four

Solutions of linear systems
Consistent, inconsistent, and dependent equations
Solution of general rational equations

34.1 solutions of linear systems: consistent equations

We have presented three methods of finding the ordered pair that is the solution to a system of two first-degree equations in two unknowns. We call the methods the **substitution method**, the **elimination method**, and the **graphical method**. We will review these procedures by solving a system of equations by each of the three methods. The system is

$$\begin{array}{ll} \text{(a)} & \left\{ \begin{array}{l} -x + y = 1 \\ x + y = 5 \end{array} \right. \\ \text{(b)} & \end{array}$$

SUBSTITUTION	ELIMINATION
$x + (1 + x) = 5$	$-x + y = 1$
$2x + 1 = 5$	$\underline{x + y = 5}$
$2x = 4$	$2y = 6$
$x = 2$	$y = 3$
and since $x + y = 5$	and since $x + y = 5$
$y = 3$	$x = 2$
solution is (2, 3)	solution is (2, 3)

On the left above we substituted for y in equation (b) the equivalent expression for y from equation (a). On the right above we added the equations algebraically and eliminated the variable x. In Figure 34.1 we show the graph of the two linear equations and note the coordinates of the intersection of the two lines. In each case, we find that the ordered pair (2, 3) is the solution to the system of equations. If a solution exists, the solution to such a system can always be found using any one of the three methods. **Equations that have a single solution are called consistent equations.**

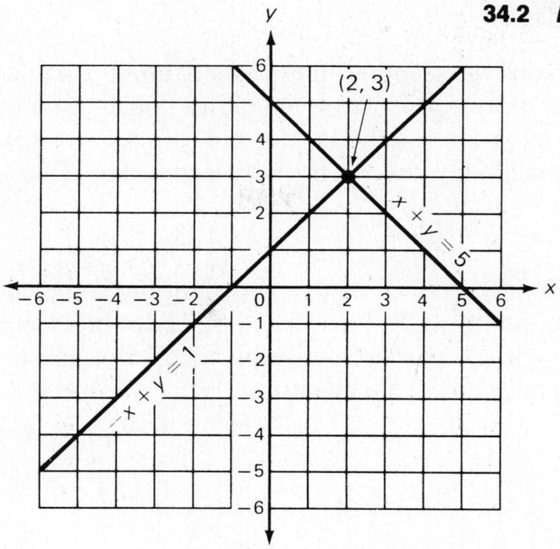

Figure 34.1

34.2 inconsistent equations

Inconsistent equations have no common solution. On the left we show a system of two linear equations and in Figure 34.2 we show the graph of the two equations.

$$\begin{cases} x + y = 2 \\ x + y = -3 \end{cases}$$

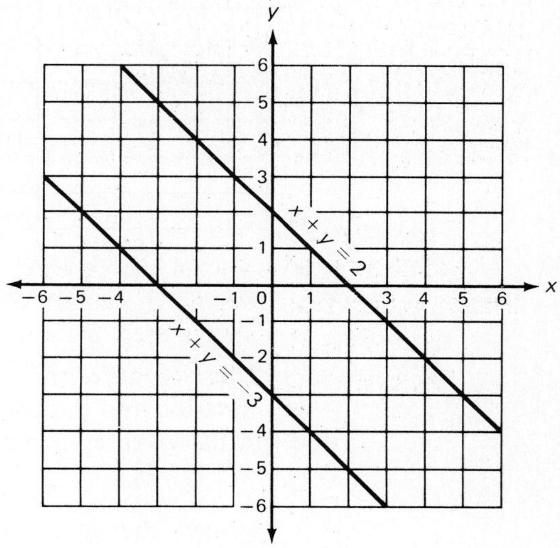

Figure 34.2

From the figure we see that the lines are parallel and do not intersect. If they do not intersect, there is no one point that lies on both of the lines and hence no ordered pair of x and y will satisfy both equations. If we try to use either the substitution method or the elimination method, we will end up with a false numerical statement. To demonstrate this result, we will attempt to solve the given system by both methods.

SUBSTITUTION	ELIMINATION
$(-3 - y) + y = 2$	$-x - y = -2$
$-3 - y + y = 2$	$x + y = -3$
$-3 = 2$ False	$0 = -5$ False

In both cases the final result is a false numerical statement.

In Lessons 13 and 17 we discussed the substitution and elimination methods in some detail. We pointed out that before either method was used, an assumption was necessary! We assumed that a value of x and y existed that would simultaneously satisfy both equations and that the symbols x and y in the equations represented these numbers. From the graph in Figure 34.2 it is obvious that our assumption was invalid, for in the graph we see that there is no point that lies on both lines and whose coordinates therefore satisfy both equations. Since our assumption was invalid, the substitution and elimination methods do not produce a solution because there is no point whose coordinates will satisfy both equations. Thus we say that the solution is the *empty set*.

34.3 dependent equations

We have defined equivalent equations to be equations that have the same solution set. If we multiply every term on both sides of an equation by the same non-zero quantity, the resulting equation is an equivalent equation to the original equation. In Figure 34.3, we have graphed the equation

$$y - x = 2$$

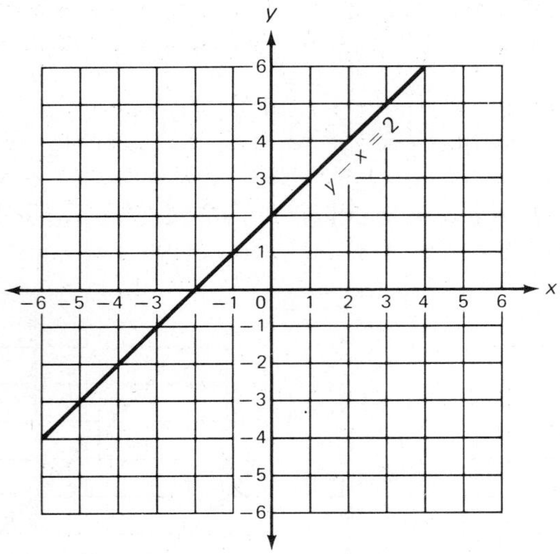

Figure 34.3

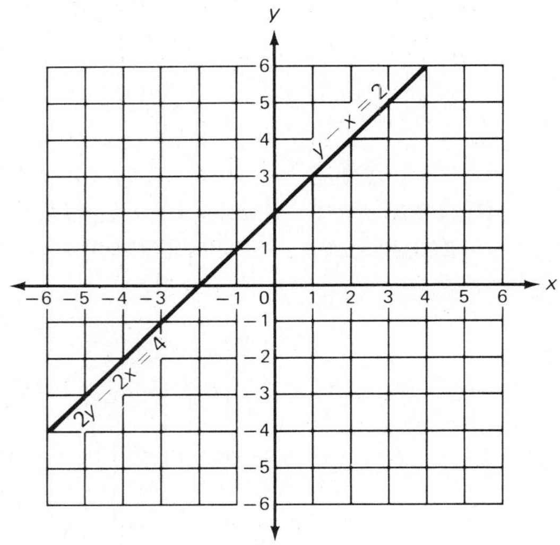

Figure 34.4

There is an infinite number of ordered pairs of x and y whose coordinates satisfy the equation and the graph of these points is the line in the figure. If we multiply every term in the given equation by some quantity, say the number 2,

$$y - x = 2 \qquad \text{multiplied by (2) yields} \qquad 2y - 2x = 4$$

we get the equation $2y - 2x = 4$, which is an equivalent equation to the original equation. Thus all ordered pairs of x and y that satisfy the original equation also satisfy the new equation and the graph of the new equation is the same as the graph of the original equation. If we are asked to find a graphical solution to a system that consists of a pair of equivalent equations such as the pair being discussed.

$$\begin{aligned} &\text{(a)} \\ &\text{(b)} \end{aligned} \quad \begin{cases} y - x = 2 \\ 2y - 2x = 4 \end{cases}$$

we find that the graph of both equations is the single line of Figure 34.4. Thus any ordered pair that is a solution to one of the equations is also a solution to the other equation. **Equivalent linear equations are called dependent equations.** If we try to find the solution to our pair of dependent equations, we find that the result is a true numerical statement. We will attempt to solve the system above by using both substitution and elimination.

SUBSTITUTION		ELIMINATION	
(a)	$y - x = 2$	(a)	$y - x = 2$
(b)	$2y - 2x = 4$	(b)	$2y - 2x = 4$
	$2(x + 2) - 2x = 4$		$-2y + 2x = -4$
	$2x + 4 - 2x = 4$		$\underline{2y - 2x = 4}$
	$4 = 4$ True		$0 = 0$ True

On the left we solved equation (a) for y and substituted this expression for y in equation (b). The result reduced to the true statement that 4 equals 4. On the right we multiplied each term in equation (a) by -2 and then added the resulting equation to equation (b). The result was the true statement that 0 equals 0. This indicates that any ordered pair of x and y that satisfies one of the equations will satisfy the other equation. Thus there is an infinite number of ordered pairs that will satisfy two dependent equations.

34.4 summary

Two linear equations in two unknowns are in one of three categories.

1. **CONSISTENT EQUATIONS**, which are equations that have a single ordered pair as a common solution. The graphs of consistent equations intersect as shown in Figure 34.5.
2. **INCONSISTENT EQUATIONS**, which are equations that have no common solution. The graphs of inconsistent equations are parallel lines as shown in Figure 34.6.
3. **DEPENDENT EQUATIONS**, which are equivalent equations, or if you will, the same equation in two different forms. The graph of two dependent equations is a single line as shown in Figure 34.7.

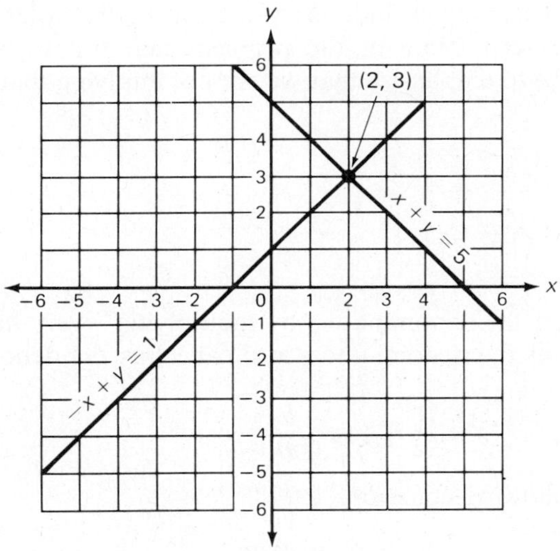

Figure 34.5

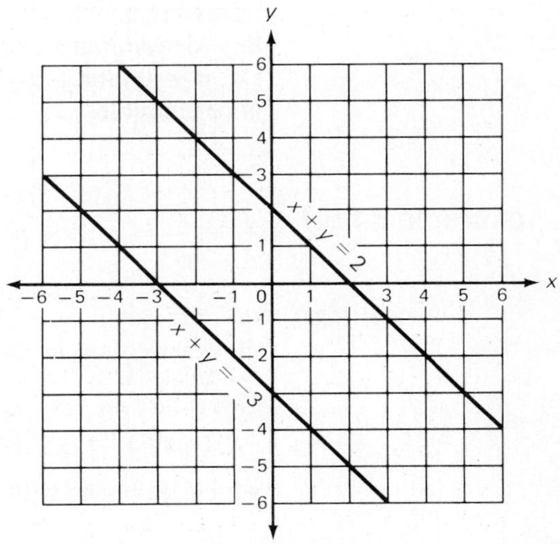

Figure 34.6

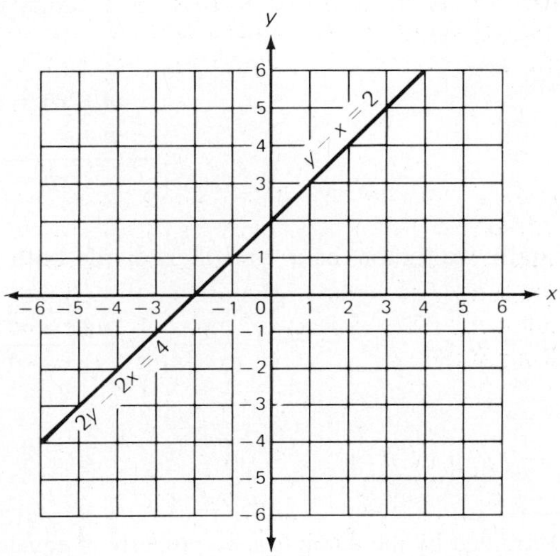

Figure 34.7

34.5 rational equations

In Lessons 19 and 28 we stated that a good way to begin the solution of an equation that contains fractions is first to eliminate the denominators by multiplying every numerator by the least common multiple of the denominators. Thus as a first step in the solution of the equation

$$\frac{y}{2} + \frac{1}{4} = \frac{y}{6}$$

we multiply each numerator by 12.

$$(12)\frac{y}{2} + (12)\frac{1}{4} = (12)\frac{y}{6}$$

Now we eliminate the denominators and complete the solution.

$$\begin{aligned} 6y + 3 &= 2y \\ 4y &= -3 \\ y &= -\tfrac{3}{4} \end{aligned}$$

The same procedure can be used to simplify any rational equation. In Lesson 28 we learned that the first step was to specify the unacceptable values of the variable, which are those values of the variable that would cause any denominator to equal zero. In the following problems we will carefully exclude the values of the variables that could make any denominator equal to zero. Note in the problem that follows we have written $(x, m \neq 0)$ and $(cx - 1 \neq 0)$ to emphasize that we are not implying that division by zero is permissible.

example 34.5.1 $\dfrac{1}{x} + \dfrac{b}{m} = c$; find m $\quad (x, m \neq 0)$.

solution We will begin by eliminating the denominators by multiplying[†] every numerator by xm, the least common multiple of the denominators and canceling the denominator,

$$(\cancel{x}m)\frac{1}{\cancel{x}} + (x\cancel{m})\frac{b}{\cancel{m}} = cxm$$

which leaves us with the following equation.

$$m + xb = cxm$$

Now we use the additive property of equality to place all terms with m's on one side and all other terms on the other side.

$$\begin{aligned} m + xb &= cxm \\ -m \qquad &\quad\; -m \\ \hline xb &= cxm - m \end{aligned}$$

Finally we factor the m and then divide both sides by the coefficient of m.

$$xb = m(cx - 1) \longrightarrow \frac{xb}{cx - 1} = \frac{m\cancel{(cx - 1)}}{\cancel{(cx - 1)}} \longrightarrow \frac{\mathbf{xb}}{\mathbf{cx - 1}} = \mathbf{m} \qquad (cx - 1 \neq 0)$$

[†]Permitted by the multiplicative property of equality.

example 34.5.2 $\frac{a}{b} + \frac{c}{d} = x$; find b $(b, d \neq 0)$.

solution As the first step we multiply each numerator by bd, the least common multiple of the denominators

$$(bd)\frac{a}{b} + (bd)\frac{c}{d} = bdx$$

and cancel the denominators

$$da + bc = bdx$$

Now we use the additive property of equality as necessary to position all terms that contain b on one side of the equation (either side) and all terms that do not contain b on the other side. We decide to position all terms that contain b on the right side of the equation by adding $-bc$ to both sides.

$$\begin{array}{rl} da + bc &= bdx \\ -bc & \quad -bc \\ \hline da \quad &= bdx - bc \end{array}$$

Now factor b on the right side

$$da = b(dx - c)$$

and as a last step divide both sides of the equation by $dx - c$.

$$\frac{da}{(dx - c)} = b\frac{\cancel{(dx - c)}}{\cancel{(dx - c)}} \longrightarrow \mathbf{\frac{da}{dx - c} = b} \qquad (dx - c \neq 0)$$

example 34.5.3 $\frac{a}{b} - c = \frac{d}{x}$, find x $(b, x \neq 0)$.

solution First we eliminate the denominators

$$(\cancel{b}x)\frac{a}{\cancel{b}} - bxc = (b\cancel{x})\frac{d}{\cancel{x}} \longrightarrow xa - bxc = bd$$

Now since all x terms are already on one side and all other terms on the other side, we factor the x and divide both sides by the coefficient of x.

$$x(a - bc) = bd \longrightarrow \frac{x\cancel{(a - bc)}}{\cancel{(a - bc)}} = \frac{bd}{(a - bc)} \longrightarrow \mathbf{x = \frac{bd}{a - bc}} \qquad (a - bc \neq 0)$$

example 34.5.4 $\frac{a}{x} - y + \frac{m}{n} = k$; find x $(x, n \neq 0)$.

solution First we eliminate the denominators.

$$(\cancel{x}n)\frac{a}{\cancel{x}} - xny + (x\cancel{n})\frac{m}{\cancel{n}} = xnk \longrightarrow na - xny + xm = xnk$$

Now we move all terms that contain an x to the right side, factor out the x and divide by the coefficient of x.

$$\begin{array}{rl} na - xny + xm &= xnk \\ +xny - xm & \quad +xny - xm \\ \hline na \quad &= xnk + xny - xm \end{array} \longrightarrow na = x(nk + ny - m)$$

$$\longrightarrow \frac{na}{(nk + ny - m)} = \frac{x\cancel{(nk + ny - m)}}{\cancel{(nk + ny - m)}} \longrightarrow \mathbf{\frac{na}{nk + ny - m}} = x \qquad (nk + ny - m \neq 0)$$

Test Your Understanding

(a) $\frac{a}{y} + b = \frac{k}{p}$; find $y (y, p \neq 0)$.

lesson checklist 34

- ☐ Consistent equations
- ☐ Inconsistent equations
- ☐ Dependent equations
- ☐ Solution of rational equations

Answers to Test Your Understanding

(a) $\frac{pa}{k - bp}$

problem set 34

1. The ratio of cowards to heroes at King Arthur's round table was 2 to 11. If Arthur had 104 knights, how many were heroes?
2. In one fiscal year, the Oil Company drilled gushers and dry holes in the ratio of 7 to 13. If they drilled 240 wells, how many dry holes did they have?
3. The ratio of losers to winners in the entering freshman class was 3 to 5. If 1440 students were in the freshman class, how many were winners?
4. Louis ran to Versailles and then walked back to town. If his running rate is 6 km/hr and his walking rate is 3 km/hr, how far was it to Versailles if the round trip took 6 hours?
5. Two cars leave Cheyenne at the same time. The northbound car has an average speed of 35 km/hr and the southbound car has an average speed of 45 km/hr. How long will it take before the two cars are 320 kilometers apart?
6. Prince Valiant bought some replacement armor for 540 florins. He bought helmets for 4 florins each and cuirasses for 6 florins each. If he bought 100 pieces of armor, how many helmets did he buy?
7. Senator Russell got 52 percent of the votes cast in his district. If he got 910 votes, how many voters cast ballots?
8. Find three consecutive even integers such that 4 times the first equals 16 times the sum of the third and the number 2.

***9.** $\dfrac{1}{x} + \dfrac{b}{m} = c$, find m.

***10.** $\dfrac{a}{b} + \dfrac{c}{d} = x$, find b.

***11.** $\dfrac{a}{b} - c = \dfrac{d}{x}$, find x.

***12.** $\dfrac{a}{x} - y + \dfrac{m}{n} = k$, find x.

Solve:

13. $\sqrt{x^2 - x - 4} - 4 = 0$

14. $\sqrt{2x + 2} - 8 = 0$

15. Find the equation of the line:

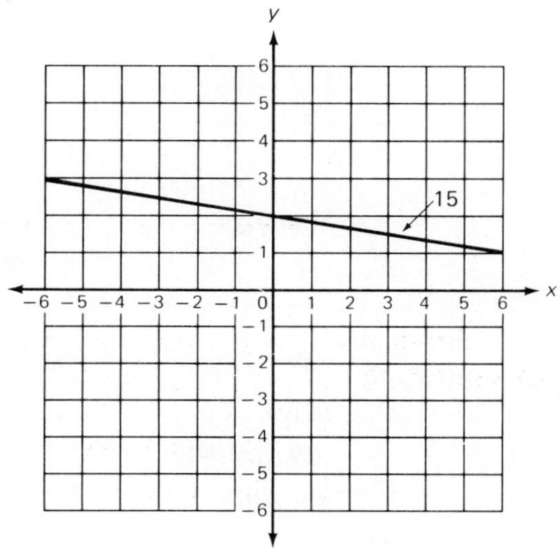

16. Find the equation of the line through $(4, -2)$ and $(1, 1)$.

17. Find the equation of the line that goes through $(2, -3)$ and is parallel to $2x + y = -3$.

18. Given that $f(x) = x - 4$, $D = \{\text{Reals}\}$; find $g(-2)$.
$g(x) = x^2 + 2$, $D = \{\text{Reals}\}$;
$\phi(x) = x^2 - 3$, $D = \{\text{Reals}\}$;

19. Which of the following represent functions? Use the vertical line test as necessary.

(a)

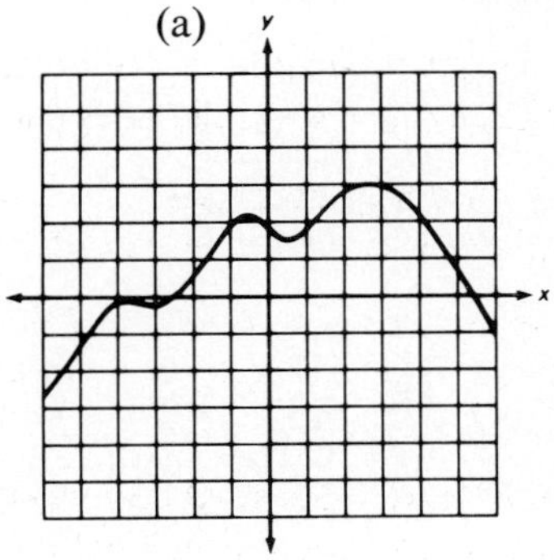

(b)

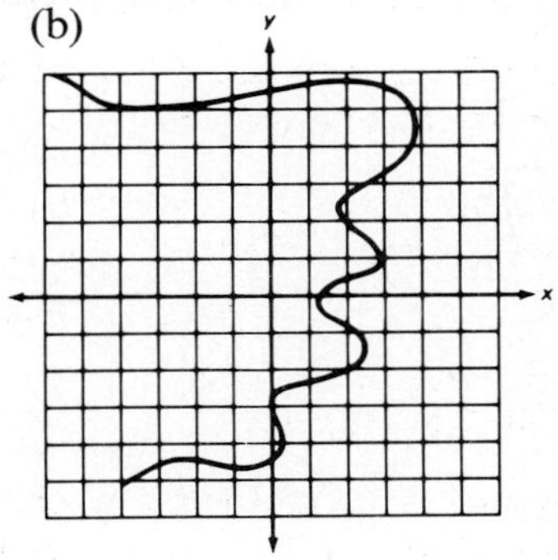

(c)

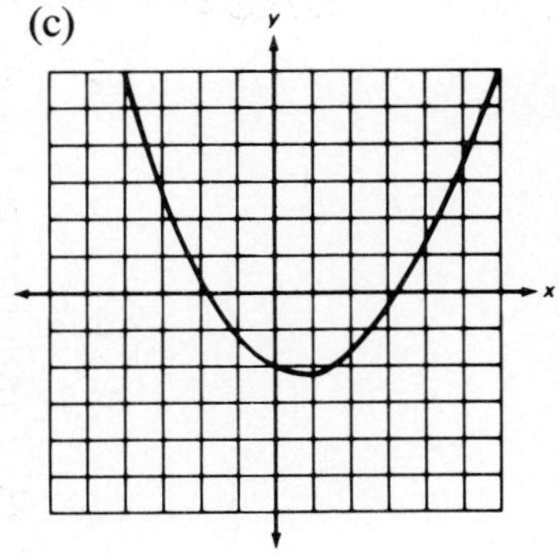

(d)
(3, 4), (5, 6), (7, 8)
(4, 5), (6, 7), (3, 9)

(e)

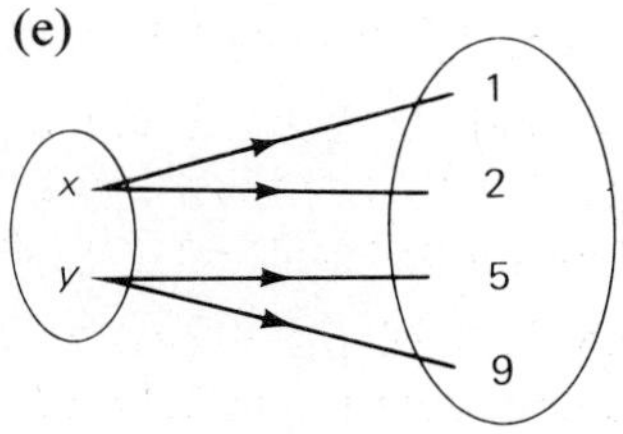

(f)

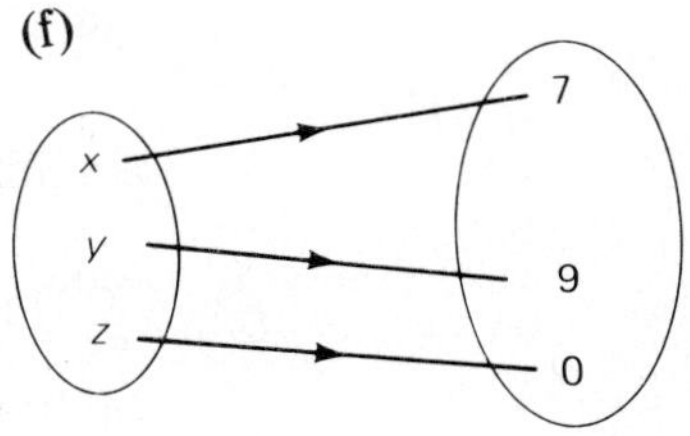

*The problems marked with an asterisk have been worked out in this lesson.

Solve:

20. $\dfrac{3a}{a + 2} = \dfrac{5}{2}$

21. $\dfrac{3x}{10x + 2} = \dfrac{2}{7}$

Graph on a number line:

22. $3 - x < 4$, $D = \{\text{Integers}\}$

23. $|x| - 6 > 1 \quad D = \{\text{Reals}\}$

24. Find h.

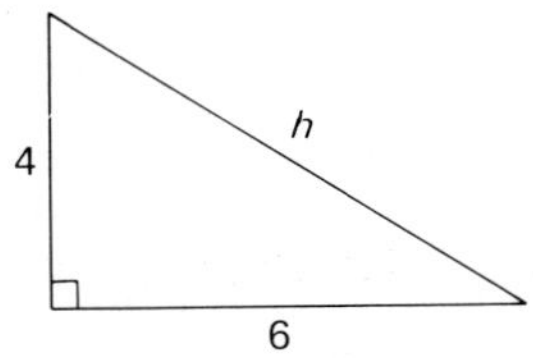

Simplify:

25. $\sqrt{12} - \sqrt{21} + \sqrt{49}$

26. $2\sqrt{5}\,(3\sqrt{5} - \sqrt{2}\,)$

27. $3\sqrt{3}\ \sqrt{2}\ \sqrt{18}$

28. Divide $7x^3 + 18x^2 - 7x + 7$ by $x + 3$.

Solve:

29. $x^2 = 5x + 14$

Simplify:

30. $\dfrac{a - b}{x + 5} \cdot \dfrac{x^2 + 3x - 10}{a^2 - b^2}$

31. $\dfrac{(.003 \times 10^8)(.02 \times 10^{-7})}{(12{,}000{,}000)}$

Use elimination:

32. $\begin{cases} 2y - 2x = -10 \\ 3y + 5x = 1 \end{cases}$

33. The set of positive integers is closed for what operations?

Solve by graphing:

34. $\begin{cases} 2x - y - 2 = 0 \\ x + 3 \end{cases}$

Simplify:

35. $\dfrac{\dfrac{a}{b} + 1}{b - \dfrac{a}{b}}$

36. $\dfrac{1 + y}{\dfrac{x}{y} + y}$

Add:

37. $\dfrac{2x}{3} - \dfrac{a}{ab^2}$

38. $\dfrac{1}{b^2a} + \dfrac{x}{a^2} - \dfrac{m}{b}$

Add like terms:

39. $3x^2yz^3 + \dfrac{2z^3}{x^2y} - \dfrac{2yz}{x^{-2}z^{-2}}$

Simplify:

40. $\dfrac{x^2 - 1}{x^2 - x - 2} \div \dfrac{x^2 - 2x + 1}{x^2 + x - 2}$

Multiply:

41. $xay\left(\dfrac{a^2}{axy^{-2}} - \dfrac{3xa^{-1}}{y}\right)$

Simplify:

42. $\dfrac{ym^2k^{-3}p}{y^{-2}m^0k^{-4}p^2}$

43. $\dfrac{abc}{(a^2)^{-1}b^0c^{-1}}$

Evaluate:

44. $(-x + a)(a + x) - x$ if $x = 2, a = -3$

Simplify:

45. $\dfrac{(-3)^{-3}}{3^{-2}}$

46. True or false: {Irrationals} $\subset$ {Reals}

47. $-\dfrac{2}{714} \in$ {what sets of numbers?}

Add:

48. $\dfrac{3x^2}{x^2 - x - 6} - \dfrac{2}{x - 3} + \dfrac{4x}{x + 2}$

lesson thirty-five

Order in the real numbers
Terminating and repeating decimal numbers

35.1 order in the real numbers

Every real number is paired with one and only one point on the number line and every point on the number line is paired with one and only one real number.

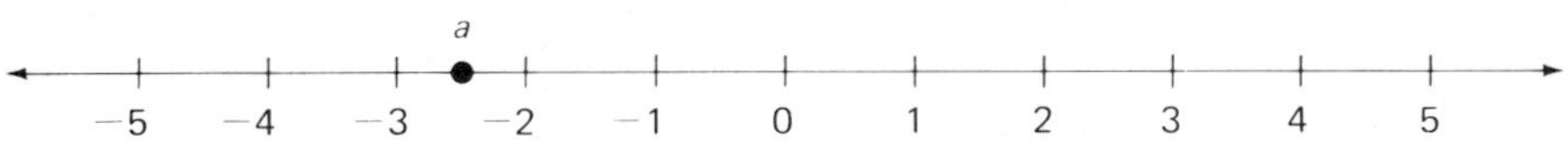

Figure 35.1

In Figure 35.1 we have graphed the location of an unspecified number, which we call a. Now if we blindly select another number on the number line and call this number b, we will find that one and only one of the following can be true about the number b.

1. The number b is the same number as the number a.
2. The number b is greater than the number a.
3. The number b is less than the number a.

While this fact is more or less self-evident and seemingly trivial, mathematicians recognize that this property of real numbers reveals that the real numbers constitute an **ORDERED SET** and this property is given the name of **TRICHOTOMY**.

> TRICHOTOMY AXIOM
>
> For any two real numbers a and b, exactly one of the following is true:
>
> $$a = b \qquad a > b \qquad a < b$$

The concept of the relative magnitudes of real numbers can be investigated graphically with the aid of the number line.

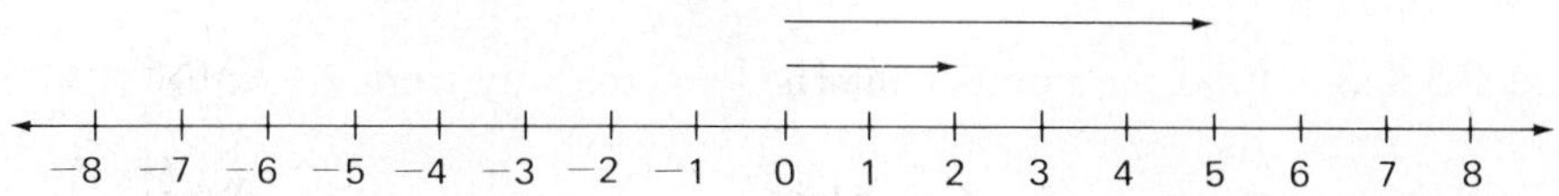

Figure 35.2

A number is said to be greater than another number if its graph on the number line lies to the right of the graph of the other number. In Figure 35.2 we see that the number 5 is 5 units to the right of the origin while the number 2 is only 2 units to the right of the origin. Thus we say that 5 is 3 greater than 2 because it lies 3 units farther to the right on the number line. **Also this figure helps to emphasize that the name of a positive number describes the distance the positive number lies to the right of the origin.**

The number line is also helpful in allowing us to visualize the problem of determining the number that has a specified relationship with respect to two other numbers.

example 35.1.1 Find the number that is $\frac{7}{10}$ of the way from 30 to 40.

solution First we draw a number line and graph the numbers 30 and 40.

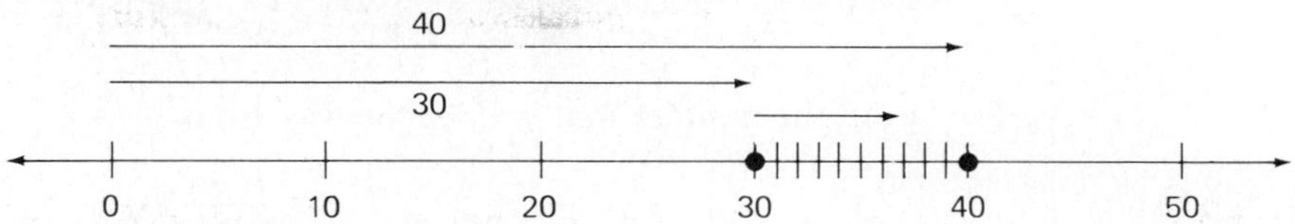

From the diagram we see that 40 is 40 units to the right of the origin and that 30 is 30 units to the right of the origin so that the distance between the numbers is $40 - 30 = 10$. Seven-tenths of 10 is 7, so the number in question lies 7 units to the right of 30 and its distance from the origin is $30 + 7 = \mathbf{37}$.

The solution may be indicated in a rather compact notation by writing

$$30 + \frac{7}{10}(40 - 30) = 37$$

example 35.1.2 Find the number that is $\frac{1}{5}$ of the distance from $\frac{1}{8}$ to $\frac{9}{11}$.

solution We begin by graphing the numbers $\frac{1}{8}$ and $\frac{9}{11}$.

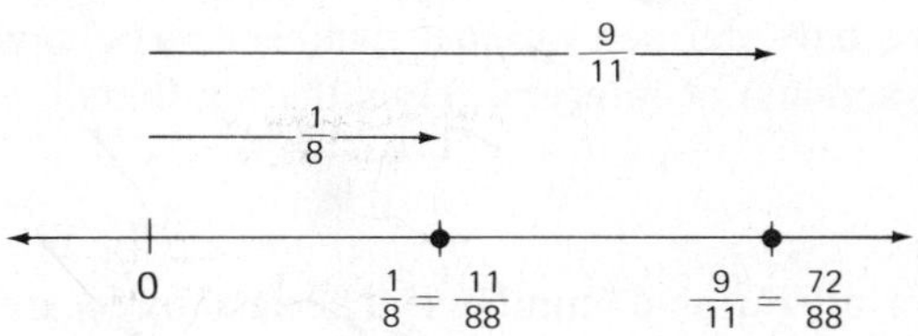

First we express $\frac{1}{8}$ and $\frac{9}{11}$ as $\frac{11}{88}$ and $\frac{72}{88}$ and find that the distance between the two numbers is $\frac{72}{88} - \frac{11}{88} = \frac{61}{88}$. One-fifth of this distance is $\frac{1}{5} \cdot \frac{61}{88} = \frac{61}{440}$. The number we wish to find is $\frac{61}{440}$ units to the right of $\frac{1}{8}$ so its total distance from the origin is

$$\frac{61}{440} + \frac{1}{8} \text{ or } \frac{61}{440} + \frac{55}{440} = \frac{116}{440} \text{ or } \frac{116 \cdot \frac{1}{4}}{440 \cdot \frac{1}{4}} = \frac{29}{110}.$$

The solution may be indicated in a compact notation by writing

$$\frac{1}{8} + \frac{1}{5}\left(\frac{9}{11} - \frac{1}{8}\right) = \frac{29}{110}$$

example 35.1.3 Find the number that is $\frac{2}{3}$ of the way from $2\frac{1}{4}$ to $3\frac{5}{6}$.

solution We began by graphing the numbers $2\frac{1}{4}$ and $3\frac{5}{6}$.

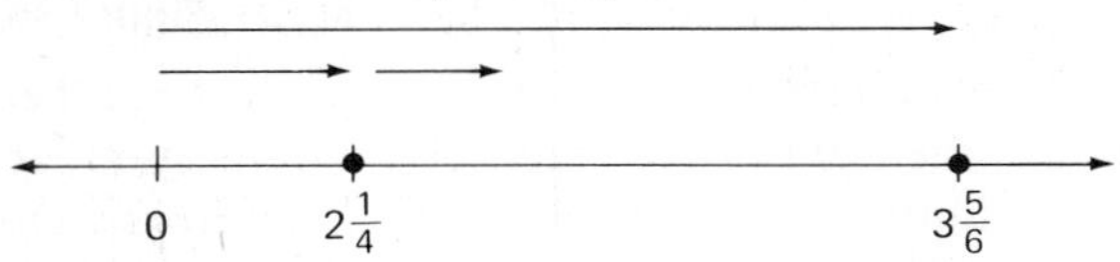

The solution may be indicated in a compact notation by writing

$$2\frac{1}{4} + \frac{2}{3}\left(3\frac{5}{6} - 2\frac{1}{4}\right)$$

$$= 2\frac{1}{4} + \frac{2}{3}\left(3\frac{5}{6} - 2\frac{1}{4}\right) = \frac{9}{4} + \frac{2}{3}\left(\frac{23}{6} - \frac{9}{4}\right) = \frac{9}{4} + \frac{2}{3}\left(\frac{92}{24} - \frac{54}{24}\right)$$

$$= \frac{9}{4} + \frac{2}{3}\cdot\frac{38}{24} = \frac{9}{4} + \frac{38}{36} = \frac{9(9)}{4(9)} + \frac{38}{36} = \frac{81}{36} + \frac{38}{36} = \mathbf{\frac{119}{36}}$$

> TEST YOUR UNDERSTANDING
>
> (a) Find the number that is $\frac{2}{3}$ of the way from $\frac{1}{11}$ to $\frac{3}{4}$.

35.2 rational numbers

We have defined rational numbers to be any numbers that can be written as a quotients (fractions) of integers. Thus the numbers

$$4 \qquad \frac{7}{10} \qquad -\frac{11}{43} \qquad \frac{150}{13}$$

are all rational numbers. The last three are already expressed as quotients of integers and the number 4 can be expressed many ways as a quotient of integers: $\frac{4}{1}$, $\frac{40}{10}$, and $\frac{-28}{-7}$ are a few.

Any terminating or repeating decimal number can be written as a quotient of integers. Thus

$$4.132 \qquad .00163 \qquad 4.1\overline{56} \qquad .61\overline{342}$$

are all rational numbers because they are either terminating or repeating decimal numbers. The first two numbers 4.132 and .00163 are classified as terminating decimal numbers because they are expressed exactly by using a finite number of digits. The line over the $\overline{56}$ and $\overline{342}$ in the numbers $4.1\overline{56}$ and $.61\overline{342}$ indicate that these digits repeat indefinitely, as

$$4.1565656565656\ldots \quad \text{and} \quad .61342342342342\ldots$$

so these numbers cannot be expressed exactly with a finite number of digits.

example 35.2.1 Show that .00134 is a rational number by expressing it as a quotient of integers.

solution The **denominator numerator same quantity theorem** says that the numerator and the denominator of an expression may be multiplied by the same quantity without changing the value of the expression. We will multiply the numerator and the denominator of .00134 by 100,000.

$$\frac{.00134}{1}\frac{(100{,}000)}{(100{,}000)} = \frac{134}{100{,}000}$$

which is a quotient of integers.

example 35.2.2 Show that 143.25 is a rational number by expressing it as a quotient of integers.

solution We will multiply by 100 over 100.

$$\frac{143.25}{1}\frac{(100)}{(100)} = \frac{14325}{100}$$

which is a quotient of integers.

example 35.2.3 Show that .2 is a rational number by expressing it as a quotient of integers.

solution If we multiply .2 by 10, the product is 2, which is an integer. But we must also divide by 10 so that the value is unchanged.

$$\frac{.2}{1}\frac{(10)}{(10)} = \frac{2}{10}$$

which is a quotient of integers.

TEST YOUR UNDERSTANDING

(b) Show that 4.0033 is a rational number by writing it as a quotient of integers.

example 35.2.4 Show that $4.\overline{16}$ is a rational number by writing it as a quotient of integers.

solution The line over the $\overline{16}$ indicates that these two digits repeat indefinitely as 4.1616161616 . . . . To write this number as a quotient of integers, we are going to have to get rid of the repeating 16's. If we write the number down twice

$$N = 4.|16|16|16|16|16| \cdots$$
$$N = 4.|16|16|16|16|16| \cdots$$

we note that the repeating digits are in groups of two. If we multiply both sides of the top equation by 100, this multiplication will shift each digit in the top number exactly 2 places to the left.

$$100N = 416.|16|16|16|16|16| \cdots$$
$$N = 4.|16|16|16|16|16| \cdots$$

The repeating 16's are still lined up one underneath the other. And now if we subtract the lower expression from the upper expression, all the repeaters will be eliminated.

$$\begin{aligned} 100N &= 416.|16|16|16|16|16| \cdots \\ N &= 4.|16|16|16|16|16| \cdots \\ \hline 99N &= 412 \end{aligned}$$

and thus

$$\frac{\cancel{99}N}{\cancel{99}} = \frac{412}{99} \longrightarrow N = \frac{412}{99}$$

which is a quotient of integers.

example 35.2.5 Show that the number $1.4\overline{321}$ is a rational number by writing it as a quotient of integers.

solution The line over the $\overline{321}$ indicates that these three digits repeat indefinitely as $1.4\underline{321}\ \underline{321}\ \underline{321}\ \underline{321}\ldots$. To write this number as a quotient of integers we are going to have to get rid of the repeating 321's. If we write the number down twice

$$\begin{aligned} N &= 1.4|321|321|321|321| \cdots \\ N &= 1.4|321|321|321|321| \cdots \end{aligned}$$

we note that the repeating digits in this number are in groups of three. If we multiply both sides of the top equation by 1000, this multiplication will shift each 321 in the top number exactly 3 places to the left.

$$\begin{aligned} 1000N &= 1432.1|321|321|321| \cdots \\ N &= 1.4|321|321|321| \cdots \end{aligned}$$

Now if we subtract the lower equation from the upper equation, the 321 repeaters are eliminated and we get

$$\begin{aligned} 1000N &= 1432.1|321|321|321|321| \cdots \\ N &= 1.4|321|321|321|321| \cdots \\ \hline 999N &= 1430.7 \end{aligned}$$

and solving for N,

$$\frac{\cancel{999}N}{\cancel{999}} = \frac{1430.7}{999} \longrightarrow N = \frac{1430.7}{999}$$

The statement of the problem asked that we write the result as a quotient of integers. The number 1430.7 is not an integer so we multiply the numerator and denominator by 10.

$$N = \frac{1430.7}{999} \longrightarrow N = \frac{1430.7\,(10)}{999\,(10)} \longrightarrow N = \frac{14{,}307}{9990}$$

Now we have a quotient of integers.

example 35.2.6 Show that $.001\overline{23}$ is a rational number by writing it as a quotient of integers.

solution We begin by writing the number down twice

$$\begin{aligned} N &= .001|23|23|23|23| \cdots \\ N &= .001|23|23|23|23| \cdots \end{aligned}$$

In this problem we will shift each digit in the upper number two places to the left by multiplying both sides of the upper equation by 100. Then we subtract the lower equation from the new upper equation.

$$\begin{aligned} 100N &= .123|23|23|23|23| \cdots \\ N &= .001|23|23|23|23| \cdots \\ \hline 99N &= .122 \end{aligned}$$

$$\frac{\cancel{99}N}{\cancel{99}} = \frac{.122}{99} \longrightarrow N = \frac{.122}{99} \longrightarrow N = \frac{.122(1000)}{99(1000)} \longrightarrow N = \frac{122}{99{,}000}$$

Note that in the last step we multiplied both the numerator and the denominator by 1000 so that .122 could be written as the integer 122.

The power of 10 that we multiply by to line up the repeating digits is equal to the number of digits that repeat. If only one digit repeats as in $1.00\overline{3}$, we multiply by 10. If two digits repeat, we multiply by 100; if 3 repeat, by 1000; if 4 repeat, by 10,000; etc.

example 35.2.7 Show that $.\overline{3}$ is a rational number by writing it as a quotient of integers.

solution We begin by writing the number down twice.

$$N = .3|3|3|3|3| \cdots$$
$$N = .3|3|3|3|3| \cdots$$

We see that we have only one repeating digit so we multiply both sides of the upper equation by 10, which will shift each of the repeating 3's one place to the left. Then we subtract.

$$\begin{aligned} 10N &= 3.|3|3|3|3|3| \cdots \\ N &= .|3|3|3|3|3| \cdots \\ \hline 9N &= 3 \end{aligned}$$

$$\frac{\cancel{9}N}{\cancel{9}} \longrightarrow N = \frac{1}{3}$$

example 35.2.8 Show that $14.2\overline{1234}$ is a rational number by writing it as a quotient of integers.

solution In this problem we have 4 repeating digits so it is necessary to multiply the top equation by 10,000.

$$\begin{aligned} 10{,}000N &= 142{,}123.4|1234|1234| \cdots \\ N &= \phantom{142{,}1}14.2|1234|1234| \cdots \\ \hline 9999N &= 142{,}109.2 \end{aligned}$$

$$\frac{\cancel{9999}N}{\cancel{9999}} = \frac{142{,}109.2}{9999} \longrightarrow N = \frac{142{,}109.2}{9999} \longrightarrow N = \frac{\mathbf{1{,}421{,}092}}{\mathbf{99{,}990}}$$

In the last step we multiplied both the numerator and the denominator by 10 so that we would get a quotient of integers.

TEST YOUR UNDERSTANDING

(c) Show that $.001\overline{233}$ is a rational number by writing it as a quotient of integers.

lesson checklist 35

- ☐ Trichotomy axiom
- ☐ Definition of greater than on the number line
- ☐ Terminating decimals as rational numbers
- ☐ Repeating decimals as rational numbers

ANSWERS TO TEST YOUR UNDERSTANDING

(a) $\frac{35}{66}$ (b) $\frac{40{,}033}{10{,}000}$ (c) $\frac{1232}{999{,}000}$

problem set 35

1. The ratio of powerboats to sailboats at the Billy Bowlegs festival was 7 to 5. If there were 204 boats at the festival, how many were powerboats?

2. The ratio of haddock to codfish in the hold of the Sister Sel was 7 to 11. If there were 12,600 lb of fish in the hold, how much did the codfish weigh?

3. When Scotty made a statement, it normally contained fact and bluster in the ratio of 5 to 3. If he spoke for 32 minutes, how many minutes were devoted to bluster?

4. Ed and Alice walked to the dock at 5 mph, jumped in the boat, and motored to Destin at 15 mph. If the total distance covered was 20 miles and the trip took 2 hours, how far did they go by boat?

5. In the race, the hare traveled at 12 km/hr while the tortoise could only manage 2 km/hr. How long did it take the hare to get 35 kilometers ahead?

6. Shotguns sold for \$75 each while rifles cost only \$35 each. The hunting club spent \$965 for 15 guns. How many rifles did they buy?

7. In 1875 in Boston, bustles sold for \$2.00 each while whalebone corsets sold for \$3.00 each. At the end of the day the store had collected \$65 from the sale of corsets and bustles and had sold 15 more bustles than corsets. How many corsets were sold?

8. Find two consecutive integers such that five times the first is 5 greater than six times the sum of the second and -3.

9. On the first day of the sale, the girls sold 20 percent of their cookies. If they sold 700 cookies, how many cookies do they have left to sell?

*10. Find the number that is $\frac{7}{10}$ of the way from 30 to 40.

*11. Find the number that is $\frac{1}{5}$ of the distance from $\frac{1}{8}$ to $\frac{9}{11}$.

Show that the following numbers are rational numbers by writing them as quotients of integers.

*12. .00134 *13. 143.25

*14. $4.\overline{16}$ *15. $1.4\overline{321}$

16. $\frac{m}{p} - \frac{c}{x} = d$; find p 17. $k - \frac{1}{c} + \frac{p}{x} = n$; find c

*The problems marked with an asterisk were worked out in this lesson.

18. $\frac{m}{x} - p = \frac{c}{d}$; find x

Solve:

19. $\sqrt{x^2 + 2x - 6} - 3 = 0$

20. $\sqrt{t + 1} = 4$

21. Find the equation of the line:

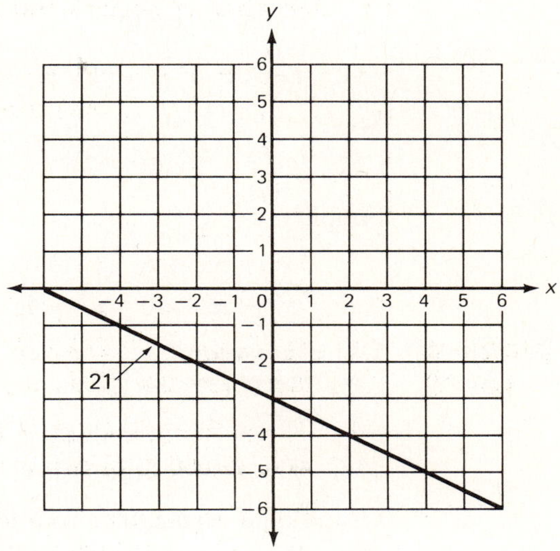

22. Find the equation of the line through $(-5, -7)$ and $(0, 2)$.

23. Find the equation of the line that passes through $(-5, -7)$ and is parallel to $3y + 2x - 5 = 0$.

24. Given that $m(x) = -x^2$, $D = \{\text{Reals}\}$; $\phi(x) = x + 4$, $D = \{\text{Reals}\}$; $k(x) = 2x - 3$, $D = \{\text{Reals}\}$; find (a) $m(-5)$. (b) $k(3)$.

25. Which of the following designate functions? Use the vertical line test as necessary.

(a)

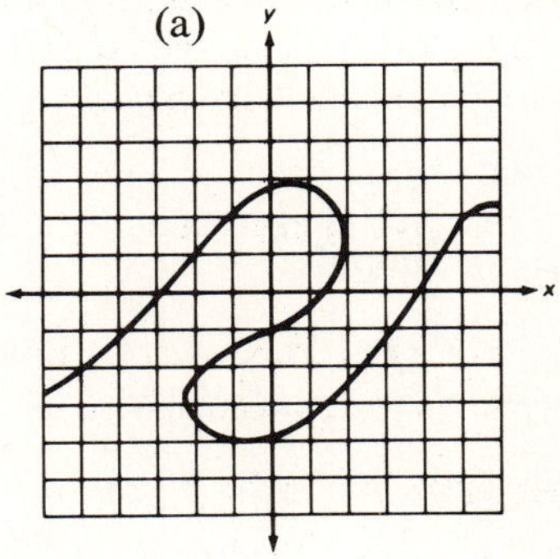

(b)

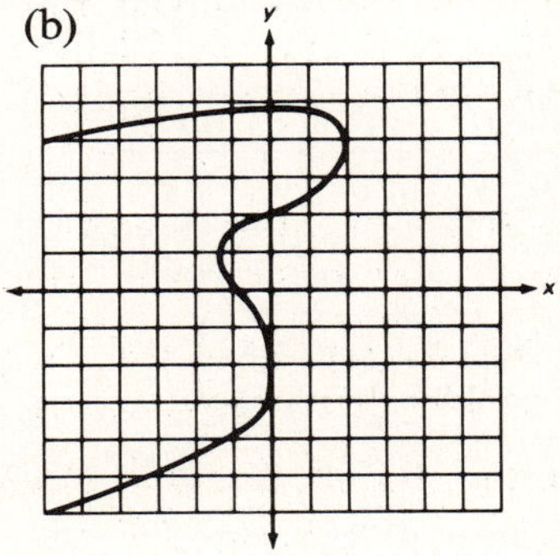

(c)

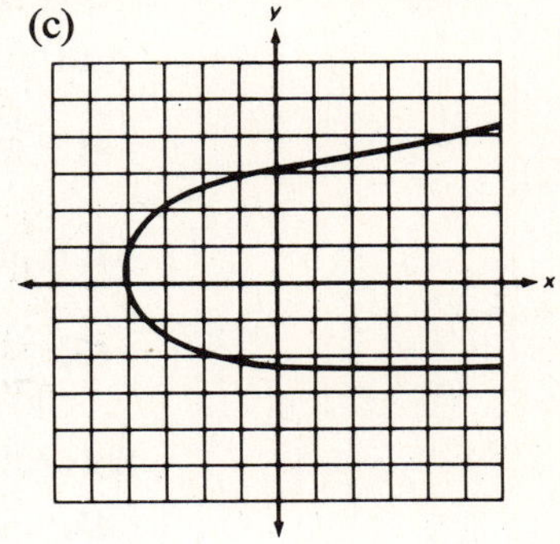

(d)
(4, 5), (3, −7), (5, 3)
(−7, 5), (7, 11), (11, 7)

(e)

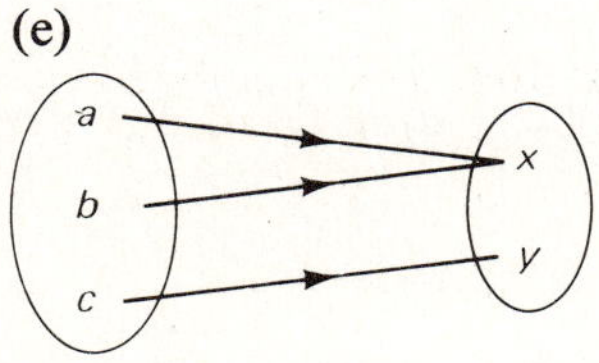

(f)

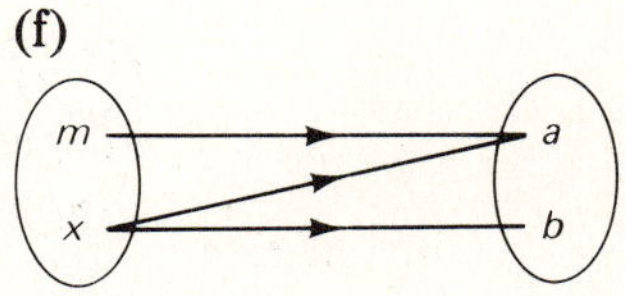

Solve:

26. $\frac{3w}{w + 2} = \frac{5}{2}$

Graph on a number line:

27. $3 - x < 5$, $D = \{\text{Reals}\}$

28. Find s.

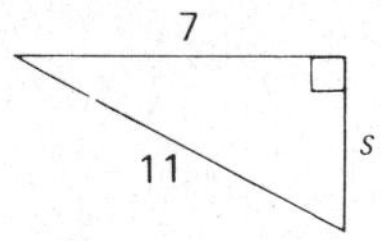

Simplify:

29. $5\sqrt{45} - 3\sqrt{20}$

30. $3\sqrt{6}\,(2\sqrt{2} - 4\sqrt{3})$

31. Divide $m^3 - 15$ by $m - 3$.

Solve:

32. $-x^2 + 2x = -24$

Simplify:

33. $\dfrac{x^2 + 5x + 6}{x^2 - x - 20} \cdot \dfrac{x^2 - 6x + 5}{x^2 + 5x + 6}$

34. $\dfrac{(450 \times 10^{-11})(.0007 \times 10^4)}{(.000009)}$

Use elimination:

35. $\begin{cases} 3x + 2y = 3 \\ x - 4y = 15 \end{cases}$

36. The set $\{1, 2, 3, 4\}$ is closed for what operations?

Graph on a rectangular coordinate system:

37. $2x + 3y - 2 = 0$

Simplify:

38. $\dfrac{a - \dfrac{1}{a + b}}{\dfrac{a}{a + b} - 1}$

39. $\dfrac{\dfrac{x}{y}}{\dfrac{1}{c}}$

Add:

40. $\dfrac{x}{m^2} + \dfrac{3}{m + 2a}$

Add like terms:

41. $\dfrac{4a^{-2}b}{c} - \dfrac{2c^{-1}}{a^2b^{-1}} + \dfrac{6a^{-2}b^{-1}}{c^{-1}}$

Multiply:

42. $4a^{-2}x\left(\dfrac{a^2}{x^{-1}} + \dfrac{8a^{-2}}{x^{-2}}\right)$

Simplify:

43. $\dfrac{(2a^{-2}y)^2xy^{-4}}{a^2yx^0y^2x^{-5}}$

Solve:

44. $-5(2x - 4) + 6 = -3x - 5 - x$

Evaluate:

45. $-x(x)(x - a) + (-a)$ if $a = -2$, $x = 5$.

Simplify:

46. -4^{-2}

47. True or false $\{\text{Whole}\} \subset \{\text{Irrationals}\}$

48. $-.0135 \in$ {what sets of numbers?}

Add:

49. $\dfrac{6x + 5}{x} - \dfrac{2}{2x^2 - x}$

lesson thirty-six

Three equations in three unknowns
More on absolute value inequalities

36.1 substitution method

In Lesson 13 we demonstrated the substitution method of solving a system of simultaneous first-degree equations in two unknowns such as the system shown here.

$$\begin{cases} 2x - y = 1 \\ x = -3y + 11 \end{cases}$$

The value of neither of the unknowns is given in the original equations. In Lesson 20 we demonstrated the use of the substitution method in solving systems of four equations in four unknowns such as

$$R_M T_M = R_W T_W \qquad R_M = 8 \qquad R_W = 2 \qquad T_W + T_M = 5$$

Although this system has four unknowns, the values of two of the unknowns are given so the system can quickly be reduced a system of two equations in two unknowns by combining the first three equations.

$$8T_M = 2T_W \qquad T_W + T_M = 5$$

Now the solution could be completed by using either substitution or elimination.

In this section we will demonstrate the use of the substitution method to initiate the solution of a system of three equations in three unknowns in which none of the values of the unknowns is given.

example 36.1.1 Solve

$$\begin{cases} \text{(a)} \quad N_P + N_N + N_D = 30 \\ \text{(b)} \quad N_P + 5N_N + 10N_D = 205. \\ \text{(c)} \quad 2N_P = N_N \end{cases}$$

solution These equations concern nickels, pennies and dimes. We note that the first two equations contain three unknowns and that the last equation gives a relationship between two of these unknowns. Equation (c) states that $2N_P$ equals N_N. We will use this in our first step, which is to substitute $2N_P$ for N_N everywhere an N_N appears in equations (a) and (b).

$$\begin{aligned}&(\text{a}')\quad N_P + (2N_P) + N_D = 30\\ &(\text{b}')\quad N_P + 5(2N_P) + 10N_D = 205\end{aligned}$$

These equations can be simplified by combining the terms that contain N_P and the result is a pair of two equations in two unknowns that can be solved by using either substitution or elimination. We will use elimination.

$$\begin{array}{llcr}(\text{a}')\ 3N_P + N_D = 30 & \longrightarrow\ (-10)\ \longrightarrow & & -30N_P - 10N_D = -300\\ (\text{b}')\ 11N_P + 10N_D = 205 & \longrightarrow\ (1)\ \longrightarrow & & 11N_P + 10N_D = 205\\ \hline & & & -19N_P \quad = -95\\ & & & \mathbf{N_P = 5}\end{array}$$

Now since $N_N = 2N_P$, we get

$$N_N = 2(5) \text{ or } \mathbf{N_N = 10}$$

and since $N_P + N_N + N_D = 30$, and we know that $N_P = 5$ and $N_N = 10$, we get

$$5 + 10 + N_D = 30 \longrightarrow 15 + N_P = 30 \longrightarrow \mathbf{N_P = 15}$$

example 36.1.2 Solve
$$\begin{array}{ll}(\text{a}) & N_C + N_P + N_E = 64\\ (\text{b}) & 2N_C + 4N_P + 10N_E = 200.\\ (\text{c}) & N_E = 2N_P\end{array}$$

solution These three equations are equations from a value problem that involves the purchase of some clips, pencils, and erasers.

Again, we see that if we begin by substituting the equivalence expressed in equation (c) into the top two equations, we can reduce the number of unknowns in the top two equations to two. We will substitute $2N_P$ for N_E in equations (a) and (b).

$$\begin{aligned}&(\text{a}')\quad N_C + N_P + (2N_P) = 64\\ &(\text{b}')\quad 2N_C + 4N_P + 10(2N_P) = 200\end{aligned}$$

Now we combine the terms that contain N_P and solve the resulting equations by using elimination.

$$\begin{array}{llcr}(\text{a}')\ N_C + 3N_P = 64 & \longrightarrow\ (-2)\ \longrightarrow & & -2N_C - 6N_P = -128\\ (\text{b}')\ 2N_C + 24N_P = 200 & \longrightarrow\ (1)\ \longrightarrow & & 2N_C + 24N_P = 200\\ \hline & & & 18N_P = 72\\ & & & \mathbf{N_P = 4}\end{array}$$

Now since

$$N_E = 2N_P \longrightarrow N_E = 2(4) \longrightarrow \mathbf{N_E = 8}$$

Now equation (a) is $N_C + N_P + N_E = 64$ and $N_P = 4$ and $N_E = 8$, so

$$N_C + 4 + 8 = 64 \longrightarrow N_C + 12 = 64 \longrightarrow \mathbf{N_C = 52}$$

TEST YOUR UNDERSTANDING

(a) Solve the system:

$$\begin{cases} N_B + N_G + N_T = 21 \\ 5N_B + 2N_G + 7N_T = 107 \\ N_T = 2N_B \end{cases}$$

36.2 more on absolute value inequalities

We have said (sections 1.7 and 27.1) that we can think of the absolute value of a number as describing the "bigness" of the number or we can think of the absolute value as representing the distance from the graph of the number to the origin. **In either case, we say that the absolute value of a non-zero real number is a positive number**[†] **and that the absolute value of zero is zero.** Some authorities use the fact that all non-zero real numbers can be paired, each with its opposite. Thus $+2$ is paired with -2, $+14$ with -14, and $+1.615$ with -1.615. **If we think of the real numbers in this way, we can say that both members of the pair have the same absolute value, which is the positive member of the pair.**

In Lesson 27 we also introduced the graphing of the solutions of absolute value inequalities. We can classify absolute value inequalities into four categories as determined by the solution sets of the inequalities.

1. The first type is an absolute value inequality whose solution set is the empty set.

   $$|x| < -6 \qquad |x| < -2 \qquad |x| < -4$$

   No real number replacement for the unknown will satisfy any of the inequalities above because the absolute value of any real number is equal to or greater than zero. Thus there is no real number whose absolute value is less than zero.
2. The second type is an absolute value inequality whose solution set is the set of real numbers.

   $$|x| > -6 \qquad |x| > -2 \qquad |x| > -4$$

   Any real number replacement for the unknown will satisfy each of the inequalities above because the absolute value of any real number is equal to or greater than zero. If the absolute value of every real number is equal to or greater than zero, certainly the absolute value is greater than any negative number.
3. The third type is an absolute value inequality whose solution set consists of those numbers whose absolute value is less than a given positive number.

   $$|x| < 3 \qquad D = \{\text{Reals}\}$$

[†]A positive number, not the positive value of a number, because numbers do not have different values. A non-zero real number is either a positive number or a negative number.

The solution of this inequality is graphed in Figure 36.1 and consists of all real numbers whose absolute value is less than 3.

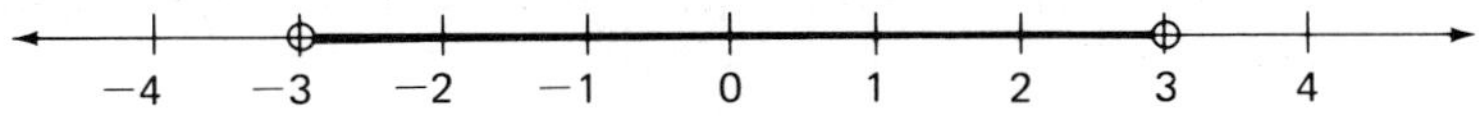

Figure 36.1

4. The fourth type is an absolute value inequality whose solution set consists of those numbers whose absolute value is greater than a given positive number.

$$|x| > 3 \qquad D = \{\text{Reals}\}$$

The solution to this inequality is graphed in Figure 36.2 and consists of all real numbers whose absolute value is greater than 3.

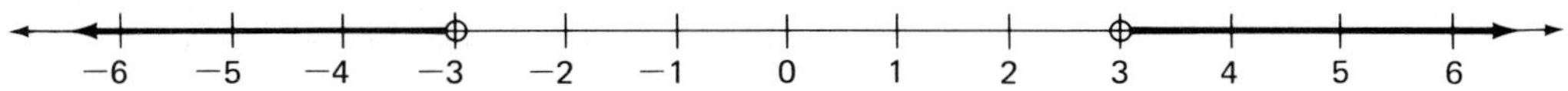

Figure 36.2

If we dismiss the first two cases, (1) and (2), in which no real number satisfies or all real numbers satisfy, we see that absolute value inequalities fall into *two categories.*

3. The solution is all numbers that are greater than one specified number **AND** also are less than another specified number.
4. The solution is all numbers that are greater than one specified number **OR** are less than another specified number.

36.3 conjunctions and disjunctions

In both (3) and (4) above we note that there are two restrictions on the numbers that can be used. In (3), both of the restrictions must be satisfied, and we use the word **and** to designate that compliance with both restrictions is required. We call a dual restriction of this type a **CONJUNCTION**. In (4) a number satisfies the stated condition if it satisfies either one of the restrictions. We use the word **or** to designate that a solution of either restriction is all that is necessary and we call this type of dual restriction a **DISJUNCTION**.

example 36.3.1 Graph $x > -1$ and $x \leq 4$, $D = \{\text{Reals}\}$.

solution The use of the word **AND** designates this dual restriction on the variable as a conjunction and indicates that **both restrictions must be satisfied.**

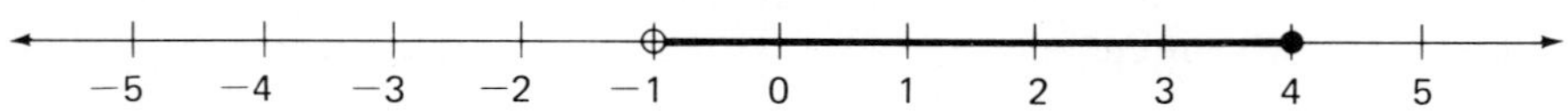

The solution designates all numbers that are greater than -1 **and** are less than or equal to 4.

example 36.3.2 Graph $-3 \leq x < 2$, $D = \{\text{Integers}\}$.

solution This notation is an alternate notation for the designation of a conjunction in that both of the restrictions must be satisfied. We can break the notation into two notations by beginning in the middle and reading to the left as "x is greater than or equal to -3." Now we go back to the middle and read to the right as "x is less than 2."

$$x \leq -3 \quad \textbf{and} \quad x < 2$$

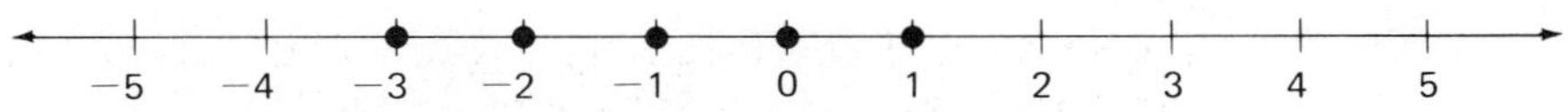

In the figure we indicate all integers that are equal to or greater than -3 and are also less than 2.

example 36.3.3 Graph $0 \le x + 2 < 5$, $D = \{\text{Reals}\}$.

solution This notation designates a conjunction but places the restrictions on $x + 2$ instead of x. To find the restrictions on x, we must simplify both of the stated conditions.

$$\begin{array}{rcl} 0 \le x + 2 & \quad \textbf{AND} \quad & x + 2 < 5 \\ -2 \quad\;\; -2 & & -2 \quad -2 \\ \hline 2 \le x & & x \quad < 3 \end{array}$$

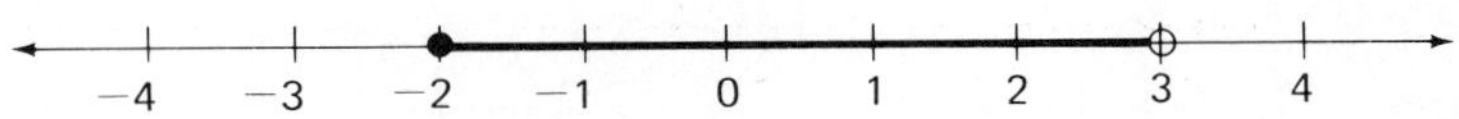

In the graph we designate all real numbers that are equal to or greater than -2 and are also less than 3.

TEST YOUR UNDERSTANDING

(b) Graph the solution of $-3 \le x - 1 \le 5$, $D = \{\text{Integers}\}$.

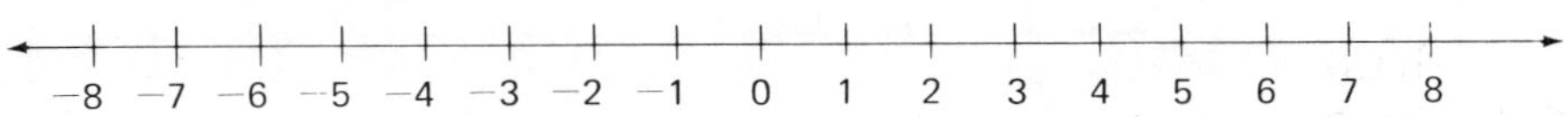

example 36.3.4 Graph $x < -1$ or $x \ge 2$, $D = \{\text{Integers}\}$.

solution **The use of the word OR between the two restrictions tells us that this dual restriction is a DISJUNCTION and is satisfied by any number that satisfies either one of the conditions.**

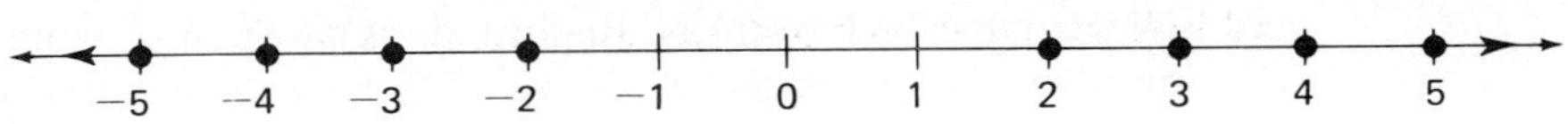

Figure 36E

In the graph we designate all integers that are less than -1 **OR** are greater than or equal to 2.

example 36.3.5 Graph $x > 15$ or $x \le 10$, $D = \{\text{Reals}\}$.

solution This dual restriction is a disjunction as indicated by the word **or**.

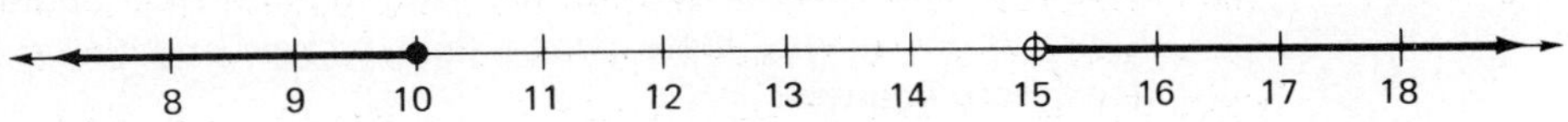

We indicate all real numbers that are less than or equal to 10 **OR** are greater than 15.

> **TEST YOUR UNDERSTANDING**
>
> (c) Graph the solution of $x \geq 3$ or $x < 0$, $D = \{\text{Integers}\}$.
>
> 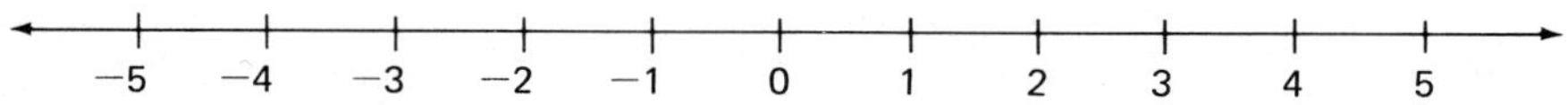
>

lesson checklist 36

- ☐ Substitution to solve three equations in three unknowns
- ☐ Absolute value
- ☐ Replacing absolute value inequality with two non-absolute value inequalities

> **ANSWERS TO TEST YOUR UNDERSTANDING**
>
> (a) $N_B = 5$, $N_G = 6$, $N_T = 10$
>
> (b) 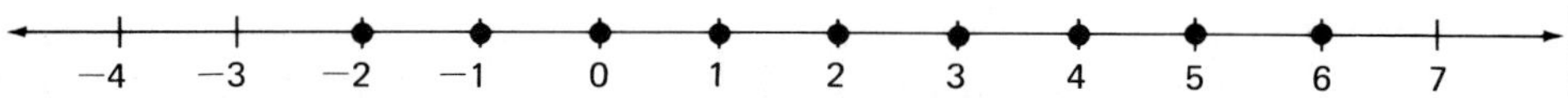
>
>
> (c) 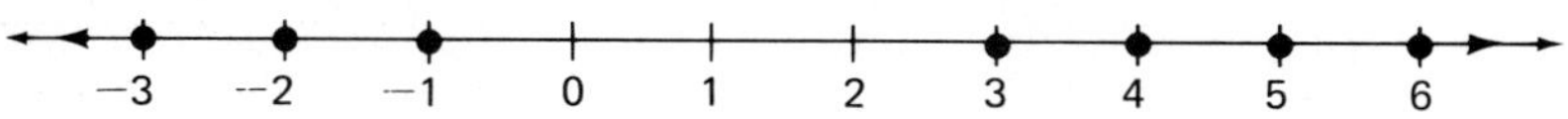
>

problem set 36

1. Robin Hood divided 567 arrows into two piles in the ratio of 4 to 3. How many arrows were in the larger pile?
2. The sum of two numbers is 24 and the ratio of the smaller to the larger is 3 to 5. What are the numbers?
3. In the battle, the ratio of English ships to Spanish ships was 5 to 3. If the total number of ships in the battle was 272, how many ships were Spanish ships?
4. At noon the armadillo left the wild kingdom and headed north at 3 km/hr. Two hours later the aardvark left the same kingdom and headed south at 5 km/hr. At what time will the animals be 18 kilometers apart?
5. At 10 a.m., Romeo left the house of the Montagues to visit Juliet who lived at the house of the Capulets. At 1 p.m. Juliet left home to go to meet Romeo. If the houses were 29 kilometers apart, if they met at 3 p.m., and if Romeo traveled at 4 km/hr, how fast did Juliet travel?
6. Stephanie threw her pennies in the sandbox and then Shannon added her nickels. If they threw in one more penny than nickels and threw in 67 cents total, how many were pennies?
7. Betty and the old witch saved 67 cents in pennies and nickels. If they had 19 coins in all, how many pennies did they have?

8. Find three consecutive even integers such that nine times the first is 14 greater than two times the sum of the third and 3.

9. Wyatt Earp had only 15% of the bullets that Big John had. If Wyatt had 27 bullets, how many did Big John have?

Solve:

***10.** $N_P + N_N + N_D = 30$
$N_P + 5N_N + 10N_D = 205$
$2N_P = N_N$

***11.** $N_C + N_P + N_E = 64$
$2N_C + 4N_P + 10N_E = 200$
$N_E = 2N_P$

Graph on a number line:

***12.** $0 \le x + 2 < 5$, $D = \{\text{Reals}\}$

***13.** $x < -1$ or $x \ge 2$, $D = \{\text{Integers}\}$

14. Find the number $\frac{1}{3}$ of the way from $\frac{1}{5}$ to $\frac{6}{7}$.

15. Show that $.001\overline{325}$ is a rational number by expressing it as a quotient of integers.

16. $\frac{x}{y} - \frac{m}{n} = p$; find n

17. $a + \frac{1}{x} + c = m$; find x

18. $\frac{x}{p} - k + \frac{1}{c} = m$; find c

Solve:

19. $\sqrt{p^2 - 4p - 1} - 2 = 0$

20. $\sqrt{2x - 4} - 6 = 0$

21. Find the equation of the line:

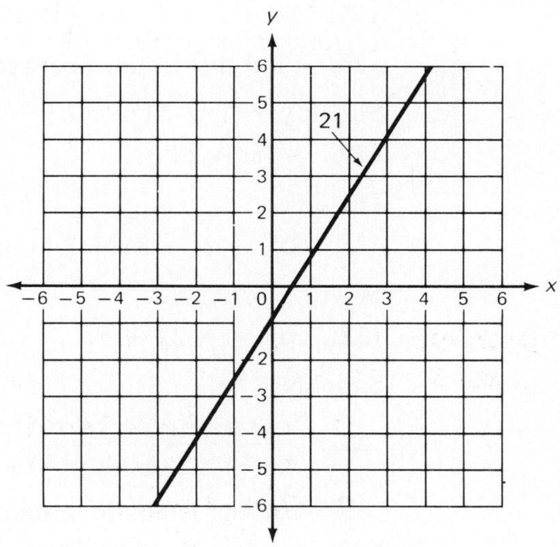

22. Find the equation of the line that goes through $(-2, 3)$ and is parallel to $x - y = -3$.

23. Find the equation of the line through $(1, 2)$ that has a slope of $\frac{-3}{2}$.

24. Given that $f = x^2 - 3$, $D = \{\text{Integers}\}$; find $f(2)$.
$\phi(x) = x + 3$, $D = \{\text{Negative integers}\}$;

---

*The problems marked with an asterisk were worked out in this lesson.

25. Which of the following designate functions? Use the vertical line test as necessary.

(a)

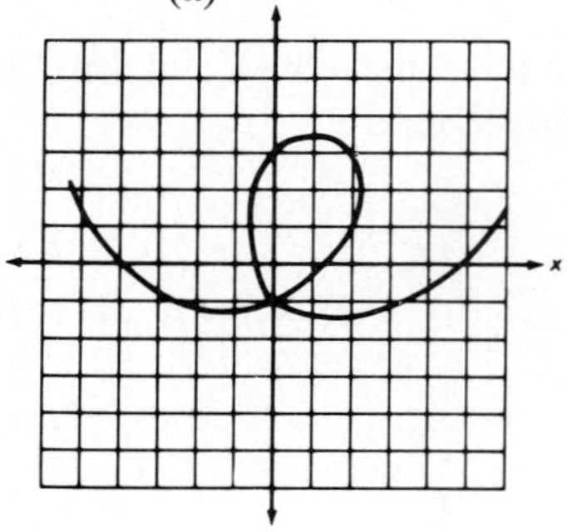

(b)

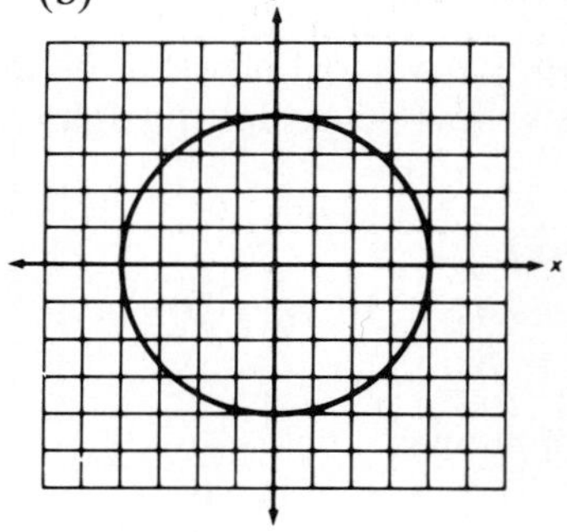

(c)

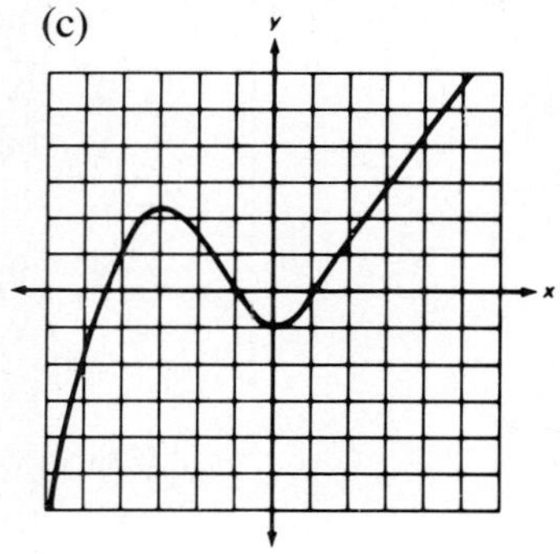

(d)

(1, 3), (3, 5), (5, 7)
(2, 4), (4, 6), (6, 8)

(e)

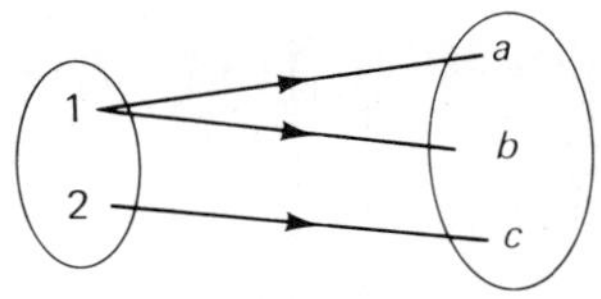

(f)

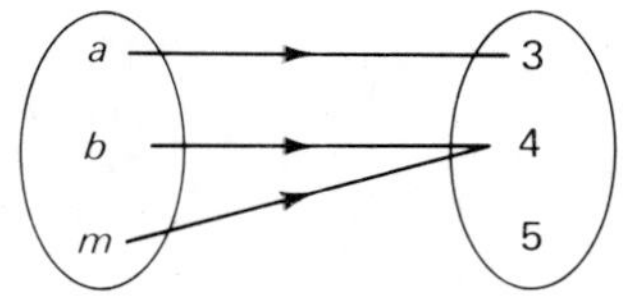

Solve:

26. $\dfrac{y+1}{2} = \dfrac{y}{3}$

Graph on a number line:

27. $|x| + 2 > 5, D = \{\text{Reals}\}$

28. Find the distance between (3, 5) and (−2, 1).

Simplify:

29. $\sqrt{18} - 2\sqrt{3}$

30. $3\sqrt{3}\,(2\sqrt{6} - \sqrt{12}\,)$

Divide:

31. $2x^3 - x^2 - 8x + 3$ by $x - 2$

Solve:

32. $3x^2 = -2x + 5$

Simplify:

33. $\dfrac{x-y}{a-1} \cdot \dfrac{3a^2 + 2a - 5}{x^2 - y^2}$

34. $\dfrac{(300{,}000)(.007 \times 10^{-5})}{(.0042 \times 10^{10})}$

35. Use the letters a, b, and c to demonstrate the associative property of multiplication.

Graph on a rectangular coordinate system:

36. $2x - y + 3 = 0$

Simplify:

37. $\dfrac{\dfrac{1}{y} + x}{y - \dfrac{x}{y}}$

38. $\dfrac{\dfrac{a}{b}}{c}$

Add:

39. $\dfrac{1}{m^2x} + \dfrac{b}{mx^2} - \dfrac{c}{1}$

Add like terms:

40. $\dfrac{3ac}{b^{-2}} + \dfrac{2ab^2}{c} - \dfrac{2b^2}{a^{-1}c^{-1}}$

Multiply:

41. $ab^2c^0x\left(\dfrac{2bc}{ab^{-1}} - \dfrac{3a^{-1}}{b^2x}\right)$

Simplify:

42. $\dfrac{a^2b^0c^{-3}}{x^3a^{-2}c^0}$

Evaluate:

43. $r - s(-s)(s - r)$ if $r = -3, s = 1$

Solve:

44. $(2x - 1) + 2(-3) = 3(x + 2)$

Graph on a number line:

45. $-3 + x > 2, D = \{\text{Integers}\}$

Simplify:

46. $\dfrac{1}{(-2)^{-3}}$

47. True or false: {Irrationals} $\subset$ {Reals}

48. $77 \in$ {what sets of numbers?}

Add:

49. $\dfrac{3}{x - 5} + \dfrac{2}{x^2 - 25}$

lesson thirty-seven

Direct and inverse variation

37.1 direct variation

In every classroom of a particular school, there are twice as many boys as there are girls. Thus, if a classroom contains 10 girls, there must be 20 boys in that classroom. We may express the relationship mathematically by writing

$$B = 2G$$

We can use this equation to find the number of boys in any classroom if we are told the number of girls in that room, or we may find the number of girls in a room if we are told the number of boys in that room.

In another school there are three times as many boys in each room as there are girls. For this school, the relationship can be expressed mathematically by writing

$$B = 3G$$

In the equation for the first school we had the variables B and G and the constant was the number 2. This last equation is the same equation except that the constant is the number 3. The general form of the relationship is

$$B = kG$$

where k represents the constant for any particular school.

We call a relationship such as this, where one variable is expressed as a constant times another variable, a direct variation or a direct proportion and the constant in the equation is called the constant of proportionality.

If we discuss a third school and are told that

1. The number of boys in any classroom **varies directly** as the number of girls in the room or
2. That the number of boys in any room **is directly proportional to** the number of girls in the room,

we have been told that the relationship between the number of boys and girls in any room in the school may be stated mathematically as

$$B = kG$$

Before we can solve any problem, we need to know the constant of proportionality for this school. If we are told the number of boys and girls in any room in the school, we can use these values in the equation to solve for k. If there are 30 boys and 5 girls in one room and we use these values for B and G in the equation, we find

$$30 = k(5) \quad \text{and thus} \quad k = 6$$

Now we have the constant for this school and may write the relationship for this school as

$$B = 6G$$

Now if we are given the number of girls in any room in this school, we can solve for the number of boys and conversely can find the number of girls if we are given the number of boys.

The key to this type of problem is recognizing the verbal statement that denotes a direct variation or a direct proportion and realizing the equation that is implied by this statement. We will give some examples here. On the left is the key verbal statement and on the right is the equation that is implied by the statement.

STATEMENT	EQUATION
The weight of a substance **varies directly** as the volume of the substance.	$W = kV$
Force **varies directly** as the current.	$F = kC$
The distance traveled **varies directly** as the time.	$D = kT$

The words **directly proportional** imply the same equation as do the words **vary directly**.

STATEMENT	EQUATION
The circumference of a circle is **directly proportional** to the length of the radius.	$C = kR$
The volume of a right circular cylinder of fixed radius is **directly proportional** to its height.	$V = kH$

In each of the relationships above, the equation contains an unknown constant k that we call the constant of proportionality. **The first step in the solution of a direct variation problem is to find the value of the constant of proportionality for that problem.** Then we can solve for the value of one unknown if we are given the value of the other unknown.

example 37.1.1 The weight of a substance varies directly as the volume of the substance. If the weight of 2 liters of the substance is 10 kg, what will be the volume of 35 kg of the substance?

solution We will solve the problem in *four steps*.

Step 1: Recognize that the words **varies directly** imply the relationship

$$W = kV$$

Step 2: Reread the problem to find the values of W and V that can be used to find the value of k. Use these values to solve for k.

$$10 = k(2) \longrightarrow k = 5$$

Step 3: Replace k in the equation with the value we have found

$$W = 5V$$

Step 4: Reread the problem to find that we are asked to find the value of V if W is 35. We replace W in the equation with 35 and solve for V.

$$35 = 5V \longrightarrow \frac{35}{5} = \frac{\cancel{5}V}{\cancel{5}} \longrightarrow \mathbf{7\ liters} = V$$

example 37.1.2 The distance traversed by a car traveling at a constant speed is directly proportional to the time spent traveling. If the car goes 75 kilometers in 5 hours, how far will it go in 7 hours?

solution

Step 1 $\quad D = kT$

Step 2 $\quad 75 = k(5) \longrightarrow k = 15$

Step 3 $\quad D = 15T$

Step 4 $\quad D = (15)(7) \longrightarrow \mathbf{D = 105\ kilometers}$

TEST YOUR UNDERSTANDING

(a) The number of atoms of oxygen varied directly as the number of atoms of hydrogen. When there were 150 atoms of oxygen, there were 300 atoms of hydrogen. How many atoms of oxygen were present when the hydrogen was increased to 750 atoms?

37.2 inverse variation

When the problem states that one variable **varies inversely** as the other variable or that the value of one variable is **inversely proportional** to the value of the other variable, an equation of the form

$$V_1 = \frac{k}{V_2}$$

is implied where k is the constant of proportionality and V_1 and V_2 are the two variables. If we look at the equations for direct variation and inverse variation

DIRECT VARIATION EQUATION	INVERSE VARIATION EQUATION
$V_1 = kV_2$	$V_1 = \frac{k}{V_2}$

we see that each equation contains two variables and a constant of proportionality. In both equations, the constant k is in the numerator! **In a direct variation equation, both variables are in the numerator; and in an inverse variation equation, one variable is in the numerator and the other variable is in the denominator!**

STATEMENT	EQUATION
The pressure of a perfect gas **varies inversely** as the volume.	$P = \frac{k}{V}$
The current **varies inversely** as the resistance.	$C = \frac{k}{R}$
The velocity is **inversely proportional** to the time.	$V = \frac{k}{T}$

Inverse variation problems are solved in the same way as the direct variation problems. First we recognize the equation implied by the statement of inverse variation. Then we find the constant of proportionality for this problem and use this constant in the equation to find the final solution.

example 37.2.1 For a constant temperature, the pressure of a perfect gas varies inversely as the volume. When the temperature of a quantity of a perfect gas is 400°K, the volume is 75 liters, and the pressure is 7 torr. What would the volume be at the same temperature if the pressure is 15 torr?

solution The temperature does not enter into the problem. The pressure varies inversely as the volume if the temperature is held constant. If the temperature is not held constant, the pressure won't necessarily vary inversely as the volume. **We don't care what the temperature is as long as it stays the same.**

Step 1 $\quad P = \frac{k}{V}$

Step 2 $\quad 7 = \frac{k}{75} \longrightarrow k = 525$

Step 3 $\quad P = \frac{525}{V}$

Step 4 $\quad 15 = \frac{525}{V} \longrightarrow V = \frac{525}{15} \longrightarrow$ $\boldsymbol{V = 35}$ **liters**

example 37.2.2 To travel a fixed distance, the velocity is inversely proportional to the time required. When the velocity is 60 km/hr, the time required is 4 hours. What would be the time required for the same distance if the velocity were increased to 80 km/hr?

solution

Step 1 $\quad V = \frac{k}{T}$

Step 2 $\quad 60 = \frac{k}{4} \longrightarrow k = 240$

Step 3 $\quad V = \frac{240}{T}$

Step 4 $\quad 80 = \frac{240}{T} \longrightarrow T = \frac{240}{80} \longrightarrow$ $\boldsymbol{T = 3}$ **hr**

example 37.2.3 For a constant volume the pressure varies directly as the temperature. If the volume of a quantity of gas is 60 cc and the pressure is 800 torr, the temperature is 400°K. If the volume is held at 60 cc and the pressure is changed to 400 torr, what is the temperature?

solution The volume does not enter into the solution. The pressure varies directly as the temperature only if the volume is held constant.

Step 1 $P = kT$
Step 2 $800 = k(400) \longrightarrow k = 2$
Step 3 $P = 2T$
Step 4 $400 = 2T \longrightarrow$ $\mathbf{T = 200°K}$

TEST YOUR UNDERSTANDING

(b) If the temperature is held constant, the volume of a fixed quantity of a perfect gas varies inversely as the pressure. When the temperature of a quantity of a perfect gas was 700°K, the pressure was 15 torr, and the volume was 100 liters. What was the volume when the pressure was increased to 50 torr if the temperature is held constant?

---

lesson checklist 37

☐ Direct variation equations

☐ Constant of proportionality

☐ Inverse variation equation

ANSWERS TO TEST YOUR UNDERSTANDING

(a) $O_x = kH$ and $k = \frac{1}{2}$ so $O_x = 375$

(b) $V = \frac{k}{P}$ and $k = 1500$ so $V = 30$ liters.

---

problem set 37

*1. The weight of a substance varies directly as the volume of the substance. If the weight of 2 cc of the substance is 10 kg, what would be the volume of 35 kg of the substance?

*2. The distance traversed by a car moving at a constant speed is directly proportional to the time spent traveling. If the car goes 75 miles in 5 hours, how far will it go in 7 hours?

*3. For a constant temperature, the pressure of a perfect gas varies inversely as the volume. When the temperature of a quantity of a perfect gas is 400°K, the

---

*The problems marked with an asterisk were worked out in this lesson.

volume is 75 liters and the pressure is 7 atm. What would the volume be at the same temperature if the pressure is 15 atm?

*4. To travel a fixed distance, the velocity is inversely proportional to the time required. When the velocity is 60 km/hr, the time required is 4 hours. What would be the time required to cover the same distance if the velocity were increased to 80 km/hr?

5. The ratio of pheasants to turkeys at the thanksgiving hunt was 7 to 2. If 207 birds were killed, how many were turkeys?

6. Fredrick headed for Lutzen at 3 km/hr. Later he increased his speed to 4 km/hr. If it was 50 km to Lutzen and the total time of travel was 15 hours, for how long did he walk at 3 km/hr?

7. Gold was worth 422 marks a gram and copper was worth 4 marks a gram. If Brother Gregory had an 8-gram mixture of gold and copper that was worth 2122 marks, how many grams of gold did he have?

8. Find three consecutive odd integers such that four times the first is 14 less than twice the sum of 2 and the third.

9. Thirty percent of the people did not like the king. If these people who did not like the king totaled 81,150, how many people lived in the kingdom?

10. Solve: $\begin{cases} N_P + N_N + N_Q = 23 \\ N_P + 5N_N + 25N_Q = 155 \\ N_P = 3N_Q \end{cases}$

Graph on a number line:

11. $-3 \leq x < 2$, $D = \{\text{Integers}\}$

12. $x > 15$ or $x \leq 10$, $D = \{\text{Reals}\}$

13. Find the number $\frac{4}{5}$ of the way from $\frac{1}{6}$ to $\frac{7}{8}$.

Show that the following numbers are rational numbers by writing them as quotients of integers.

14. $12.31\overline{685}$

15. $1.00\overline{326}$

16. $\frac{x}{a} + m - \frac{1}{c} = p$, find a

17. $m + \frac{a}{x} + \frac{3}{y} = 6$, find x

Solve:

18. $\sqrt{x^2 + 6x + 12} - 2 = 0$

19. $\sqrt{x + 6} - 8 = 0$

20. Find the equation of the line:

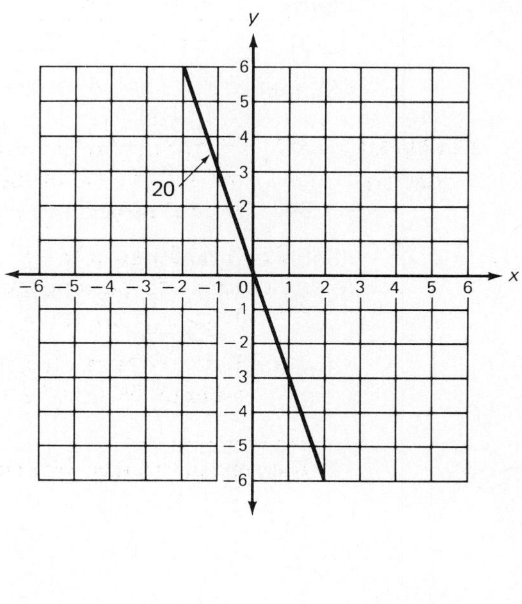

21. Find the equation of the line through $(3, -5)$ and $(-6, -2)$.

22. Find the equation of the line through $(-4, -6)$ that is parallel to $3x + y - 2 = 0$.

23. Given that $p(x) = 2x - 5$ $D = \{\text{Reals}\}$; find (a) $p(5)$.
$f(x) = x + 2$, $D = \{\text{Reals}\}$; (b) $\theta(0)$.
$\theta(x) = 4x + 2$, $D = \{\text{Reals}\}$;

24. Which of the following designate functions? Use the vertical line test.

(a)

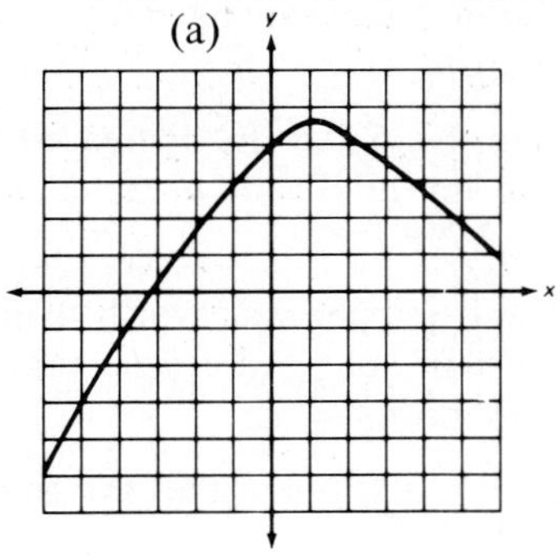

(b)

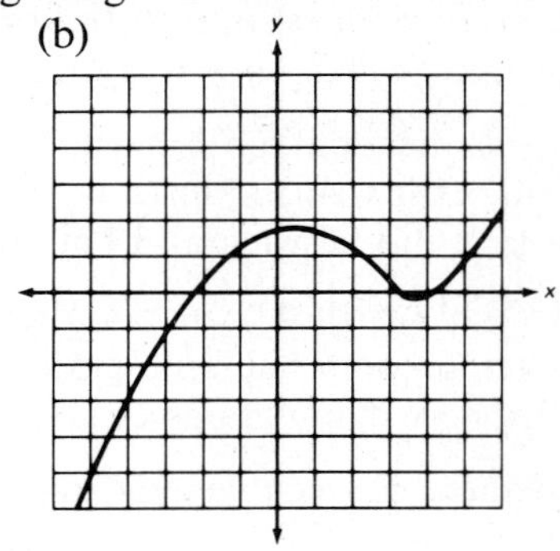

(c)

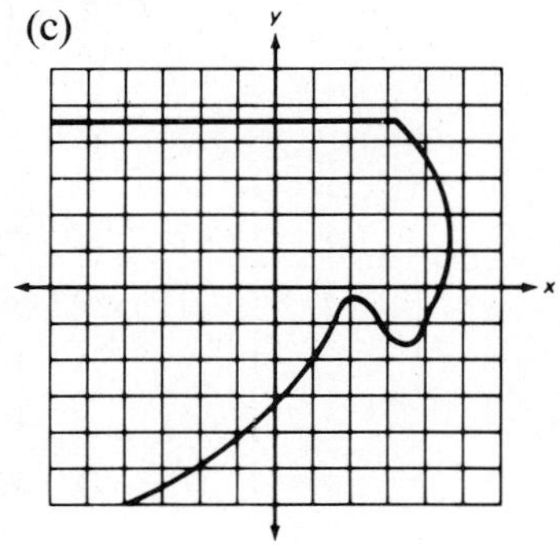

(d)
(4, 2), (5, 7), (6, 11)
(2, 5), (7, 6), (3, 7)

(e)

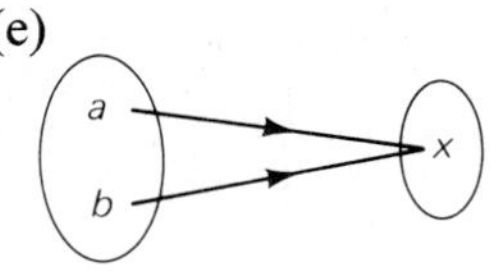

(f)

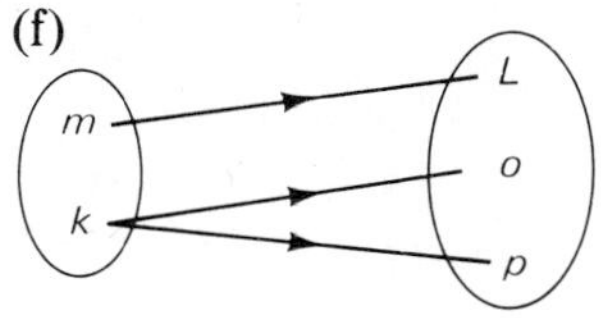

Solve:

25. $\dfrac{7p - 2}{4p + 13} = \dfrac{6}{5}$

Graph on a number line:

26. $-x + 2 < 5$, $D = \{\text{Reals}\}$

27. Find x.

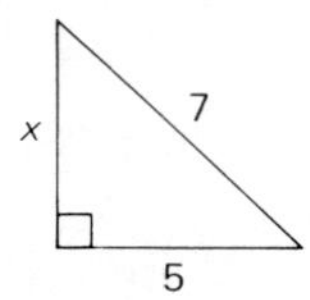

Simplify:

28. $\sqrt{48} - \sqrt{75}$

29. $2\sqrt{15}(3\sqrt{5} - 2\sqrt{3})$

Divide:

30. $3x^2 - 4x + 11$ by $x + 2$

Solve:

31. $-x^2 + 9x = -22$

Simplify:

32. $\dfrac{x^2 + 2x - 15}{x^2 + x - 12} \cdot \dfrac{x^2 + 3x - 4}{x^2 + 7x + 10}$

33. $\dfrac{(.005 \times 10^{21})(4{,}000{,}000)}{(200{,}000 \times 10^{-11})}$

34. The set of negative even integers is closed for what operations?

Solve by graphing:

35. $\begin{cases} 2x - y + 3 = 0 \\ -3 = y \end{cases}$

Simplify:

36. $\dfrac{a - \dfrac{x}{y}}{\dfrac{b}{y} - a}$

37. $\dfrac{1}{\dfrac{c}{a}}$

Add:

38. $\dfrac{p}{x^4} - \dfrac{3}{x - a}$

Add like terms:

39. $\dfrac{2a^2b}{x} - \dfrac{3x^{-1}a^2}{b^{-1}} + 2a^2bx - \dfrac{5bx}{a^{-2}}$

Multiply:

40. $\dfrac{2a^{-2}}{b}\left(\dfrac{4b}{a^2} + \dfrac{3a^4}{b^{-3}}\right)$

Simplify:

41. $\dfrac{(m^{-2}x^2)^{-2}m^{-4}x^0}{x^{-5}m^4x^3m}$

Solve:

42. $5(x + 3 - 2x) = 5 + 2x$

Evaluate:

43. $x - (-y)(-y)(-y + x)$ if $y = -2$ and $x = 4$

Add:

44. $\dfrac{x}{ab^2} + \dfrac{y}{b^2} - \dfrac{z}{a^2b}$

Simplify:

45. $(-3)^{-3} \cdot 3^4$

46. True or false: {Natural} $\in$ {Reals}

47. $4\sqrt{2} \in$ {what sets of numbers?}

Add:

48. $\dfrac{x}{x^2 + 2x + 1} - \dfrac{3}{x + 1}$

lesson thirty-eight

Linear inequalities
More on direct and inverse variation
Rational equations that lead to quadratic equations

38.1 linear inequalities

In Figure 38.1 we have graphed the line whose equation is $y = \frac{1}{2}x + 1$. We see that this line divides the set of all the points of the plane into three mutually exclusive subsets:

1. The set of points that lie on the line.
2. The set of points that lie above the line (region A).
3. The set of points that lie below the line (region B).

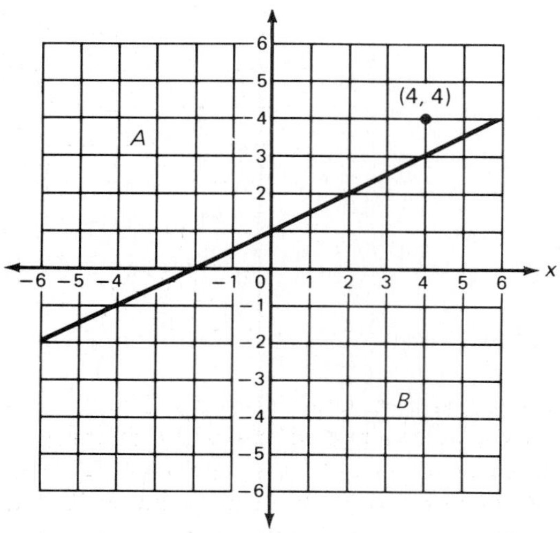

Figure 38.1

It can be shown that the coordinates of any point in the plane either will satisfy the equation of a given line or will satisfy one of the two linear inequalities that define the regions on either side of the line. For this particular line, we can say that the coordinates of any point in the plane will satisfy one and only one of the following:

$$y > \tfrac{1}{2}x + 1, \qquad y = \tfrac{1}{2}x + 1, \qquad y < \tfrac{1}{2}x + 1$$

The points in the regions denoted by A and B above do not lie on the line and the coordinates of any point in region A or region B will satisfy one and only one of the inequalities. To see which of these inequalities defines the region above the line, we will choose a test point that clearly lies above the line and test the coordinates of this point in both inequalities. We choose the point (4, 4).

$y > \frac{1}{2}x + 1$	$y < \frac{1}{2}x + 1$
$4 > \frac{1}{2}(4) + 1$	$4 < \frac{1}{2}(4) + 1$
$4 > 2 + 1$	$4 < 2 + 1$
$4 > 3$ True	False

Thus the coordinates of all points above the line satisfy the inequality $y > \frac{1}{2}x + 1$. If we choose a point below the line in region B, we can show that all points in this region satisfy the other inequality, $y < \frac{1}{2}x + 1$. We choose the point (0, 0).

$$y > \tfrac{1}{2}x + 1 \quad \longrightarrow \quad 0 < \tfrac{1}{2}(0) + 1 \quad \longrightarrow \quad 0 < 1 \quad \text{True}$$

We indicate in Figure 38.2 that the coordinates of all points above the line satisfy the inequality $y > \frac{1}{2}x + 1$ and that the coordinates of all points below the line satisfy the inequality $y < \frac{1}{2}x + 1$.

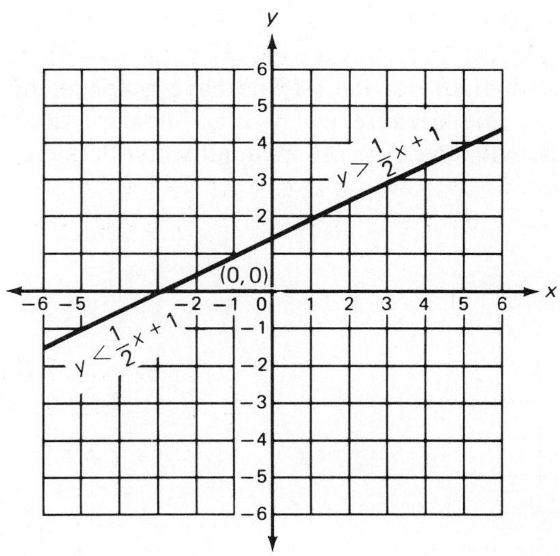

Figure 38.2

example 38.1.1 Graph the inequality $y > -x + 2$.

solution **We first graph the line** $y = -x + 2$ in the figure at the left page 366, top. We show the line as a dotted line to indicate that the points on the line do not satisfy the stated inequality, which uses a *greater than* symbol. Had the inequality used an *equal to or greater than* symbol, the line would have been drawn as a solid line. We choose the point (0, 0) as a test point because it clearly lies on one side of the line and also because it is the easiest test point to use.

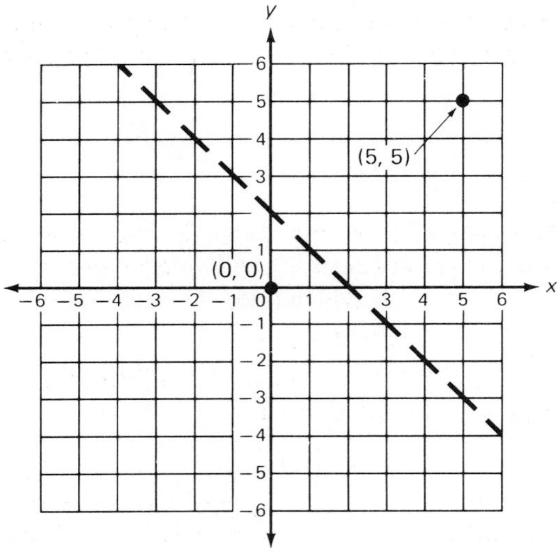

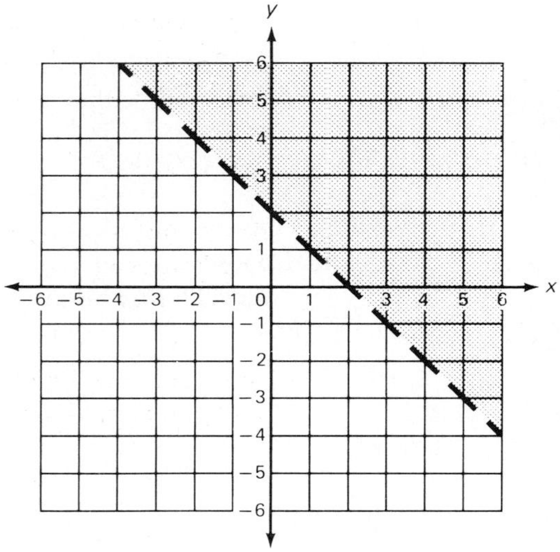

$$0 > -0 + 2$$
$$0 > 2 \quad \text{False}$$

The coordinates of this point do not satisfy the inequality so the coordinates of any point above the line must satisfy the inequality. We will use the point (5, 5).

$$5 > -5 + 2$$
$$5 > -3 \quad \text{True}$$

We indicate that the coordinates of all points above the line satisfy the given condition by shading the region above the line in the figure at the right.

example 38.1.2 Graph $y \leq 2x - 1$.

solution In the figure at the left we have graphed the line $y = 2x - 1$. We show the line as a solid line because we wish to indicate that the points on the line satisfy the stated inequality because the inequality symbol here is read from left to right as *equal to or less than*.

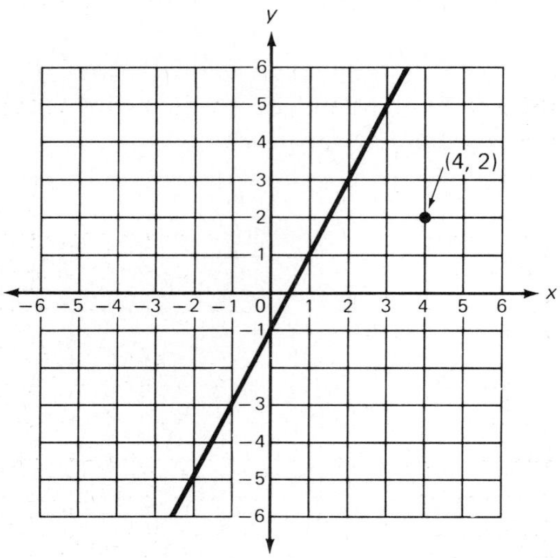

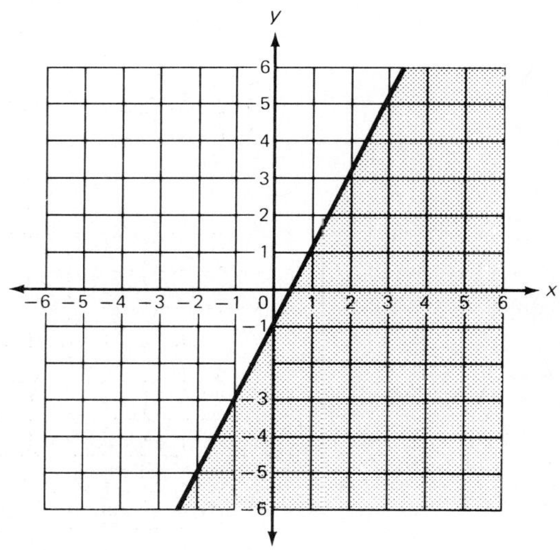

We could use the test point (0, 0), but it is rather close to the line. To be sure we have a point that is well on one side of the line, we choose the point (4, 2).

$$2 < 2(4) - 1$$
$$2 < 7 \quad \text{True}$$

Thus the coordinates of the points on the line or in the region to the right of the line satisfy the stated inequality. We indicate the solution by shading in this region in the other figure.

example 38.1.3 Graph $y > -2$.

solution In the figure at the left we have graphed the line $y = -2$. We show the line as a dotted line because the points on the line do not satisfy the stated inequality. We choose (0, 0) as the test point.

$$0 > -2 \quad \text{True}$$

We shade the region above the line in the next figure to indicate that the coordinates of any point above the line will satisfy the stated inequality.

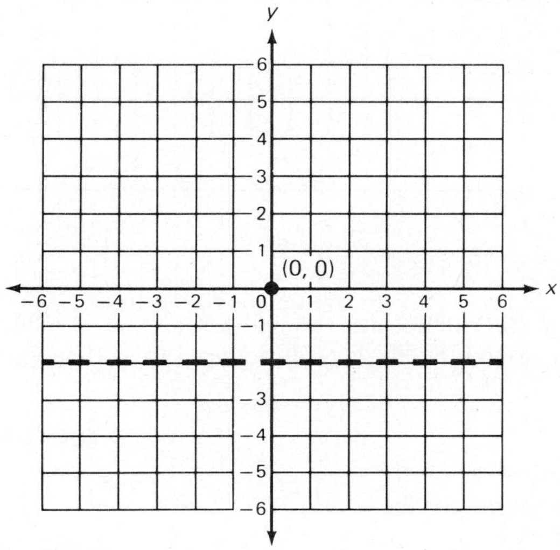

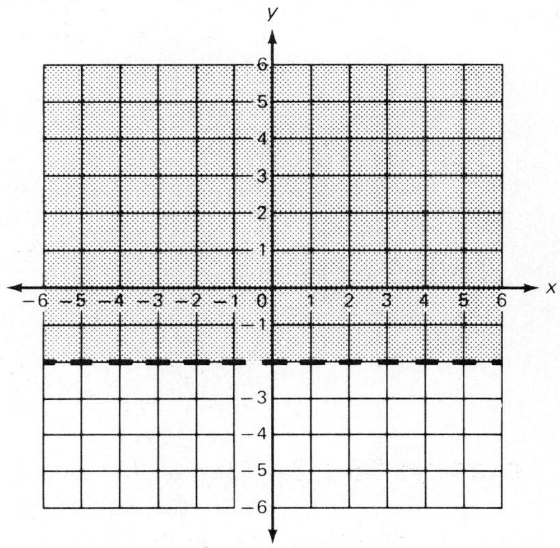

example 38.1.4 Graph $x < 3$.

solution In the figure at the left we graph the equation $x = 3$. We draw the line as a dotted line because points on the line do not satisfy $x < 3$, the given inequality. We will use the point (0, 0) as our test point.

$$0 < 3 \quad \text{True}$$

We shade in the region to the left of the dotted line in the figure at the right to indicate that all points in this region will satisfy the inequality $x < 3$.

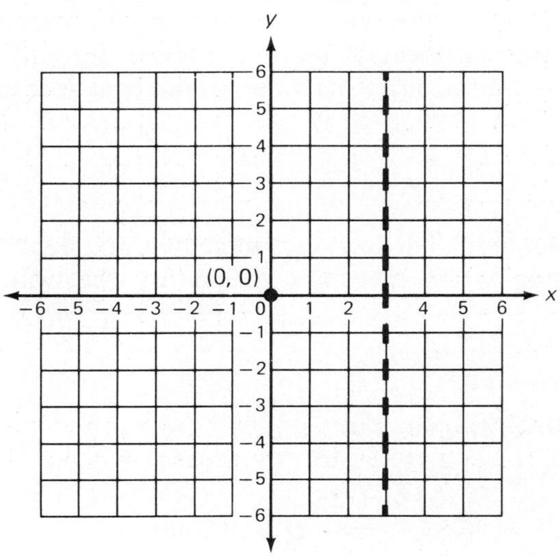

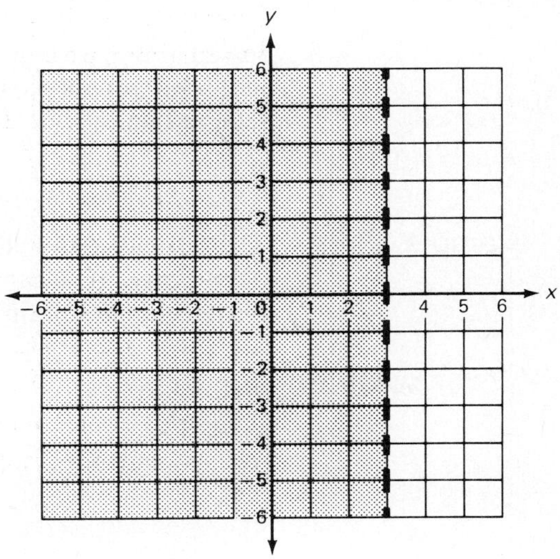

TEST YOUR UNDERSTANDING

(a) Graph the solution to $y > -2x + 3$.

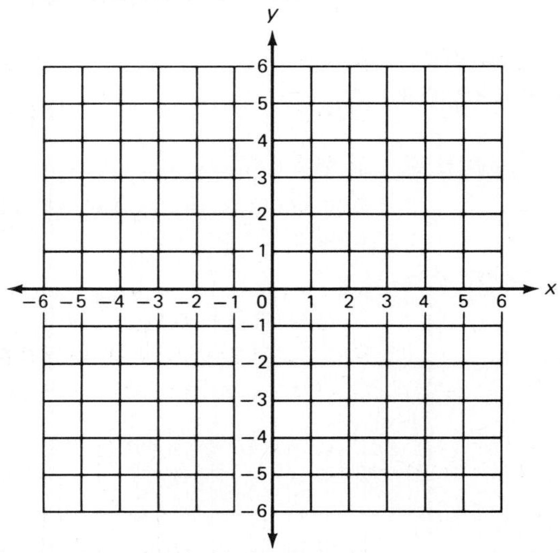

38.2 variation

Direct and inverse variation statements are not always simple statements of direct or inverse variation. Often one variable will vary directly as the other variable squared or the other variable cubed.

STATEMENT	IMPLIED EQUATION
The weight of a body varies inversely with the square of the distance to the center of the earth.	$W = \dfrac{k}{D^2}$
The distance required to stop is directly proportional to the square of the velocity.	$D = kV^2$
The price of a diamond varies directly as the square of its weight.	$P = kW^2$
The strength of the field is inversely proportional to the cube of the radius.	$S = \dfrac{k}{R^3}$

These problems are solved in the same way that simple variation problems are solved. First we recognize the statement of the implied variation and write down the indicated equation. Then we find k, insert its value in the equation, and solve for the required unknown.

example 38.2.1 The distance required for an automobile to stop is directly proportional to the square of its velocity. If a car can stop in 200 meters at 20 km/hr, what will be the required distance at 28 km/hr?

solution

Step 1 $\quad D = kV^2$

Step 2 $\quad 200 = k(20)^2 \longrightarrow k = .5$

Step 3 $\quad D = .5V^2$

Step 4 $\quad D = .5(28)^2 \longrightarrow$ $\mathbf{D = 392}$ **meters**

example 38.2.2 The distance a body falls varies directly as the square of the time that it falls. If it falls 144 meters in 3 seconds, how far will it fall in 10 seconds?

solution Step 1 $D = kt^2$

Step 2 $144 = k(3)^2 \longrightarrow k = 16$

Step 3 $D = 16t^2$

Step 4 $D = 16(10)^2 \longrightarrow \mathbf{D = 1600\ meters}$

example 38.2.3 The weight of a body on or above the surface of the earth varies inversely with the square of the distance from the body to the center of the earth. If a body weighs 10,000 lb at a distance of 5000 miles from the center of the earth, how much would it weigh 50,000 miles from the center of the earth?

solution Step 1 $W = \dfrac{k}{D^2}$

Step 2 $10{,}000 = \dfrac{k}{(5000)^2} \longrightarrow 10{,}000(5000)^2 = k \longrightarrow 25 \times 10^{10} = k$

Step 3 $W = \dfrac{25 \times 10^{10}}{D^2}$

Step 4 $W = \dfrac{25 \times 10^{10}}{(50{,}000)^2} \longrightarrow W = \dfrac{25 \times 10^{10}}{25 \times 10^8} \longrightarrow \mathbf{W = 100\ lb}$

TEST YOUR UNDERSTANDING

(b) The price of a diamond varies directly as the square of its weight. If a 2-carat diamond costs $8000, what is the price of a 3-carat diamond?

38.3 quadratic equations

We have been solving rational equations by multiplying the numerator of every term in the equation by the least common multiple of the denominators and then eliminating all the denominators as we show here.

$$\frac{x}{4} - \frac{5}{6} = \frac{x}{2} \longrightarrow 12 \cdot \frac{x}{4} - 12 \cdot \frac{5}{6} = 12 \cdot \frac{x}{2} \longrightarrow 3x - 10 = 6x$$

$$\longrightarrow -3x = 10 \longrightarrow \mathbf{x = -\frac{10}{3}}$$

As we learned in Lesson 28, when we encounter rational equations such as

$$\frac{2}{3n} = \frac{2}{n + 4}$$

we must use care in stating the solution or solutions because some values of the variable cannot be solutions to the equation because some equations by their structure restrict the permissible values of the variable. In the equation

$$\frac{2}{3n} = \frac{2}{n + 4}$$

we see that the numbers 0 and -4 are not permissible values of n, for if we let n equal zero,

$$\frac{2}{3n} = \frac{2}{n + 4} \longrightarrow \frac{2}{3 \cdot 0} = \frac{2}{0 + 4} \longrightarrow \frac{2}{0} = \frac{2}{4}$$

we find we are trying to divide by zero and if we let $n = -4$

$$\frac{2}{3n} = \frac{2}{n + 4} \longrightarrow \frac{2}{3(-4)} = \frac{2}{-4 + 4} \longrightarrow \frac{2}{-6} = \frac{2}{0}$$

we also indicate a division by zero, which of course has no meaning.

Often the solution of a rational equation requires the solution of a quadratic equation as we see in the following problems.

example 38.3.1 Solve $\frac{r + 1}{r - 1} = \frac{2}{r(r - 1)}$.

solution ($r \neq 0, 1$) We begin by multiplying every term by $r(r - 1)$ so that we can cancel the denominators.

$$r\cancel{(r - 1)}\frac{r + 1}{\cancel{r - 1}} = \cancel{r(r - 1)}\frac{2}{\cancel{r(r - 1)}} \longrightarrow r^2 + r = 2$$

Now we write the quadratic equation in standard form, factor, and solve.

$$\longrightarrow r^2 + r - 2 = 0 \longrightarrow (r + 2)(r - 1) = 0 \longrightarrow \boldsymbol{r = -2, 1}$$

but 1 is not an acceptable value for r. Thus $\boldsymbol{r = -2}$.

example 38.3.2 Solve $1 + \frac{2}{b - 1} = \frac{2}{b^2 - b}$.

solution As the first step it is advisable to factor the denominator of the term $\frac{2}{b^2 - b}$.

$$1 + \frac{2}{b - 1} = \frac{2}{b(b - 1)}$$

Now we see that the unacceptable values of the variable are 0 and 1 ($b \neq 0, 1$). The least common multiple of the denominators is $b(b - 1)$. So we eliminate the denominators by multiplying every numerator by $b(b - 1)$ and canceling.

$$b(b - 1) \cdot 1 + b\cancel{(b - 1)}\frac{2}{\cancel{b - 1}} = \cancel{b(b - 1)}\frac{2}{\cancel{b(b - 1)}} \longrightarrow b^2 - b + 2b = 2$$

Now we write the quadratic equation in standard form, factor, and solve.

$$\longrightarrow b^2 + b - 2 = 0 \longrightarrow (b + 2)(b - 1) = 0 \longrightarrow \boldsymbol{b = -2, 1}$$

but 1 is not an acceptable value for b so our solution is $\boldsymbol{b = -2}$.

example 38.3.3 Solve $\dfrac{y-3}{2} = \dfrac{4y}{y+3}$.

solution $(y \neq -3)$. We begin by multiplying every numerator by $2(y + 3)$ and canceling the denominators.

$$\cancel{2}(y+3)\frac{(y-3)}{\cancel{2}} = 2\cancel{(y+3)}\frac{4y}{\cancel{y+3}}$$

$$\longrightarrow \quad (y+3)(y-3) = 2(4y) \quad \longrightarrow \quad y^2 - 9 = 8y$$

Now we write the quadratic in standard form, factor, and solve.

$$\longrightarrow \quad y^2 - 8y - 9 = 0 \quad \longrightarrow \quad (y-9)(y+1) = 0 \quad \longrightarrow \quad \mathbf{y = 9, -1}$$

Both values are acceptable.

example 38.3.4 Solve $\dfrac{x-5}{8x} = \dfrac{3}{x+5}$.

solution $(x \neq 0, -5)$. We begin by multiplying every numerator by $8x(x + 5)$ and canceling the denominators.

$$\cancel{(8x)}(x+5)\frac{(x-5)}{\cancel{8x}} = (8x)\cancel{(x+5)}\frac{3}{\cancel{x+5}} \quad \longrightarrow \quad (x+5)(x-5) = 8x(3)$$

Now we write the quadratic equation in standard form, factor, and solve.

$$\longrightarrow \quad x^2 - 25 = 24x \quad \longrightarrow \quad x^2 - 24x - 25 = 0 \quad \longrightarrow \quad (x-25)(x+1) = 0$$

$$\longrightarrow \quad \mathbf{x = 25, -1}$$

Neither of these numbers was excluded as a solution when we began the problem so both are solutions.

TEST YOUR UNDERSTANDING

(c) Solve: $\dfrac{x+3}{4} = \dfrac{2x}{x-3}$

lesson checklist 38

- ☐ Linear inequality
- ☐ Test point
- ☐ Direct square variation
- ☐ Inverse square variation
- ☐ Rational equations with quadratic solutions

ANSWERS TO TEST YOUR UNDERSTANDING

(a)

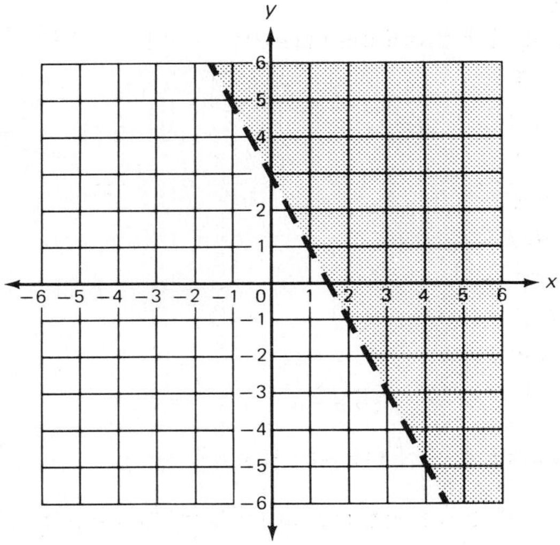

(b) $18,000
(c) 9, −1

problem set 38

*1. The distance required for an automobile to stop is directly proportional to the square of its velocity. If a car can stop in 200 ft at 20 mph, what will be the required distance at 28 mph?

*2. The weight of a body on or above the surface of the earth varies inversely with the square of the distance from the body to the center of the earth. If a body weighs 10,000 lb at a distance of 5000 miles from the center of the earth, how much would it weigh 50,000 miles from the center of the earth?

3. Francois had 42 salamanders. If the value of the large salamanders was 5 francs and the value of the small salamanders was 3 francs and their total value was 186 francs, how many of the salamanders were small ones?

4. The ratio of pikemen to archers at the battle of Agincourt was 7 to 4. If the English army had 3850 pikemen and archers, how many were archers?

5. The mustang had a 700-meter headstart on the buckaroo. If the speed of the mustang was 7 meters a second and that of the buckaroo was 11 meters a second, how long did it take the buckaroo to catch the mustang?

6. Miss Kate and Miss Willy lived in towns 280 miles apart. At 8 a.m. Miss Kate left her home to visit Miss Willy. At 9 a.m. Miss Willy left her home to visit Miss Kate. If both traveled at 40 mph, what time did they meet?

7. Woody saved dimes and Gwenn saved nickels. Between them they had 35 coins whose value was $2.00. How many dimes did Woody have?

8. Find three consecutive integers such that if the first is increased by 100, it is 69 greater than three times the sum of the third and the number 5.

*The problems marked with an asterisk were worked out in this lesson.

9. Forty percent of Harry Vardon's rounds were sub-par rounds. If he had 240 sub-par rounds in 3 years, how many total rounds of golf did he play in that length of time?

Graph on a rectangular coordinate system:

*10. $y > -x + 2$ *11. $y \leq 2x - 1$ *12. $y \geq -2$

Solve:

*13. $\frac{x-5}{8x} = \frac{3}{x+5}$ 14. $\frac{x+3}{4} = \frac{2x}{x-3}$

15. $N_P + N_N + N_D = 15$
$N_P + 5N_N + 10N_D = 83$
$N_N = 2N_D$

Graph on a number line:

16. $-4 < x - 2 < 4$, $D = \{\text{Reals}\}$

17. Find the number $\frac{2}{3}$ of the way from $\frac{2}{5}$ to $\frac{5}{8}$.

18. Show that $1.0\overline{345}$ is a rational number by writing it as a quotient of integers.

19. $\frac{x}{m} - p + \frac{k}{a} = c$, find a.

Solve:

20. $\sqrt{x-3} - 5 = 0$

21. Find the equation of the line:

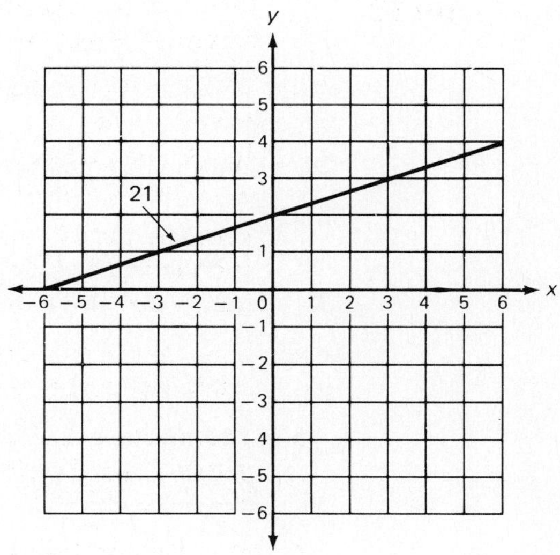

22. Find the equation of the line that goes through $(4, -5)$ and is parallel to $4x + 3y - 7 = 0$.

23. Find the equation of the line through $(-2, 5)$ that has a slope of $-\frac{2}{3}$.

24. Given that $f(x) = 4x + 2$, $D = \{\text{Reals}\}$
$m(x) = 5x - 3$, $D = \{\text{Reals}\}$
$\phi(x) = 2x - 8$, $D = \{\text{Reals}\}$

find (a) $f(0)$.
(b) $\phi(-5)$.

25. Which of the following designate functions? Use the vertical line test as necessary.

(a)

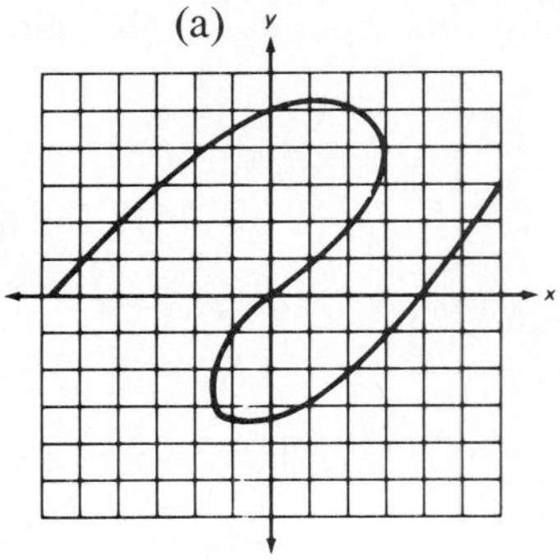

(b)

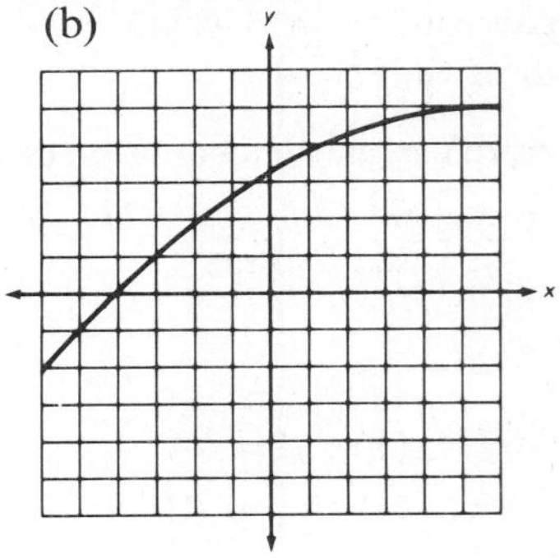

(c)

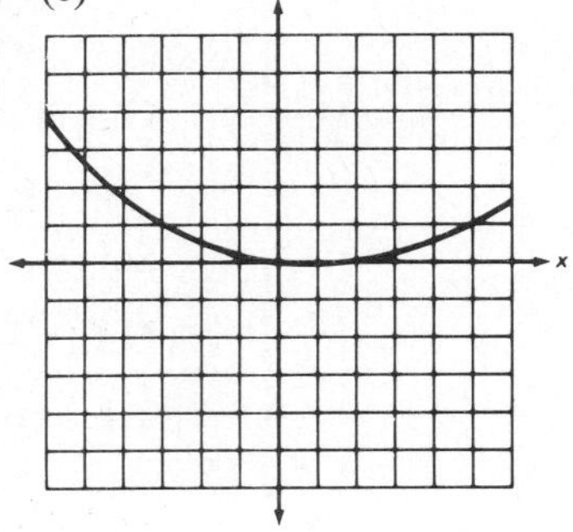

(d)

(4, 2), (5, 7), (6, 11)
(2, 5), (7, 6), (8, 14)

(e)

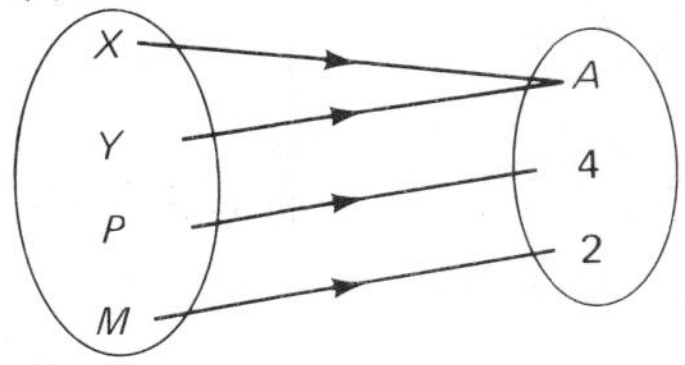

(f)

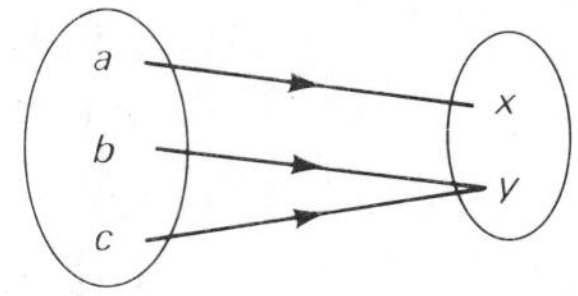

Graph on a number line:

26. $-3 - x < -2$, $D = \{\text{Reals}\}$

27. $-|x| < 2$, $D = \{\text{Reals}\}$

28. Find m.

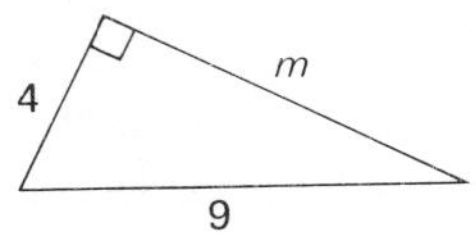

29. Find the distance between $(-4, 2)$ and $(5, 3)$.

Simplify:

30. $3\sqrt{98} - \sqrt{147}$

31. $2\sqrt{21}\,(\sqrt{7} + 3\sqrt{3}\,)$

32. $2\sqrt{2}\,(\sqrt{2} - 4)$

33. Divide $x^3 + 2x - 5$ by $x + 5$.

Solve:

34. $-x^2 = 4x - 21$

Simplify:

35. $\dfrac{k^2 + k}{k^2} \cdot \dfrac{3k - 3}{k^2 - 1}$

36. $\dfrac{(.00007 \times 10^{20})(30{,}000)}{(14{,}000 \times 10^{-30})}$

37. Use the letters x, y, and z to state the distributive property.

Graph on a rectangular coordinate system:

38. $y - 2x + 2 = 0$

Simplify:

39. $\dfrac{a + \dfrac{1}{x - y}}{\dfrac{a}{x - y} + 1}$

40. $\dfrac{\dfrac{1}{a + b}}{\dfrac{1}{a}}$

Add:

41. $\dfrac{m}{x^2c} + \dfrac{b + y}{x^3c} + \dfrac{k}{c^3}$

Add like terms:

42. $\dfrac{a^2c^{-2}}{b^4} - \dfrac{3b^{-2}}{b^2c^2} + \dfrac{7a^2}{b^4c^2}$

Multiply:

43. $\dfrac{2a^2}{x^4}\left(x^4a^{-2} - \dfrac{3x^{-4}}{a^2}\right)$

Simplify:

44. $\dfrac{a^2(x^{-2}y)^{-3}y^3}{a^{-2}x^2yax^2}$

Solve:

45. $-5(2x + 6) - (-4) = 5(-x - 2)$

Evaluate:

46. $-a(-a - b) + (-b)$ if $a = 2, b = -7$

47. True or false: {Reals} $\subset$ {Irrationals}

48. $-6 \in$ {what sets of numbers?}

Add:

49. $\dfrac{x + 1}{x} + \dfrac{3}{x^2 - 2x}$

lesson thirty-nine

Difference of two squares theorem
Completing the square

39.1 difference of two squares

To review the difference of two squares theorem, we note that if $p^2 = q^2$, we can use the additive property of equality to justify adding $-q^2$ to both sides of the equation.

$$\begin{array}{rcl} p^2 & = & q^2 \\ -q^2 & & -q^2 \\ \hline p^2 - q^2 & = & 0 \end{array}$$

Now we factor the left side of the equation and use the zero factor theorem to find that $p = q$ or $p = -q$.

$$p^2 - q^2 = 0 \longrightarrow (p + q)(p - q) = 0 \longrightarrow p = q, -q$$

> DIFFERENCE OF TWO SQUARES THEOREM
>
> If p and q are real numbers and if $p^2 = q^2$, then
>
> $p = q \quad \text{or} \quad p = -q$

We review the use of this theorem by finding the values of the unknowns that satisfy the following equations.

	EQUATION	SOLUTION
(a)	$p^2 = q^2$	$p = \pm q$
(b)	$x^2 = 4$	$x = \pm 2$

	EQUATION	SOLUTION
(c)	$m^2 = 3$	$m = \pm\sqrt{3}$
(d)	$(x + 2)^2 = 3$	$x + 2 = \pm\sqrt{3}$ so $x = -2 \pm \sqrt{3}$
(e)	$(x - 3)^2 = 5$	$x - 3 = \pm\sqrt{5}$ so $x = 3 \pm \sqrt{5}$

39.2 completing the square

The values of x that satisfy $x^2 + 5x + 6 = 0$ can be found by factoring this quadratic equation and using the zero factor theorem.

$$x^2 + 5x + 6 = 0 \longrightarrow (x + 3)(x + 2) = 0 \longrightarrow \mathbf{x = -3, -2}$$

If we wish to find the replacement values of x that will satisfy the quadratic equation

$$x^2 - x - 5 = 0$$

we are stumped because this equation cannot be factored over the integers. It is possible, however, to solve this equation or any quadratic equation by rewriting the equation in the form $(x + a)^2 = k$ and then using the difference of two squares theorem to write that $x + a = \pm\sqrt{k}$ or that $x = -a \pm \sqrt{k}$.

To understand the development of this method of solution, it is helpful to observe the relationship between the terms of a particular binomial and the terms of the square of the same binomial.

BINOMIAL	BINOMIAL SQUARED
$x + 3$	$x^2 + 6x + 9$
$x - 3$	$x^2 - 6x + 9$
$x + 5$	$x^2 + 10x + 25$
$x - 5$	$x^2 - 10x + 25$
$x + 7$	$x^2 + 14x + 49$
$x - 7$	$x^2 - 14x + 49$

In each of the expressions above there is a definite relationship between the constant term of the binomial and the two constants of the trinomial. In every example the coefficient of the second term of the trinomial is twice the constant term of the binomial and the third term of the trinomial is the square of the constant term of the binomial. But more important is the relationship between the constants in the trinomial. **In each case the last term is the square of one half of the coefficient of x in the middle term.** In the first trinomial

$$x^2 + 6x + 9$$

we note that 9 is $\left(\frac{6}{2}\right)^2$. In the middle trinomial

$$x^2 + 10x + 25$$

we note that 25 is $\left(\frac{10}{2}\right)^2$. In the last trinomial

$$x^2 - 14x + 49$$

we note that 49 is $\left(\frac{-14}{2}\right)^2$.

We can use the patterns pointed out above to help us write a given quadratic equation in the form $(x + a)^2 = k$, which can be solved for x by using the difference of two squares theorem. This procedure is called completing the square.

We always begin with the given quadratic equation written in standard form. Thus if we are given the quadratic equation

$$-2x^2 = -x - 5$$

we first write it in standard form as

$$2x^2 - x - 5 = 0$$

Now we will demonstrate the procedure of completing the square.

Step 1: Divide every term in the equation by the coefficient of the x^2 term so that the resulting coefficient of the x^2 term will be 1.

$$\frac{2x^2}{2} - \frac{x}{2} - \frac{5}{2} = 0 \longrightarrow x^2 - \frac{1}{2}x - \frac{5}{2} = 0$$

Step 2: Move the constant term to the right side of the equation.

$$\begin{aligned} x^2 - \frac{1}{2}x - \frac{5}{2} &= 0 \\ + \frac{5}{2} \quad &+ \frac{5}{2} \qquad \text{added } \frac{5}{2} \text{ to both sides} \\ \hline x^2 - \frac{1}{2}x \qquad &= \frac{5}{2} \end{aligned}$$

Step 3: Square the product of the coefficient of x and $\frac{1}{2}$

$$\left(-\frac{1}{2}\cdot\frac{1}{2}\right)^2 = \frac{1}{16}$$

and add the result of $\frac{1}{16}$ to both sides of the equation.

$$x^2 - \frac{1}{2}x + \frac{1}{16} = \frac{5}{2} + \frac{1}{16}$$

Step 4: Write the left side as the square of a binomial and simplify the right side.

$$\left(x - \frac{1}{4}\right)^2 = \frac{41}{16}$$

Step 5: Use the difference of two squares theorem to complete the solution.

$$x - \frac{1}{4} = \pm\sqrt{\frac{41}{16}} \longrightarrow x = \frac{1}{4} \pm \sqrt{\frac{41}{16}} \longrightarrow \mathbf{x = \frac{1}{4} \pm \frac{\sqrt{41}}{4}}$$

example 39.2.1 Solve $x^2 + 3x - 5 = 0$ by completing the square.

solution (1) Since the coefficient of the x^2 term is already 1, the first step is omitted.

(2)
$$\begin{aligned} x^2 + 3x - 5 &= 0 \\ +5 \quad &+5 \\ \hline x^2 + 3x \qquad &= 5 \end{aligned}$$

(3)
$$\left(3 \cdot \frac{1}{2}\right)^2 = \frac{9}{4}, \text{ so add } \frac{9}{4} \text{ to both sides.}$$

$$x^2 + 3x + \frac{9}{4} = 5 + \frac{9}{4}$$

Now write the left side as a binomial squared and simplify the right side.

(4)
$$\left(x + \frac{3}{2}\right)^2 = \frac{29}{4}$$

Use the difference of two squares theorem to complete the solution.

(5)
$$x + \frac{3}{2} = \pm\sqrt{\frac{29}{4}} \longrightarrow x = -\frac{3}{2} \pm \sqrt{\frac{29}{4}} \longrightarrow \mathbf{x = -\frac{3}{2} \pm \frac{\sqrt{29}}{2}}$$

example 39.2.2 Solve $x^2 + 5x - 8 = 0$ by completing the square.

solution (1) Not required as the coefficient of the x^2 term is already 1.

(2)
$$\begin{array}{r} x^2 + 5x - 8 = 0 \\ +8 \quad +8 \\ \hline x^2 + 5x \quad = 8 \end{array}$$

(3)
$$\left(5 \cdot \frac{1}{2}\right)^2 = \frac{25}{4}, \text{ so add } \frac{25}{4} \text{ to both sides.}$$
$$x^2 + 5x + \frac{25}{4} = 8 + \frac{25}{4}$$

Now write the left side as a binomial squared and simplify the right side.

(4)
$$\left(x + \frac{5}{2}\right)^2 = \frac{57}{4}$$

Use the difference of two squares theorem to complete the solution.

(5)
$$x + \frac{5}{2} = \pm\sqrt{\frac{57}{4}} \longrightarrow x = -\frac{5}{2} \pm \sqrt{\frac{57}{4}} \longrightarrow \mathbf{x = -\frac{5}{2} \pm \frac{\sqrt{57}}{2}}$$

example 39.2.3 Solve $2x^2 + \frac{1}{3}x - 6 = 0$ by completing the square.

solution (1) $2x^2 + \frac{1}{3}x - 6 = 0 \longrightarrow x^2 + \frac{1}{6}x - 3 = 0$

(2)
$$\begin{array}{r} x^2 + \frac{1}{6}x - 3 = 0 \\ +3 \quad +3 \\ \hline x^2 + \frac{1}{6}x \quad = 3 \end{array}$$

(3)
$$\left(\frac{1}{6} \cdot \frac{1}{2}\right)^2 = \frac{1}{144}, \text{ so add } \frac{1}{144} \text{ to both sides.}$$
$$x^2 + \frac{1}{6}x + \frac{1}{144} = 3 + \frac{1}{144}$$

Now write the left side as a binomial squared and simplify the right side.

(4)
$$\left(x + \frac{1}{12}\right)^2 = \frac{433}{144}$$

Now use the difference of two squares theorem to complete the solution.

(5)
$$x + \frac{1}{12} = \pm\sqrt{\frac{433}{144}} \longrightarrow \mathbf{x = -\frac{1}{12} \pm \frac{\sqrt{433}}{12}}$$

TEST YOUR UNDERSTANDING

(a) Solve: $x^2 + 3x - 8 = 0$ by completing the square.

---

lesson checklist 39

- ☐ Difference of two squares theorem
- ☐ Completing the square

ANSWER TO TEST YOUR UNDERSTANDING

(a) $x = -\frac{3}{2} \pm \frac{\sqrt{41}}{2}$

problem set 39

1. The number of days required to build a pyramid varies inversely as the number of men who work on the pyramid. If 6000 men can complete a pyramid in 8 years, how many men would it take to do the job in 3 years?

2. The value of the emerald varies directly as the square of its weight. If a 2-carat emerald has a value of 3000 dollars, what would be the value of a 3-carat emerald?

3. The size of the weight required to balance the load varies inversely as the distance of the weight from the fulcrum. If a 60-kg weight at a distance of 4 meters balances the load, what weight would be required at a distance of $\frac{1}{2}$ meter?

4. The ratio of fats to leans was 2 to 5. If there were a total of 343 fats and leans, how many were fats?

5. John left town at 9 a.m. and headed north at 45 km/hr. Mary Esther left in pursuit at 10 a.m. and she drove 60 km/hr. At what time did Mary Esther catch John?

6. To kill time before the ball game, Sal jogged to Times Square and rode back on the subway. If he jogged at 10 km/hr and the speed of the subway was 40 km/hr, how far did he jog if he was gone for 3 hours?

7. Pembroke bought 30 globs. The large globs cost $5 each and the very large ones cost $15 each. If he spent $250 for the lot, how many large globs did he buy?

8. Find three consecutive even integers such that if 14 is added to twice the first, the sum is one-half of three times the sum of the third and the number 2.

9. At the party, green dresses and blue dresses were in the ratio of 2 to 5. If 154 girls attended, how many wore blue dresses?

Solve by completing the square:

***10.** $x^2 + 3x - 5 = 0$ ***11.** $x^2 + 5x - 8 = 0$ ***12.** $2x^2 + \frac{1}{3}x - 6 = 0$

Graph on a rectangular coordinate system:

13. $y > -x - 2$

Solve:

14. $\frac{x - 4}{5x} = \frac{3}{x + 4}$ **15.** $\frac{x + 2}{3} = \frac{x}{x - 2}$ **16.** $N_P + N_N + N_D = 15$; $N_P + 5N_N + 10N_D = 63$; $N_P = 2N_D$

Graph on a number line:

17. $x + 3 > 1$ or $x + 3 < -1$, $D = \{\text{Reals}\}$

---

*The problems marked with an asterisk were worked in this lesson.

18. Find the number $\frac{3}{5}$ of the way from $\frac{4}{5}$ to $\frac{8}{9}$.

19. $\frac{1}{a} + c = b$; find a.

20. $\frac{x}{y} - \frac{c}{k} = m$, find k.

Solve:

21. $\sqrt{x^2 - 3x - 12} - 4 = 0$

22. Find the equation of the line:

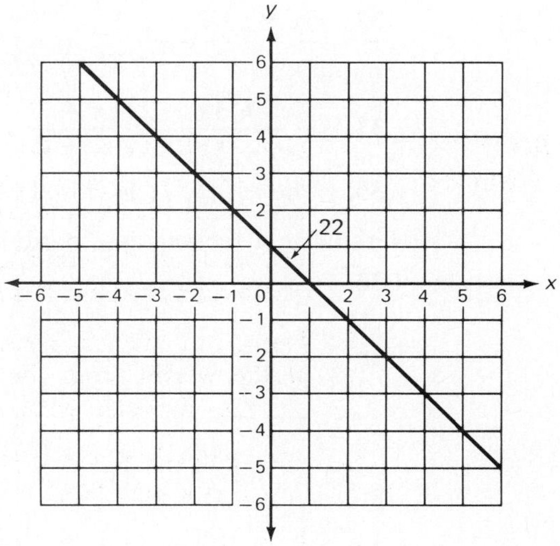

23. Find the equation of the line that goes through $(-3, 2)$ and is parallel to $2x + 3y = 7$.

24. Find the equation of the line through $(2, 4)$ and $(-3, -5)$.

25. Given that $f(x) = x^2 - 3$, $D = \{\text{Reals}\}$; find $f(4)$.
$g(x) = x^2 + 5$, $D = \{\text{Reals}\}$;
$\phi(x) = x - 1$, $D = \{\text{Reals}\}$;

26. Which of the following designate functions? Use the vertical line test as necessary.

(a)

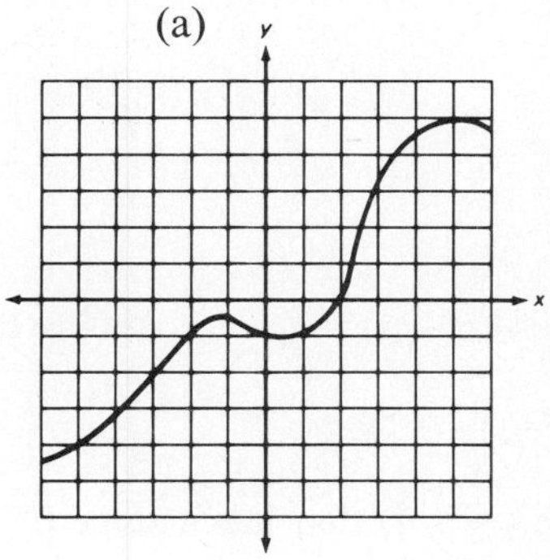

(b)

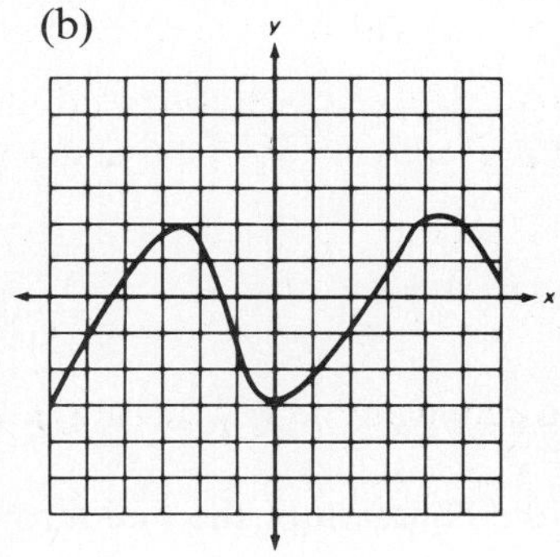

(c)

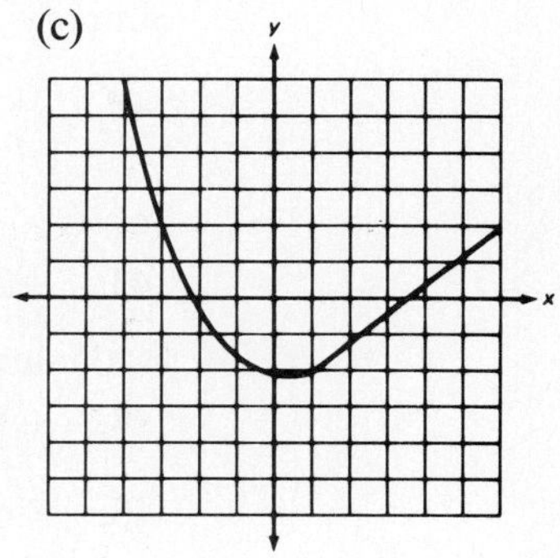

(d)

(6, 7), (3, 2), (5, 1)
(7, k), (5, 9), (8, 2)

(e)

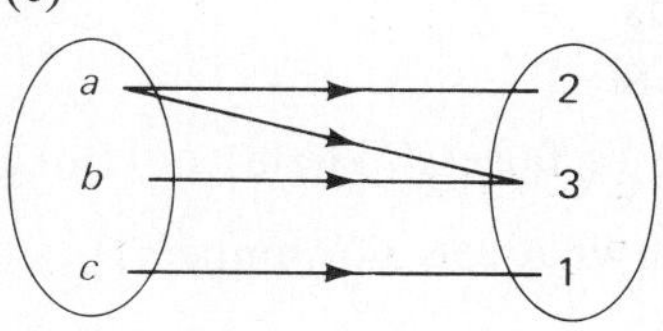

(f)

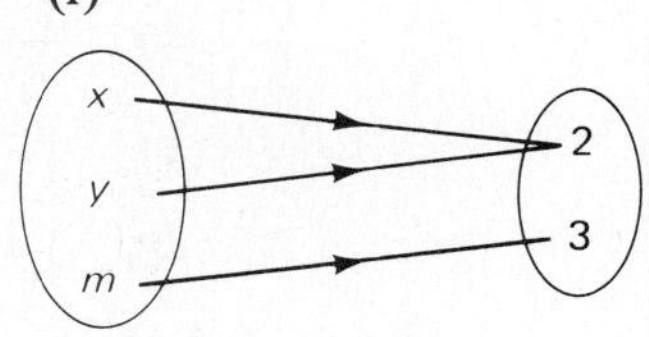

Graph on a number line:

27. $-3 - x < 4$, $D = \{\text{Reals}\}$

28. Find p.

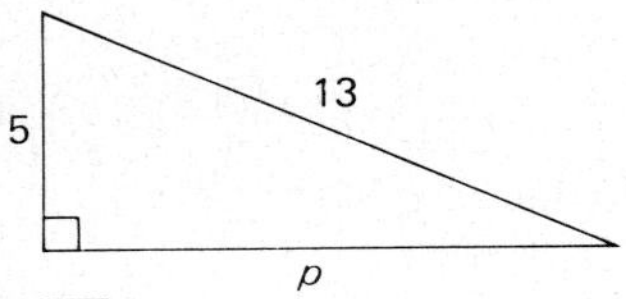

Simplify:

29. $\sqrt{25} - \sqrt{18}$

30. $3\sqrt{8}\,(2\sqrt{2} - 4\sqrt{12}\,)$

Divide

31. $2x^3 + 7x^2 + x - 13$ by $x + 2$

Solve:

32. $5x = -2x^2 + 3$

Simplify:

33. $\dfrac{2x^2 + 3x - 2}{2x^2 + 5x - 3} \cdot \dfrac{x + 3}{x + 2}$

34. $\dfrac{(.00005 \times 10^{-4})(400{,}000)}{.00002 \times 10^{17}}$

35. The set $\{0, 1\}$ is closed for what operations?

Graph on a rectangular coordinate system:

36. $y + x - 5 = 2$

Simplify:

37. $\dfrac{1 + \dfrac{1}{x}}{7}$

38. $\dfrac{\dfrac{a}{b} + 1}{\dfrac{x}{b} + 2}$

Add:

39. $\dfrac{1}{x} - \dfrac{a}{x + b}$

Add like terms:

40. $3xyz^2 - \dfrac{3xz^2}{y^{-1}} - \dfrac{2xy}{z^{-2}}$

Multiply:

41. $2a^2yb\left(\dfrac{a^2}{by^0} - \dfrac{5y^0}{b^{-1}a^2}\right)$

Simplify:

42. $\dfrac{x^{-3}y^3p^{-5}k}{b^2tx^3p^{-1}}$

Solve:

43. $-2(x + 3) + (-x) = 2(x - 1)$

Evaluate:

44. $-1 + (-b)(a - b)$ if $a = -3, b = -1$

45. Find the equation of the line through $(-2, 4)$ that has a slope of $-\dfrac{2}{3}$.

Simplify:

46. $\dfrac{(-3)^{-2}}{-6^2}$

47. True or false: $\{\text{Whole}\} \subset \{\text{Integers}\}$

48. $3 \in \{\text{what sets of numbers?}\}$

Add:

49. $\dfrac{4}{x^2 - 5x + 6} - \dfrac{6x}{x - 3}$

lesson forty

Quadratic formula

40.1 quadratic formula

In Lesson 21 we learned how to find solutions to quadratic equations by factoring over the integers and using the zero factor theorem. In Lesson 39 we learned how to use the method of completing the square to find solutions of quadratic equations that we cannot factor over the integers. **If we write a quadratic equation in standard form using *a*, *b*, and *c*, to represent the constants and complete the square, we can find a formula that can be used to find the values of *x* that will satisfy any given quadratic equation.**

$ax^2 + bx + c = 0$	general form of quadratic equation
$x^2 + \frac{b}{a}x + \frac{c}{a} = 0$	divide by a (multiplicative property of equality)
$x^2 + \frac{b}{a}x = -\frac{c}{a}$	add $-\frac{c}{a}$ to both sides (additive property of equality)
$x^2 + \frac{b}{a}x + \frac{b^2}{4a^2} = \frac{b^2}{4a^2} - \frac{c}{a}$	add $\left(\frac{b}{2a}\right)^2 = \frac{b^2}{4a^2}$ to both sides (additive property of equality)
$\left(x + \frac{b}{2a}\right)^2 = \frac{b^2 - 4ac}{4a^2}$	simplification of both sides
$x + \frac{b}{2a} = \pm\sqrt{\frac{b^2 - 4ac}{4a^2}}$	if $p^2 = q^2$, $p = \pm q$
$x + \frac{b}{2a} = \pm\frac{\sqrt{b^2 - 4ac}}{\sqrt{4a^2}}$	quotient of square roots theorem

$$x = -\frac{b}{2a} \pm \frac{\sqrt{b^2 - 4ac}}{2a} \qquad \text{additive property of equality}$$

$$x = \frac{-b \pm \sqrt{b^2 - 4ac}}{2a} \qquad \text{addition of rational expressions}$$

40.2 use of the quadratic formula

By completing the square on $ax^2 + bx + c = 0$, we have derived the **QUADRATIC FORMULA**

$$x = \frac{-b \pm \sqrt{b^2 - 4ac}}{2a} \qquad (a \neq 0)$$

This formula expresses x in terms of the constants a, b, and c of the equation. If we wish to use this formula to find the values of x that satisfy a particular quadratic equation, it is necessary to compare the particular quadratic equation with the equation $ax^2 + bx + c$ to determine the values of a, b, and c.

example 40.2.1 Use the quadratic formula to determine the roots of the equation $3x^2 + 2x - 7 = 0$.

solution The general form of the quadratic equation is $ax^2 + bx + c = 0$. If we write our equation and the general equation one over the other

$$ax^2 + bx + c = 0$$
$$3x^2 + 2x - 7 = 0$$

we see that 3 corresponds to a, 2 corresponds to b, and -7 corresponds to c. If we use these numbers as replacements for a, b, and c in the quadratic formula, we find

$$x = \frac{-b \pm \sqrt{b^2 - 4ac}}{2a} \longrightarrow x = \frac{-(2) \pm \sqrt{(2)^2 - 4(3)(-7)}}{2(3)}$$

$$\longrightarrow \frac{-2 \pm \sqrt{4 + 84}}{6} \longrightarrow \frac{-2 \pm \sqrt{88}}{6} \longrightarrow \frac{-2 \pm 2\sqrt{22}}{6} \longrightarrow \frac{-1 \pm \sqrt{22}}{3}$$

Thus the two real numbers that will satisfy the given equation are

$$x = \frac{-1 + \sqrt{22}}{3} \qquad x = \frac{-1 - \sqrt{22}}{3}$$

We can verify that these numbers are solutions by using them as replacements for x in the original equation

USING $\frac{-1 + \sqrt{22}}{3}$

$$3\left(\frac{-1 + \sqrt{22}}{3}\right)^2 + 2\left(\frac{-1 + \sqrt{22}}{3}\right) - 7 = 0$$

$$\frac{3(23 - 2\sqrt{22})}{9} + \frac{-2 + 2\sqrt{22}}{3} - 7 = 0$$

$$\frac{23}{3} - \frac{2\sqrt{22}}{3} - \frac{2}{3} + \frac{2\sqrt{22}}{3} - 7 = 0$$

$$\frac{21}{3} - 7 = 0$$

$7 - 7 = 0$ True

USING $\frac{-1 - \sqrt{22}}{3}$

$$3\left(\frac{-1 - \sqrt{22}}{3}\right)^2 + 2\left(\frac{-1 - \sqrt{22}}{3}\right) - 7 = 0$$

$$\frac{3(23 + 2\sqrt{22})}{9} + \frac{-2 - 2\sqrt{22}}{3} - 7 = 0$$

$$\frac{23}{3} + \frac{2\sqrt{22}}{3} - \frac{2}{3} - \frac{2\sqrt{22}}{3} - 7 = 0$$

$$\frac{21}{3} - 7 = 0$$

$7 - 7 = 0$ True

example 40.2.2 Use the quadratic formula to determine the roots of $-6 = -x^2 + x$.

solution We begin by writing the given equation in standard form as

$$x^2 - x - 6 = 0$$

Now we compare this equation to the equation $ax^2 + bx + c = 0$ to determine the numbers that correspond to a, b, and c.

$$ax^2 + bx + c = 0$$
$$x^2 - x - 6 = 0$$

We see that $a = 1$, $b = -1$, and $c = -6$.

$$x = \frac{-b \pm \sqrt{b^2 - 4ac}}{2a} \longrightarrow x = \frac{-(-1) \pm \sqrt{(-1)^2 - 4(1)(-6)}}{2(1)}$$

$$\longrightarrow x = \frac{1 \pm \sqrt{1 + 24}}{2} \longrightarrow x = \frac{1 \pm \sqrt{25}}{2} \longrightarrow x = \frac{1 \pm 5}{2} \longrightarrow \mathbf{x = 3, -2}$$

example 40.2.3 Use the quadratic formula to find the values of x that satisfy the equation $-x = 7 - x^2$.

solution We begin by writing the given equation in standard form and comparing it to the equation $ax^2 + bx + c = 0$

$$ax^2 + bx + c = 0$$
$$x^2 - x - 7 = 0$$

We see that $a = 1$, $b = -1$, and $c = -7$.

$$x = \frac{-b \pm \sqrt{b^2 - 4ac}}{2a} \longrightarrow x = \frac{-(-1) \pm \sqrt{(-1)^2 - 4(1)(-7)}}{2(1)}$$

$$\longrightarrow x = \frac{1 \pm \sqrt{1 + 28}}{2} \longrightarrow \mathbf{x = \frac{1 \pm \sqrt{29}}{2}}$$

example 40.2.4 Use the quadratic formula to find the roots of $5x^2 = 2x + 1$.

solution Again we begin by writing the given equation in standard form and comparing it to the equation $ax^2 + bx + c = 0$.

$$ax^2 + bx + c = 0$$
$$5x^2 - 2x - 1 = 0$$

We see that $a = 5$, $b = -2$, and $c = -1$.

$$x = \frac{-b \pm \sqrt{b^2 - 4ac}}{2a} \longrightarrow x = \frac{-(-2) \pm \sqrt{(-2)^2 - 4(5)(-1)}}{2(5)}$$

$$\longrightarrow x = \frac{2 \pm \sqrt{4 + 20}}{10} \longrightarrow x = \frac{2 \pm \sqrt{24}}{10}$$

$$\longrightarrow x = \frac{2 \pm 2\sqrt{6}}{10} \longrightarrow \mathbf{x = \frac{1 \pm \sqrt{6}}{5}}$$

TEST YOUR UNDERSTANDING

(a) Solve $2x^2 - 3x - 8 = 0$ by using the quadratic formula.

lesson checklist 40

- ☐ Derivation of the quadratic formula
- ☐ Use of the quadratic formula in solving equations

ANSWER TO TEST YOUR UNDERSTANDING

(a) $\dfrac{3 \pm \sqrt{9 + 64}}{4} = \dfrac{3}{4} \pm \dfrac{\sqrt{73}}{4}$

problem set 40

1. If the temperature is held constant, the pressure of a perfect gas varies inversely as the volume. If the pressure is 40 torr when the volume is 60 cc, what is the volume at the same temperature when the pressure is 30 torr?
2. The cost to one person varies inversely as the number of people sharing the cost. If 5 people pay \$48 each for house rent, how much rent would be paid by each person if 3 more people moved in to share the rent?
3. If the volume is held constant, the temperature of a perfect gas varies directly as the pressure. A sample of gas has temperature of 400°K at a pressure of 4 atmospheres and a volume of 70 cc. If the volume is held constant, what is the temperature when the pressure is increased to 6 atmospheres?
4. The ratio of floor space devoted to hardware and software at the computer conference was 5 to 2. If 14,700 square meters were used at convention hall, how much of this floor space was devoted to a display of hardware?
5. Bobbie headed for San Mateo at 30 km/hr. Two hours later John realized she had left her purse and he headed after her at 60 km/hr. How long did it take John to catch Bobbie?
6. Mike walks into the country at 4 km/hr and catches a ride back in a truck that goes 20 km/hr. If he was gone for 6 hours, how far did he walk into the country?
7. Jeni and Denny bought sparkplugs and condensers at the Auto Haus. If sparkplugs cost \$1.50 each and condensers cost \$1.25 each and if they paid \$57.50 for 40 parts, how many sparkplugs did they buy?

---

*The problems marked with an asterisk were worked out in this example.

8. Find four consecutive integers such that four times the sum of the first and 12 is 13 greater than five times the fourth.

9. Seventeen percent of the people at the convention bought peanuts. If 1258 bags of peanuts were sold, how many people attended the convention?

*10. Begin with $ax^2 + bx + c = 0$ and derive the quadratic formula.

Use the quadratic formula to solve:

*11. $3x^2 + 2x - 7 = 0$ *12. $-6 = -x^2 + x$ *13. $5x^2 = 2x + 1$

Solve by completing the square:

14. $2x^2 - 2x - 4 = 0$

Graph on a rectangular coordinate system:

15. $y < -2x + 3$

Solve:

16. $\frac{x-6}{8} = \frac{2x}{x+6}$

17. $N_P + N_N + N_D = 19$
$N_P + 5N_N + 10N_D = 104$
$N_N = 2N_D$

Graph on a number line:

18. $-2 < x + 3 < 2$, $D = \{\text{Reals}\}$

19. Find the number $\frac{1}{4}$ of the way from $\frac{2}{5}$ to $\frac{8}{9}$.

20. Show that $1.0\overline{23}$ is a rational number by writing it as a quotient of integers.

21. $\frac{x}{a} + \frac{y}{b} - c = d$, find a.

Solve:

22. $x^2 - 2x - 6 - 9 = 0$

23. Find the equation of the line:

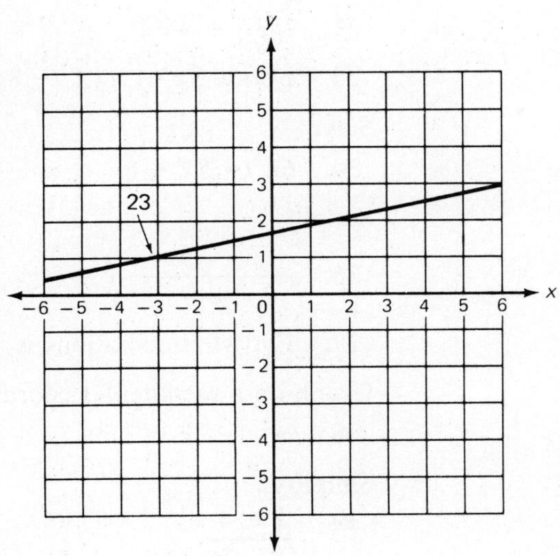

24. Find the equation of the line that goes through $(4, -2)$ and is parallel to $x - 2y = 3$.

25. Find the equation of the line through $(4, 3)$ that has a slope of $-\frac{1}{2}$.

26. Given that $f(x) = x - 5, \quad D = \{\text{Reals}\}$ find $g(3)$.
$g(x) = x^2 - 7, \quad D = \{\text{Reals}\}$
$\phi(x) = x + 2, \quad D = \{\text{Reals}\}$

27. Which of the following designate functions? Use the vertical line test as necessary.

(a)

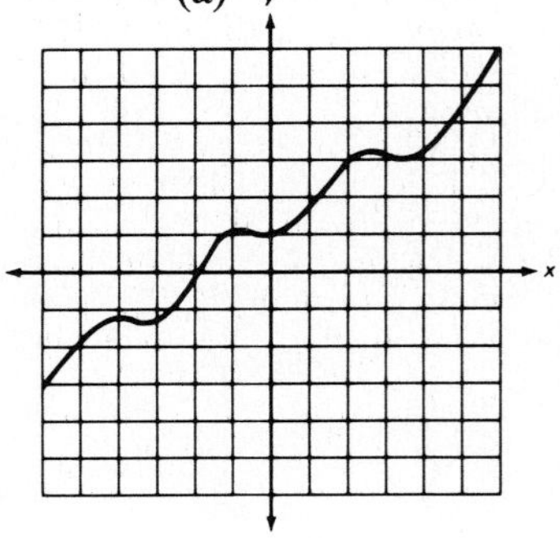

(b)

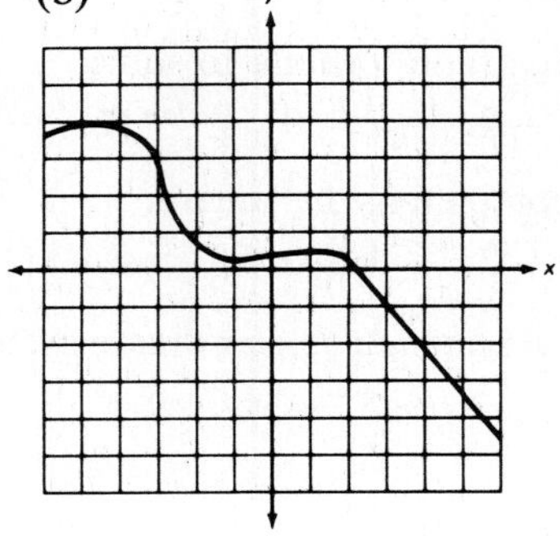

(c)

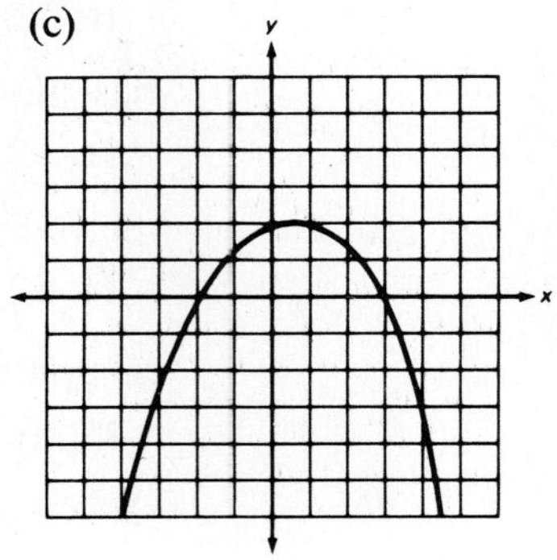

(d)
(3, 1), (2, 1), (4, 1)
(5, 6), (7, 3), (9, 2)

(e)

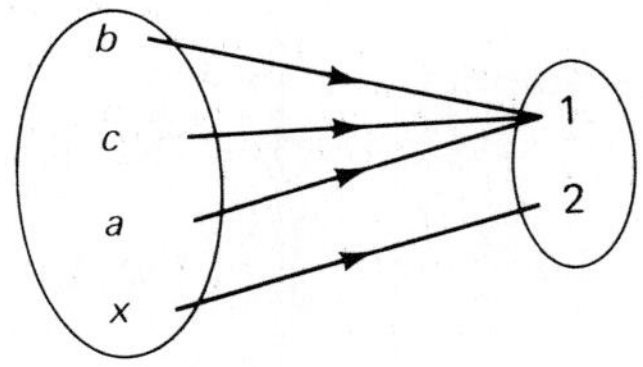

(f)

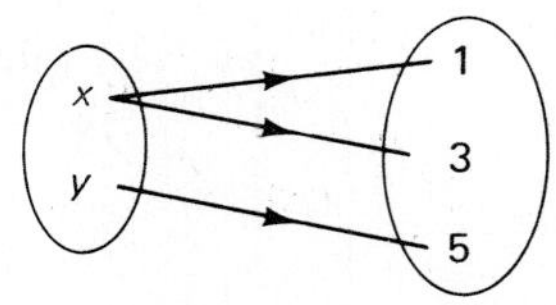

Graph on a number line:

28. $-7 + x \geq 2, D = \{\text{Reals}\}$

29. $2 + |x| \leq 3, \quad D = \{\text{Reals}\}$

30. Find the distance between (3, 5) and (−2, −1).

Simplify:

31. $\sqrt{32} + 2\sqrt{2}$

32. $2\sqrt{6}\,(5\sqrt{2} - 3\sqrt{3}\,)$

33. Divide $3x^3 + 11x^2 - 8x + 6$ by $x - 3$.

Solve:

34. $6x^2 = 5x + 6$

Simplify:

35. $\dfrac{y^2 - 9}{y^2 - y - 6} \cdot \dfrac{y - 3}{y + 3}$

36. $\dfrac{(.004 \times 10^{-6})(500{,}000)}{.002 \times 10^{16}}$

37. For what operations is the set of real numbers closed?

Graph on a rectangular coordinate system:

38. $-y - 2x = 3$

Simplify:

39. $\dfrac{t - 4}{1 - \dfrac{2}{t}}$

40. $\dfrac{\dfrac{1}{x}}{\dfrac{x}{y}}$

Add:

41. $\dfrac{x}{a} + \dfrac{y}{a + b}$

Add like terms:

42. $\dfrac{3a^2b}{y^{-3}} - \dfrac{2a^0b^{-1}y^3}{a^{-2}} + \dfrac{y^3b}{a^{-2}}$

Multiply:

43. $a^2xy^3\left(\dfrac{2x^{-2}}{a^2y^{-3}} + \dfrac{3x^0a^{-2}}{y^3}\right)$

Simplify:

44. $\dfrac{x^2y^3b^0m^{-4}}{m^{-4}b^{-3}y^2x^4}$

Solve:

45. $-3(x - 2) + (-x) = 7(x - 2)$

Simplify:

46. $\dfrac{4^{-1}}{-2^{-4}}$

47. True or false: {Natural} $\subset$ {Whole}

48. $\dfrac{\sqrt{15}}{2} \in$ {what sets of numbers?}

Add:

49. $\dfrac{7x^2}{x^2 - 9} - \dfrac{3x + 2}{x + 3}$

lesson forty-one

Slope-intercept method of graphing a line
Double-intercept method of graphing a line
Linear inequalities

41.1 slope-intercept method of graphing a line

We have been graphing a line by writing the equation of the line in the slope-intercept form and finding the values of y that the equation pairs with several different values of x. For instance, to graph $3x + 5y = 10$, we would first write the equation in slope-intercept form by solving for y.

$$3x + 5y = 10 \longrightarrow 5y = -3x + 10 \longrightarrow y = -\frac{3}{5}x + 2$$

Next we make a table and insert convenient values of x,

x	0	5	-5
y			

and now use the equation to find the corresponding values of y.

When $x = 0$

$$y = -\frac{3}{5}(0) + 2$$
$$y = 0 + 2$$
$$y = 2$$

When $x = 5$

$$y = -\frac{3}{5}(5) + 2$$
$$y = -3 + 2$$
$$y = -1$$

When $x = -5$

$$y = -\frac{3}{5}(-5) + 2$$
$$y = 3 + 2$$
$$y = 5$$

Now we insert these values in the table, graph the ordered pairs that we have found and draw the line indicated by these points.

x	0	5	-5
y	2	-1	5

If we use a procedure called the slope-intercept method, we can graph the line accurately and in less time than the method just described required. We remember that the two constants in the slope-intercept form of the equation are the values of the slope of the

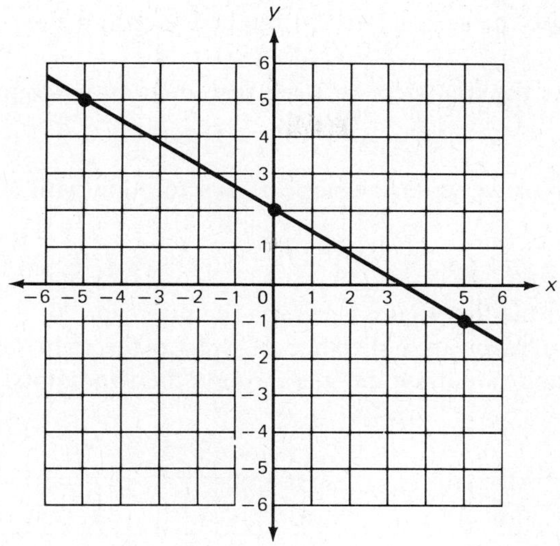

Figure 41.1

line and the intercept of the line. To demonstrate this method we will use the equation of the line for Fig. 41.1, which is

$$y = -\frac{3}{5}x + 2$$

We alter the form of this equation slightly by writing the slope in the form of a fraction **that has a positive denominator**. The numerator will be positive or negative as dictated by the sign of the slope.

$$y = \frac{-3}{+5}x + 2$$

Now we will graph the line in three steps as shown below. As the first step, we graph the intercept (0, 2) in Figure 41.2.

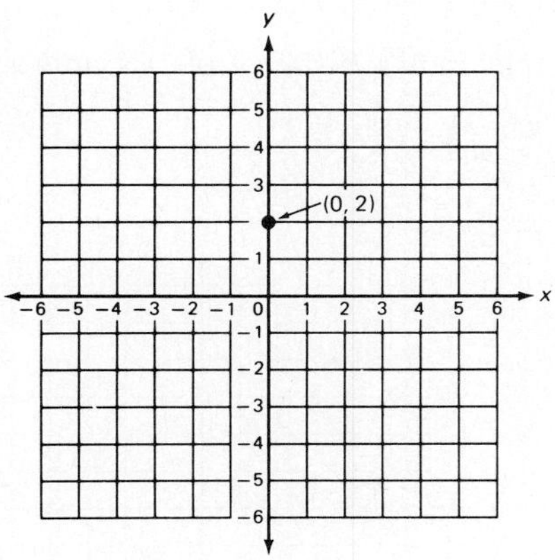

Figure 41.2

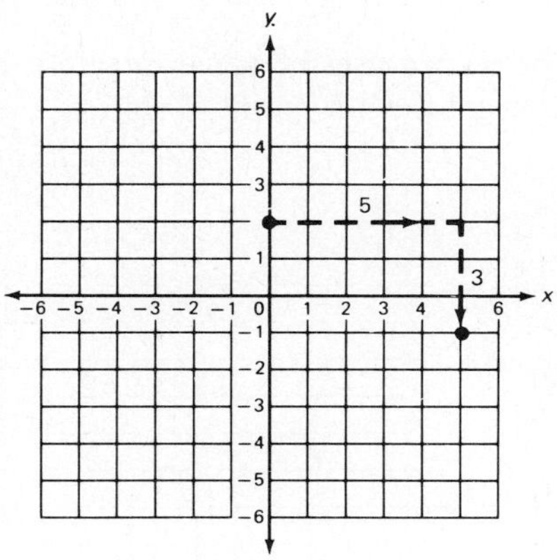

Figure 41.3

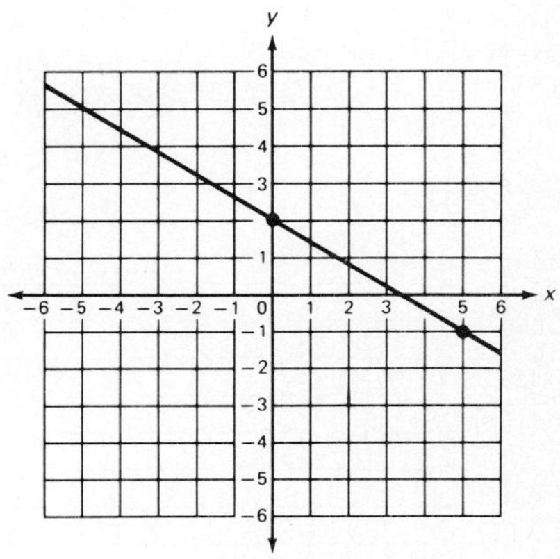

Figure 41.4

Now, **from the intercept**, we move to the right (in the positive x direction) a distance equal to the value of the denominator of the slope, which in our case is 5, and then move up or down the distance indicated by the value of the numerator of the slope, in this case -3 or down 3. We graph this new point in Figure 41.3. Now we draw the line through the two points in Fig. 41.4.

We note that the graph is the same as before but the procedure used was uncomplicated and required less time.

example 41.1.1 Use the slope-intercept method to graph the equation $x - 2y = 4$.

solution As the first step we write the equation in slope-intercept form by solving for y.

$$x - 2y = 4 \longrightarrow -2y = -x + 4 \longrightarrow 2y = x - 4 \longrightarrow y = \tfrac{1}{2}x - 2$$

Now we write the slope $\frac{1}{2}$ as a fraction with a positive denominator.

$$y = \frac{+1}{+2}x - 2$$

In the first figure we graph the intercept, $(0, -2)$. In the second figure we move from the intercept an x distance of $+2$ (to the right) and a y distance of $+1$ (up). In the third figure we draw the line through the two points.

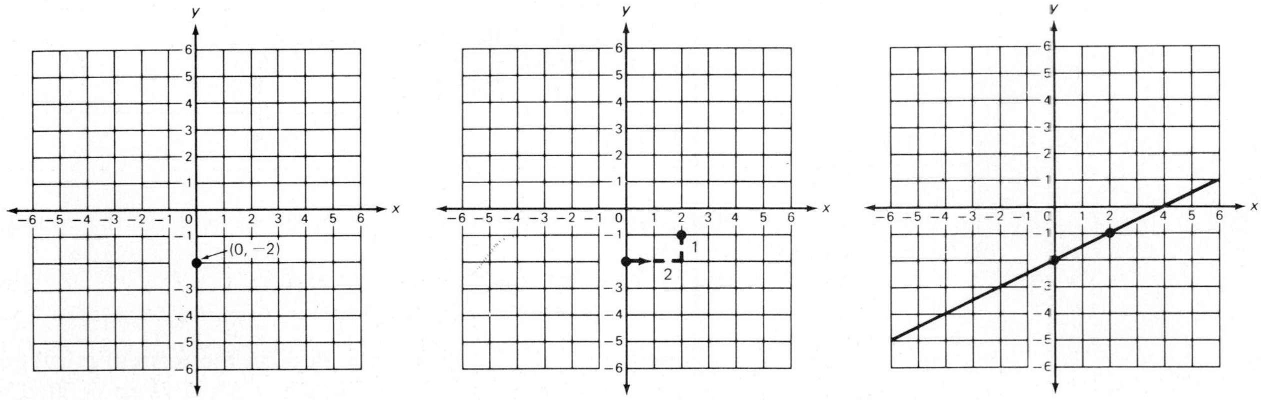

When the points are close together as in this case, it is difficult to draw the indicated line accurately. To get another point, we multiply the denominator and the numerator of the slope by a convenient integer and use the new form of the slope to get the second point. For the line under discussion, we will multiply the slope by $\frac{2}{2}$ and get

$$\frac{+1}{+2} \cdot \frac{(2)}{(2)} \longrightarrow \frac{+2}{+4}$$

In the figures below we use the same intercept but move an x distance of $+4$ and a y distance of $+2$ to get the new point.

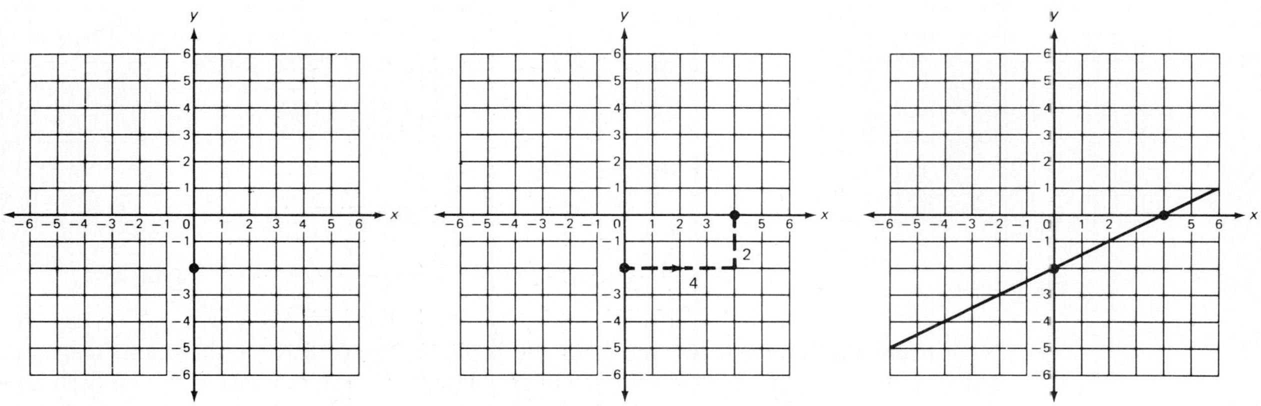

TEST YOUR UNDERSTANDING

(a) Use the slope-intercept method, to graph the equation
$$4y + 3x = 8.$$

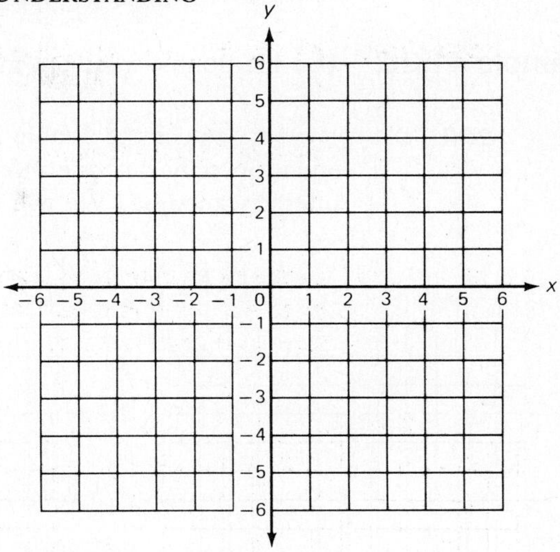

41.2 double-intercept method of graphing a line

The point where a line crosses the y-axis is called the y-intercept of the line and the point where a line crosses the x-axis is called the x-intercept of the line. In section 41.1 we used the y-intercept and the slope to enable a quick graph of the line. An even quicker method is the double-intercept method, which works if the line has two easily graphable intercepts. By letting $x = 0$, we can find the y-intercept; by letting $y = 0$, we can find the x-intercept.

example 41.2.1 Use the double-intercept method to graph the equation $2y + 3x = 6$.

solution If we let x equal zero,

$$2y + 3(0) = 6 \longrightarrow 2y = 6 \quad \text{or} \quad \mathbf{y = 3}$$

If we let y equal 0,

$$2(0) + 3x = 6 \longrightarrow \mathbf{x = 2}$$

In the accompanying figure we graph the points (0, 3) and (2, 0) and draw the line.

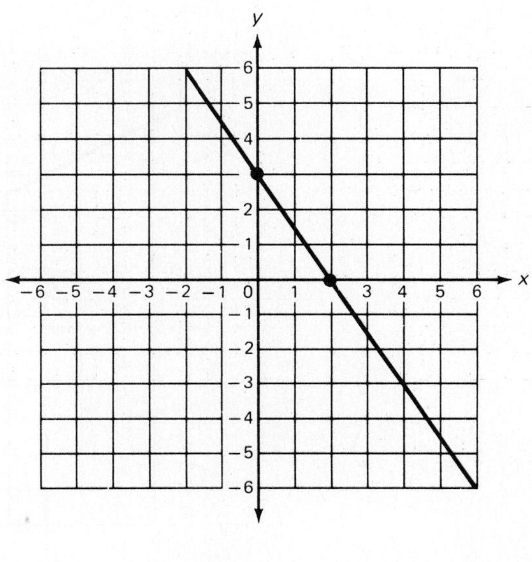

The double-intercept method will not work if the intercept is the origin or if one intercept falls off the graph.

example 41.2.2 Use the double-intercept method to graph $y = -3x$.

solution If we try to use the double-intercept method, we let $x = 0$ and find that $y = 0$. Thus this equation only has one intercept, which is the origin and thus we can't use the double-intercept method. We will complete the graph by using the slope-intercept method.

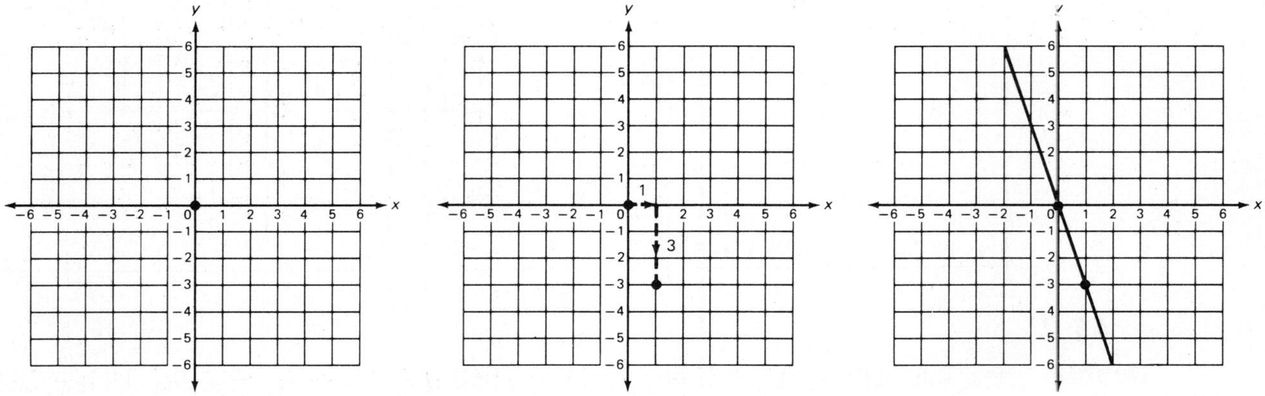

$$y = \frac{-3}{+1}x$$

example 41.2.3 Use the double-intercept method to graph the line $2y + x - 8 = 0$.

solution If $x = 0$,

$$2y + 0 - 8 = 0 \quad \text{or} \quad y = 4$$

If $y = 0$,

$$2(0) + x - 8 = 0 \quad \text{or} \quad x = 8$$

Thus the intercepts are (0, 4) and (8, 0). If we use larger graph paper or change the scale on the graphs we have been using, we can graph both points. In this case we choose to graph the point (8, 0) even though it falls outside the border of our graph paper.

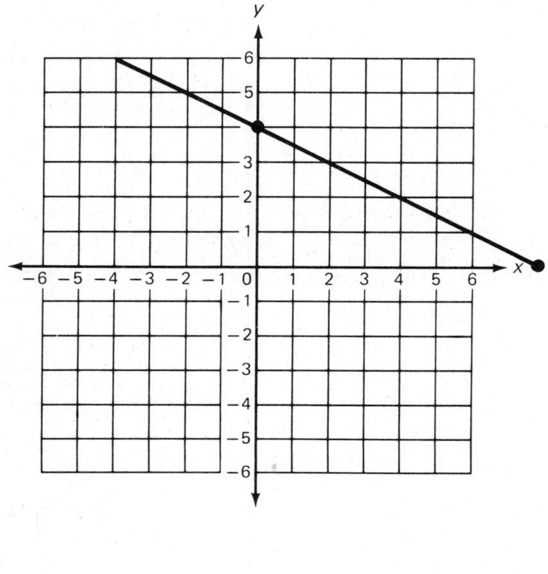

TEST YOUR UNDERSTANDING

(b) Use the double-intercept method to graph the equation

$$4y + 3x = 8.$$

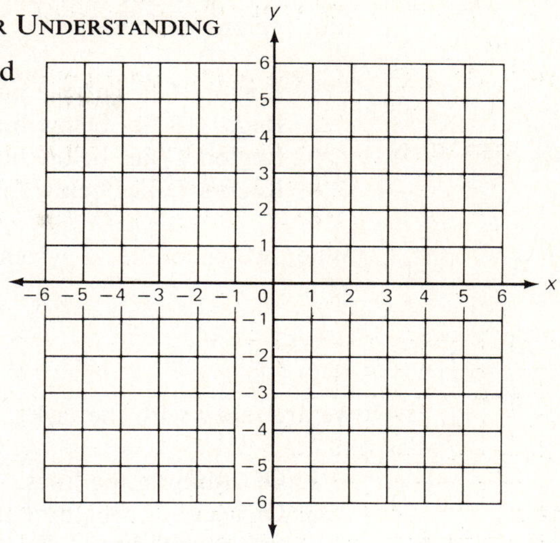

41.3 linear inequalities

A line divides the points in the plane into three disjoint subsets. These are the points on the line and the points that fall in the half planes on either side of the line. In Figure 41.5 we have graphed the line $y = x - 2$ and indicated that the coordinates of any point above the line will satisfy the inequality $y > x - 2$ and that the coordinates of any point below the line will satisfy the inequality $y < x - 2$.

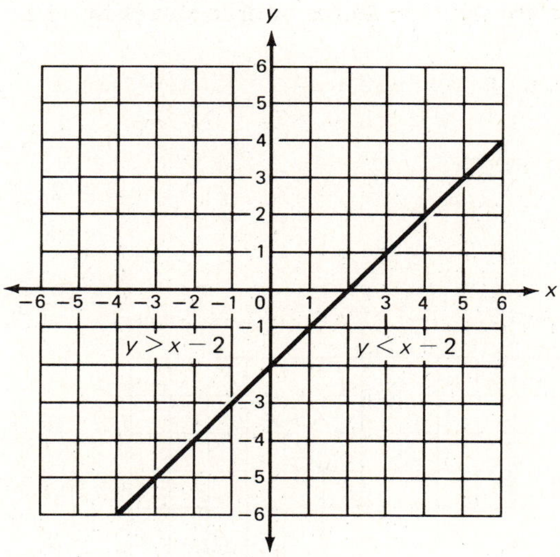

Figure 41.5

If two lines are designated, they intersect or do not intersect. If the lines intersect, they divide the points in the plane that are not on the lines into four disjoint subsets. The four subsets are indicated here by the letters A, B, C, and D.

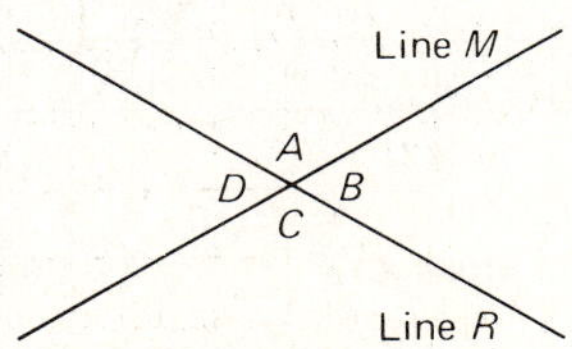

Figure 41.6

If we call the lines line M and line R as shown in Figure 41.6, we can define the four regions as follows:

1. Region A lies above both lines.
2. Region B lies below line M and above line R.
3. Region C lies below both lines.
4. Region D lies below line R and above line M.

When we encounter a system of linear inequalities that designate intersecting lines such as

$$\begin{cases} y \geq x + 2 \\ y \leq -x - 1 \end{cases}$$

we are faced with the task

1. Of graphing the lines.
2. Of deciding which of the four regions of the graph satisfy the condition stated by the inequalities.

The double-intercept method will work for both lines and give us a quick accurate graph.

For $y = x + 2$	For $y = -x - 1$
when $x = 0, y = 2 \longrightarrow (0, 2)$	when $x = 0, y = -1 \longrightarrow (0, -1)$
when $y = 0, x = -2 \longrightarrow (-2, 0)$	when $y = 0, x = -1 \longrightarrow (-1, 0)$

Next we use these ordered pairs to graph the lines as shown in Figure 41.7. We will use the test point (0, 0) for both inequalities

$$\text{For } y \geq x + 2$$
$$0 \geq 0 + 2$$
$$0 \geq 2 \quad \text{False}$$

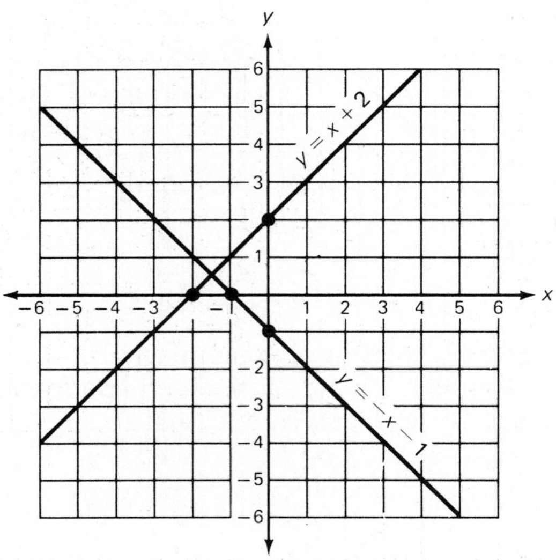

Figure 41.7

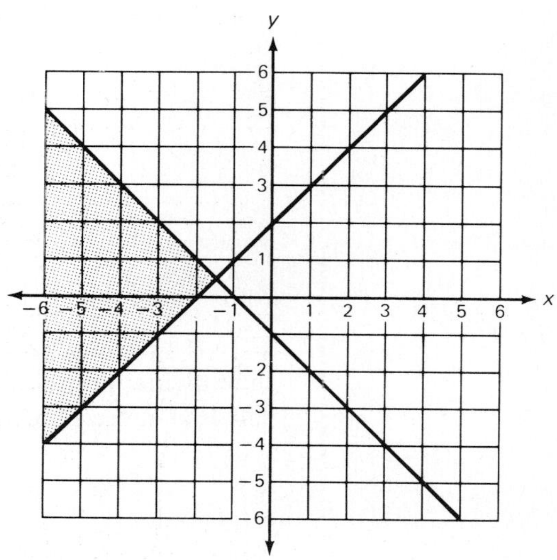

Figure 41.8

Thus the points that satisfy this inequality lie on or above the line $y = x + 2$.

$$\text{For } y \leq -x - 1$$
$$0 \leq -(0) - 1$$
$$0 \leq -1 \quad \text{True}$$

Thus the points that satisfy this inequality lie on or below the line $y = -x - 1$.
In Figure 41.8 we designate the area defined by the two inequalities.

example 41.3.1 Graph the solution to this system of linear inequalities on a rectangular coordinate system.

$$\begin{cases} 2y - x + 2 > 0 \\ y + 5x - 5 < 0 \end{cases}$$

solution First we graph the lines. We will use the double intercept method.

For $2y - x + 2 = 0$			For $y + 5x - 5 = 0$		
when $x = 0, y =$	$\longrightarrow$	$(0, -1)$	when $x = 0, y = 5$	$\longrightarrow$	$(0, 5)$
when $y = 0, x = 2$	$\longrightarrow$	$(2, 0)$	when $y = 0, x = 1$	$\longrightarrow$	$(1, 0)$

We graph the lines in the figure at the left as dotted lines to indicate that the coordinates of the points on the lines do not satisfy the stated inequalities. Again we choose the test point (0, 0).

For $2y - x + 2 > 0$	For $y + 5x - 5 < 0$
$2(0) - (0) + 2 > 0$	$(0) + 5(0) - 5 < 0$
$2 > 0$ True	$-5 < 0$ True

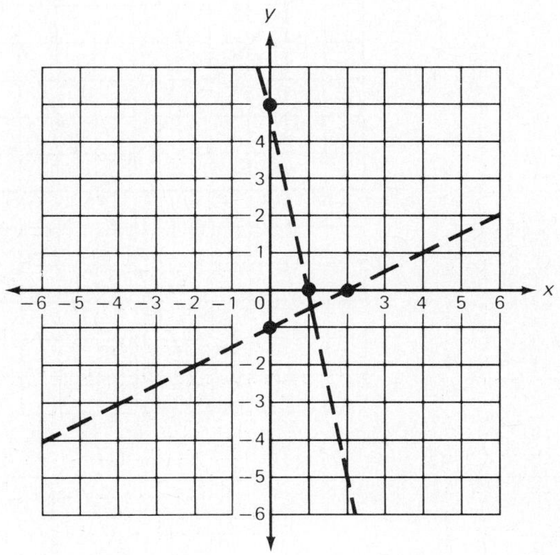

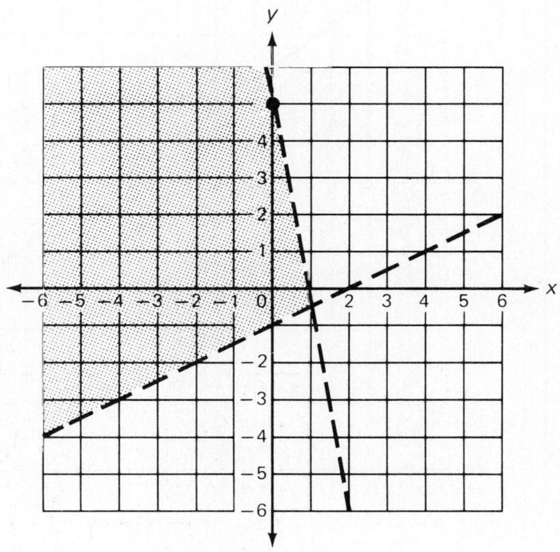

The coordinates of any of the points in the area designated in the figure at the right will satisfy both of the given linear inequalities.

TEST YOUR UNDERSTANDING

(c) Graph the solution to the system:

$$\begin{cases} y > -x + 1 \\ y < 2x - 2 \end{cases}$$

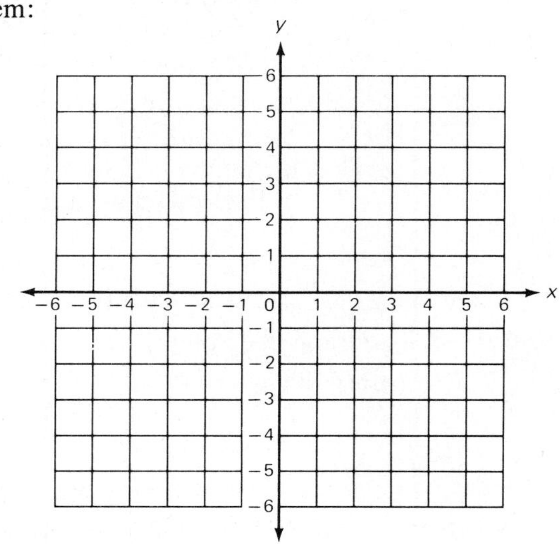

lesson checklist 41

- ☐ Slope-intercept method of graphing a line
- ☐ Double-intercept method of graphing a line
- ☐ Solution of systems of linear inequalities

ANSWERS TO TEST YOUR UNDERSTANDING

(a) and (b)

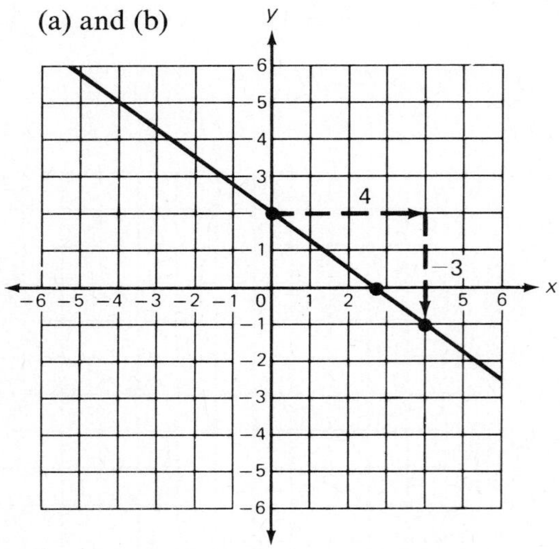

(c)

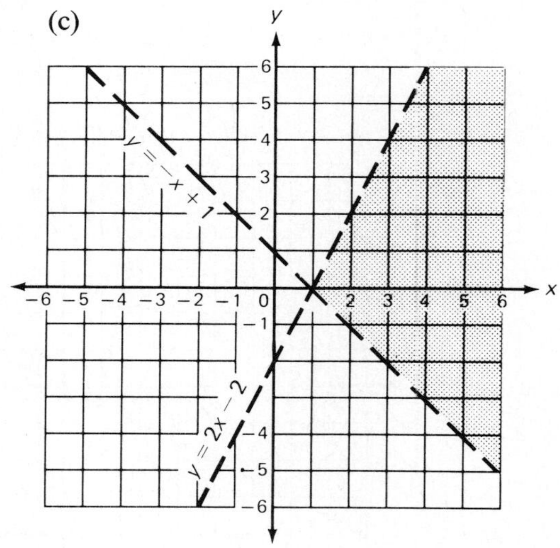

problem set 41

1. For a constant temperature, the volume of a perfect gas varies inversely with the pressure. When the pressure is 4 atmospheres, the volume of a quantity of the gas is 80 cc. At the same temperature, what is the volume of this gas at a pressure of 10 atmospheres?

2. If the pressure of a perfect gas is held constant, the volume varies directly as the temperature. At a pressure of 40 torr, the temperature of an amount of the gas was 700°K and the volume was 80 cc. At the same pressure, what is the temperature of this gas if the volume is increased to 100 cc?

3. The ratio of mosquitoes to other bugs in the swamp was 2 to 7. If there were 2,430,000 bugs and mosquitoes in the swamp, how many were mosquitoes?

4. The ratio of field goals to missed shots on the Junior High basketball team was 2 to 5. If 84 shots were taken by the team, how many field goals were made?

5. At 6 a.m. the good fairy began her trip north at 400 km/hr. Three hours later the bad fairy headed south at 700 km/hr. At what time were they 5600 kilometers apart?

6. The bad guy who ran at 6 meters per second had a 50-meter start on the good guy. How long did it take the good guy to catch the bad guy if the good guy ran at 11 meters per second?

7. Karen had pennies, Buzz had nickels, and Debby had quarters. Their total worth was $8.00 and they had 52 coins in all. How many quarters did Debby have if she had twice as many coins as Karen?

8. Find four consecutive even integers such that five times the first is 16 greater than twice the sum of the fourth and the number 4.

9. Willis had 12,500 of them before Billie Anne took 6 percent. How many of them did Billie Anne take?

*10. Graph on a rectangular coordinate system: $y \geq x + 2$, $y \leq -x + 1$

Use the quadratic formula to find the roots of:

11. $4x^2 - x - 8 = 0$

12. $x^2 + 2x - 7 = 0$

Solve by completing the square:

13. $2x^2 - x - 5 = 0$

Graph on a rectangular coordinate system:

14. $y \leq 2x - 2$

Solve:

15. $\dfrac{x + 4}{1} = \dfrac{15x}{x - 4}$

16. $N_P + N_N + N_D = 22$
$N_P + 5N_N + 10N_D = 165$
$N_D = 3N_P$

Graph on a number line:

17. $-3 < x - 3 < 3$, $D = \{\text{Reals}\}$

18. Find the number $\frac{2}{9}$ of the way from $\frac{4}{5}$ to $\frac{7}{8}$.

19. Show that $.006\overline{123}$ is a rational number by writing it as a quotient of integers.

20. $x + \dfrac{1}{c} - \dfrac{y}{a} = m$, find c.

Solve:

21. $\sqrt{x - 3} = 7$

---

*The problems marked with an asterisk were worked out in this lesson.

22. Find the equation of the line:

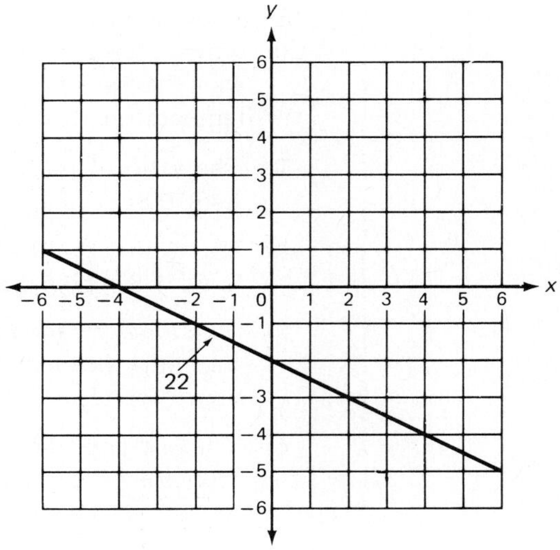

23. Find the equation of the line that goes through $(-2, 4)$ and is parallel to $-x - 2y = 3$.

24. Find the equation of the line through $(3, -1)$ and $(-3, 3)$.

25. Given that $f(x) = x^2 + 11$, $D = \{\text{Reals}\}$ find $g(1)$
$g(x) = x - 5$, $D = \{\text{Reals}\}$

26. Which of the following designate functions? Use the vertical line test as necessary.

(a)

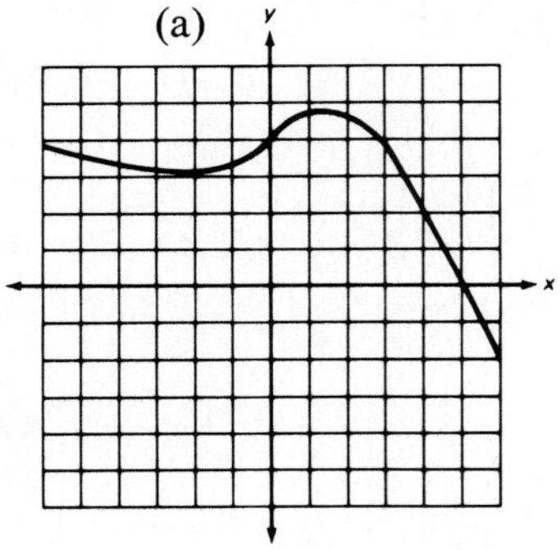

(b)

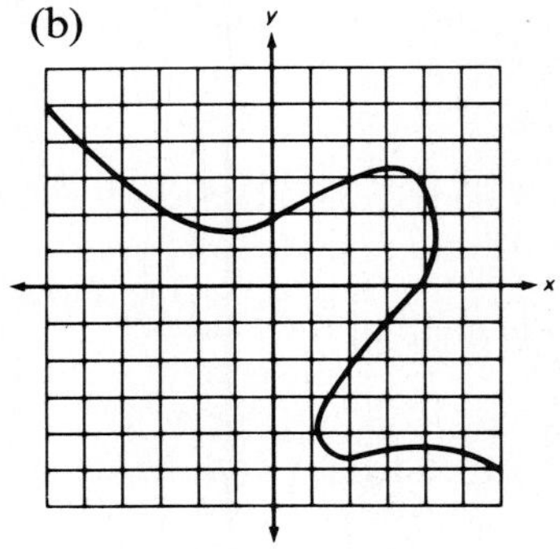

(c)

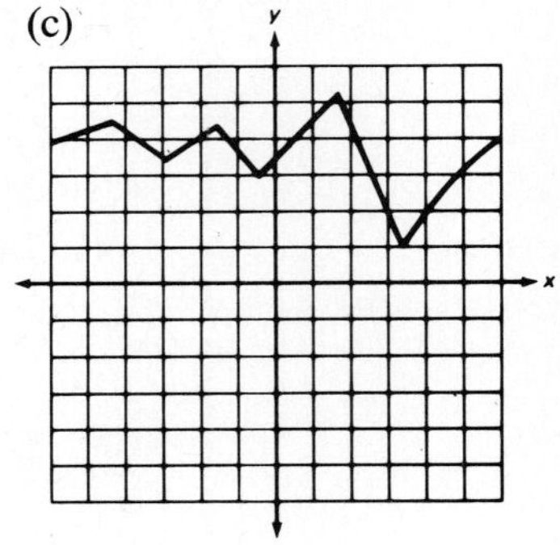

(d)
(1, 3), (7, 9), (8, 2)
(9, 7), (3, 1), (2, 8)

(e)

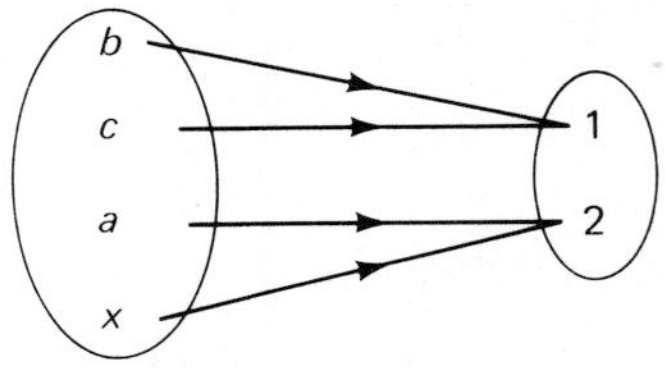

(f)

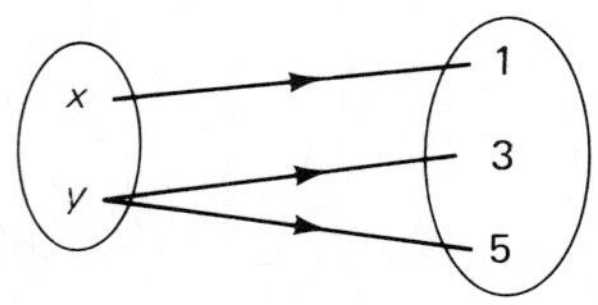

Graph on a number line:

27. $-7 - x < 2$, $D = \{\text{Reals}\}$

28. $|y| - 5 < 0$, $D = \{\text{Reals}\}$

29. Find p.

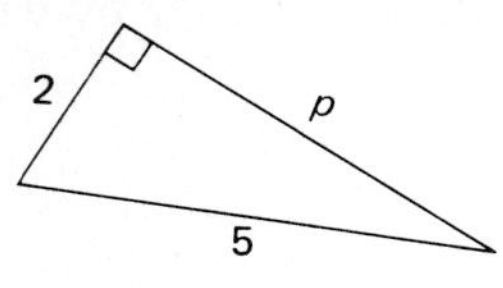

Simplify:

30. $\sqrt{12} - 8\sqrt{6} + 2\sqrt{3}$

31. $2\sqrt{2}(\sqrt{3} + \sqrt{12})$

32. Divide $3x^3 + 11x^2 - 8x + 9$ by $x - 3$.

Solve:

33. $x^2 = -2x + 3$

Simplify:

34. $\dfrac{x^2 - 4}{y^2 - 1} \cdot \dfrac{y - 1}{x + 2}$

35. $\dfrac{(7000 \times 10^{-5})(.08 \times 10^4)}{(.00028 \times 10^{-12})}$

36. Use the letters a and b to illustrate the commutative property of addition.

Graph on a rectangular coordinate system:

37. $\begin{cases} y - 2x - 2 \le 0 \\ y + 2 < 0 \end{cases}$

Simplify:

38. $\dfrac{a + \dfrac{1}{b}}{y - \dfrac{a}{b}}$

39. $\dfrac{\dfrac{c}{x} - 1}{\dfrac{c}{x} + 1}$

Add:

40. $\dfrac{3}{a^2} - \dfrac{1}{a + b}$

41. $\dfrac{1}{a^2x} - \dfrac{m}{ax} + b$

Add like terms:

42. $\dfrac{xzy^2}{a} + \dfrac{3az}{xy^2} - \dfrac{2a^{-1}yz}{x^{-1}y^{-1}}$

Multiply:

43. $3x^2yz^2\left(\dfrac{1}{x^2y^0z^2} + \dfrac{2y^{-1}z^2}{x^{-2}}\right)$

Simplify:

44. $\dfrac{a^2b^0c^{-3}x}{a^3b^{-2}c^{-2}x^2}$

Evaluate:

45. $a + (b - a) - (-b)$ if $a = -3$, $b = 2$

Add:

46. $\dfrac{x + 6}{x - 6} - \dfrac{x}{x^2 - 36}$

appendix one

Supplementary factoring

A-1.1 supplementary factoring problems

problem set A-1.1a Trinomials with a lead coefficient of one.

Factor. Remember that as a first step it is sometimes necessary to rewrite trinomials in descending powers of the variable.

1. $x^2 + 6x - 16$	**2.** $6x + x^2 + 9$	**3.** $6x - 27 + x^2$
4. $x^2 - 6x - 16$	**5.** $x^2 + 9 - 6x$	**6.** $x^2 - 6x - 27$
7. $-20 - p + p^2$	**8.** $p^2 - p - 12$	**9.** $-6 + p^2 - p$
10. $p^2 + p - 20$	**11.** $p - 12 + p^2$	**12.** $p^2 + p - 6$
13. $x^2 + 33 + 14x$	**14.** $p^2 - 13p + 36$	**15.** $m^2 - m - 30$
16. $11n + n^2 + 18$	**17.** $x^2 + 27 + 12x$	**18.** $x^2 + 90 - 19x$
19. $x^2 + x - 132$	**20.** $a^2 + 90 - 47a$	**21.** $10m + m^2 + 16$
22. $-48 - 8n + n^2$	**23.** $y^2 - 15y + 56$	**24.** $p^2 - 55 - 6p$
25. $12t + 35 + t^2$	**26.** $y^2 + 50 + 51y$	**27.** $x^2 + 12 - 7x$
28. $77 - 18r + r^2$	**29.** $m^2 + 21m + 90$	**30.** $55 + v^2 + 16v$
31. $s^2 + 15 + 8s$	**32.** $a^2 - 40 + 3a$	**33.** $b^2 - 24 - 5b$
34. $-63 - 2h + h^2$	**35.** $-30 - 13x + x^2$	**36.** $w^2 + 22 - 13w$
37. $-110 + k^2 + k$	**38.** $d^2 - 16 - 6d$	**39.** $-36 + 9c + c^2$
40. $x^2 - 8x - 33$	**41.** $-99 + r^2 - 2r$	**42.** $-90 - x + x^2$

43. $w^2 - 96 - 4w$
44. $-15 - 2p + p^2$
45. $w^2 + 8w + 7$
46. $x^2 + 25 + 10x$
47. $x^2 + 9 + 10x$
48. $x^2 + 16 + 10x$
49. $x^2 + 25 - 10x$
50. $p^2 + 9 - 10p$
51. $x^2 + 16 - 8x$

problem set A-1.1b Difference of two squares

Factor. Rearrange as necessary before factoring.

1. $x^2 - y^2$
2. $-y^2 + x^2$
3. $9x^2y^2 - z^2$
4. $-z^2 + 16x^2y^2$
5. $-81 + r^2$
6. $c^2 - d^2$
7. $-a^2 + 121$
8. $36 - a^2b^2$
9. $4z^2 - 1$
10. $-9 + x^2$
11. $64x^2 - y^2$
12. $16 - 9y^2$
13. $25r^2 - 9s^2$
14. $-25p^2 + 4$
15. $-1 + 4x^2$
16. $-r^2s^2 + 25$
17. $36 - 49p^2$
18. $4x^2y^2 - r^2$
19. $m^2 - 16n^2$
20. $.09x^2 - 4$
21. $1 - 9n^2$
22. $c^2 - .64$
23. $-d^2 + .81$
24. $-\frac{1}{9} + x^2$
25. $16a^2 - 25b^2$
26. $-25a^2y^2 + 16x^2$
27. $9a^2x^2 - p^2$

problem set A-1.1c Trinomials and difference of two squares with and without common factors

Factor. As the first step rearrange the expressions as necessary and then begin by factoring the greatest common factor if one exists.

1. $4x^2 - y^2$
2. $8x^2 - 2y^2$
3. $12a^2x^2 - 3a^2y^2$
4. $x^2 - x - 6$
5. $abx^2 - abx - 6ab$
6. $(x - 1)x^2 - (x - 1)x - 6(x - 1)$
7. $-a^2 + y^2$
8. $-9a^2 + 9y^2$
9. $-9a^2 + 36y^2$
10. $(x - a)x^2 - (x - a)16$
11. $(x - a)p^2 - (x - a)p - 20(x - a)$
12. $x^4y^2 - 9x^4$
13. $6x^3 - 12x^2 - 90x$
14. $3x^2 + 45 + 24x$
15. $10x - 75 + 5x^2$
16. $-a^2y^2 + a^4$
17. $-a^2y^2 + 4a^4$
18. $6m^2 + 126m + 540$
19. $-15b^2 - 72b + 3b^3$
20. $-x^2 - 3x + 40$
21. $(x + 2)x^2 - 5x(x + 2) - 24(x + 2)$
22. $(x - a)x^2 - 6x(x - a) - 27(x - a)$
23. $-s^3 - 8s^2 - 15s$
24. $-24w^2 - 44 - 26w$
25. $p^2 + 9p + 20$
26. $-p^2 - 9p - 20$
27. $-2p^2 - 18p - 40$
28. $-60 - 27p - 3p^2$
29. $ax^2 + 6a + 7ax$
30. $p^2x^3 + 9px - 10px^2$
31. $kpx^2 + 25kp - 10kpx$
32. $(a + b)x^2 + 2x(a + b) - 24(a + b)$
33. $(x + 2y)x^2 - 7x(x + 2y) + 10(x + 2y)$

34. $-4x^2 - 4 - 8x$

35. $-3x^2y^2 + 75$

36. $4ap + 5apx + apx^2$

37. $5 - 20x^2$

38. $a^2x^3 + 4a^2x - 4a^2x^2$

39. $7x^2 - 42 - 7x$

40. $-25x^2 + m^2p^2$

41. $5m^2 - 80n^2$

42. $a^2x^2 - m^2p^2$

43. $-16 + 49x^2y^2$

44. $-20 + 5x^2$

45. $63 + 3x^2 + 30x$

46. $(x - 1)x^2 + x(x - 1) - 12(x - 1)$

47. $(p + c)x^2 - (p + c)x - 12(p + c)$

48. $-30x - 3x^2 - 63$

49. $-3x^2 + 9x + 30$

50. $15x - 5x^2 + 50$

51. $-2x^2 + 20x - 50$

52. $-75 + 30x - 3x^2$

53. $21a^2 + 10a^2x + a^2x^2$

54. $-8x - 2x^2 + 42$

55. $6a^2x + 3a^2x^2 - 24a^2$

56. $8mx - 7m - mx^2$

57. $-16 + 49x^2$

58. $-32 + 98x^2$

59. $32a - 98x^2a$

60. $16x^2y^2 - 9m^2$

61. $16ax^2y^2 - 9am^2$

62. $n^2 - 16 - 6n$

63. $2 - 4z + 2z^2$

64. $-9x^2 + 6x + 3x^3$

65. $4x^2 - 48 - 16x$

66. $-5x^3 - 25x + 30x^2$

67. $16x - 2x^2 - 14$

68. $-ax^2 + 11ax - 10a$

69. $-ax^2 + 2a + ax$

70. $bx^2 - 3b - 2bx$

71. $-4c + cx^2 - 3cx$

72. $-dx^2 + 4dx + 5d$

73. $6m^2 - 36 - 30m$

74. $7kx + 8k - kx^2$

75. $9x + 10 - x^2$

76. $-2p^2 + p^2x + p^2x^2$

77. $-3k^2 + 12$

78. $10a + 3ax - ax^2$

79. $bx^2 - 14b - 5bx$

80. $-kx^2 + 18k + 7kx$

81. $x^3 - 9x^2 - 22x$

82. $-8p^2 - p^3 - 15p$

83. $4x^2 - 2x^3 + 16x$

84. $-30x - 63 - 3x^2$

85. $4ax - ax^2 + 21a$

86. $-a^3 - 24a + 5a^2$

87. $3by - 4b + by^2$

88. $-2x^2 - 8 - 10x$

89. $-8c^2 - 7c^2x + c^2x^2$

90. $-8b - 6b^2 - b^3$

91. $-3x^2 + 36 - 3x$

92. $12ax + 7ax^2 + ax^3$

93. $-3x^2 + 48$

94. $-a^2k^3 + 16k$

95. $-5x^2 + 80$

96. $-x^2a^2 - 7a^2x - 12a^2$

97. $b^2x^2 - 20b^2 - b^2x$

98. $acx^2 - 20ac + acx$

A-1.2 trinomials of the form $x^2 + 8xy + 15y^2$

To be able to factor quickly and effectively, it is necessary to memorize the forms of different types of factorable polynomials. If we are asked to factor the polynomial

$$16x^2 - 9$$

we recognize that this form is the difference of two squares and can be factored as

$$(4x + 3)(4x - 3)$$

If we had not recognized the form of the expression, we would have had great difficulty in factoring. Another form of a factorable polynomial that we must learn to recognize is

$$x^2 + 4xy - 12y^2$$

This polynomial begins with an x^2 term, ends with a y^2 term, and the middle term is xy. These polynomials also have the same format.

$$x^2 + 5ax + 6a^2 \qquad y^2 - xp - 6p^2 \qquad a^2 - 5pax + 6p^2x^2$$

We factor these polynomials by recognizing the format of the variables and realizing that the constants are the same as those in the more familiar trinomial that has only one variable.

$$\begin{array}{ll} x \; +3 & x \; +3a \\ \underline{x \; +2} & \underline{x \; +2a} \\ x^2 + 3x & x^2 + 3ax \\ \underline{\quad + 2x + 6} & \underline{\quad + 2ax + 6a^2} \\ x^2 + 5x + 6 & x^2 + 5ax + 6a^2 \end{array}$$

We note that the constants in the multiplications are the same and that the only difference is that the second multiplication also involves the variable a.

example A-1.2a Factor $x^2 - xp - 6p^2$

solution We recognize the form of the polynomial is similar to $x^2 - x - 6$ which factors as

$$(x - 3)(x + 2)$$

Thus the polynomial $x^2 - xp - 6p^2$ factors as

$$(x - 3p)(x + 2p)$$

We check by multiplying:

$$(x - 3p)(x + 2p) = x^2 + 2xp - 3xp - 6p^2 = x^2 - xp - 6p^2 \quad \text{Check}$$

example A-1.2b Factor $a^2 - 5pax + 6p^2x^2$

solution This expression is similar to the expression $a^2 - 5a + 6$ which equals $(a - 3)(a - 2)$. In the polynomial $a^2 - 5pax + 6p^2x^2$, the variables are a and px so we factor as

$$a^2 - 5pax + 6p^2x^2 = (a - 3px)(a - 2px)$$

problem set A-1.2

Factor. Rearrange the terms as necessary.

1. $x^2 + 6xp - 16p^2$
2. $k^2 + 6ka + 9a^2$
3. $p^2 + 6xp - 27x^2$
4. $-px + p^2 - 20x^2$
5. $k^2 - 12m^2 - km$
6. $m^2 - 27z^2 - 6mz$
7. $pm - 20m^2 + p^2$
8. $a^2 - 12m^2 + am$
9. $-6k^2m^2 + pkm + p^2$
10. $11nb + n^2 + 18b^2$
11. $x^2 + 27a^2 + 12ax$
12. $x^2a^2 + 90b^2 - 21xab$

13. $-48k^2 - 8nk + n^2$
14. $y^2p^2 - 15ypx + 56x^2$
15. $p^2 - 55a^2 - 6pa$
16. $77n^2 - 18rn + r^2$
17. $m^2 + 21mn + 90n^2$
18. $55x^2p^2 + v^2 + 16vpx$
19. $-63a^2 - 2ka + k^2$
20. $-30x^2y^2 - 13axy + a^2$
21. $w^2 + 22k^2 - 13wk$
22. $x^2y^2 - 8xya - 33a^2$
23. $-99x^2 + r^2 - 2rx$
24. $-90p^2 - xp + x^2$
25. $x^2 + 25p^2r^2 + 10xpr$
26. $x^2 + 9a^2y^2 + 10xay$
27. $x^2 + 16p^2m^2 + 10pmx$
28. $x^2 + 25a^2k^2 - 10xak$
29. $p^2 + 9m^2 - 10pm$
30. $x^2 + 16a^2m^2 - 8xam$
31. $x^2m^2 + 33p^2 + 14xmp$
32. $p^2 - 12px + 36x^2$
33. $m^2x^2 - mxa - 30a^2$
34. $x^2 + xp - 132p^2$
35. $a^2x^2 + 90m^2 - 47amx$
36. $10mxp + m^2 + 16x^2p^2$
37. $12tm + 35m^2 + t^2$
38. $y^2x^2 + 51a^2 + 52axy$
39. $x^2 + 12m^2n^2 - 7mxn$

A-1.3 factoring by grouping

Sometimes it is necessary to factor polynomials that are the product of two binomials such as

$$(x - 4)(a + b)$$

When we multiply these binomials and look at the product

$$(x - 4)(a + b) = xa + xb - 4a - 4b$$

we see that it consists of 4 terms. Thus again, we encounter a factorable polynomial that can be factored only if we recognize the form.

example A-1.3a Factor $xya - 4a + xyb - 4b$

solution We recognize the four term format and note that the first two terms have a as a factor and that the last two terms have b as a factor. We begin by using parentheses to **group** these terms.

$$(xya - 4a) + (xyb - 4b)$$

Now we factor a from the first group and b from the second.

$$a(xy - 4) + b(xy - 4)$$

Lastly, we recognize that both of these terms have $xy - 4$ as a factor, so we factor this expression.

$$(xy - 4)(a + b)$$

example A-1.3b Factor $ac + 2ad + 2bc + 4bd$

solution We recognize the form and note that the first and third terms have a common factor of c, and that the second and fourth terms have a common factor of $2d$; so we rearrange the terms and use parentheses as follows:

$$(ac + 2bc) + (4bd + 2ad)$$

Now we factor these terms as

$$c(a + 2b) + 2d(2b + a)$$

and complete the problem by factoring $2b + a$. The final result is

$$(a + 2b)(c + 2d)$$

There is no set procedure for rearranging the terms so that the expression can be factored. Adeptness comes after much practice and frustration. Hopefully the problems in this problem set will be of help in improving your ability to do this type of factoring. The first six do not require rearranging, but many of the rest do require rearranging as the first step.

problem set A-1.3

Factor. Rearrange the terms as required as the first step.

1. $ac - ad + bc - bd$
2. $ab + 4a + 2b + 8$
3. $ab + ac + xb + xc$
4. $2mx - 3m + 2pcx - 3pc$
5. $4k - kxy + 4pc - pcxy$
6. $ac - axy + dc - dxy$
7. $ab + 15 + 5a + 3b$
8. $ay + xy + ac + xc$
9. $3mx - 2p + 3px - 2m$
10. $kx - 15 - 5k + 3x$
11. $xpc + 4b + bpc + 4x$
12. $acb - ack + 2b - 2k$
13. $p^2c - ab + p^2b - ac$
14. $ax^2 - ca + 3x^2 - 3c$
15. $2y + mx^3 + my + 2x^3$
16. $4ab + 4c + abx + cx$
17. $km^2 + 2c - 2m^2 - kc$
18. $6a - xya - xyb + 6b$
19. $abx - 2yc + xc - 2yab$
20. $4xm + abn - abm - 4xn$
21. $pa + 2qc + pc + 2qa$
22. $2ka - 15 + 10k - 3a$
23. $6x - 2py + 12y - px$
24. $za - 2kc + zc - 2ka$
25. $4mx - mxc + 4b - bc$
26. $axk + aky + 2x + 2y$
27. $2am - 4m + 6a - 12$
28. $6xy - 18 + pxy - 3p$
29. $2m - 3kp - 3km + 2p$
30. $ma - 2cp - 2m + cpa$
31. $xa - 2by - bx + 2ya$
32. $3ak + 3ap + 2k + 2p$
33. $amx - bmx + 4a - 4b$
34. $cmx - 2pcx + 6p - 3m$
35. $2kp - ka - 6p + 3a$
36. $xa - xb - a + b$

A-1.4 advanced trinomial factoring

The text concentrates on factoring trinomials such as

$$x^2 - x - 6$$

in which the coefficient of the x^2 term is 1 and on factoring trinomials that can be reduced to this form by factoring a common factor. Trinomials whose lead coefficient is not 1 can be formed by multiplying binomials as we see here.

$$\begin{array}{rr} 2x & +3 \\ x & -5 \\ \hline 2x^2 + 3x & \\ -10x & -15 \\ \hline 2x^2 - 7x & -15 \end{array} \qquad \begin{array}{rr} 3x & +2 \\ 2x & -3 \\ \hline 6x^2 + 4x & \\ -9x & -6 \\ \hline 6x^2 - 5x & -6 \end{array}$$

We note that the first term of the trinomial is the product of the first two terms of the binomials and that the last term of the trinomial is the product of the last terms of the binomials, **but alas, the coefficient of the middle term of the trinomial is no longer the sum of the last two terms of the binomials.** The middle term is the sum of the product of the first term of the first binomial and the last term of the second binomial and the product of the last term of the first binomial and the first term of the second binomial. It is easier to see this if we write the original indicated multiplication in horizontal form and note that the middle term is the sum of the products of the means and the extremes.†

extremes

$$(2x + 3)(x - 5) = 2x^2 - 7x - 15$$

means

extremes

$$(3x + 2)(2x - 3) = 6x^2 - 5x - 6$$

means

product of means	=	$3x$
product of extremes	=	$-10x$
sum	=	$-7x$

product of means	=	$4x$
product of extremes	=	$-9x$
sum	=	$-5x$

example A-1.4 Factor $-7x - 15 + 2x^2$

solution We begin by writing the trinomial in descending powers of the variable

$$2x^2 - 7x - 15$$

Now to factor $2x^2 - 7x - 15$, we note that the product of the first terms of the binomials is $2x^2$; the product of the last terms of the binomials is -15; and the middle term is the sum of the products of the means and extremes. Since the term $2x^2$ is the product of the first terms of the binomial we write

$$(2x \qquad)(x \qquad)$$

Now, the pairs of integral factors of -15 are $(+15, -1)$, $(-15, 1)$, $(5, -3)$ and $(-5, 3)$. Now we must try each pair **twice** and see what middle term will result in each case.

for $(15, -1)$	$(2x + 15)(x - 1)$	middle term is $13x$
	$(2x - 1)(x + 15)$	middle term is $29x$
for $(5, -3)$	$(2x + 5)(x - 3)$	middle term is $-x$
	$(2x - 3)(x + 5)$	middle term is $7x$
for $(-15, 1)$	$(2x - 15)(x + 1)$	middle term is $-13x$
	$(2x + 1)(x - 15)$	middle term is $-29x$
for $(-5, 3)$	$(2x - 5)(x + 3)$	middle term is x
	$(2x + 3)(x - 5)$	middle term is $-7x$

†Mathematicians sometimes use the word "mean" to mean middle and the word "extreme" to mean end. Thus the mean terms in the multiplication shown are the middle terms, and the extreme terms are the end terms.

We will use the last entry because the sum of the products of the means and the extremes is $-7x$. Thus we see that $2x^2 - 7x - 15$ can be factored over the integers as $(2x + 3)(x - 5)$.

problem set A-1.4

Factor the following trinomials. Always begin by writing the trinomial in descending powers of the variable and then factoring the greatest common factor if one exists.

1. $-7x - 15 + 2x^2$
2. $-14x - 30 + 4x^2$
3. $-5x^2 - 2x + 3x^3$
4. $-10x - 4 + 6x^2$
5. $8x + 4 + 3x^2$
6. $24x^2 + 9x^3 + 12x$
7. $2mp^2 - 3pm - 5m$
8. $8 + 18x + 4x^2$
9. $8x^2 + 6x + 3x^3$
10. $-8mx + 10m^2x - 2x$
11. $2z^2 + 13z + 15$
12. $28s^2 + 30s + 6s^3$
13. $12px^2 - 39px + 9p$
14. $-11a + 15 + 2a^2$
15. $14x^2 + 12x + 4x^3$
16. $14ax + 6ax^2 + 4a$
17. $x^2 - 15x + 2x^3$
18. $4 + 12x + 8x^2$
19. $11m^2 + 2m + 5m^3$
20. $16s^2 + 6s^3 + 10s$
21. $17x^2 + 4x + 4x^3$
22. $9xm + 2x^2m + 9m$
23. $6b + 20b^2 + 6b^3$
24. $-14k + 6k^2 - 12$
25. $-3p + 6p^2 - 30$
26. $-8x - 14x^2 + 4x^3$
27. $32mk + 10km^2 + 6k$
28. $11ps + 3ps^2 + 10p$
29. $-2x + 8x^3 + 6x^2$
30. $12n + 11n^2 + 2n^3$

answers to problem sets

answers to problem set A-1.1a

1. $(x + 8)(x - 2)$
2. $(x + 3)(x + 3)$
3. $(x + 9)(x - 3)$
4. $(x - 8)(x + 2)$
5. $(x - 3)(x - 3)$
6. $(x - 9)(x + 3)$
7. $(p - 5)(p + 4)$
8. $(p - 4)(p + 3)$
9. $(p - 3)(p + 2)$
10. $(p + 5)(p - 4)$
11. $(p + 4)(p - 3)$
12. $(p + 3)(p - 2)$
13. $(x + 3)(x + 11)$
14. $(p - 4)(p - 9)$
15. $(m - 6)(m + 5)$
16. $(n + 2)(n + 9)$
17. $(x + 3)(x + 9)$
18. $(x - 10)(x - 9)$
19. $(x + 12)(x - 11)$
20. $(a - 2)(a - 45)$
21. $(m + 2)(m + 8)$
22. $(n - 12)(n + 4)$
23. $(y - 8)(y - 7)$
24. $(p - 11)(p + 5)$
25. $(t + 5)(t + 7)$
26. $(y + 50)(y + 1)$
27. $(x - 4)(x - 3)$
28. $(r - 7)(r - 11)$
29. $(m + 15)(m + 6)$
30. $(v + 5)(v + 11)$
31. $(s + 5)(s + 3)$
32. $(a + 8)(a - 5)$
33. $(b - 8)(b + 3)$
34. $(h - 9)(h + 7)$
35. $(x - 15)(x + 2)$
36. $(w - 2)(w - 11)$
37. $(k + 11)(k - 10)$
38. $(d - 8)(d + 2)$
39. $(c + 12)(c - 3)$
40. $(x - 11)(x + 3)$
41. $(r - 11)(r + 9)$
42. $(x - 10)(x + 9)$

43. $(w - 12)(w + 8)$
44. $(p - 5)(p + 3)$
45. $(w + 7)(w + 1)$
46. $(x + 5)(x + 5)$
47. $(x + 9)(x + 1)$
48. $(x + 2)(x + 8)$
49. $(x - 5)(x - 5)$
50. $(p - 1)(p - 9)$
51. $(x - 4)(x - 4)$

answers to problem set A-1.1b

1. $(x + y)(x - y)$
2. $(x + y)(x - y)$
3. $(3xy + z)(3xy - z)$
4. $(4xy + z)(4xy - z)$
5. $(r + 9)(r - 9)$
6. $(c + d)(c - d)$
7. $(11 + a)(11 - a)$
8. $(6 - ab)(6 + ab)$
9. $(2z + 1)(2z - 1)$
10. $(x + 3)(x - 3)$
11. $(8x + y)(8x - y)$
12. $(4 + 3y)(4 - 3y)$
13. $(5r + 3s)(5r - 3s)$
14. $(2 + 5p)(2 - 5p)$
15. $(2x + 1)(2x - 1)$
16. $(5 + rs)(5 - rs)$
17. $(6 - 7p)(6 + 7p)$
18. $(2xy + r)(2xy - r)$
19. $(m - 4n)(m + 4n)$
20. $(.3x + 2)(.3x - 2)$
21. $(1 - 3n)(1 + 3n)$
22. $(c + .8)(c - .8)$
23. $(.9 + d)(.9 - d)$
24. $(x + \frac{1}{3})(x - \frac{1}{3})$
25. $(4a + 5b)(4a - 5b)$
26. $(4x + 5ay)(4x - 5ay)$
27. $(3ax + p)(3ax - p)$

answers to problem set A-1.1c

1. $(2x + y)(2x - y)$
2. $2(2x + y)(2x - y)$
3. $3a^2(2x + y)(2x - y)$
4. $(x - 3)(x + 2)$
5. $ab(x - 3)(x + 2)$
6. $(x - 1)(x - 3)(x + 2)$
7. $(y + a)(y - a)$
8. $9(y + a)(y - a)$
9. $9(2y + a)(2y - a)$
10. $(x - a)(x + 4)(x - 4)$
11. $(x - a)(p - 5)(p + 4)$
12. $x^4(y + 3)(y - 3)$
13. $6x(x - 5)(x + 3)$
14. $3(x + 5)(x + 3)$
15. $5(x + 5)(x - 3)$
16. $a^2(a + y)(a - y)$
17. $a^2(2a + y)(2a - y)$
18. $6(m + 15)(m + 6)$
19. $3b(b - 8)(b + 3)$
20. $(-x - 8)(x - 5)$
21. $(x + 2)(x - 8)(x + 3)$
22. $(x - a)(x - 9)(x + 3)$
23. $-s(s + 3)(s + 5)$
24. $-2(w + 2)(w + 11)$
25. $(p + 5)(p + 4)$
26. $(-p - 5)(p + 4)$
27. $-2(p + 4)(p + 5)$
28. $-3(p + 5)(p + 4)$
29. $a(x + 6)(x + 1)$
30. $px(x - 9)(x - 1)$
31. $kp(x - 5)(x - 5)$
32. $(a + b)(x + 6)(x - 4)$
33. $(x + 2y)(x - 5)(x - 2)$
34. $-4(x + 1)(x + 1)$
35. $3(5 + xy)(5 - xy)$
36. $ap(x + 4)(x + 1)$
37. $5(1 - 2x)(1 + 2x)$
38. $a^2x(x - 2)(x - 2)$
39. $7(x - 3)(x + 2)$
40. $(mp + 5x)(mp - 5x)$
41. $5(m - 4n)(m + 4n)$
42. $(ax + mp)(ax - mp)$
43. $(7xy + 4)(7xy - 4)$
44. $5(x + 2)(x - 2)$

45. $3(x + 3)(x + 7)$
46. $(x - 1)(x + 4)(x - 3)$
47. $(p + c)(x - 4)(x + 3)$
48. $-3(x + 7)(x + 3)$
49. $-3(x - 5)(x + 2)$
50. $-5(x - 5)(x + 2)$
51. $-2(x - 5)(x - 5)$
52. $-3(x - 5)(x - 5)$
53. $a^2(x + 7)(x + 3)$
54. $-2(x + 7)(x - 3)$
55. $3a^2(x - 2)(x + 4)$
56. $-m(x - 7)(x - 1)$
57. $(7x - 4)(7x + 4)$
58. $2(7x + 4)(7x - 4)$
59. $2a(4 - 7x)(4 + 7x)$
60. $(4xy - 3m)(4xy + 3m)$
61. $a(4xy + 3m)(4xy - 3m)$
62. $(n - 8)(n + 2)$
63. $2(z - 1)(z - 1)$
64. $3x(x - 2)(x - 1)$
65. $4(x - 6)(x + 2)$
66. $-5x(x - 5)(x - 1)$
67. $-2(x - 7)(x - 1)$
68. $-a(x - 10)(x - 1)$
69. $-a(x - 2)(x + 1)$
70. $b(x - 3)(x + 1)$
71. $c(x - 4)(x + 1)$
72. $-d(x - 5)(x + 1)$
73. $6(m - 6)(m + 1)$
74. $-k(x - 8)(x + 1)$
75. $(-1)(x - 10)(x + 1)$
76. $p^2(x + 2)(x - 1)$
77. $-3(k + 2)(k - 2)$
78. $-a(x - 5)(x + 2)$
79. $b(x - 7)(x + 2)$
80. $-k(x - 9)(x + 2)$
81. $x(x - 11)(x + 2)$
82. $-p(p + 3)(p + 5)$
83. $-2x(x - 4)(x + 2)$
84. $-3(x + 3)(x + 7)$
85. $-a(x - 7)(x + 3)$
86. $-a(a - 8)(a + 3)$
87. $b(y + 4)(y - 1)$
88. $-2(x + 1)(x + 4)$
89. $c^2(x - 8)(x + 1)$
90. $-b(b + 2)(b + 4)$
91. $-3(x + 4)(x - 3)$
92. $ax(x + 4)(x + 3)$
93. $-3(x + 4)(x - 4)$
94. $k(4 + ak)(4 - ak)$
95. $-5(x + 4)(x - 4)$
96. $-a^2(x + 4)(x + 3)$
97. $b^2(x - 5)(x + 4)$
98. $ac(x + 5)(x - 4)$

answers to problem set A-1.2

1. $(x + 8p)(x - 2p)$
2. $(k + 3a)(k + 3a)$
3. $(p + 9x)(p - 3x)$
4. $(p - 5x)(p + 4x)$
5. $(k - 4m)(k + 3m)$
6. $(m - 9z)(m + 3z)$
7. $(p + 5m)(p - 4m)$
8. $(a + 4m)(a - 3m)$
9. $(p + 3mk)(p - 2mk)$
10. $(n + 2b)(n + 9b)$
11. $(x + 3a)(x + 9a)$
12. $(xa - 15b)(xa - 6b)$
13. $(n - 12k)(n + 4k)$
14. $(yp - 8x)(yp - 7x)$
15. $(p - 11a)(p + 5a)$
16. $(r - 7n)(r - 11n)$
17. $(m + 15n)(m + 6n)$
18. $(v + 11px)(v + 5px)$

19. $(k - 9a)(k + 7a)$
20. $(a - 15xy)(a + 2xy)$
21. $(w - 2k)(w - 11k)$
22. $(xy - 11a)(xy + 3a)$
23. $(r - 11x)(r + 9x)$
24. $(x - 10p)(x + 9p)$
25. $(x + 5pr)(x + 5pr)$
26. $(x + 9ay)(x + ay)$
27. $(x + 2pm)(x + 8pm)$
28. $(x - 5ak)(x - 5ak)$
29. $(p - 9m)(p - m)$
30. $(x - 4am)(x - 4am)$
31. $(xm + 3p)(xm + 11p)$
32. $(p - 6x)(p - 6x)$
33. $(mx - 6a)(mx + 5a)$
34. $(x + 12p)(x - 11p)$
35. $(ax - 2m)(ax - 45m)$
36. $(m + 8xp)(m + 2xp)$
37. $(t + 5m)(t + 7m)$
38. $(xy + 51a)(xy + a)$
39. $(x - 4mn)(x - 3mn)$

answers to problem set A-1.3

1. $(a + b)(c - d)$
2. $(a + 2)(b + 4)$
3. $(a + x)(b + c)$
4. $(m + pc)(2x - 3)$
5. $(k + pc)(4 - xy)$
6. $(a + d)(c - xy)$
7. $(a + 3)(b + 5)$
8. $(a + x)(y + c)$
9. $(m + p)(3x - 2)$
10. $(k + 3)(x - 5)$
11. $(pc + 4)(x + b)$
12. $(ac + 2)(b - k)$
13. $(p^2 - a)(c + b)$
14. $(a + 3)(x^2 - c)$
15. $(m + 2)(x^3 + y)$
16. $(ab + c)(4 + x)$
17. $(k - 2)(m^2 - c)$
18. $(6 - xy)(a + b)$
19. $(x - 2y)(ab + c)$
20. $(m - n)(4x - ab)$
21. $(p + 2q)(a + c)$
22. $(2k - 3)(a + 5)$
23. $(6 - p)(x + 2y)$
24. $(z - 2k)(a + c)$
25. $(mx + b)(4 - c)$
26. $(ak + 2)(x + y)$
27. $(m + 3)(2a - 4)$
28. $(6 + p)(xy - 3)$
29. $(m + p)(2 - 3k)$
30. $(m + cp)(a - 2)$
31. $(x + 2y)(a - b)$
32. $(3a + 2)(k + p)$
33. $(mx + 4)(a - b)$
34. $(cx - 3)(m - 2p)$
35. $(k - 3)(2p - a)$
36. $(x - 1)(a - b)$

answers to problem set A-1.4

1. $(2x + 3)(x - 5)$
2. $2(2x + 3)(x - 5)$
3. $x(3x + 1)(x - 2)$
4. $2(3x + 1)(x - 2)$
5. $(3x + 2)(x + 2)$
6. $3x(3x + 2)(x + 2)$
7. $m(2p - 5)(p + 1)$
8. $2(2x + 1)(x + 4)$
9. $x(3x + 2)(x + 3)$
10. $2x(5m + 1)(m - 1)$
11. $(2z + 3)(z + 5)$
12. $2s(3s + 5)(s + 3)$
13. $3p(4x - 1)(x - 3)$
14. $(2a - 5)(a - 3)$
15. $2x(2x + 3)(x + 2)$
16. $2a(3x + 1)(x + 2)$
17. $x(2x - 5)(x + 3)$
18. $4(2x + 1)(x + 1)$
19. $m(5m + 1)(m + 2)$
20. $2s(3s + 5)(s + 1)$
21. $x(4x + 1)(x + 4)$
22. $m(2x + 3)(x + 3)$
23. $2b(3b + 1)(b + 3)$
24. $2(3k + 2)(k - 3)$
25. $3(2p - 5)(p + 2)$
26. $2x(2x + 1)(x - 4)$
27. $2k(5m + 1)(m + 3)$
28. $p(3s + 5)(s + 2)$
29. $2x(4x - 1)(x + 1)$
30. $n(2n + 3)(n + 4)$

appendix two

Decimals and fractions

A-2.1 introduction

This appendix consists of a brief review of multiplication and division of decimal numbers; addition, subtraction, multiplication, and division of fractions; conversions of improper fractions to mixed numbers and conversions from fractions to decimals. One homework problem set with answers is included for students who feel that they need a quick review.

A-2.2 multiplication of decimal numbers

Decimal numbers are multiplied in the same way whole numbers are multiplied. The decimal point in the product is placed so that the product has as many digits to the right of the decimal point as did both of the numbers multiplied.

example A-2.2.1 Multiply: (a) 4.16 by 7.5 and (b) .00416 by .75.

solution

```
(a)    4.16          (b)    .000416
        7.5                     .75
      -----               ---------
       2080                    2080
      2912                    2912
      -----               ---------
     31.200               .00031200
```

In (a) 4.16 has *two* digits to the right of the decimal point and 7.5 has *one* digit to the right of the decimal point. Two plus one equals three so in the product the decimal point is placed three places to the left of the final digit. In (b) .000416 has six digits to the right of the decimal point and .75 has two digits to the right of the decimal point. Six plus two equals eight so in the product the decimal point is placed eight places to the left of the final digit. Note that three zeros had to be added in front to provide the proper spacing.

Observe the placement of the decimal point in the multiplications shown here.

```
(a)      .0416      (b)    4.16     (c)    714.621    (d)   .000011
         .0002              200              .053                25
    ----------           ------        ----------         -------
    .00000832            832.00           2143863                55
                                         3573105                22
                                       ----------         -------
                                        37.874913           .000275
```

A-2.3 division of decimal numbers

If the divisor is a whole number, place the decimal in the quotient just above the decimal point of the dividend.

example A-2.3.1 Divide 75.35 by 72.

solution We record the numbers for division as we would if we were dividing whole numbers.

```
      .
77 | 75.35
```

The only difference is that before we begin, we place the decimal point of the quotient (answer) just above the decimal point in the dividend. Then we divide as if both numbers were whole numbers.

```
     1.046
72 | 75.35
     72
     --
      3 35
      2 88
     -----
        470
        432
        ---
         38
```

We could continue the division until the remainder is zero (which will not necessarily occur), or we can stop when we reach the required number of decimal places. We will write the answer above correct to two decimal places as

1.05

If both the divisor and dividend have internal decimal points, the problem is just a little more complicated. The most straightforward method of handling the decimal point is to move the decimal in the divisor as necessary to make the divisor a whole number. Then we make a corresponding adjustment in the position of the decimal point in the dividend.

example A-2.3.2 Divide 40.756 by 3.25 to at least two decimal places.

solution We first write the numbers down in the standard form for division.

```
3.25 | 40.756
```

Now we place a small inverted v called a *caret* to mark the place where the decimal point was in the divisor and move the decimal point to the end of the number.

```
3‸25. ) 40.756
```

Now we move the decimal point of the dividend the same number of places as we moved the decimal point in the divisor, which in this case is two places. Again, we use the caret to mark the old position of the decimal point.

```
3‸25. ) 40‸75.6
```

Now we place a decimal point in the space just above the new decimal point in the dividend. This will be our decimal point for the answer (which we call a quotient).

```
                .
3‸25. ) 40‸75.6
```

Now we divide using the same procedure that we use to divide whole numbers.

```
          12.5403
3‸25. ) 40‸75.6
        32 5
        ----
         8 25
         6 50
         ----
         1 75 6
         1 62 5
         ------
           13 10
           13 00
           -----
              1000
               975
              ----
                25
```

It doesn't look like this one is going to come out even so we will write the answer correct to two decimal places as

12.54

example A-2.3.3 Divide .000631 by .014.

solution

```
          .04507
‸014. ) ‸000.631
            56
            --
             71
             70
             --
              100
               98
              ---
                2
```

The answer to four decimal places is

.0451

A-2.4 simplification of fractions

The denominator and the numerator of a fraction can be multiplied or divided by the same non-zero number without changing the value of the fraction. We use this rule to help us reduce fractions to lowest terms.

example A-2.4.1 Simplify $\frac{8}{12}$.

solution We see that both numbers are divisible by 2 and we divide each number by 2:

$$\frac{8}{12} = \frac{8 \div 2}{12 \div 2} = \frac{4}{6}$$

Now we see that both are still divisible by 2 and we divide each number by 2:

$$\frac{4}{6} = \frac{4 \div 2}{6 \div 2} = \frac{2}{3}$$

No whole number will divide both 2 and 3 evenly so our simplification is complete.

It is not necessary to show each step as we did above. We can do the divisions in our head and make such marks on the paper as are required to help us keep track of what we have done.

example A-2.4.2 Reduce $\frac{42}{56}$ to lowest terms.

solution We first divide by 2 and then by 7.

$$\frac{\overset{\overset{3}{\cancel{21}}}{\cancel{42}}}{\underset{\underset{4}{\cancel{28}}}{\cancel{56}}} = \frac{3}{4}$$

example A-2.4.3 Reduce $\frac{75}{15}$ to lowest terms.

solution We see that 15 and 75 are both divisible by 5 so we divide them both by 5.

$$\frac{\overset{15}{\cancel{75}}}{\underset{3}{\cancel{15}}}$$

Now we see that 15 and 3 are both divisible by 3 and we divide by 3.

$$\frac{\overset{5}{\cancel{15}}}{\underset{1}{\cancel{3}}} = \mathbf{5} \quad \text{so} \quad \frac{\mathbf{75}}{\mathbf{15}} = \mathbf{5}$$

example A-2.4.4 Reduce $\frac{44}{66}$ to lowest terms.

solution We first divide by 11 and then by 2.

$$\frac{\overset{4}{\cancel{44}}}{\underset{6}{\cancel{66}}} = \frac{\overset{2}{\cancel{4}}}{\underset{3}{\cancel{6}}} = \frac{\mathbf{2}}{\mathbf{3}}$$

example A-2.4.5 Reduce $\frac{42}{84}$ to lowest terms.

solution We divide by 2 and then by 7 and 3.

$$\frac{\overset{21}{\cancel{42}}}{\underset{42}{\cancel{84}}} = \frac{\overset{3}{\cancel{21}}}{\underset{6}{\cancel{42}}} = \frac{\overset{1}{\cancel{3}}}{\underset{2}{\cancel{6}}} = \mathbf{\frac{1}{2}}$$

A-2.5 multiplication of fractions

To multiply two fractions, we multiply the numerators to form the numerator of the product and we multiply the denominators to form the denominator of the product.

example A-2.5.1 Multiply $\frac{3}{4}$ by $\frac{8}{11}$.

solution

$$\frac{3}{4} \times \frac{8}{11} = \frac{3 \cdot 8}{4 \cdot 11} = \frac{24}{44}$$

and we can simplify $\frac{24}{44}$ by dividing both numbers by 4.

$$\frac{\overset{6}{\cancel{24}}}{\underset{11}{\cancel{44}}} = \mathbf{\frac{6}{11}}$$

example A-2.5.2 Multiply $\frac{5}{7}$ by $\frac{11}{13}$.

solution

$$\frac{5}{7} \times \frac{11}{13} = \frac{5 \times 11}{7 \times 13} = \mathbf{\frac{55}{91}}$$

A-2.6 division of fractions

We divide fractions by multiplying both the numerator and the denominator of the fraction by the reciprocal of the denominator.

example A-2.6.1 Simplify $\dfrac{\frac{4}{3}}{\frac{7}{11}}$.

solution We will multiply the denominator by $\frac{11}{7}$. Now $\frac{11}{7}$ times $\frac{7}{11}$ equals 1 so we have eliminated the denominator. Now we must multiply $\frac{4}{3}$ by the same thing so we don't change the value of the expression.

$$\frac{\frac{4}{3} \cdot \frac{11}{7}}{\frac{7}{11} \cdot \frac{11}{7}} = \frac{\frac{44}{21}}{1} = \mathbf{\frac{44}{21}}$$

We see that the answer is the reciprocal of the denominator times the numerator. We will work the problem again.

$$\frac{\frac{4}{3}}{\frac{7}{11}} = \frac{4}{3} \times \frac{11}{7} = \mathbf{\frac{44}{21}}$$

example A-2.6.2 Simplify $\dfrac{\frac{7}{9}}{\frac{19}{8}}$.

solution We invert the divisor and multiply.

$$\frac{7}{9} \cdot \frac{8}{19} = \frac{7 \times 8}{9 \times 19} = \mathbf{\frac{56}{171}}$$

example A-2.6.3 Simplify $\frac{7}{9} \div \frac{19}{8}$.

solution This is the same problem as the last one, but in a different form. So to simplify, we turn the division problem into a multiplication problem by inverting the divisor and multiplying.

$$\frac{7}{9} \div \frac{19}{8} = \frac{7}{9} \times \frac{8}{19} = \mathbf{\frac{56}{171}}$$

A-2.7 addition and subtraction of fractions

If fractions have the same denominators, they are added or subtracted by adding and subtracting the numerators as required and recording the sum over a single denominator.

example A-2.7.1 Simplify $\frac{10}{8} - \frac{3}{8} + \frac{2}{8}$.

solution

$$\frac{10}{8} - \frac{3}{8} + \frac{2}{8} = \frac{10 - 3 + 2}{8} = \mathbf{\frac{9}{8}}$$

example A-2.7.2 Simplify $\frac{748}{963} + \frac{10}{963}$.

solution

$$\frac{748}{963} + \frac{10}{963} = \frac{748 + 10}{963} = \mathbf{\frac{758}{963}}$$

If the fractions do not have the same denominator, it is necessary to change one or both of the fractions by multiplying the denominator and the numerator by the same quantity.

example A-2.7.3 Add $\frac{3}{8} + \frac{2}{5} + \frac{5}{4}$.

solution We choose as the new denominator the number 40, which is the least common multiple of the original denominators. Then we will multiply the denominator of each fraction by the required number so that the product is 40. Then we multiply the numerator by the same number.

To change $\frac{3}{8}$ to a fraction whose denominator is 40, we multiply the numerator and the denominator by 5.

$$\frac{3}{8} = \frac{3 \times 5}{8 \times 5} = \frac{15}{40}$$

To change $\frac{2}{5}$ to a fraction whose denominator is 40, we multiply by 8 over 8.

$$\frac{2}{5} = \frac{2 \times 8}{5 \times 8} = \frac{16}{40}$$

To change $\frac{5}{4}$ to a fraction whose denominator is 40, we multiply by $\frac{10}{10}$.

$$\frac{5}{4} = \frac{5 \times 10}{4 \times 10} = \frac{50}{40}$$

Now instead of adding

$$\frac{3}{8} + \frac{2}{5} + \frac{5}{4}$$

we can add the fractions in their new form as

$$\frac{15}{40} + \frac{16}{40} + \frac{50}{40} = \mathbf{\frac{81}{40}}$$

example A-2.7.4 Add $\frac{7}{11} + \frac{5}{3}$.

solution The least common denominator is 33.

$$\frac{7}{11} + \frac{5}{3} = \frac{7 \times 3}{11 \times 3} + \frac{5 \times 11}{3 \times 11} = \frac{21}{33} + \frac{55}{33} = \mathbf{\frac{76}{33}}$$

example A-2.7.5 Subtract $\frac{7}{21} - \frac{3}{14}$.

solution The least common denominator is 42.

$$\frac{7}{21} - \frac{3}{14} = \frac{7 \times 2}{21 \times 2} - \frac{3 \times 3}{14 \times 3} = \frac{14}{42} - \frac{9}{42} = \mathbf{\frac{5}{42}}$$

A-2.8 converting fractions to decimal numbers

To convert a fraction to a decimal number, we perform the indicated division.

example A-2.8.1 Write $\frac{3}{4}$ as a decimal number.

solution We perform the indicated division.

$$\frac{3}{4} = 4\overline{)3.0} \quad = .75$$

$$\begin{array}{r} .75 \\ 4\,\overline{)\,3.0} \\ 2\,8 \\ \hline 2\,0 \\ 2\,0 \\ \hline \end{array}$$

and find that the decimal representation of $\frac{3}{4}$ is .75.

example A-2.8.2 Write $\frac{34}{11}$ as a decimal number.

solution We divide 34 by 11.

$$\begin{array}{r} 3.0909 \\ 11\,\overline{)\,34} \\ 33 \\ \hline 1\,00 \\ 99 \\ \hline 100 \\ 99 \\ \hline \end{array}$$

Thus $\frac{34}{11}$ written as a decimal rounded off to two places is 3.09.

A-2.9 writing mixed numbers as fractions

To write a mixed number as a fraction, we multiply the whole part by the denominator of the fraction, add the numerator and record over the original denominator.

example A-2.9.1 Write $4\frac{3}{5}$ as a fraction.

solution $$4\frac{3}{5} = \frac{4 \times 5 + 3}{5} = \frac{23}{5}$$

example A-2.9.2 Write $7\frac{3}{11}$ as a fraction.

solution $$7\frac{3}{11} = \frac{7 \times 11 + 3}{11} = \frac{80}{11}$$

example A-2.9.3 Write $4\frac{7}{8}$ as a decimal number.

solution We will convert $4\frac{7}{8}$ to a fraction and then divide.

$$4\frac{7}{8} = \frac{8 \times 4 + 7}{8} = \frac{39}{8} \rightarrow \begin{array}{r} 4.875 \\ 8\overline{)39.000} \\ \underline{32} \\ 7\,0 \\ \underline{6\,4} \\ 60 \\ \underline{56} \\ 40 \\ \underline{40} \end{array}$$

So $4\frac{7}{8}$ written as a decimal is **4.875**.

problem set appendix 2

Multiply:

*1. 4.16 by 7.5 *2. .000416 by .75

3. 72.51 by .083 4. .0713 by 7005.216

Divide:

*5. 75.35 by 72 *6. 40.756 by 3.25 *7. .000631 by .014

8. .8432 by .00615 9. .000802 by .00041 10. 76.35 by .005

Simplify:

*11. $\frac{8}{12}$ *12. $\frac{42}{56}$ *13. $\frac{75}{15}$

---

*The problems marked with an asterisk were worked out in this appendix.

*14. $\frac{44}{66}$ 15. $\frac{39}{65}$ 16. $\frac{35}{55}$

17. $\frac{60}{80}$ 18. $\frac{290}{610}$ 19. $\frac{148}{188}$

20. $\frac{175}{325}$ 21. $\frac{78}{142}$

Multiply:

*22. $\frac{3}{4} \times \frac{8}{11}$ *23. $\frac{5}{7} \times \frac{11}{13}$ 24. $\frac{71}{83} \times \frac{7}{11}$

25. $6 \times \frac{4}{3}$ 26. $\frac{11}{3} \times 42$ 27. $\frac{6}{11} \times \frac{13}{21}$

Divide as indicated:

*28. $\dfrac{\frac{4}{3}}{\frac{7}{11}}$ *29. $\dfrac{\frac{7}{9}}{\frac{19}{8}}$ *30. $\frac{7}{9} \div \frac{19}{8}$

31. $\frac{3}{5} \div \frac{11}{7}$ 32. $\frac{8}{13} \div \frac{1}{5}$ 33. $\frac{7}{13} \div \frac{2}{3}$

Simplify:

*34. $\frac{10}{8} - \frac{3}{8} + \frac{2}{8}$ *35. $\frac{3}{8} + \frac{2}{5} + \frac{5}{4}$ *36. $\frac{7}{11} + \frac{5}{3}$

*37. $\frac{7}{21} - \frac{3}{14}$ 38. $\frac{15}{11} - \frac{2}{5}$ 39. $\frac{7}{13} + \frac{8}{7}$

40. $\frac{5}{11} - \frac{1}{19}$ 41. $\frac{62}{11} + \frac{5}{3}$ 42. $\frac{105}{63} - \frac{5}{7}$

Write the decimal equivalent of the following fractions and mixed numbers:

*43. $\frac{3}{4}$ *44. $\frac{34}{11}$ *45. $4\frac{7}{8}$

46. $6\frac{11}{5}$ 47. $5\frac{2}{7}$ 48. $6\frac{11}{13}$

49. $8\frac{1}{16}$ 50. $4\frac{1}{11}$

answers to problem set appendix 2

1. 31.2 2. .000312 3. 6.01833 4. 499.4719008 5. 1.046
6. 12.54 7. .45071 8. 137.1057 9. 1.9561 10. 15,270
11. $\frac{2}{3}$ 12. $\frac{3}{4}$ 13. 5 14. $\frac{2}{3}$ 15. $\frac{3}{5}$ 16. $\frac{7}{11}$ 17. $\frac{3}{4}$
18. $\frac{29}{61}$ 19. $\frac{37}{47}$ 20. $\frac{7}{13}$ 21. $\frac{39}{71}$ 22. $\frac{6}{11}$ 23. $\frac{55}{21}$
24. $\frac{497}{913}$ 25. 8 26. 154 27. $\frac{78}{231}$ 28. $\frac{44}{21}$ 29. $\frac{56}{171}$
30. $\frac{56}{171}$ 31. $\frac{21}{55}$ 32. $\frac{40}{13}$ 33. $\frac{21}{26}$ 34. $\frac{9}{8}$ 35. $\frac{81}{40}$
36. $\frac{76}{33}$ 37. $\frac{5}{42}$ 38. $\frac{43}{55}$ 39. $\frac{153}{91}$ 40. $\frac{84}{209}$ 41. $\frac{241}{33}$
42. $\frac{60}{63}$ 43. .75 44. 3.091 45. 4.875 46. 8.2 47. 5.286
48. 6.846 49. 8.0625 50. 4.09

appendix three

Units and the metric system

A-3.1 English units

Throughout recorded history we find evidence that man has realized the requirement for a system of units that could be used to help him describe basic quantities such as length, area, volume, weight, and time. In the beginning, we believe that local societies developed their own system of units, and as communications between societies improved, one system of units would come to be used over a large area such as a country or several countries.

Distances or lengths were often measured by reference to the human body. The biblical cubit was the length of the forearm from the elbow to the end of the middle finger. In the Bible Noah was instructed to "make thee an arc of gopher wood . . . . The length of the arc shall be three hundred cubits, the breadth of it fifty cubits, and the height of it thirty cubits."

The Romans used the length of a man's foot as the basic unit of distance. They defined the pace to be five Roman feet and the mile to be 1000 Roman paces.

The Romans also had units of measure for volumes and for weights and the total Roman system of units was introduced to what is now modern Europe as an adjunct of the Roman occupation of that area about the time of Christ. When the Romans withdrew (by 300 A.D.) the natives used the Roman system augmented by local systems. Units were not standard from country to country or even from town to town. It was not unusual to find merchants in one marketplace using two or even three systems of weights and measures as required by the particular transaction under consideration.

The basic unit of length in England came to be the foot. The yard originally was defined as the distance from the king's nose to the end of his middle finger and later was defined to be equal to three feet. The inch was at one time the length of the last joint of the king's thumb and later was defined to be one-twelvth of a foot. In the sixteenth century an English law was promulgated that defined a mile to consist of 5280 feet.

In a similar haphazard and unplanned fashion, the English systems of weights and volume measures evolved. The relationships between some of these English units are shown here.

LENGTH

12 in. equals 1 ft
3 ft equals 1 yard
$5\frac{1}{2}$ yards equals 1 rod
5280 ft equals 1 mile

LIQUID MEASURE

4 gills equals 1 pint
2 pints equals 1 quart
4 quarts equals 1 gallon
42 gallons equals 1 barrel

HOUSEHOLD MEASURES

3 teaspoons equals 1 tablespoon
16 tablespoons equals 1 cup
2 cups equals 1 pint

WEIGHT

3.086 grains equals 1 carat
480 grains equals 1 ounce
16 ounces equals 1 pound

DRY MEASURES

2 pints equals 1 quart
16 pints equals 1 peck
4 pecks equal 1 bushel
33.6003125 cubic in. equals 1 pint

AREA MEASURES

144 sq in. equals 1 sq ft
9 sq ft equals 1 sq yard
4840 sq yards equals 1 acre
640 acres equals 1 square mile

Given the information above, we ask two questions:

1. How many teaspoons are there in a barrel?
2. How many square inches are there in a square mile?

Of course, the answer to both questions is: "don't be silly." The computations required are time-consuming and no one wants to do the work required to find the answers. If this confusion exists within one standardized system of units, we can imagine the confusion that existed in Europe in the late eighteenth century when the use of several non-standardized systems of units was prevalent.

A-3.2 history of the metric system

For centuries, scholars, merchants, and the man on the street had all realized the need for a simple and accurate system of weights and measures, and articles had been written on the subject in most European countries. But the systems of government in Europe, mostly headed by kings and princes, were lethargic and resisted change. A catalyst was needed to permit the adoption of new units or new anything.

In 1787 King Louis XVI of France found that the country was bankrupt and he requested a session of the Estates-General (The French Congress) be held to pass new tax laws. This request set in motion the series of events that we call the French Revolution and among other things caused Louis to "lose his head" in 1793. During the revolution everything associated with the old regime was suspect and change was the order of the day. Everyone was addressed as "Citizen" rather than as Monsieur or Madame as had been the custom. The names of the months were changed and the first day of the year was changed to 22 September and was called the 1st of Vendémiaire. The proponents of a new simple system of weights and measures found the climate of change to their liking and were able to persuade the government to request in 1790 that the French Academy of Sciences develop such a system. The system was developed as requested and in 1799 a law was passed that made the system official.

In 1840 in France another law was passed that provided severe penalties for using any other system of weights and measures. All the major nations of the world were cosigners of the Treaty of the Meter in 1875, which established the International Bureau of Weights and Measures at Sèvres, France, and provided that the governing body is to meet every six years to consider problems in weights and measures. Over the years this bureau has made slight changes in the system.

England began converting to the metric system in the 1960's and the United States is the last major holdout. Conversion by law in this country, however, has been decreed with the major companies who have been "going metric" on their own as the biggest supporters. General Motors, Caterpillar, IBM, Hughes Aircraft, and International Harvester are some of the companies that have decided to go metric on their own and are in the process of a total conversion to the metric system.

A-3.3 the metric system

The proper name for the metric system is the Système International d'Unités, or S.I. for short. It has seven base units, which are the **METER** for length, the **KILOGRAM** (1000 grams) for mass, the **SECOND** for time, the **AMPERE** for electric current, degrees **KELVIN** for temperature, the **CANDELA** for light intensity, and the **MOL** or **MOLE** for amount of substance.

Any of these base units can be multiplied by positive powers of 10 by attaching the proper prefix (from the Greek), as we will demonstrate here by using the meter.

	Prefix	*Unit*		*Abbreviation*
1 meter	–	meter	10^0	m
10 meters	deca	decameter	10^1	dkm
100 meters	hecto	hectometer	10^2	hm
1000 meters	kilo	kilometer	10^3	km
1,000,000 meters	mega	megameter	10^6	Mm
1,000,000,000 meters	giga	gigameter	10^9	Gm

The prefixes *tera*, *peta*, and *exa* are used to indicate multiplication by 10^{12}, 10^{15}, and 10^{18}, respectively, but are seldom encountered.

Any of the base units can be multiplied by negative powers of 10 by attaching the proper prefix (from the Latin) as we show here.

	Prefix	*Unit*		*Abbreviation*
1 meter	–	meter	10^0	m
.1 meter	deci	decimeter	10^{-1}	dm
.01 meter	centi	centimeter	10^{-2}	cm
.001 meter	milli	millimeter	10^{-3}	mm
.000001 meter	micro	micrometer	10^{-6}	μm
.000000001 meter	nano	nanometer	10^{-9}	nm

The prefixes pico, femto, and alto are used to designate multiplication by 10^{-12}, 10^{-15}, and 10^{-18}, respectively, but are seldom encountered.

These prefixes can be used with any of the base units to designate a unit that is some integral power of 10 times the base unit. For instance, to designate an illumination that is 1000 times as bright as one candela we would write

one kilocandela

To designate a unit of mass that is one one-millionth of a gram, we would write

one microgram

Basic Unit of Length: The Meter

The **meter** was originally defined to be one ten-thousandth of the distance from the North Pole to the Equator on a great circle through Paris. This definition has now been replaced with a definition that relates the length of the meter to the wavelengths of the element krypton.

A METER IS A YARD—PLUS A LITTLE EXTRA.

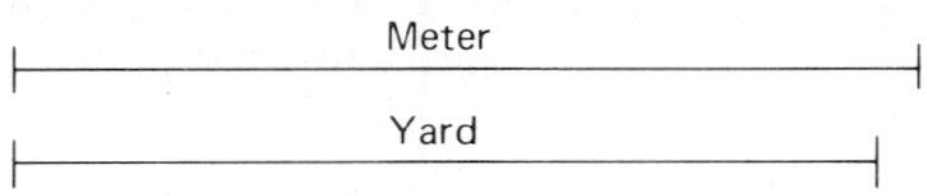

one meter $\approx$ 1.093613 yards

Basic Unit of Mass: The Kilogram

The **kilogram** equals 1000 grams and is defined to be the base unit although it and other mass units are expressed products of grams and powers of 10. To the physicist the kilogram is a unit of mass and a NEWTON is a unit of force or weight. Unfortunately, the merchant and the man in the street use the kilogram as a unit of weight although this is incorrect. If we use the kilogram *incorrectly* as a unit of weight, we can say

A KILOGRAM IS TWO POUNDS—PLUS A LITTLE EXTRA.

1 kilogram $\approx$ 2.204623 pounds

Basic Units of Area: Length Squared

All area is defined by squares whose sides are some power of 10 times the meter as we show here.

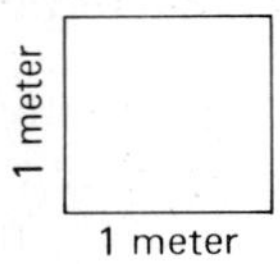

1 square meter (or m^2)

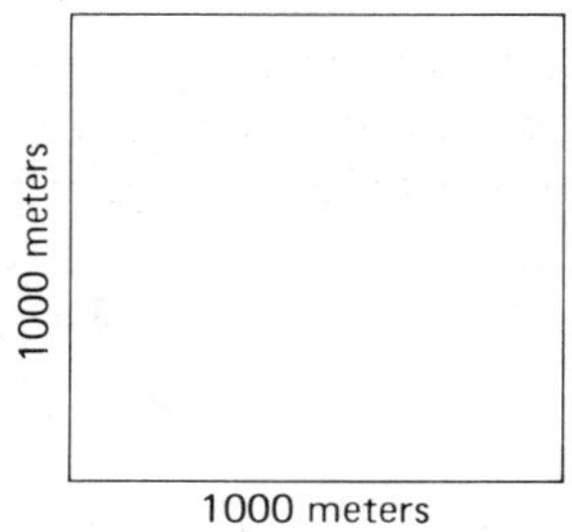

1 square kilometer (or km^2)

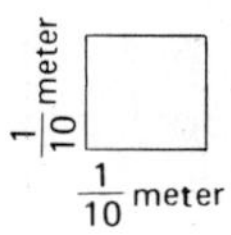

1 square decimeter (or dm^2)

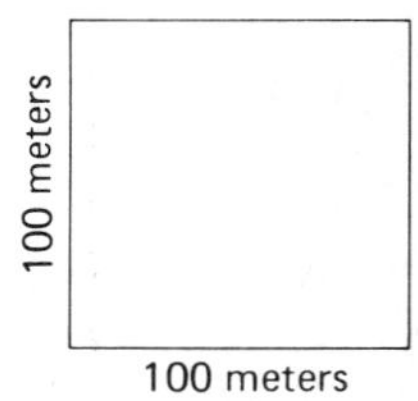

1 square hectometer or 1 hectare

Basic Units of Volume: Length Cubed

The units of volume are expressed in terms of cubes each of whose sides is the same basic metric length, as

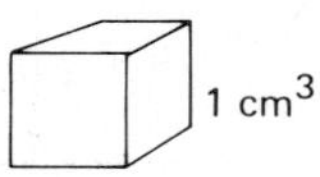

Each side of cube is 1 centimeter.

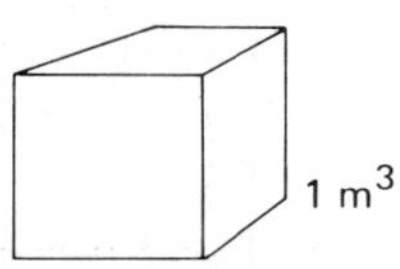

Each side of cube is 1 meter.

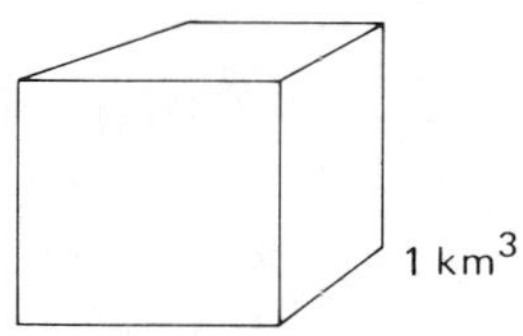

Each side of cube is 1 kilometer.

We introduce a new unit: the **LITER**

1 liter equals 1000 cubic centimeters

A LITER IS ONE QUART—PLUS A LITTLE EXTRA.

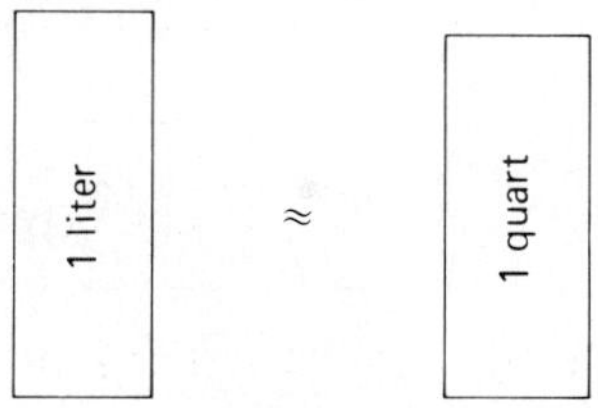

1 liter ≈ 1.056 quarts

TEMPERATURE: DEGREES KELVIN OR DEGREES CELSIUS (CENTIGRADE)

The three major reference temperatures in any system of measuring temperature are absolute zero, the freezing temperature of water, and the boiling temperature of water. In the metric system the temperature difference between the freezing and the boiling points of water is divided into 100 units called **degrees**. When we speak of degrees Kelvin, we are designating the number of these degrees above absolute zero. When we speak of the number of degrees Celsius (formerly centigrade), we are designating the number of these degrees above the freezing temperature of water. Thus any temperature in degrees Kelvin is 273.15 greater than the same temperature in degrees Celsius. Thus a temperature 10 degrees above the freezing point of water could be designated as either

10 degrees Celsius or 283.15 degrees Kelvin

The corresponding English units of temperature in degrees Fahrenheit and degrees Rankine are shown on the left. To convert from degrees Fahrenheit to degrees Celsius, either of the following equations may be used.

$$°F = \frac{9}{5}°C + 32 \qquad °C = (°F - 32)\frac{5}{9}$$

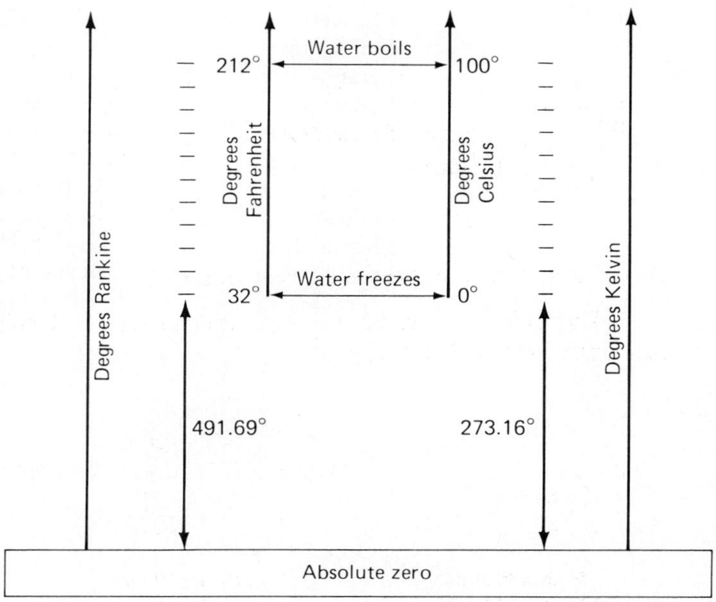

METRIC EQUIVALENTS

1 millimeter	=0.0394 inch
1 centimeter	=0.3937 inch
1 meter	=39.3708 inches
1 kilometer	=1093.63 yards
1 inch	=25.4 millimeters
1 foot	=0.3048 meter
1 yard	=0.9144 meter
1 mile	=1.6093 kilometers
1 international nautical mile	=1.852 kilometers
1 square centimeter	=0.155 square inch
1 square meter	=1.196 square yards
1 acre	=4840 square yards
1 hectare	=2.471 acres
1 square inch	=6.452 square centimeters
1 square foot	=9.290 square decimeters
1 square yard	=0.836 square meter
1 acre	=0.405 hectare
1 square mile	=259 hectares
1 cubic centimeter	=0.061 cubic inch
1 cubic meter	=1.308 cubic yards
1 cubic inch	=16.387 cubic centimeters
1 cubic foot	=0.028 cubic meter
1 cubic yard	=0.765 cubic meter
1 liter	=1.056 quart
1 gill	=118.291 milliliters
1 pint	=0.473 liter
1 quart	=0.946 liter
1 gallon	=3.785 liters
1 bushel	=35.23 liters
1 milligram	=0.015 grain
1 gram	=15.432 grains
1 kilogram	=2.2046 pounds
1 metric ton	=1.1 tons
1 grain	=0.0648 gram
1 ounce	=28.350 grams
1 pound	=0.4536 kilograms
1 hundredweight	=45.359 kilograms
1 ton	=909.1 kilograms
	=0.907 metric ton

Table of squares and square roots

$$2m^3xy^2p + 3pxy^2m^3 - 10xy^2m^3p + yx^2_{3_3}p$$

$$-5m^3pxy^2 + m^3px^2y$$

TABLE OF SQUARES AND SQUARE ROOTS

n	n^2	$\sqrt{n}$	n	n^2	$\sqrt{n}$	n	n^2	$\sqrt{n}$	n	n^2	$\sqrt{n}$	n	n^2	$\sqrt{n}$	n	n^2	$\sqrt{n}$
1.00	1.0000	1.00000	1.50	2.2500	1.22474	2.00	4.0000	1.41421	2.50	6.2500	1.58114	3.00	9.0000	1.73205	3.50	12.2500	1.87083
1.01	1.0201	1.00499	1.51	2.2801	1.22882	2.01	4.0401	1.41774	2.51	6.3001	1.58430	3.01	9.0601	1.73494	3.51	12.3201	1.87350
1.02	1.0404	1.00995	1.52	2.3104	1.23288	2.02	4.0804	1.42127	2.52	6.3504	1.58745	3.02	9.1204	1.73781	3.52	12.3904	1.87617
1.03	1.0609	1.01489	1.53	2.3409	1.23693	2.03	4.1209	1.42478	2.53	6.4009	1.59060	3.03	9.1809	1.74069	3.53	12.4609	1.87883
1.04	1.0816	1.01980	1.54	2.3716	1.24097	2.04	4.1616	1.42829	2.54	6.4516	1.59374	3.04	9.2416	1.74356	3.54	12.5316	1.88149
1.05	1.1025	1.02470	1.55	2.4025	1.24499	2.05	4.2025	1.43178	2.55	6.5025	1.59687	3.05	9.3025	1.74642	3.55	12.6025	1.88414
1.06	1.1236	1.02956	1.56	2.4336	1.24900	2.06	4.2436	1.43527	2.56	6.5536	1.60000	3.06	9.3636	1.74929	3.56	12.6736	1.88680
1.07	1.1449	1.03441	1.57	2.4649	1.25300	2.07	4.2849	1.43875	2.57	6.6049	1.60312	3.07	9.4249	1.75214	3.57	12.7449	1.88944
1.08	1.1664	1.03923	1.58	2.4964	1.25698	2.08	4.3264	1.44222	2.58	6.6564	1.60624	3.08	9.4864	1.75499	3.58	12.8164	1.89209
1.09	1.1881	1.04403	1.59	2.5281	1.26095	2.09	4.3681	1.44568	2.59	6.7081	1.60935	3.09	9.5481	1.75784	3.59	12.8881	1.89473
1.10	1.2100	1.04881	1.60	2.5600	1.26491	2.10	4.4100	1.44914	2.60	6.7600	1.61245	3.10	9.6100	1.76068	3.60	12.9600	1.89737
1.11	1.2321	1.05357	1.61	2.5921	1.26886	2.11	4.4521	1.45258	2.61	6.8121	1.61555	3.11	9.6721	1.76352	3.61	13.0321	1.90000
1.12	1.2544	1.05830	1.62	2.6244	1.27279	2.12	4.4944	1.45602	2.62	6.8644	1.61864	3.12	9.7344	1.76635	3.62	13.1044	1.90263
1.13	1.2769	1.06301	1.63	2.6569	1.27671	2.13	4.5369	1.45945	2.63	6.9169	1.62173	3.13	9.7969	1.76918	3.63	13.1769	1.90526
1.14	1.2996	1.06771	1.64	2.6896	1.28062	2.14	4.5796	1.46287	2.64	6.9696	1.62481	3.14	9.8596	1.77200	3.64	13.2496	1.90788
1.15	1.3225	1.07238	1.65	2.7225	1.28452	2.15	4.6225	1.46629	2.65	7.0225	1.62788	3.15	9.9225	1.77482	3.65	13.3225	1.91050
1.16	1.3456	1.07703	1.66	2.7556	1.28841	2.16	4.6656	1.46969	2.66	7.0756	1.63095	3.16	9.9856	1.77764	3.66	13.3956	1.91311
1.17	1.3689	1.08167	1.67	2.7889	1.29228	2.17	4.7089	1.47309	2.67	7.1289	1.63401	3.17	10.0489	1.78045	3.67	13.4689	1.91572
1.18	1.3924	1.08628	1.68	2.8224	1.29615	2.18	4.7524	1.47648	2.68	7.1824	1.63707	3.18	10.1124	1.78326	3.68	13.5424	1.91833
1.19	1.4161	1.09087	1.69	2.8561	1.30000	2.19	4.7961	1.47986	2.69	7.2361	1.64012	3.19	10.1761	1.78606	3.69	13.6161	1.92094
1.20	1.4400	1.09545	1.70	2.8900	1.30384	2.20	4.8400	1.48324	2.70	7.2900	1.64317	3.20	10.2400	1.78885	3.70	13.6900	1.92354
1.21	1.4641	1.10000	1.71	2.9241	1.30767	2.21	4.8841	1.48661	2.71	7.3441	1.64621	3.21	10.3041	1.79165	3.71	13.7641	1.92614
1.22	1.4884	1.10454	1.72	2.9584	1.31149	2.22	4.9284	1.48997	2.72	7.3984	1.64924	3.22	10.3684	1.79444	3.72	13.8384	1.92873
1.23	1.5129	1.10905	1.73	2.9929	1.31529	2.23	4.9729	1.49332	2.73	7.4529	1.65227	3.23	10.4329	1.79722	3.73	13.9129	1.93132
1.24	1.5376	1.11355	1.74	3.0276	1.31909	2.24	5.0176	1.49666	2.74	7.5076	1.65529	3.24	10.4976	1.80000	3.74	13.9876	1.93391
1.25	1.5625	1.11803	1.75	3.0625	1.32288	2.25	5.0625	1.50000	2.75	7.5625	1.65831	3.25	10.5625	1.80278	3.75	14.0625	1.93649
1.26	1.5876	1.12250	1.76	3.0976	1.32665	2.26	5.1076	1.50333	2.76	7.6176	1.66132	3.26	10.6276	1.80555	3.76	14.1376	1.93907
1.27	1.6129	1.12694	1.77	3.1329	1.33041	2.27	5.1529	1.50665	2.77	7.6729	1.66433	3.27	10.6929	1.80831	3.77	14.2129	1.94165
1.28	1.6384	1.13137	1.78	3.1684	1.33417	2.28	5.1984	1.50997	2.78	7.7284	1.66733	3.28	10.7584	1.81108	3.78	14.2884	1.94422
1.29	1.6641	1.13578	1.79	3.2041	1.33791	2.29	5.2441	1.51327	2.79	7.7841	1.67033	3.29	10.8241	1.81384	3.79	14.3641	1.94679
1.30	1.6900	1.14018	1.80	3.2400	1.34164	2.30	5.2900	1.51658	2.80	7.8400	1.67332	3.30	10.8900	1.81659	3.80	14.4400	1.94936
1.31	1.7161	1.14455	1.81	3.2761	1.34536	2.31	5.3361	1.51987	2.81	7.8961	1.67631	3.31	10.9561	1.81934	3.81	14.5161	1.95192
1.32	1.7424	1.14891	1.82	3.3124	1.34907	2.32	5.3824	1.52315	2.82	7.9524	1.67929	3.32	11.0224	1.82209	3.82	14.5924	1.95448
1.33	1.7689	1.15326	1.83	3.3489	1.35277	2.33	5.4289	1.52643	2.83	8.0089	1.68226	3.33	11.0889	1.82483	3.83	14.6689	1.95704
1.34	1.7956	1.15758	1.84	3.3856	1.35647	2.34	5.4756	1.52971	2.84	8.0656	1.68523	3.34	11.1556	1.82757	3.84	14.7456	1.95959
1.35	1.8225	1.16190	1.85	3.4225	1.36015	2.35	5.5225	1.53297	2.85	8.1225	1.68819	3.35	11.2225	1.83030	3.85	14.8225	1.96214
1.36	1.8496	1.16619	1.86	3.4596	1.36382	2.36	5.5696	1.53623	2.86	8.1796	1.69115	3.36	11.2896	1.83303	3.86	14.8996	1.96469
1.37	1.8769	1.17047	1.87	3.4969	1.36748	2.37	5.6169	1.53948	2.87	8.2369	1.69411	3.37	11.3569	1.83576	3.87	14.9769	1.96723
1.38	1.9044	1.17473	1.88	3.5344	1.37113	2.38	5.6644	1.54272	2.88	8.2944	1.69706	3.38	11.4244	1.83848	3.88	15.0544	1.96977
1.39	1.9321	1.17898	1.89	3.5721	1.37477	2.39	5.7121	1.54596	2.89	8.3521	1.70000	3.39	11.4921	1.84120	3.89	15.1321	1.97231
1.40	1.9600	1.18322	1.90	3.6100	1.37840	2.40	5.7600	1.54919	2.90	8.4100	1.70294	3.40	11.5600	1.84391	3.90	15.2100	1.97484
1.41	1.9881	1.18743	1.91	3.6481	1.38203	2.41	5.8081	1.55242	2.91	8.4681	1.70587	3.41	11.6281	1.84662	3.91	15.2881	1.97737
1.42	2.0164	1.19164	1.92	3.6864	1.38564	2.42	5.8564	1.55563	2.92	8.5264	1.70880	3.42	11.6964	1.84932	3.92	15.3664	1.97990
1.43	2.0449	1.19583	1.93	3.7249	1.38924	2.43	5.9049	1.55885	2.93	8.5849	1.71172	3.43	11.7649	1.85203	3.93	15.4449	1.98242
1.44	2.0736	1.20000	1.94	3.7636	1.39284	2.44	5.9536	1.56205	2.94	8.6436	1.71464	3.44	11.8336	1.85472	3.94	15.5236	1.98494
1.45	2.1025	1.20416	1.95	3.8025	1.39642	2.45	6.0025	1.56525	2.95	8.7025	1.71756	3.45	11.9025	1.85742	3.95	15.6025	1.98746
1.46	2.1316	1.20830	1.96	3.8416	1.40000	2.46	6.0516	1.56844	2.96	8.7616	1.72047	3.46	11.9716	1.86011	3.96	15.6816	1.98997
1.47	2.1609	1.21244	1.97	3.8809	1.40357	2.47	6.1009	1.57162	2.97	8.8209	1.72337	3.47	12.0409	1.86279	3.97	15.7609	1.99249
1.48	2.1904	1.21655	1.98	3.9204	1.40712	2.48	6.1504	1.57480	2.98	8.8804	1.72627	3.48	12.1104	1.86548	3.98	15.8404	1.99499
1.49	2.2201	1.22066	1.99	3.9601	1.41067	2.49	6.2001	1.57797	2.99	8.9401	1.72916	3.49	12.1801	1.86815	3.99	15.9201	1.99750
1.50	2.2500	1.22474	2.00	4.0000	1.41421	2.50	6.2500	1.58114	3.00	9.0000	1.73205	3.50	12.2500	1.87083	4.00	16.0000	2.00000
n	n^2	$\sqrt{n}$	n	n^2	$\sqrt{n}$	n	n^2	$\sqrt{n}$	n	n^2	$\sqrt{n}$	n	n^2	$\sqrt{n}$	n	n^2	$\sqrt{n}$

TABLE OF SQUARES AND SQUARE ROOTS (Cont.)

n	n^2	$\sqrt{n}$	n	n^2	$\sqrt{n}$	n	n^2	$\sqrt{n}$	n	n^2	$\sqrt{n}$	n	n^2	$\sqrt{n}$	n	n^2	$\sqrt{n}$
4.00	16.0000	2.00000	4.50	20.2500	2.12132	5.00	25.0000	2.23607	5.50	30.2500	2.34521	6.00	36.0000	2.44949	6.50	42.2500	2.54951
4.01	16.0801	2.00250	4.51	20.3401	2.12368	5.01	25.1001	2.23830	5.51	30.3601	2.34734	6.01	36.1201	2.45153	6.51	42.3801	2.55147
4.02	16.1604	2.00499	4.52	20.4304	2.12603	5.02	25.2004	2.24054	5.52	30.4704	2.34947	6.02	36.2404	2.45357	6.52	42.5104	2.55343
4.03	16.2409	2.00749	4.53	20.5209	2.12838	5.03	25.3009	2.24277	5.53	30.5809	2.35160	6.03	36.3609	2.45561	6.53	42.6409	2.55539
4.04	16.3216	2.00998	4.54	20.6116	2.13073	5.04	25.4016	2.24499	5.54	30.6916	2.35372	6.04	36.4816	2.45764	6.54	42.7716	2.55734
4.05	16.4025	2.01246	4.55	20.7025	2.13307	5.05	25.5025	2.24722	5.55	30.8025	2.35584	6.05	36.6025	2.45967	6.55	42.9025	2.55930
4.06	16.4836	2.01494	4.56	20.7936	2.13542	5.06	25.6036	2.24944	5.56	30.9136	2.35797	6.06	36.7236	2.46171	6.56	43.0336	2.56125
4.07	16.5649	2.01742	4.57	20.8849	2.13776	5.07	25.7049	2.25167	5.57	31.0249	2.36008	6.07	36.8449	2.46374	6.57	43.1649	2.56320
4.08	16.6464	2.01990	4.58	20.9764	2.14009	5.08	25.8064	2.25389	5.58	31.1364	2.36220	6.08	36.9664	2.46577	6.58	43.2964	2.56515
4.09	16.7281	2.02237	4.59	21.0681	2.14243	5.09	25.9081	2.25610	5.59	31.2481	2.36432	6.09	37.0881	2.46779	6.59	43.4281	2.56710
4.10	16.8100	2.02485	4.60	21.1600	2.14476	5.10	26.0100	2.25832	5.60	31.3600	2.36643	6.10	37.2100	2.46982	6.60	43.5600	2.56905
4.11	16.8921	2.02731	4.61	21.2521	2.14709	5.11	26.1121	2.26053	5.61	31.4721	2.36854	6.11	37.3321	2.47184	6.61	43.6921	2.57099
4.12	16.9744	2.02978	4.62	21.3444	2.14942	5.12	26.2144	2.26274	5.62	31.5844	2.37065	6.12	37.4544	2.47386	6.62	43.8244	2.57294
4.13	17.0569	2.03224	4.63	21.4369	2.15174	5.13	26.3169	2.26495	5.63	31.6969	2.37276	6.13	37.5769	2.47588	6.63	43.9569	2.57488
4.14	17.1396	2.03470	4.64	21.5296	2.15407	5.14	26.4196	2.26716	5.64	31.8096	2.37487	6.14	37.6996	2.47790	6.64	44.0896	2.57682
4.15	17.2225	2.03715	4.65	21.6225	2.15639	5.15	26.5225	2.26936	5.65	31.9225	2.37697	6.15	37.8225	2.47992	6.65	44.2225	2.57876
4.16	17.3056	2.03961	4.66	21.7156	2.15870	5.16	26.6256	2.27156	5.66	32.0356	2.37908	6.16	37.9456	2.48193	6.66	44.3556	2.58070
4.17	17.3889	2.04206	4.67	21.8089	2.16102	5.17	26.7289	2.27376	5.67	32.1489	2.38118	6.17	38.0689	2.48395	6.67	44.4889	2.58263
4.18	17.4724	2.04450	4.68	21.9024	2.16333	5.18	26.8324	2.27596	5.68	32.2624	2.38328	6.18	38.1924	2.48596	6.68	44.6224	2.58457
4.19	17.5561	2.04695	4.69	21.9961	2.16564	5.19	26.9361	2.27816	5.69	32.3761	2.38537	6.19	38.3161	2.48797	6.69	44.7561	2.58650
4.20	17.6400	2.04939	4.70	22.0900	2.16795	5.20	27.0400	2.28035	5.70	32.4900	2.38747	6.20	38.4400	2.48998	6.70	44.8900	2.58844
4.21	17.7241	2.05183	4.71	22.1841	2.17025	5.21	27.1441	2.28254	5.71	32.6041	2.38956	6.21	38.5641	2.49199	6.71	45.0241	2.59037
4.22	17.8084	2.05426	4.72	22.2784	2.17256	5.22	27.2484	2.28473	5.72	32.7184	2.39165	6.22	38.6884	2.49399	6.72	45.1584	2.59230
4.23	17.8929	2.05670	4.73	22.3729	2.17486	5.23	27.3529	2.28692	5.73	32.8329	2.39374	6.23	38.8129	2.49600	6.73	45.2929	2.59422
4.24	17.9776	2.05913	4.74	22.4676	2.17715	5.24	27.4576	2.28910	5.74	32.9476	2.39583	6.24	38.9376	2.49800	6.74	45.4276	2.59615
4.25	18.0625	2.06155	4.75	22.5625	2.17945	5.25	27.5625	2.29129	5.75	33.0625	2.39792	6.25	39.0625	2.50000	6.75	45.5625	2.59808
4.26	18.1476	2.06398	4.76	22.6576	2.18174	5.26	27.6676	2.29347	5.76	33.1776	2.40000	6.26	39.1876	2.50200	6.76	45.6976	2.60000
4.27	18.2329	2.06640	4.77	22.7529	2.18403	5.27	27.7729	2.29565	5.77	33.2929	2.40208	6.27	39.3129	2.50400	6.77	45.8329	2.60192
4.28	18.3184	2.06882	4.78	22.8484	2.18632	5.28	27.8784	2.29783	5.78	33.4084	2.40416	6.28	39.4384	2.50599	6.78	45.9684	2.60384
4.29	18.4041	2.07123	4.79	22.9441	2.18861	5.29	27.9841	2.30000	5.79	33.5241	2.40624	6.29	39.5641	2.50799	6.79	46.1041	2.60576
4.30	18.4900	2.07364	4.80	23.0400	2.19089	5.30	28.0900	2.30217	5.80	33.6400	2.40832	6.30	39.6900	2.50998	6.80	46.2400	2.60768
4.31	18.5761	2.07605	4.81	23.1361	2.19317	5.31	28.1961	2.30434	5.81	33.7561	2.41039	6.31	39.8161	2.51197	6.81	46.3761	2.60960
4.32	18.6624	2.07846	4.82	23.2324	2.19545	5.32	28.3024	2.30651	5.82	33.8724	2.41247	6.32	39.9424	2.51396	6.82	46.5124	2.61151
4.33	18.7489	2.08087	4.83	23.3289	2.19773	5.33	28.4089	2.30868	5.83	33.9889	2.41454	6.33	40.0689	2.51595	6.83	46.6489	2.61343
4.34	18.8356	2.08327	4.84	23.4256	2.20000	5.34	28.5156	2.31084	5.84	34.1056	2.41661	6.34	40.1956	2.51794	6.84	46.7856	2.61534
4.35	18.9225	2.08567	4.85	23.5225	2.20227	5.35	28.6225	2.31301	5.85	34.2225	2.41868	6.35	40.3225	2.51992	6.85	46.9225	2.61725
4.36	19.0096	2.08806	4.86	23.6196	2.20454	5.36	28.7296	2.31517	5.86	34.3396	2.42074	6.36	40.4496	2.52190	6.86	47.0596	2.61916
4.37	19.0969	2.09045	4.87	23.7169	2.20681	5.37	28.8369	2.31733	5.87	34.4569	2.42281	6.37	40.5769	2.52389	6.87	47.1969	2.62107
4.38	19.1844	2.09284	4.88	23.8144	2.20907	5.38	28.9444	2.31948	5.88	34.5744	2.42487	6.38	40.7044	2.52587	6.88	47.3344	2.62298
4.39	19.2721	2.09523	4.89	23.9121	2.21133	5.39	29.0521	2.32164	5.89	34.6921	2.42693	6.39	40.8321	2.52784	6.89	47.4721	2.62488
4.40	19.3600	2.09762	4.90	24.0100	2.21359	5.40	29.1600	2.32379	5.90	34.8100	2.42899	6.40	40.9600	2.52982	6.90	47.6100	2.62679
4.41	19.4481	2.10000	4.91	24.1081	2.21585	5.41	29.2681	2.32594	5.91	34.9281	2.43105	6.41	41.0881	2.53180	6.91	47.7481	2.62869
4.42	19.5364	2.10238	4.92	24.2064	2.21811	5.42	29.3764	2.32809	5.92	35.0464	2.43311	6.42	41.2164	2.53377	6.92	47.8864	2.63059
4.43	19.6249	2.10476	4.93	24.3049	2.22036	5.43	29.4849	2.33024	5.93	35.1649	2.43516	6.43	41.3449	2.53574	6.93	48.0249	2.63249
4.44	19.7136	2.10713	4.94	24.4036	2.22261	5.44	29.5936	2.33238	5.94	35.2836	2.43721	6.44	41.4736	2.53772	6.94	48.1636	2.63439
4.45	19.8025	2.10950	4.95	24.5025	2.22486	5.45	29.7025	2.33452	5.95	35.4025	2.43926	6.45	41.6025	2.53969	6.95	48.3025	2.63629
4.46	19.8916	2.11187	4.96	24.6016	2.22711	5.46	29.8116	2.33666	5.96	35.5216	2.44131	6.46	41.7316	2.54165	6.96	48.4416	2.63818
4.47	19.9809	2.11424	4.97	24.7009	2.22935	5.47	29.9209	2.33880	5.97	35.6409	2.44336	6.47	41.8609	2.54362	6.97	48.5809	2.64008
4.48	20.0704	2.11660	4.98	24.8004	2.23159	5.48	30.0304	2.34094	5.98	35.7604	2.44540	6.48	41.9904	2.54558	6.98	48.7204	2.64197
4.49	20.1601	2.11896	4.99	24.9001	2.23383	5.49	30.1401	2.34307	5.99	35.8801	2.44745	6.49	42.1201	2.54755	6.99	48.8601	2.64386
4.50	20.2500	2.12132	5.00	25.0000	2.23607	5.50	30.2500	2.34521	6.00	36.0000	2.44949	6.50	42.2500	2.54951	7.00	49.0000	2.64575
n	n^2	$\sqrt{n}$	n	n^2	$\sqrt{n}$	n	n^2	$\sqrt{n}$	n	n^2	$\sqrt{n}$	n	n^2	$\sqrt{n}$	n	n^2	$\sqrt{n}$

TABLE OF SQUARES AND SQUARE ROOTS (Cont.)

n	n^2	$\sqrt{n}$	n	n^2	$\sqrt{n}$	n	n^2	$\sqrt{n}$	n	n^2	$\sqrt{n}$	n	n^2	$\sqrt{n}$	n	n^2	$\sqrt{n}$
7.00	49.0000	2.64575	7.50	56.2500	2.73861	8.00	64.0000	2.82843	8.50	72.2500	2.91548	9.00	81.0000	3.00000	9.50	90.2500	3.08221
7.01	49.1401	2.64764	7.51	56.4001	2.74044	8.01	64.1601	2.83019	8.51	72.4201	2.91719	9.01	81.1801	3.00167	9.51	90.4401	3.08383
7.02	49.2804	2.64953	7.52	56.5504	2.74226	8.02	64.3204	2.83196	8.52	72.5904	2.91890	9.02	81.3604	3.00333	9.52	90.6304	3.08545
7.03	49.4209	2.65141	7.53	56.7009	2.74408	8.03	64.4809	2.83373	8.53	72.7609	2.92062	9.03	81.5409	3.00500	9.53	90.8209	3.08707
7.04	49.5616	2.65330	7.54	56.8516	2.74591	8.04	64.6416	2.83549	8.54	72.9316	2.92233	9.04	81.7216	3.00666	9.54	91.0116	3.08869
7.05	49.7025	2.65518	7.55	57.0025	2.74773	8.05	64.8025	2.83725	8.55	73.1025	2.92404	9.05	81.9025	3.00832	9.55	91.2025	3.09031
7.06	49.8436	2.65707	7.56	57.1536	2.74955	8.06	64.9636	2.83901	8.56	73.2736	2.92575	9.06	82.0836	3.00998	9.56	91.3936	3.09192
7.07	49.9849	2.65895	7.57	57.3049	2.75136	8.07	65.1249	2.84077	8.57	73.4449	2.92746	9.07	82.2649	3.01164	9.57	91.5849	3.09354
7.08	50.1264	2.66083	7.58	57.4564	2.75318	8.08	65.2864	2.84253	8.58	73.6164	2.92916	9.08	82.4464	3.01330	9.58	91.7764	3.09516
7.09	50.2681	2.66271	7.59	57.6081	2.75500	8.09	65.4481	2.84429	8.59	73.7881	2.93087	9.09	82.6281	3.01496	9.59	91.9681	3.09677
7.10	50.4100	2.66458	7.60	57.7600	2.75681	8.10	65.6100	2.84605	8.60	73.9600	2.93258	9.10	82.8100	3.01662	9.60	92.1600	3.09839
7.11	50.5521	2.66646	7.61	57.9121	2.75862	8.11	65.7721	2.84781	8.61	74.1321	2.93428	9.11	82.9921	3.01828	9.61	92.3521	3.10000
7.12	50.6944	2.66833	7.62	58.0644	2.76043	8.12	65.9344	2.84956	8.62	74.3044	2.93598	9.12	83.1744	3.01993	9.62	92.5444	3.10161
7.13	50.8369	2.67021	7.63	58.2169	2.76225	8.13	66.0969	2.85132	8.63	74.4769	2.93769	9.13	83.3569	3.02159	9.63	92.7369	3.10322
7.14	50.9796	2.67208	7.64	58.3696	2.76405	8.14	66.2596	2.85307	8.64	74.6496	2.93939	9.14	83.5396	3.02324	9.64	92.9296	3.10483
7.15	51.1225	2.67395	7.65	58.5225	2.76586	8.15	66.4225	2.85482	8.65	74.8225	2.94109	9.15	83.7225	3.02490	9.65	93.1225	3.10644
7.16	51.2656	2.67582	7.66	58.6756	2.76767	8.16	66.5856	2.85657	8.66	74.9956	2.94279	9.16	83.9056	3.02655	9.66	93.3156	3.10805
7.17	51.4089	2.67769	7.67	58.8289	2.76948	8.17	66.7489	2.85832	8.67	75.1689	2.94449	9.17	84.0889	3.02820	9.67	93.5089	3.10966
7.18	51.5524	2.67955	7.68	58.9824	2.77128	8.18	66.9124	2.86007	8.68	75.3424	2.94618	9.18	84.2724	3.02985	9.68	93.7024	3.11127
7.19	51.6961	2.68142	7.69	59.1361	2.77308	8.19	67.0761	2.86182	8.69	75.5161	2.94788	9.19	84.4561	3.03150	9.69	93.8961	3.11288
7.20	51.8400	2.68328	7.70	59.2900	2.77489	8.20	67.2400	2.86356	8.70	75.6900	2.94958	9.20	84.6400	3.03315	9.70	94.0900	3.11448
7.21	51.9841	2.68514	7.71	59.4441	2.77669	8.21	67.4041	2.86531	8.71	75.8641	2.95127	9.21	84.8241	3.03480	9.71	94.2841	3.11609
7.22	52.1284	2.68701	7.72	59.5984	2.77849	8.22	67.5684	2.86705	8.72	76.0384	2.95296	9.22	85.0084	3.03645	9.72	94.4784	3.11769
7.23	52.2729	2.68887	7.73	59.7529	2.78029	8.23	67.7329	2.86880	8.73	76.2129	2.95466	9.23	85.1929	3.03809	9.73	94.6729	3.11929
7.24	52.4176	2.69072	7.74	59.9076	2.78209	8.24	67.8976	2.87054	8.74	76.3876	2.95635	9.24	85.3776	3.03974	9.74	94.8676	3.12090
7.25	52.5625	2.69258	7.75	60.0625	2.78388	8.25	68.0625	2.87228	8.75	76.5625	2.95804	9.25	85.5625	3.04138	9.75	95.0625	3.12250
7.26	52.7076	2.69444	7.76	60.2176	2.78568	8.26	68.2276	2.87402	8.76	76.7376	2.95973	9.26	85.7476	3.04302	9.76	95.2576	3.12410
7.27	52.8529	2.69629	7.77	60.3729	2.78747	8.27	68.3929	2.87576	8.77	76.9129	2.96142	9.27	85.9329	3.04467	9.77	95.4529	3.12570
7.28	52.9984	2.69815	7.78	60.5284	2.78927	8.28	68.5584	2.87750	8.78	77.0884	2.96311	9.28	86.1184	3.04631	9.78	95.6484	3.12730
7.29	53.1441	2.70000	7.79	60.6841	2.79106	8.29	68.7241	2.87924	8.79	77.2641	2.96479	9.29	86.3041	3.04795	9.79	95.8441	3.12890
7.30	53.2900	2.70185	7.80	60.8400	2.79285	8.30	68.8900	2.88097	8.80	77.4400	2.96648	9.30	86.4900	3.04959	9.80	96.0400	3.13050
7.31	53.4361	2.70370	7.81	60.9961	2.79464	8.31	69.0561	2.88271	8.81	77.6161	2.96816	9.31	86.6761	3.05123	9.81	96.2361	3.13209
7.32	53.5824	2.70555	7.82	61.1524	2.79643	8.32	69.2224	2.88444	8.82	77.7924	2.96985	9.32	86.8624	3.05287	9.82	96.4324	3.13369
7.33	53.7289	2.70740	7.83	61.3089	2.79821	8.33	69.3889	2.88617	8.83	77.9689	2.97153	9.33	87.0489	3.05450	9.83	96.6289	3.13528
7.34	53.8756	2.70924	7.84	61.4656	2.80000	8.34	69.5556	2.88791	8.84	78.1456	2.97321	9.34	87.2356	3.05614	9.84	96.8256	3.13688
7.35	54.0225	2.71109	7.85	61.6225	2.80179	8.35	69.7225	2.88964	8.85	78.3225	2.97489	9.35	87.4225	3.05778	9.85	97.0225	3.13847
7.36	54.1696	2.71293	7.86	61.7796	2.80357	8.36	69.8896	2.89137	8.86	78.4996	2.97658	9.36	87.6096	3.05941	9.86	97.2196	3.14006
7.37	54.3169	2.71477	7.87	61.9369	2.80535	8.37	70.0569	2.89310	8.87	78.6769	2.97825	9.37	87.7969	3.06105	9.87	97.4169	3.14166
7.38	54.4644	2.71662	7.88	62.0944	2.80713	8.38	70.2244	2.89482	8.88	78.8544	2.97993	9.38	87.9844	3.06268	9.88	97.6144	3.14325
7.39	54.6121	2.71846	7.89	62.2521	2.80891	8.39	70.3921	2.89655	8.89	79.0321	2.98161	9.39	88.1721	3.06431	9.89	97.8121	3.14484
7.40	54.7600	2.72029	7.90	62.4100	2.81069	8.40	70.5600	2.89828	8.90	79.2100	2.98329	9.40	88.3600	3.06594	9.90	98.0100	3.14643
7.41	54.9081	2.72213	7.91	62.5681	2.81247	8.41	70.7281	2.90000	8.91	79.3881	2.98496	9.41	88.5481	3.06757	9.91	98.2081	3.14802
7.42	55.0564	2.72397	7.92	62.7264	2.81425	8.42	70.8964	2.90172	8.92	79.5664	2.98664	9.42	88.7364	3.06920	9.92	98.4064	3.14960
7.43	55.2049	2.72580	7.93	62.8849	2.81603	8.43	71.0649	2.90345	8.93	79.7449	2.98831	9.43	88.9249	3.07083	9.93	98.6049	3.15119
7.44	55.3536	2.72764	7.94	63.0436	2.81780	8.44	71.2336	2.90517	8.94	79.9236	2.98998	9.44	89.1136	3.07246	9.94	98.8036	3.15278
7.45	55.5025	2.72947	7.95	63.2025	2.81957	8.45	71.4025	2.90689	8.95	80.1025	2.99166	9.45	89.3025	3.07409	9.95	99.0025	3.15436
7.46	55.6516	2.73130	7.96	63.3616	2.82135	8.46	71.5716	2.90861	8.96	80.2816	2.99333	9.46	89.4916	3.07571	9.96	99.2016	3.15595
7.47	55.8009	2.73313	7.97	63.5209	2.82312	8.47	71.7409	2.91033	8.97	80.4609	2.99500	9.47	89.6809	3.07734	9.97	99.4009	3.15753
7.48	55.9504	2.73496	7.98	63.6804	2.82489	8.48	71.9104	2.91204	8.98	80.6404	2.99666	9.48	89.8704	3.07896	9.98	99.6004	3.15911
7.49	56.1001	2.73679	7.99	63.8401	2.82666	8.49	72.0801	2.91376	8.99	80.8201	2.99833	9.49	90.0601	3.08058	9.99	99.8001	3.16070
7.50	56.2500	2.73861	8.00	64.0000	2.82843	8.50	72.2500	2.91548	9.00	81.0000	3.00000	9.50	90.2500	3.08221	10.00	100.0000	3.16228
n	n^2	$\sqrt{n}$	n	n^2	$\sqrt{n}$	n	n^2	$\sqrt{n}$	n	n^2	$\sqrt{n}$	n	n^2	$\sqrt{n}$	n	n^2	$\sqrt{n}$

Answers to problem sets

problem set 1

1. The symbols written on the paper form a numeral that we use to represent the idea of the number.
2. The negative counterparts of each positive number and also the number zero.
3. None. *Every* real number has a home on the number line.
4. $\{1, 2, 3, \cdots\}$ **5.** $\{\cdots, -3, -2, -1, 0, 1, 2, 3, \cdots\}$
6. $\{0, 1, 2, 3, \cdots\}$
7. The *result* of a division. **8.** The *result* of an addition.
9. If two or more quantities are multiplied, each quantity is called a *factor*.
10. Multiplicative property of zero, $4 \cdot 0 = 0$; additive property of zero, $4 + 0 = 4$.
11. Multiplicative property of one, $4 \cdot 1 = 4$.
12. The *result* of a multiplication of factors. **13.** 5 **14.** 9
15. 25 **16.** -18

17.

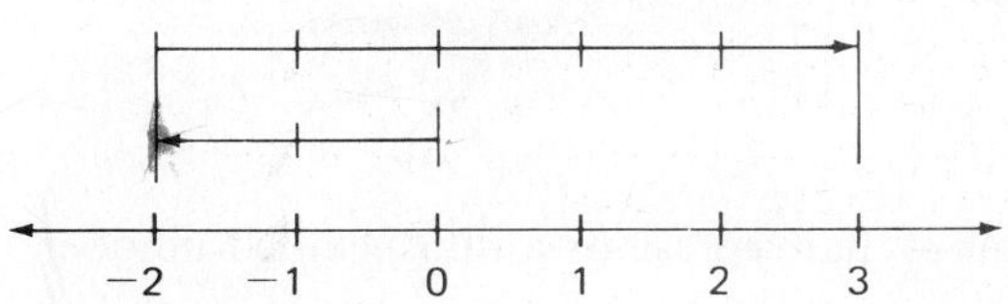

18.

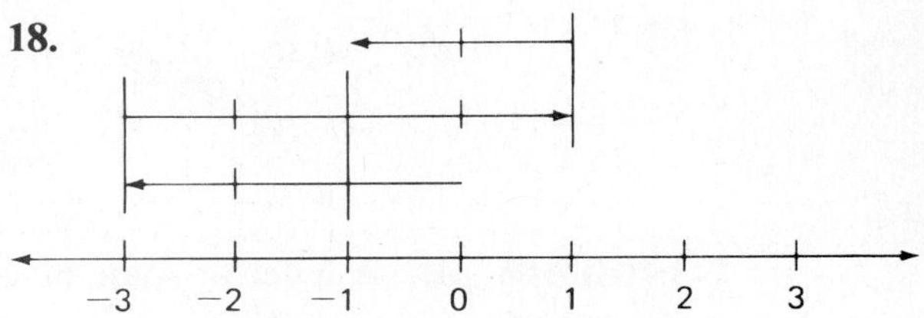

19. 2 **20.** -3 **21.** -14 **22.** -18 **23.** -18 **24.** -5

25. -19 **26.** 10 **27.** -10 **28.** -8 **29.** 15 **30.** -27

31. 36 **32.** −28 **33.** 17 **34.** −13 **35.** −14
36. 30 **37.** −33 **38.** 7 **39.** 14 **40.** 7 **41.** −5
42. −10

problem set 2

1. Additive inverse **2.** 1 **3.** Multiplicative inverse
4. A number is an idea. A numeral is the symbol used to represent the idea.
5. $\{0, 1, 2, 3, \cdots\}$ **6.** The result of a division.
7. If two or more quantities are multiplied, each quantity is called a *factor*.
8. 8 **9.** An operation that can be performed using *only two* quantities.
10. $4 + 3 = 7$ also $3 + 4 = 7$ **11.** $4 \cdot 3 = 12$ also $3 \cdot 4 = 12$
12. −3 **13.** 5 **14.** −10 **15.** 10 **16.** −15 **17.** −8
18. −12 **19.** 6 **20.** 20 **21.** 15 **22.** −144 **23.** 120
24. 10 **25.** −2 **26.** −1 **27.** 44 **28.** +5 **29.** 51
30. 3 **31.** −4 **32.** −7 **33.** 1 **34.** −18 **35.** 6 **36.** 12
37. −6 **38.** −6 **39.** −15 **40.** −12 **41.** −5 **42.** −14
43. 11 **44.** 5 **45.** −1 **46.** 4 **47.** −14 **48.** 17

problem set 3

1. A negative number. The product of an *odd* number of negative factors is negative.
2. A negative number. The product of an *odd* number of negative factors is negative.
3. Reciprocal **4.** $\{\cdots, -3, -2, -1, 0, 1, 2, 3, \cdots\}$ **5.** Yes
6. Yes. Any number that can be located on the number line is a real number.
7. $-\frac{3}{2}, -\frac{1}{4}$ **8.** 0 **9.** −8 **10.** −7 **11.** 14 **12.** 5
13. −51 **14.** 6 **15.** −13 **16.** 2 **17.** $-\frac{1}{13}$ **18.** $-\frac{6}{13}$
19. −17 **20.** 7 **21.** −4 **22.** 15 **23.** 24 **24.** −1
25. $-4(6 + 2 \cdot 2) + 6 = -40 + 6 = \mathbf{-34}$ **26.** 1
27. $-8(7 - 9 + 7) = -8(5) = \mathbf{-40}$ **28.** −4 **29.** 7 **30.** 20
31. 13 **32.** −32 **33.** 40 **34.** −19 **35.** 100 **36.** −181
37. −291 **38.** −228 **39.** 35 **40.** −10 **41.** −721 **42.** 490
43. −540 **44.** 16

problem set 4

1. A letter or some other symbol that represents an unspecified number.
2. $\{0, 1, 2, 3, \cdots\}$ **3.** The result of a division.
4. If two or more quantities are multiplied, each of the quantities is said to be a factor of the resulting product.

5. $0 \cdot 15 = 0, 0 + 15 = 15$ **6.** $7 \cdot 1 = 7$

7. An operation in which only two numbers can be used at one time.

8. $4 + 3 = 7$ also $3 + 4 = 7$ **9.** $4 \cdot 3 = 12$ also $3 \cdot 4 = 12$

10. 0 **11.** 10 **12.** -8 **13.** -2 **14.** -88 **15.** 46

16. $-5 - (20) = -$ **25** **17.** $-2[-6(12) + 3] = -2(-69) =$ **138**

18. $[6(16)] - 5 = (96) - 5 =$ **91** **19.** $-8(4 + 12) = -$ **128**

20. $-14 + 22 + 4 - 18 = -$ **6** **21.** $-5(2 - 9 - 6) =$ **65**

22. $-3[7(20) + 2] = -$ **426** **23.** $96 - 11 + 5 =$ **90** **24.** $3 + 7 =$ **10**

25. $4(7 + 4 + 5) =$ **64** **26.** 10 **27.** -7

28. $\frac{45}{-11} = -\frac{45}{11}$ **29.** $\frac{70}{78} = \frac{35}{39}$ **30.** $\frac{-41}{-4} = \frac{41}{4}$ **31.** $\frac{19}{-1} = -19$

32. 0 **33.** -29 **34.** 0 **35.** 384 **36.** 10 **37.** -43

38. 37 **39.** -46 **40.** -139 **41.** -84 **42.** -14

problem set 5

1. $a(b + c) = ab + ac$ **2.** $4 + 3 = 7$ and $3 + 4 = 7$

3. $15 + 0 = 15, \quad 15 \cdot 0 = 0$

4. This expression consists of only one term so there is no third term.

5. None, all integers are real numbers. **6.** Reciprocal

7. Positive. The product of an *even number* of negative numbers is a positive number.

8. $-\frac{7}{4}$

9. A letter or a symbol used to represent an unspecified number.

10. When the terms have the same variables in the same form so that the terms represent the same number regardless of the non-zero values assigned to the variables.

11. $x - 2xy + mx$ **12.** $-4mxy + 2pyx$ **13.** $3x + 2y - 5xy + 11$

14. $-5pyx - 2ym - 5$ **15.** $xy - 5y + x + 11$

16. $x + 3y + 5 + 2xy + 6mz - p$

17. $4x^2yp^5 + xy^2p^5 + x^2y^5p$ **18.** $-5xy^2m^3p + x^2ym^3p$

19. $4x^2y - 3x^2y^2$ **20.** x^7y^5 **21.** $x^5y^5m^5$ **22.** x^4y^3

23. $m^6p^6x^4$ **24.** -23 **25.** -216 **26.** 24 **27.** $ab + ac$

28. $xmp - xmqc$ **29.** $xya + xyb - xyc$ **30.** $pxym - pxyk - pxyc$

31. $kmp - kn - kq$ **32.** $-2[4 - (-9)] = -$ **26**

33. $2[3(5 - 2) + 4] = 2(13) =$ **26**

34. $-3\{4(-2 - 4)[4 - (-2)] + 2\} =$ **426**

35. $-2 + 4(-6 + 2) = -2 + 4(-4) = -$ **18** **36.** $-6 - 4[-3(6 + 3)] =$ **102**

37. $\{4[-3(-8)]\} = \{4[24]\} =$ **96** **38.** $-7[4 - 12(-3)] = -7(4 + 36) = -$ **280**

39. $-11 + 2 - 18 = -$ **27** **40.** 24 **41.** 41 **42.** 150 **43.** -100

44. -210 **45.** 56 **46.** 106

problem set 6

1. Yes, but an algebraic expression is not necessarily a numerical expression.
2. If $a = b$, then $c + a = c + b$. 3. If $a = b$, then $ca = cb$.
4. $\{\cdots, -3, -2, -1, 0, 1, 2, 3, \cdots\}$ 5. $2 + 3 = 3 + 2$
6. None. *Every* real number can be paired with a unique point on the number line.
7. The result of a division. 8. $1 \cdot 33 = 33$ 9. $4 \cdot 7 = 7 \cdot 4$
10. $a(b + c) = ab + ac$ 11. Multiplicative inverse
12. A letter or a symbol used to stand for an unspecified number.
13. 4 14. -5 15. 3 16. $\frac{4}{5}$ 17. $\frac{10}{3}$ 18. -6
19. $xy^3 - x^3yz$ 20. $4ax^3 - 8x^2$ 21. $4x^5y^4 - 20x^2y^3$
22. $4x^3y^3z - 8x^4y^4z^2 + 4cx^2y^3z^2$ 23. $x^3y^7m^2$ 24. p^8x^5m
25. $m^6x^5y^6$ 26. x^7m^5y 27. $7x^2pm + 2pa$ 28. $3x^2a + 2a^2x$
29. $-5k^2pa + 5k^2p$ 30. $5x^2pk$ 31. $5p^2mx - 2pxym$
32. $10a^2x^3 + 2a^3x^2$ 33. $-3[-2 - 5(9)] = -3(-2 - 45) = 141$
34. $2 - 3(-6) = \mathbf{20}$ 35. $-3 - 8(12) = \mathbf{-99}$
36. $-5[4 - 3(-7) + 2] = -5(27) = \mathbf{-135}$
37. $4(-4 + 6) + 6 + 4 = 8 + 6 + 4 = \mathbf{18}$
38. $5[10(-10 - 5)] = 5(-150) = \mathbf{-750}$ 39. $(5 + 3)(-3 + 5) = \mathbf{16}$
40. $2 + (5)(-2) + 5 = -3$ 41. $4[-3(-3 + 4)] = 4(-3) = \mathbf{-12}$
42. 18 43. -15 44. -6 45. -69 46. 27
47. $\frac{35}{2}$ 48. $\frac{12}{7}$ 49. 62 50. -16 51. -416

problem set 7

1. $4 \cdot 3 = 12$, $3 \cdot 4 = 12$ 2. $\{0, 1, 2, 3, \cdots\}$ 3. None
4. The result of a division 5. None
6. Multiplicative property of one, $33 \cdot 1 = 33$.
7. Multiplicative and additive properties of zero, $4 \cdot 0 = 0$, $4 + 0 = 4$.
8. If two or more quantities are to be multiplied, each of the quantities is said to be a factor of the resultant product.
9. $x^{-2}m^3$ 10. $p^{-3}x^{-2}m$ 11. $z^3m^5p^{-4}$ 12. y^3 13. m^6
14. $y^2x - 1$ 15. $p^{-6}y^2 - y^3$ 16. $x^5p^{-3}z - 4$
17. $1 + 4x^2y^{-2}$ 18. $4x^5y^5 - 16y^5 + 8$ 19. $1 + x^2y^2$ 20. $\frac{7}{8}$
21. $-\frac{5}{2}$ 22. $\frac{11}{3}$ 23. $\frac{6}{5}$ 24. $b = \frac{3}{8}$ 25. $x = -\frac{9}{5}$ 26. $b = \frac{2}{3}$
27. 2 28. $3a^3x^3 + a^2x^4$ 29. $10m^2p^2x + 3m^2xp + 2mx^2p^2$
30. $6xa^2 + 2ax - 3a^2x^2$ 31. $6a^2xp^2$ 32. $8a^2xz^4m$ 33. 27
34. 11 35. -6 36. 9 37. 4 38. -7 39. -77
40. -64 41. -32 42. 34 43. -10 44. -28 45. -108

46. $\frac{1}{27}$ **47.** $\frac{1}{9}$ **48.** $m = \frac{20}{3}$ **49.** $m = 225$ **50.** $p = \frac{21}{10}$

problem set 8

1. A natural number greater than 1 that is evenly divisible only by itself and the number 1.
2. No, 14 is divisible by both 7 and 2.
3. See definition in Lesson 8.
4. $\{\cdots, -3, -2, -1, 0, 1, 2, 3, \cdots\}$ **5.** $a(b + c) = ab + ac$
6. Because we have *defined* x^{-n} and $\dfrac{1}{x^n}$ to have the same meaning.
7. If a, b, and c are real numbers and if $a = b$, then $c + a = c + b$.
8. Not necessarily. $x + 4$ is an algebraic expression but not a numerical expression.
9. The result of addition.
10. The magnitude or size of the number discounting the sign preceding the number.

11. $2 \cdot 2 \cdot 2 \cdot 2 \cdot 5$ **12.** $3 \cdot 7 \cdot 7$ **13.** $2 \cdot 2 \cdot 2 \cdot 5 \cdot 5$ **14.** $2 \cdot 7 \cdot 7$

15. $2 \cdot 2 \cdot 3 \cdot 3$ **16.** $5 \cdot 5 \cdot 7$ **17.** $a(b + c)$ **18.** $a^2xm^2(ax + m - a^2x^2)$

19. $2a^2bz^3(2ab^3 + z)$ **20.** $2a^2x^2(3 + ax + 2a^2x)$ **21.** $m^2y(3mxy + 1)$

22. $3a^2x^2m(2a - m)$ **23.** $\dfrac{xa}{y^2} - \dfrac{bx}{y}$ **24.** $\dfrac{x^4}{y^2} - \dfrac{3x^2y^2}{m}$

25. $\dfrac{axp^2}{z} - \dfrac{2m^5p^4}{z} = \dfrac{axp^2 - 2m^5p^4}{z}$ **26.** $3x^{-2}y^{-2}m - 6$

27. $ax^{-5}p^2y^2 - 3x^{-3} + 6$ **28.** $\frac{1}{16}$ **29.** 3 **30.** $\frac{1}{18}$ **31.** 4

32. 2 **33.** 0 **34.** $2k^2x + 4axk^2$ **35.** $3abc - 2a^2bc^2$

36. $3mpx$ **37.** 13 **38.** 11 **39.** 12 **40.** -55

41. -1 **42.** 8 **43.** 8 **44.** 3 **45.** -16 **46.** 16 **47.** $-\frac{1}{16}$

48. $\frac{1}{16}$ **49.** -8 **50.** $x = \frac{66}{5}$

problem set 9

1. Yes. Divisible evenly only by 23 and 1. **2.** $7 + 11 = 11 + 7$

3. $a(b + c) = ab + ac$ **4.** Reciprocal

5.

6.
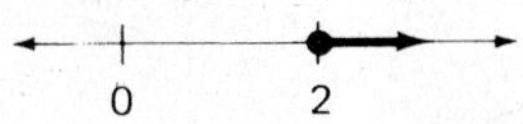

7.
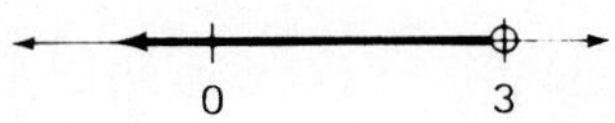

8.
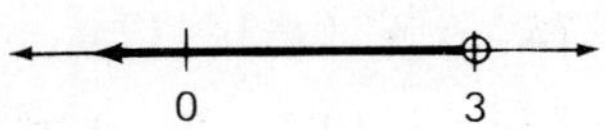

9.

10. $\dfrac{y^4z^4}{x^6}$ **11.** $\dfrac{a}{b^3c}$ **12.** $\dfrac{m^5}{k^{10}p^3}$ **13.** z^7p **14.** $5a^4x^2m$

15. $\dfrac{3b^4}{a^3}$ **16.** $\dfrac{8x^2}{y}$ **17.** $4abc^2 - 6acb^2$ **18.** $\dfrac{4a^2bc}{x}$

19. $\dfrac{y^5}{x^4} - \dfrac{x^2}{y}$ **20.** $ap^2 - 4k^2p^4$ **21.** $\dfrac{4p^2}{k} - 5$ **22.** $2 \cdot 5 \cdot 5 \cdot 5$

23. $3 \cdot 7 \cdot 5 \cdot 5$ **24.** $2 \cdot 3 \cdot 3 \cdot 3$ **25.** $3 \cdot 3 \cdot 5 \cdot 13$ **26.** $2 \cdot 3 \cdot 7 \cdot 11$

27. $x^2by(b^2 - 4b^2y - 1)$ **28.** $2mb^3x^3(mx^2 + x^3 - b^3)$

29. 14 **30.** 14 **31.** $-\frac{11}{2}$ **32.** $\frac{20}{9}$ **33.** -3 **34.** 1

35. $\frac{1}{6}$ **36.** $3x^3 + 2x^2$ **37.** $-2m - 32n$ **38.** $6x^3 - 4x^2 - 11x$

39. m^6 **40.** $\frac{y^2x^2}{m^2}$ **41.** 8 **42.** 8 **43.** $\frac{1}{3}$ **44.** 3

45. $-\frac{20}{3}$ **46.** 36 **47.** 9 **48.** 48 **49.** $\frac{451}{2}$ **50.** $\frac{451}{2}$

51. -16 **52.** -186

problem set 10

1. $y = \frac{1}{6}x - \frac{1}{6}z + \frac{2}{3}$ **2.** $y = \frac{4}{3}x + 2$ **3.** $p = \frac{4}{3}a + \frac{5}{3}$

4. $x = -\frac{3}{4}y + 1$ **5.** $n = \frac{1}{3}m - \frac{1}{6}a + \frac{2}{3}$ **6.** $y = -\frac{1}{2}x + \frac{1}{2}$

7. $\frac{3b + 8}{4b}$ **8.** $\frac{8c + 2b + bc}{2bc}$ **9.** $\frac{2m + 8c - c^3}{2c^3}$

10. $\frac{pb - 2ab + 4c}{4b}$ **11.** $\frac{12ac^2 + 9c + 4m}{12c^4}$

12. 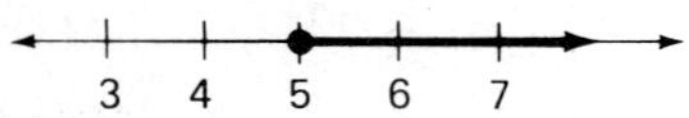

13. 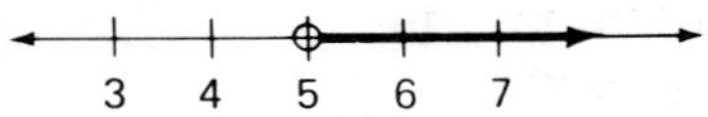

14. 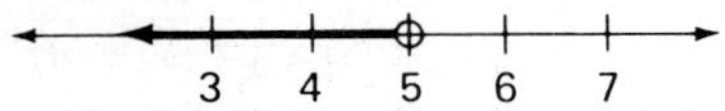

15. 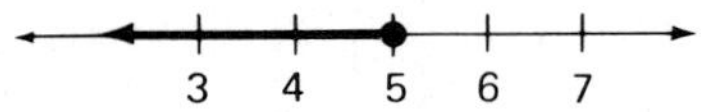

16. m^3b^7 **17.** $4ma^4$ **18.** 64 **19.** x^4y^4 **20.** $p^7z^6y^5$

21. $\frac{x^8}{4}$ **22.** $x^6y^{-6}m^{-1}$ **23.** $x^2m^2y^4$ **24.** $\frac{ac^4d^2}{b}$

25. m^2p^2 **26.** $-3x^5y - 2xy^5$ **27.** $-3a^2xm^{-5}$ **28.** $5px^{-4}$

29. 600 **30.** 1200 **31.** $k = -\frac{52}{5}$

32. $m = \frac{17}{6}$ **33.** $x = \frac{14}{5}$ **34.** $p = 32$ **35.** -18 **36.** -6

37. -28 **38.** -10 **39.** $xc(c - b)$ **40.** $ab^2c^2(abc^2 - 2 + 3ac)$

41. 45 **42.** 45 **43.** $\frac{1}{25}$ **44.** $\frac{1}{25}$ **45.** 18

46. 12 **47.** $\frac{27}{5}$ **48.** $\frac{27}{5}$ **49.** -45 **50.** -15

problem set 11

1. $3x^3 + 3x^2 + 6$ **2.** $2x^4 + 2x^3 - 3x^2 + 3$ **3.** $12x^2 + 7x - 10$

4. $4x^3 + 2x^2 + 14x - 8$ **5.** $12x^2 - 14x - 10$ **6.** $y = -2x + \frac{5}{2}$

7. $y = -x - 1$ **8.** $y = \frac{1}{3}ax + \frac{1}{3}m - 2$ **9.** $\frac{bmc + x}{mc^2}$ or $\frac{b}{c} + \frac{x}{mc^2}$

10. $\frac{mc + x + dmc^2}{mc^2}$ **11.** $\frac{ac + bmx}{cmx}$ **12.** $\frac{amc^2 + bc^2 + d}{mc^2}$

13. 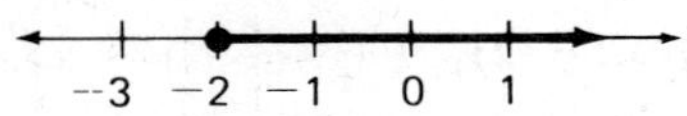

14. 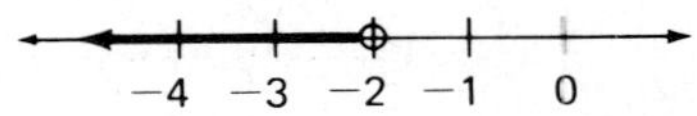

15. 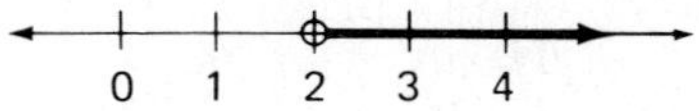

16. 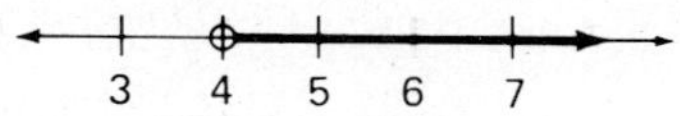

17. 9 **18.** $\dfrac{a^4xc^4}{b^3}$ **19.** $\dfrac{xk}{m^3a^4p^2}$ **20.** $\dfrac{x^4}{m^2p^3}$ **21.** $\dfrac{m^6}{p^9}$

22. $\dfrac{x^2}{p^5m^6}$ **23.** $\dfrac{x^3p^4}{y^3}$ **24.** $\dfrac{1}{y^2m^3}$ **25.** $\dfrac{x^2m^2}{p^2}$

26. $4x^5y^{-2} - 4x^{-2}y^5$ **27.** $-2x^2p^5y^{-1} + 2x^2p^5y$

28. $\dfrac{ab^3m}{c} - \dfrac{ax}{p}$ **29.** 1800 **30.** 3150

31. $x = 10$ **32.** $y = -2$ **33.** $p = 12$

34. $k = \frac{1}{2}$ **35.** 56 **36.** -7 **37.** -15 **38.** -180

39. $m^2x^2y^3(x^2y^3 - 3m)$ **40.** $x^2y^2z^3(yz^2 + m)$ **41.** 21 **42.** 7

43. $\frac{1}{49}$ **44.** $-\frac{1}{49}$ **45.** 28 **46.** 98 **47.** 49 **48.** $\frac{98}{3}$

49. 134 **50.** -714

problem set 12

1. 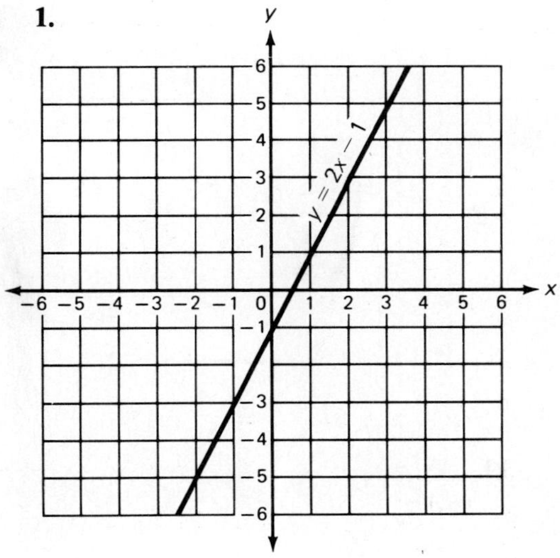

2. 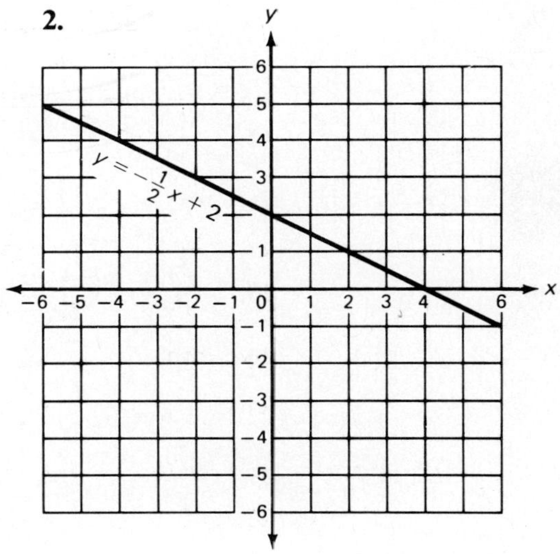

3. 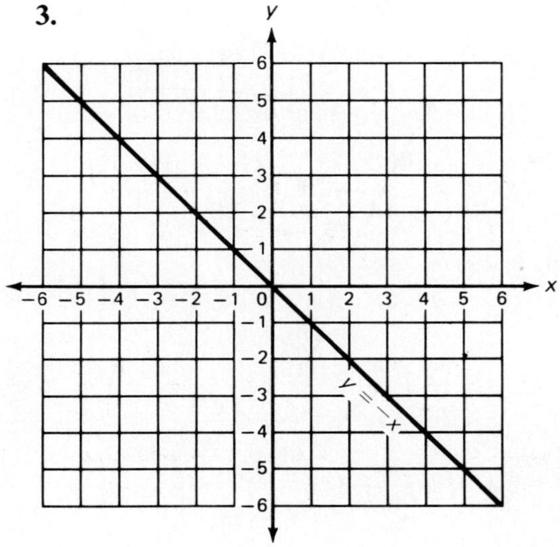

4. 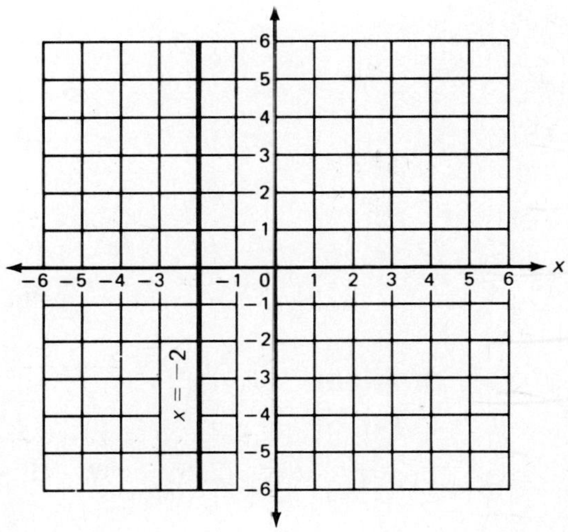

5.

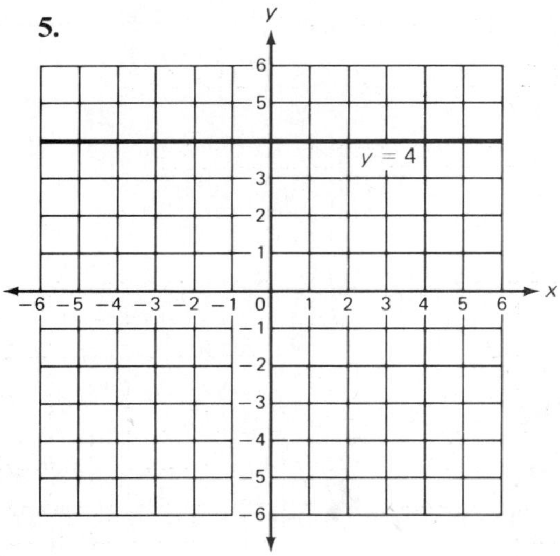

6. $3x^2 + 3x - 11$ 7. $\dfrac{bx + 2x + 3b}{abx}$

8. $\dfrac{m^2ay + mbx + ca^2y}{ma^2y}$

9. $\dfrac{ax + c + 4mx^2}{mx^2}$ 10. $\dfrac{x^2az + 2yz + 2am}{2az}$ 11. $y = -\frac{4}{5}x + \frac{6}{5}$

12. $y = \frac{7}{3}x + \frac{2}{3}a + \frac{7}{3}$ 13. $y = -2x + 1$ 14. $3x^2 - 10x - 8$

15. $2x^3 + 6x^2 - 6x - 20$ 16. $\dfrac{ax + ay + bx}{x^2 + xy}$ 17. $\dfrac{4x^2 + ax + ac + bc}{ax + bx}$

18. $\dfrac{ma + mb + cx + dmx}{mx}$ 19. $\dfrac{ka - xkb - xkc + db + dc + mb + mc}{kb + kc}$

20. 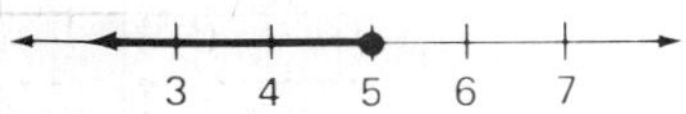 21. 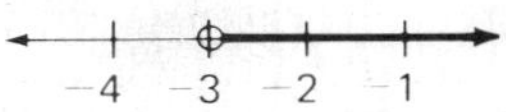 22. 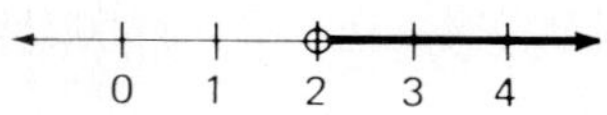

23. $\dfrac{x^2m^2}{p^2}$ 24. 1 25. $\dfrac{x}{m^6a^2p^2}$ 26. $\dfrac{4a^2bc}{x}$

27. $m^2x^{-1} + 5m^{-2}x^{-1}$ 28. $\dfrac{-x}{ap} + \dfrac{2b^2x}{a^2p^2}$ 29. $x = -17$

30. $x = \frac{36}{5}$ 31. $x = 3$ 32. $z = -1$ 33. 24

34. -22 35. -7 36. -87 37. $mx^2y(1 - 3myz)$

38. $p^2x^2(4p^2x - 3m)$ 39. 3 40. 3 41. $3 \cdot 3 \cdot 11$

42. $2 \cdot 2 \cdot 2 \cdot 3 \cdot 7$ 43. -16 44. 16 45. $\frac{1}{16}$ 46. $-\frac{1}{16}$

47. 6 48. -6 49. -6 50. -132

problem set 13

1. $\dfrac{x^4z^8}{y^{10}}$ 2. $\dfrac{a^{10}z^{50}}{b^{10}}$ 3. $16x^2y^2m^4$ 4. $\dfrac{y^4p^{10}}{9x^2}$ 5. $\dfrac{y^{60}}{x^{30}z^{15}}$

6. (2, 3) 7. (1, −5) 8. (−1, 3)

9. (4, −4) **10.** (1, 1) **11.** $-x^2 + 4x + 6$

12. $x^2 - x + 7$ **13.** $\dfrac{ax - bc}{bx}$ **14.** $\dfrac{ax - cbx + mb}{bx}$

15. $\dfrac{8 + 2x - ma}{2a}$ **16.** $\dfrac{xm - 6m + 6n + km^2 - kmn}{m^2 - mn}$

17.

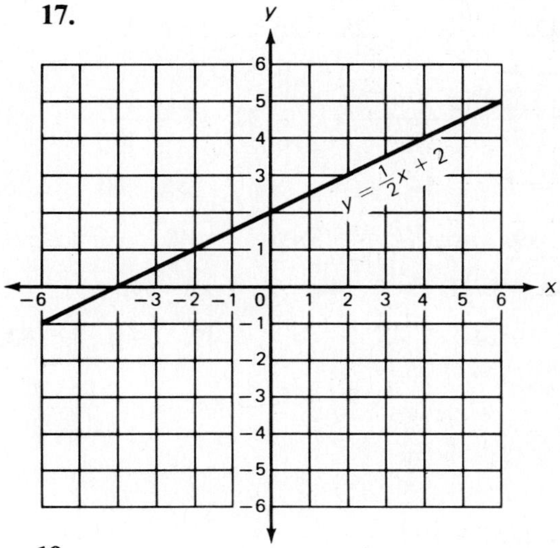

18.

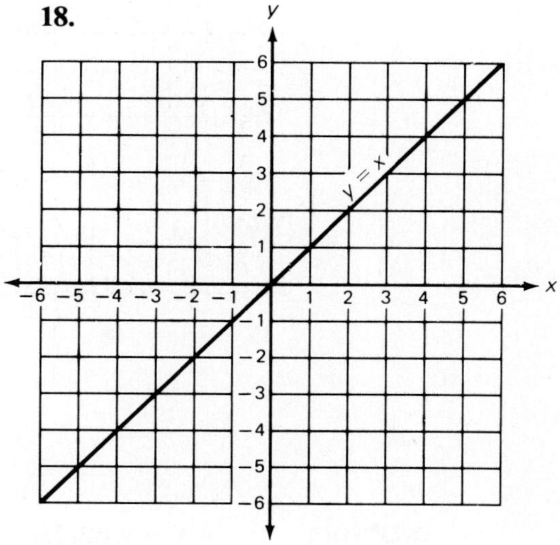

19.

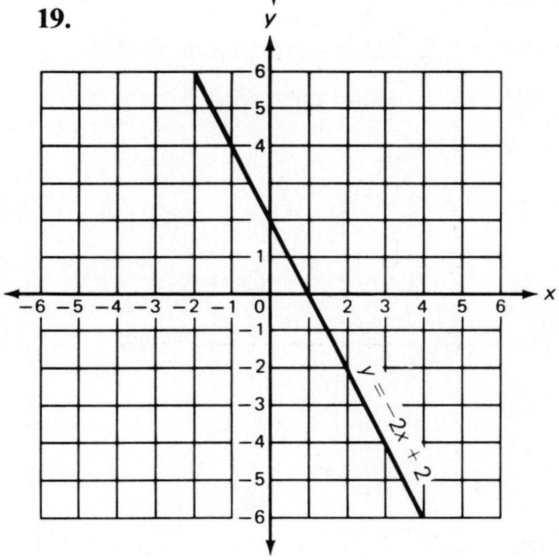

20.

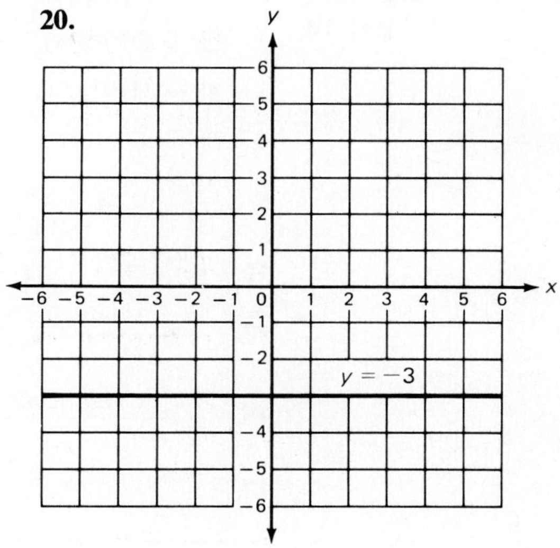

21.

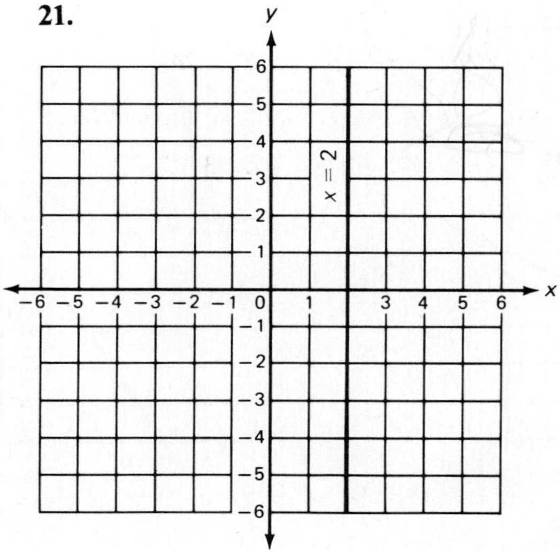

22. 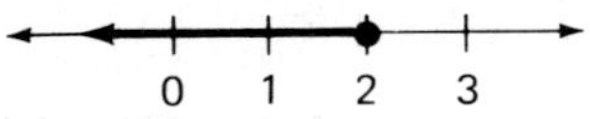

23. 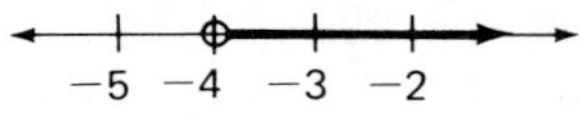

24. 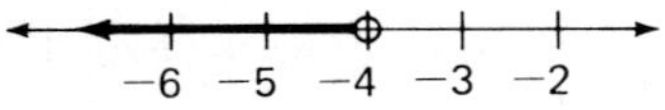

25. x^4 **26.** xm **27.** $\frac{b}{c^3p^6}$ **28.** $x^2y^3 + 2x^2y^{-3}$

29. $\frac{3x^5yp}{m} - \frac{4xy^5p}{m} \longrightarrow \frac{3x^5yp - 4y^5xp}{m}$ **30.** $1 - \frac{2ax}{pm}$ **31.** $m = \frac{7}{10}$

32. $x = -\frac{69}{10}$ **33.** $x = 1$ **34.** $p = -99$ **35.** 60 **36.** 33

37. -4 **38.** -6 **39.** $mxp^2(kx^3 - 3mp)$ **40.** $xyz^3(x^4z - yp)$

41. 32 **42.** 15 **43.** $2 \cdot 3 \cdot 5$ **44.** $2 \cdot 2 \cdot 2 \cdot 19$

45. $\frac{2}{9}$ **46.** $\frac{1}{9}$ **47.** -24 **48.** $-\frac{65}{2}$ **49.** $-\frac{65}{2}$ **50.** 68

problem set 14

1. $WN = (.14)(75) \longrightarrow WN = 10.5$

2. $20 = (WP)(80) \longrightarrow WP = .25$, which corresponds to 25%.

3. $48 = (WP)(64) \longrightarrow WP = .75$, which corresponds to 75%.

4. $68 = .04(WN) \longrightarrow WN = 1700$

5. $\frac{1}{x^2}$ **6.** $\frac{x^4}{9y^8z^2}$ **7.** $\frac{x^5z^3}{y^3}$ **8.** $\frac{x}{p^2}$ **9.** $(-1, 3)$

10. $(3, -3)$ **11.** $(4, -2)$ **12.** $-2x^2 - 6x + 8$

13. $\frac{ac^2 + mb}{bc^2}$ **14.** $\frac{ak - mkb - mkc + b + c}{kb + kc}$ **15.** $\frac{a - cb}{b}$

16. $\frac{ac^2 + mb + mc}{c^2(b + c)}$

17. 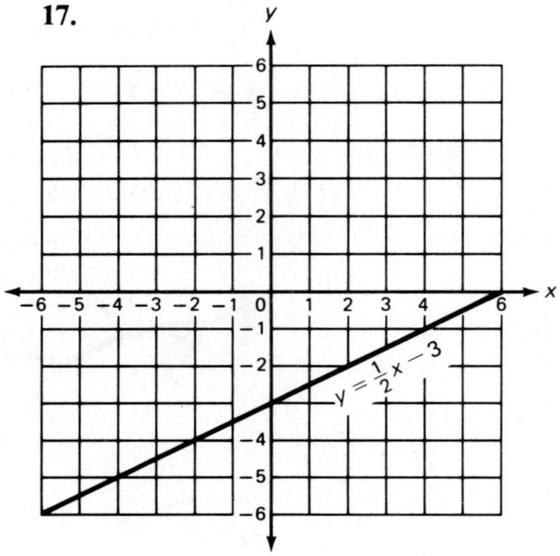

18. 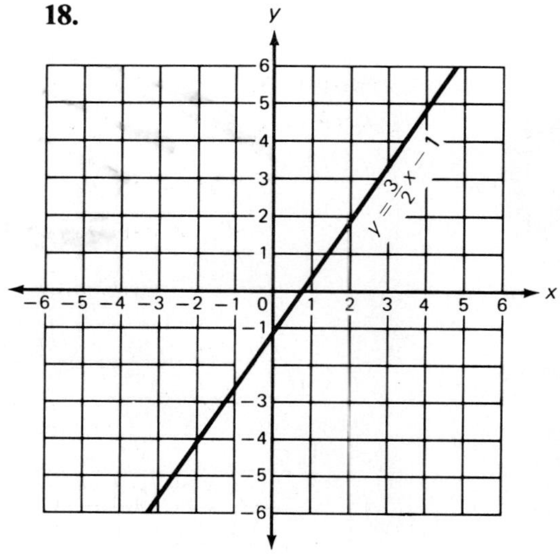

19.

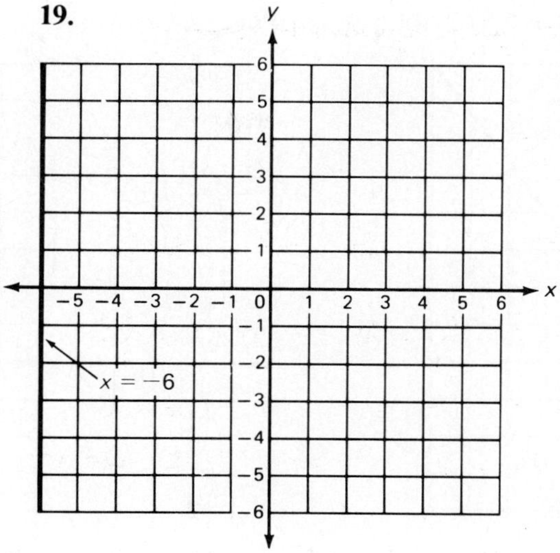

20.

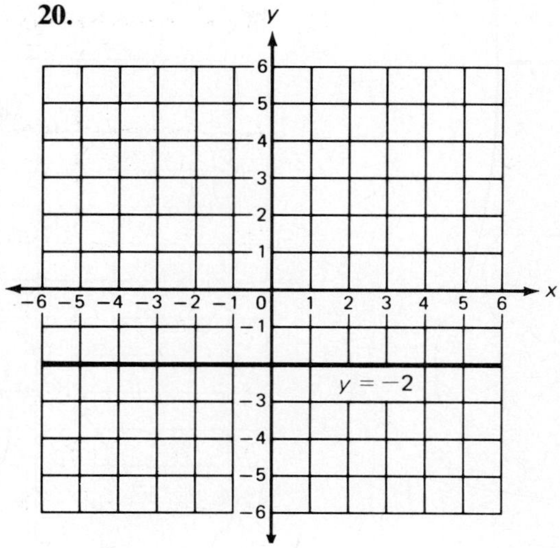

21.

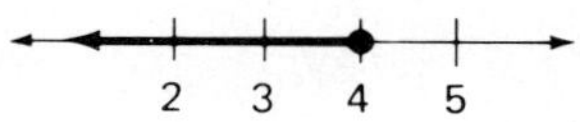

22.

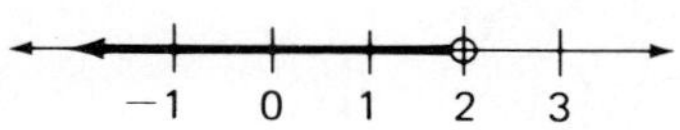

23.

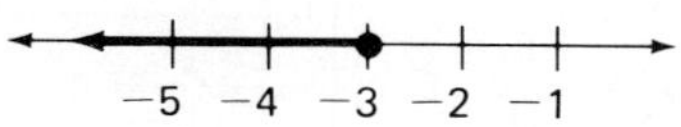

24. $4x^2 - 10x - 6$ **25.** $2x^4 - x^3zp^{-1}$

26. $4x^2ym^{-3} + 2y^2x^2m^{-1} - yx^{-2}m^{-3}$

27. $2ax^2m$ **28.** $\frac{1}{z^4} + \frac{5}{p^2}$ **29.** $\frac{x^4}{p^3y^5} + \frac{6x^2}{p^2}$ **30.** 61 **31.** 2

32. $\frac{7}{2}$ **33.** $\frac{42}{23}$ **34.** $-\frac{4}{7}$ **35.** 5 **36.** 2

37. -7 **38.** -2196 **39.** -12 **40.** $mpx(p - mz + 4mp^2x^3)$

41. $abc(a - b + 3zab)$ **42.** 60 **43.** 6 **44.** $2 \cdot 2 \cdot 3 \cdot 3$

45. $5 \cdot 5 \cdot 7$ **46.** $\frac{ad}{bc}$ **47.** $\frac{a}{bc}$ **48.** $\frac{b}{c}$ **49.** ac **50.** $\frac{a}{bx}$

problem set 15

1. $2N + 13 = 75 \longrightarrow \mathbf{N = 31}$ **2.** $2N - 13 = 137 \longrightarrow \mathbf{N = 75}$

3. $6 + (4N + 5N) = 123 \longrightarrow \mathbf{N = 13}$ **4.** $N = -N + 14 \longrightarrow \mathbf{N = 7}$

5. $3N = 5N + 2 \longrightarrow \mathbf{N = -1}$ **6.** $PV = (.04)(3000) \longrightarrow \mathbf{PV = 120}$

7. $35 = (WP)(175) \longrightarrow WP = .2 \longrightarrow$ **20%**

8. $120 = (.15)(N) \longrightarrow \mathbf{N = 800}$ **9.** (3, 2) **10.** (7, 4)

11. (4, −5) **12.** $-x^3 + 5x^2$ **13.** $\frac{bx + cx - mby - mcy + y}{by + cy}$

14. $\frac{ca + cx - m}{mc}$ **15.** $\frac{kd + kx + cmd + cmx - m}{md + mx}$

16. $\frac{a + mxy + mxc - dx^3y - dx^3c}{x^3y + x^3c}$ **17.** $\frac{xc^2p - m^3}{m^3c^3p}$

18.

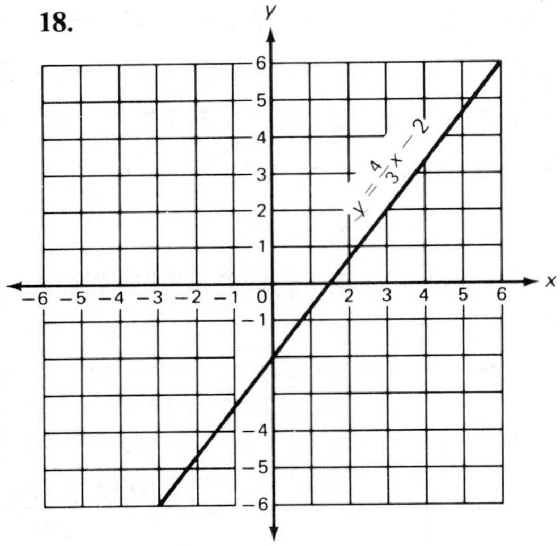

19.

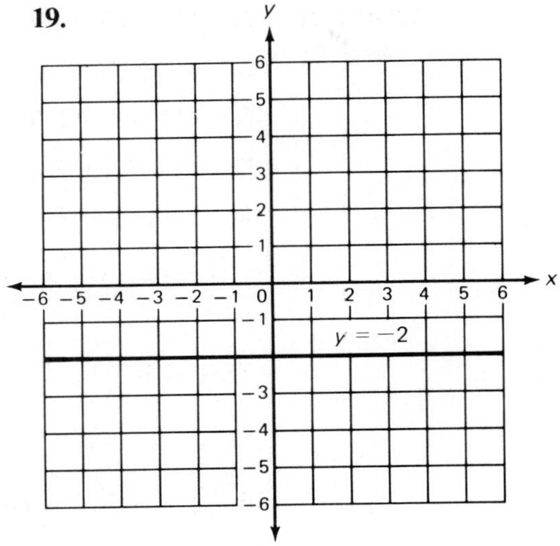

20.

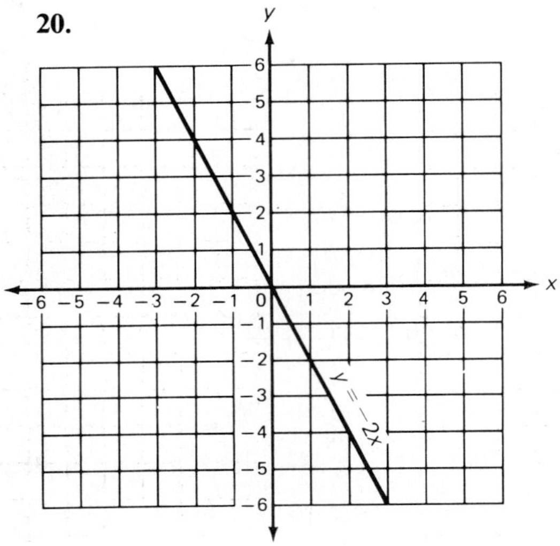

21.

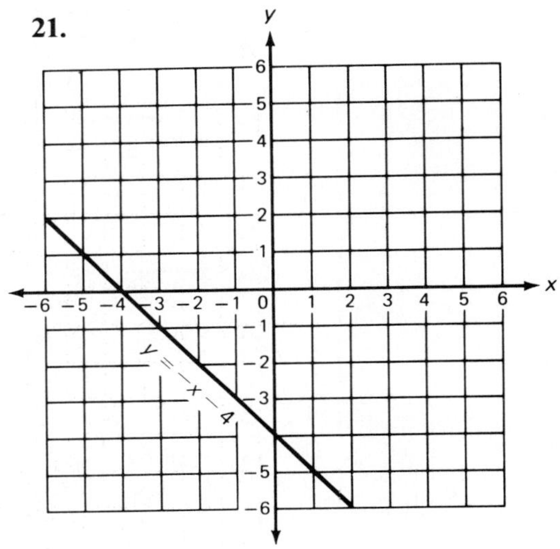

22.

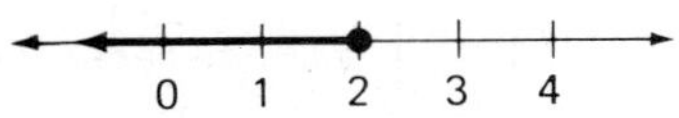

23.

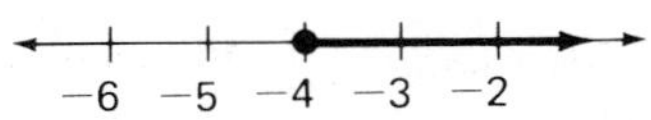

24.

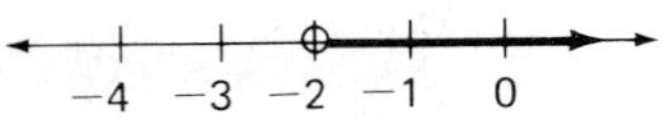

25. $3x^2 + 2x - 8$ **26.** $x^3 + 5x^2 + 10x + 8$

27. $\dfrac{y^3p^2}{x}$ **28.** $\dfrac{-3x^2}{y}$ **29.** $\dfrac{1}{x^3y^6}$ **30.** 1 **31.** $\dfrac{y + x}{y}$

32. $a^2b^{-2}x + 4a^{-2}b^2x$ **33.** $6a^2b^{-1} - 8ab^{-2}$ **34.** $2m^2x^4 + 4mb^{-1}$

35. 14 **36.** -42 **37.** -4 **38.** 9 **39.** 29 **40.** $x = -\frac{2}{3}$

41. $p = 3$ **42.** $m = \frac{3}{2}$ **43.** $\dfrac{a}{cx}$ **44.** $\dfrac{ac}{x}$ **45.** $a + b$

46. $a - b$ **47.** -125

48. False. The set of natural numbers is a subset of the set of real numbers.

49. Natural, whole, integers, rational, real. **50.** 792

problem set 16

1. $27 + 2N = 51 \longrightarrow$ $\mathbf{N = 12}$
2. $11 + 3N = -13 \longrightarrow$ $\mathbf{N = -8}$
3. $3N + 27 = 21 \longrightarrow$ $\mathbf{N = -2}$
4. $N = -N - 54 \longrightarrow$ $\mathbf{N = -27}$
5. $WN = (.05)(20) \longrightarrow$ $\mathbf{WN = 1}$
6. $75 = (.2)(WN) \longrightarrow$ $\mathbf{WN = 375}$
7. $25 = (WP)(200) \longrightarrow WP = .125 \longrightarrow$ **12.5%**
8. $5\sqrt{2}$
9. $10\sqrt{2}$
10. $7\sqrt{3}$
11. $6\sqrt{3}$
12. 

13. 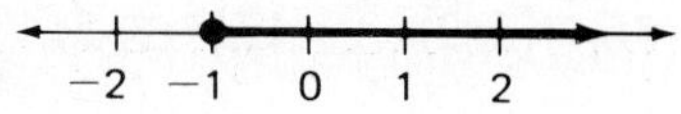

14. 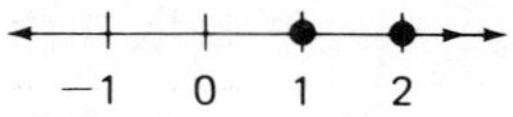

15. 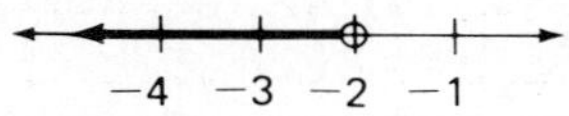

16. (2, 5)
17. (1, 2)
18. (9, 26)
19. (−4, 2)
20. $2x^3 - 3x^2 - 7x - 6$
21. $x^2 + 2x + 3$
22. $\dfrac{amc + anc + b}{a^3c}$
23. $\dfrac{my + cxy - mx}{xy}$
24. $\dfrac{x^3a + x^3b - c^2}{cx^2}$
25. $\dfrac{da - xbd - xd^2 - b - d}{d(b + d)}$
26. 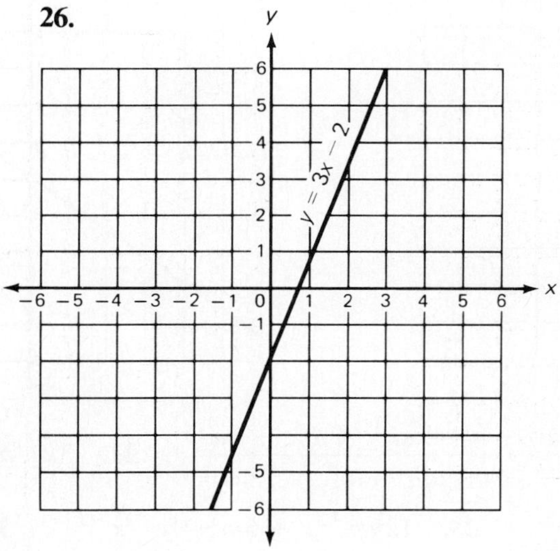

27. 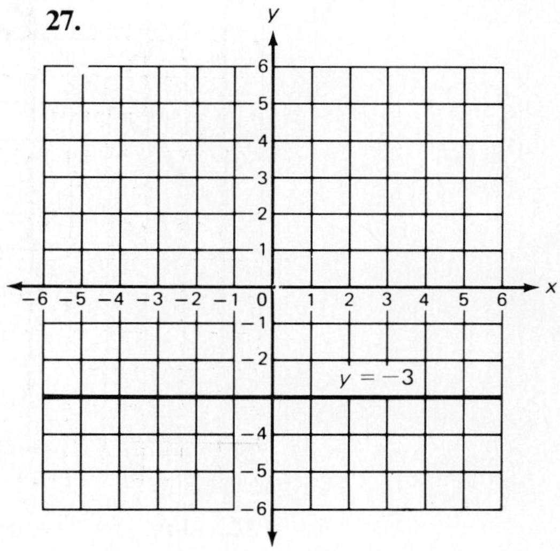

28. $2\sqrt{2}$
29. $4\sqrt{2}$
30. $3\sqrt{3}$
31. $\dfrac{y + 1}{y}$
32. $\dfrac{y^4}{z^2}$
33. y^2x^6
34. $\dfrac{a^4}{x^6}$
35. -2
36. $\dfrac{1}{y^2}$
37. $m^{-5}x^4y^2 + 3m^5y^2x^4 - 2m^{-2}y^{-2}x^4$
38. $3p^{-3}y^2m^{-1} - 3x^2m^{-2}$
39. $p^2y^3z^5 - 3p^{-1}$
40. -37
41. 36
42. 8
43. 5
44. -30
45. $x = \frac{17}{2}$
46. $x = 7$
47. $\dfrac{1}{x + y}$
48. $\dfrac{mxy}{n}$
49. $\dfrac{xp^2}{q}$
50. True
51. Irrational, real

problem set 17

1. $2N = 15 + 3N \longrightarrow$ $\mathbf{N = -15}$
2. $54 + 3N = 4N - 14 \longrightarrow$ $\mathbf{N = 68}$
3. $3N = -N - 32 \longrightarrow$ $\mathbf{N = -8}$
4. $2(N + 4) = 38 \longrightarrow$ $\mathbf{N = 15}$

5. $(N - 4)2 = 48 \longrightarrow N = 28$ 6. $42 = (.05)(WN) \longrightarrow WN = 840$

7. $W = (.08)(400) \longrightarrow W = 32$

8. $42 = (WP)(105) \longrightarrow WP = .4 \longrightarrow 40\%$

9. $4\sqrt{2}$ 10. $5\sqrt{2}$ 11. $3\sqrt{3} - 9\sqrt{2} - 18\sqrt{5}$ 12. $4\sqrt{2}$

13. (2, 3) 14. (5, −3) 15. (−2, −3)

16. (2, 3) 17. (3, 3) 18. $\dfrac{xc}{ya}$ 19. $\dfrac{ac}{b}$

20. $\dfrac{a}{mx}$ 21. $\dfrac{x+1}{x-1}$ 22. $\dfrac{x+1}{7x}$ 23. $\dfrac{1}{x-1}$ 24. $\dfrac{a+b}{x+4b}$

25. 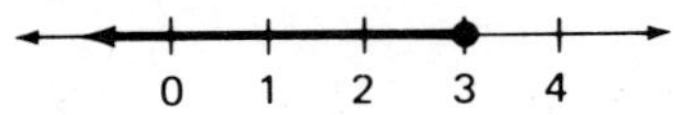

26. 

27. 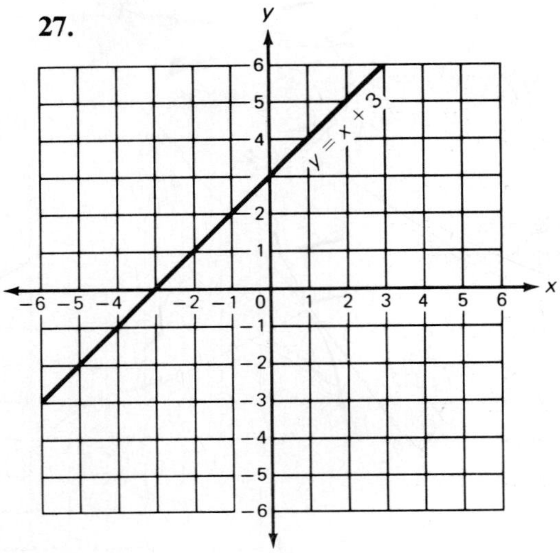

28. 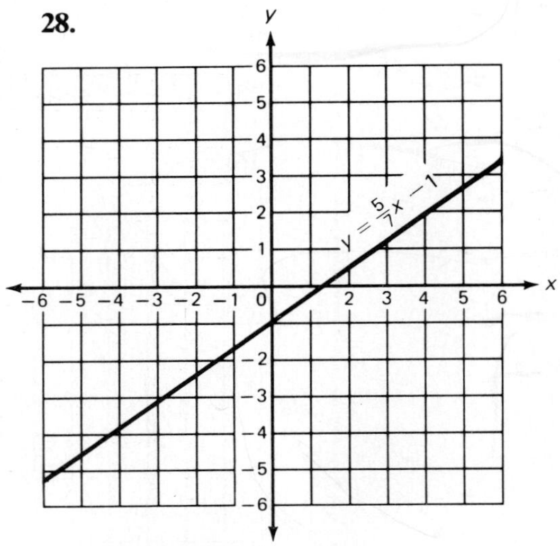

29. $\dfrac{xyb + m}{y^2b}$ 30. $\dfrac{a^2 + ab + ab^3c + b^4c + bd}{ab^3 + b^4}$ 31. $3k^{-4}p^9m^2 - 3k^{-4}m^2$

32. $4a^2p^{-2}y^{-4} + a^2p^{-2}y^3$ 33. $12mp^{-2}y + 6m^3p^{-6}y^{-2}x^{-3}$ 34. $x^6 - 5p$

35. 1 36. $-x^3 + 3x^2 - x$ 37. −10 38. 24 39. $ka(am - 3kp)$

40. −32 41. 5 42. $x = -\frac{43}{4}$ 43. $x = -\frac{14}{5}$

44. $-\frac{1}{27}$ 45. −27 46. $\frac{1}{27}$ 47. −81 48. True

49. Integers, rational, real. 50. −192

problem set 18

1. $(8 + N)2 = 36,\ N = 10$ 2. $4N + 14 = 30,\ N = 4$

3. $6(N + 4) = 42,\ N = 3$ 4. $S = (.15)80,\ S = 12$ lb

5. $6 = (WP)(120),\ WP = .05$ (5 percent) 6. $6 = .15(JM),\ JM = 40$

7. $9\sqrt{3} - 3\sqrt{2} + 6\sqrt{5}$ 8. $-5\sqrt{2}$ 9. 1, −2 10. −4, 5

11. 2, −3 12. $\dfrac{4}{ax}$ 13. $\dfrac{a-b}{4b-a}$ 14. $-\dfrac{1}{16}$ 15. 36

16. −36 17. 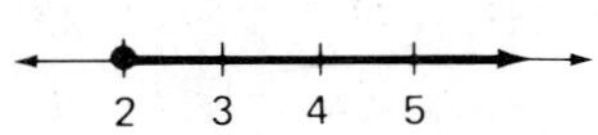 18. irrational, real

19. 3.1×10^{-4} **20.** 3.16×10^{-11} **21.** 3.16×10^{3}

22. 4.78×10^{4} **23.** 4.78×10^{-3} **24.** 4.78×10^{11} **25.** $(x - 15)(x + 1)$

26. $(x + y)(x - y)$ **27.** $x(x^2 + 6x + 5) = x(x + 5)(x + 1)$

28. $(x - 2)(x - 3)$ **29.** $(5y + 6a)(5y - 6a)$ **30.** $4bx(x - 5)(x + 4)$

31. $(x + 1)(x + 5)$ **32.** $(7m + a)(7m - a)$ **33.** $(x + 4)(-x + 5)$

34. $(x + 3)(x - 8)$ **35.** $(4x + 5a)(4x - 5a)$ **36.** $-3x(x + 6)(x - 4)$

37. $(x + 7)(x - 2)$ **38.** $5(a + x)(a - x)$ **39.** $m(x - 1)(x + 5)(x + 2)$

40. $(x - 3)(x - 8)$ **41.** $(a + 3)(a - 3)$ **42.** $b^2(x + 5)(x - 3)$

43. $(x - 1)(x - 5)$ **44.** $2(3x + 2a)(3x - 2a)$ **45.** $(x + 3)(x + 4)(x - 3)$

46.

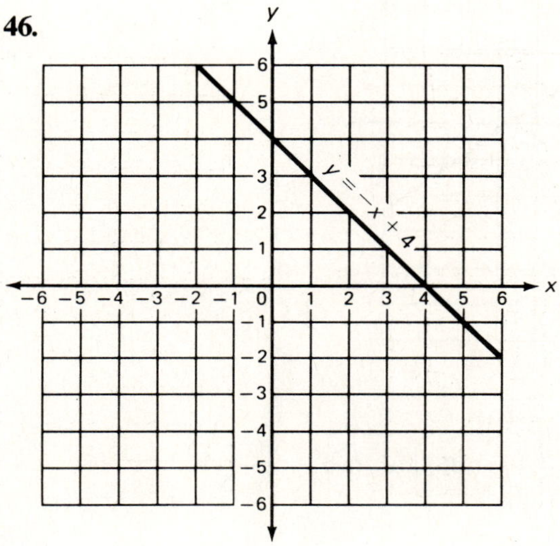

47. $\dfrac{bm + ax + bx + ay + by}{b(a + b)}$ **48.** $3 - 6y^2$ **49.** x^4z^{12}

problem set 19

1. $\underbrace{N} + \underbrace{N + 2} = 146$, $N = 72$, $N + 1 = 73$, $N + 2 = 74$

2. $\underbrace{2N} + \underbrace{2N + 2} = \underbrace{2N + 4} + (-10)$, $N = -4$, $\underbrace{2N = -8}$, $\underbrace{2N + 2 = -6}$, $\underbrace{2N + 4 = -4}$

3. $\underbrace{2N + 1} + \underbrace{2N + 3} = \underbrace{2N + 5} - 15$, $N = -7$, $\underbrace{2N + 1 = -13}$, $\underbrace{2N + 3 = -11}$, $\underbrace{2N + 5 = -9}$

4. $(3N + 7)4 = -20$, $N = -4$ **5.** $549 = (.18)(OP)$, $OP = \$3050$

6. $6 = (WP)(48)$, $WP = 12.5\%$ **7.** m **8.** am **9.** a **10.** ams

11. mas **12.** $-\frac{3}{4}$ **13.** 2 **14.** 7 **15.** 4.56×10^{-15}

16. 4.56×10^{-4} **17.** 4.56×10^{7} **18.** 4.56×10^{8}

19. $x(x^2 - 6x + 9) = x(x - 3)(x - 3)$ **20.** $(t - 5)(t + 5)$

21. $(2t + 5)(2t - 5)$ **22.** $(x - 3)(x + 4)$

23. $x(x^2 - x - 12) = x(x - 4)(x + 3)$ 24. $pxm(m - 4 + pm)$

25. $\frac{15}{8}$ 26. -2 27. 16 28. -30 29. 12 30. 16

31. $10\sqrt{2}$ 32. $2\sqrt{3}$ 33. $-4\sqrt{3}$ 34. (2, 2) 35. (3, 4)

36. (2, −4) 37. (3, 4) 38. $\frac{ac}{b}$ 39. $\frac{ac}{b}$ 40. $\frac{t^2 + 1}{t^2 - 1}$

41. $\frac{b^2 - a}{b^2 + a}$ 42. 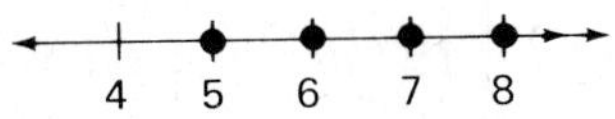

43.

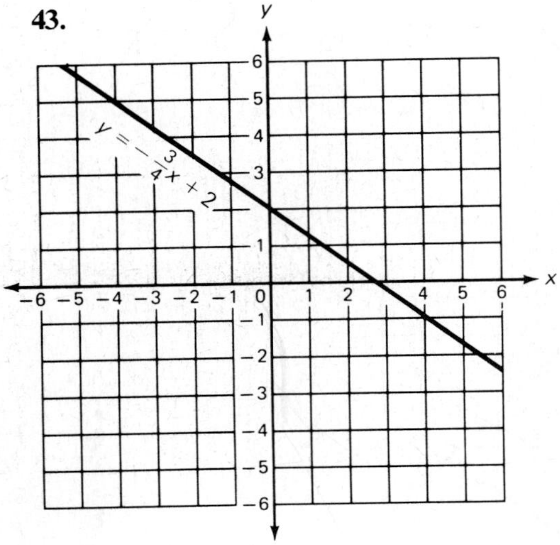

44. $\frac{x + x^2(b + c) + d}{x^3}$ 45. $x^{-3}y^2z - \frac{4x^{-3}y^2}{z}$ 46. $4 + \frac{12x^2}{y^4}$

47. -108

48. False. No irrational number is also a rational number and vice versa.

49. Rational, real

problem set 20

1. $N + N + 3 = 37$, $N = 17$, integers = 17, 18, 19, 20
2. $2N + 3(2N + 4) = 44$, $N = 4$, integers = 8, 10, 12
3. $2N + 3 + 2(2N + 5) = 19$, $N = 1$, integers = 3, 5, 7
4. $(N - 7)5 = 2N - 2$, $N = 11$
5. $180 = (.16)(AC)$, $AC = 1125$ 6. $320 = (WP)(800)$, $WP = .4$ (40 percent)
7. $R_A = 40$ 8. $R_F = 45$, $R_E = 60$ 9. $T_M = 1$, $T_W = 4$
10. 6×10^{13} 11. 4×10^{-9} 12. 2×10^{10}
13. 1×10^{-19} 14. Multiplication 15. Addition
16. $m = 20$ 17. $x = -66$ 18. $n = \frac{17}{7}$ 19. $x(x - 5)(x + 4)$
20. $(x - 7)(x - 5)$ 21. $x(x + 7)(x - 5)$ 22. $(x - 2)(x - 4)$
23. $(x - 2)(x - 1)$ 24. $(x + 6)(x - 6)$ 25. $(5x + 6)(5x - 6)$

26. $bp(x-5)(x+4)$ **27.** $\dfrac{4yz}{x^4}$ **28.** $\dfrac{xz^{10}}{y}$ **29.** $12 - \dfrac{16y^3}{x}$

30. $\dfrac{4}{y} + m$ **31.** $x = \frac{77}{8}$ **32.** -48 **33.** -7

34.

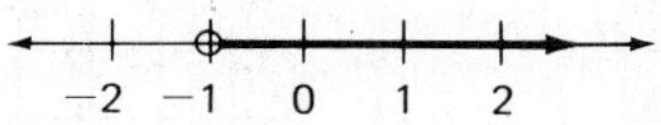

35.

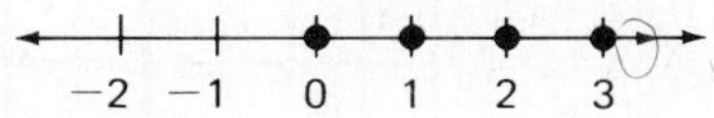

36.

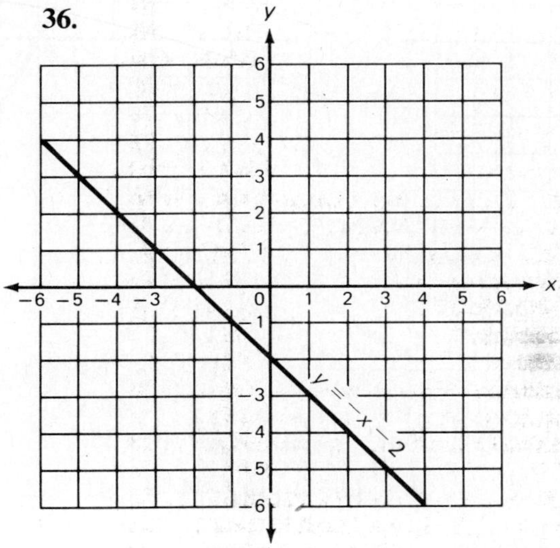

37.

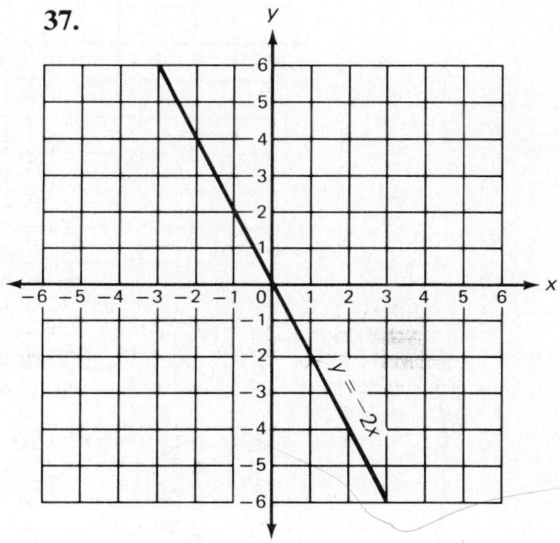

38. $\dfrac{ax + bx + cx^2 + 1}{x^2}$ **39.** $\dfrac{ab + ax + bc}{b^2 + bx}$ **40.** $\dfrac{ab+1}{mb-1}$

41. $\dfrac{\dfrac{ax-1}{x}}{\dfrac{1-ax}{x}} = \dfrac{ax-1}{1-ax} = \dfrac{ax-1}{(-1)(ax-1)} = -1$ **42.** $(2, -4)$ **43.** $\left(\frac{15}{11}, \frac{-3}{11}\right)$

44. $\dfrac{a}{ab + b^2}$ **45.** $\dfrac{xc}{y+m}$ **46.** $15\sqrt{2}$ **47.** $6\sqrt{3} - 9\sqrt{2} + 3\sqrt{5}$

48. 23 **49.** -12 **50.** $4x^2y^5m^{-2} + 5x^2m^{-5}y$

problem set 21

1. $5(2N+1) = \underbrace{2N+5} + (-32)$, $N = -4$, integers $= -7, -5, -3$

2. $(\underbrace{N} + \underbrace{N+1})2 = N + 3 + 32$, $N = 11$, integers $= 11, 12, 13, 14$

3. $3(2N) + 8 = 5(2N+4)$, $N = -3$, integers $= -6, -4, -2$

4. $(2N+5)3 = 57$, $N = 7$ **5.** $65 = (WP)(1300)$, $WP = .05$ (5 percent)

6. $8 = .2T_B$, $TB = \$40$ **7.** $20\sqrt{6} + 72$ **8.** 144

9. $24 + 20\sqrt{2}$ **10.** $20\sqrt{3} - 10\sqrt{15}$ **11.** $15\sqrt{5} - 100$

12. $24\sqrt{3} - 15\sqrt{2}$ **13.** $-12 + 15\sqrt{2}$ **14.** $-2\sqrt{2}$ **15.** $21\sqrt{2}$

16. $22\sqrt{3}$ **17.** $-2\sqrt{5}$ **18.** $T_H = 2$, $T_S = 2$

19. $R_F = 60$, $R_S = 50$ **20.** $T_M = 1$, $T_R = 4$

21. 2×10^{-30} **22.** $y = -\frac{4}{3}x + 3$ **23.** $y = \frac{5}{9}x - 1.8$

24. Addition and multiplication **25.** $x = \frac{26}{3}$ **26.** $x = -\frac{21}{2}$

27.

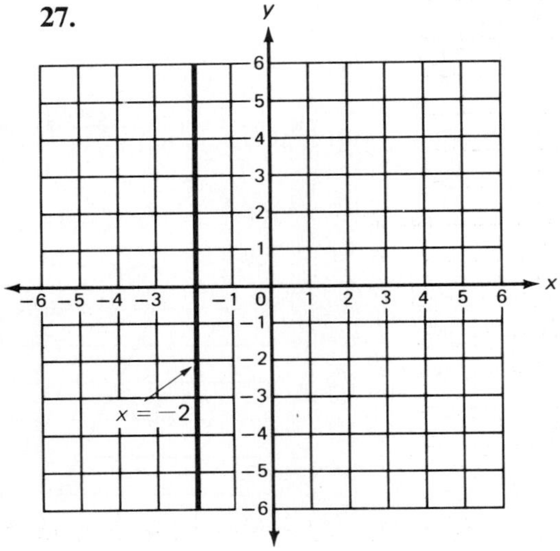

28.

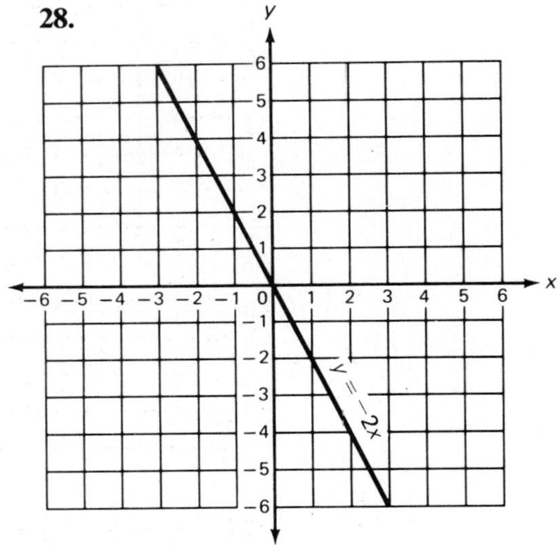

29. $\dfrac{2ax + bx + ay}{x^2 + xy}$ **30.** $\dfrac{a + cxa + cxb}{x^2(a + b)}$ **31.** $5a^{-2}yb^{-1} - 2a^{-2}y^{-1}b^{-1}$

32. $4x^{-5}b - 16x^{-4}$ **33.** $2b^4 - 6a^8$ **34.** k^7m^2 **35.** $a^4m^4y^2$ **36.** 13

37. $a = \frac{16}{13}$ **38.** $x = -35$ **39.** 17 **40.** 44 **41.** $(2, -4)$

42. $(3, -5)$ **43.** $\dfrac{a}{b(a + b)}$

44. $\dfrac{a - b}{b - a} = \dfrac{a - b}{-a + b} = \dfrac{(-1)(-a + b)}{-a + b} = -1$ **45.** $\dfrac{x - 1}{ay + 1}$

46. $(x + 3y)(x - 3y)$ **47.** $7(x + 3y)(x - 3y)$ **48.** $ap^2(x + 2y)(x - 2y)$

49. $mb(x + 5)(x - 4)$ **50.** $2(x + 3)(x + 3)$

problem set 22

1. $(N + 5)13 = 208 \longrightarrow$ $\mathbf{N = 11}$ **2.** $(.15)m = 33 \longrightarrow$ $\mathbf{m = 220}$

3. $(D + 4)12 = D + 136 \longrightarrow$ $\mathbf{D = 8}$

4. $(WP)(1100) = 55 \longrightarrow$ $WP = .05$ or 5 percent

5. $(A + 5)8 = A + 54 \longrightarrow$ $\mathbf{A = 2}$ **6.** $(.18)(1500) = K \longrightarrow$ $\mathbf{K = 270}$

7. $3x^8 - 2x^4 + x^2$ **8.** $3x^2 + 7x + 26 + \dfrac{128}{x - 3}$

9. $3x^2 + 9x + 25 + \dfrac{80}{x - 3}$ **10.** $2x^2 - 10x + 49 - \dfrac{258}{x + 5}$

11. $-3, 6$ **12.** $\frac{5}{2}, -\frac{5}{2}$ **13.** $-8, 7$ **14.** $\frac{97}{3}$ **15.** -65

16. $\dfrac{m^2b + b^2 + ma + mb}{b(a + b)}$ **17.** $\dfrac{x^2a + (x + y)(b + c)}{x^2(x + y)}$

18. $\dfrac{3y}{a^2x^2} + \dfrac{2yx^2}{a^2}$ **19.** $\dfrac{8a^2}{x^4}$ **20.** $y = -\frac{1}{3}x + 2$ **21.** $y = 3$

22., **23.**

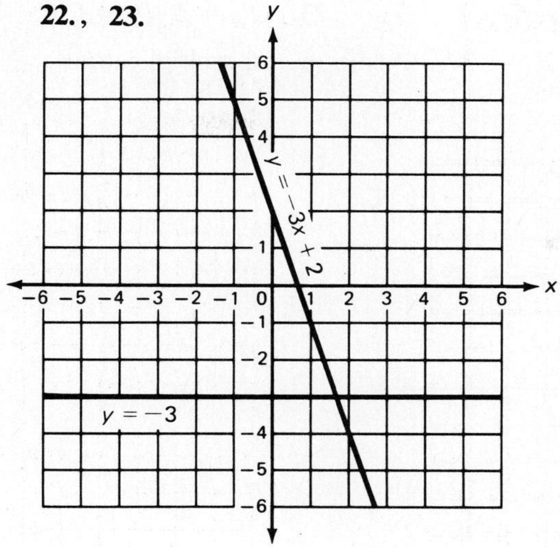

24. $T_g = 5, \quad T_B = 8$ **25.** $T_N = 16, \quad T_K = 8$ **26.** 2×10^{-48}

27. 1×10^{-20} **28.** $60 - 8\sqrt{3}$

29. $16\sqrt{3} - 12\sqrt{6}$ **30.** $-14\sqrt{3}$ **31.** $-5\sqrt{2}$ **32.** $\dfrac{3z^5x^4}{p^4} + \dfrac{2}{p^{11}}$

33. $\dfrac{k^3}{x^8p^7y^5}$ **34.** $\dfrac{x^9a^2}{y^{13}}$ **35.** $\dfrac{x^{12}}{y^5}$ **36.** 23 **37.** $\frac{22}{9}$

38. -3 **39.** 6 **40.** 20 **41.** $\frac{1}{25}$ **42.** $-\frac{1}{25}$ **43.** -25

44. False. The rationals are not a *member* of the reals. The rationals are a subset of the reals.

45. Rational, reals **46.** $(2, -3)$ **47.** $\dfrac{x}{y - 1}$ **48.** $\dfrac{x + y}{xy + a}$

problem set 23

1. $N_N + N_D = 28, \quad 5N_N + 10N_D = 195, \quad N_N = 17, \quad N_D = 11$

2. $N_F + N_T = 52, \quad 30N_F + 25N_T = 1360, \quad N_F = 12, \quad N_T = 40$

3. $20N_C + 45N_E = 1375, \quad N_C = N_E + 20, \quad N_E = 15, \quad N_C = 35$

4. $25N_Q + 10N_D = 445, \quad N_D + 8 = N_Q, \quad N_D = 7, \quad N_Q = 15$

5. $N_C + N_N = 92, \quad N_C = N_N + 6, \quad N_N = 43, \quad N_C = 49$

6. $N_B + N_G = 29, \quad N_B = N_G + 7, \quad N_B = 18, \quad N_G = 11$

7. $3(2N + 1) = \underbrace{2N + 5} + 2, \quad N = 1$, integers = 3, 5, 7

8. $300 = (.96)(TW), \quad TW = \$312.50$

9.

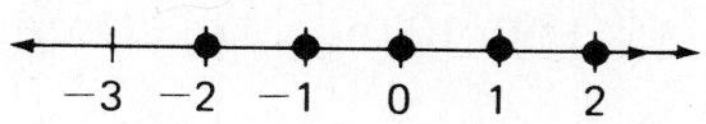

10.

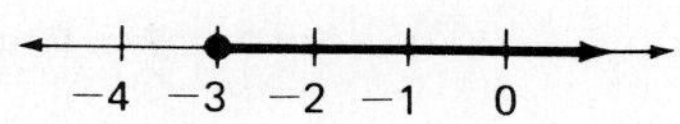

11. $12 - 8\sqrt{6}$ **12.** $18 + 9\sqrt{6}$ **13.** $70 - 15\sqrt{14}$ **14.** $6\sqrt{15} - 45$

15. $9\sqrt{2}$ **16.** $6\sqrt{6} - 10\sqrt{5}$ **17.** $x^2 - 6x + 25 - \dfrac{105}{x + 4}$

18. $x^2 + 4x + 16 + \dfrac{59}{x - 4}$ **19.** $x = 6, -6$ **20.** $x = -4, -3$

21. $x = -6, 2$ **22.** $x = -7, -3$ **23.** $T_B = 2, T_T = 6$ **24.** 2×10^{-2}

25. $y = 2x$ **26.** $y = 2$ **27.** Addition and multiplication **28.** $x = -\frac{19}{6}$

29.

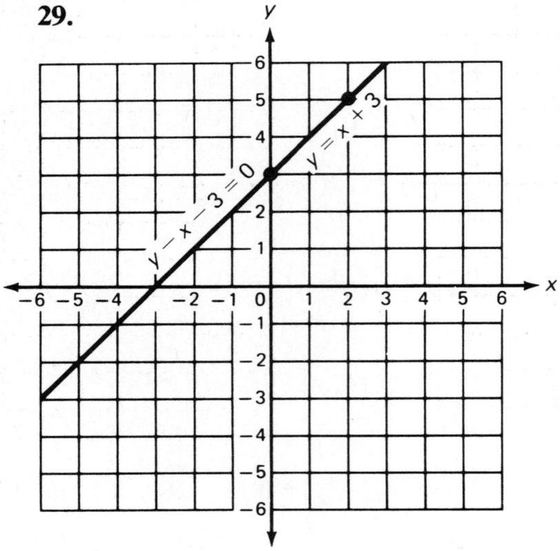

30. $\dfrac{y + a + bx^2}{x^2(m + c)}$ **31.** $3a^2xym^{-1} - 3a^2ym^{-1}$ **32.** $1 + a^{-1}x^{-3}$

33. $b^4 + 4a^2$ **34.** m^2k^6 **35.** $a^7b^{-4}c^{-7}$ **36.** 28 **37.** $x = 2$

38. $m = \frac{1}{2}$ **39.** 18 **40.** 17 **41.** $\frac{1}{4}$ **42.** 4 **43.** 4

44. False. The set of whole numbers is a subset of the set of integers. $\subset$ means *subset*, while $\in$ means *is a member of*.

45. Irrational, real

problem set 24

1. $25N_Q + 10N_D = 320, \quad N_D + 3 = N_Q, \quad N_Q = 10, \quad N_D = 7$

2. $N_7 + N_9 = 14, \quad 7N_7 + 9N_9 = 110, \quad N_7 = 8, \quad N_9 = 6$

3. $L + S = 20, \quad L = S + 8, \quad L = 14, \quad S = 6$

4. (a) $(a + b) + c = a + (b + c)$; (b) $ab = ba$; (c) $a(b + c) = ab + ac$

5. $(\underbrace{2N} + \underbrace{2N + 2})2 = \underbrace{2N + 4} + 48, \quad N = 8,$ integers = 16, 18, 20

6. (a) 0; (b) 1; (c) $\dfrac{1}{a}$

7. $762 = (.08)(N), \quad N = 9525$ **8.** $N = (.14)(1200), \quad N = 168$

9.

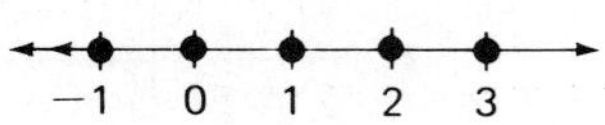

10. $12 + 9\sqrt{10}$ **11.** $12\sqrt{6} + 12\sqrt{3}$

12. $6\sqrt{3} - 36$ **13.** $28\sqrt{2}$ **14.** $48\sqrt{15}$ **15.** $x^2 - 5x + 20 - \dfrac{70}{x + 4}$

16. $x^3 + x^2 + x - 2 + \dfrac{3}{x - 1}$ **17.** $+\frac{6}{5}, -\frac{6}{5}$ **18.** $x = 3, 2$

19. $x = 7, 4$ **20.** $x = 10, -5$ **21.** $T_L = 14, T_S = 6$ **22.** 1×10^{-19},

23. $y = -1.9x - 2$ **24.** $x = -2$ **25.** Multiplication **26.** $x = \frac{24}{13}$

27.

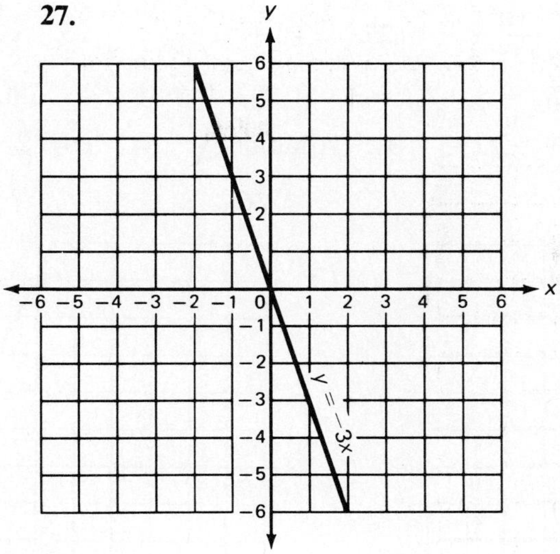

28.

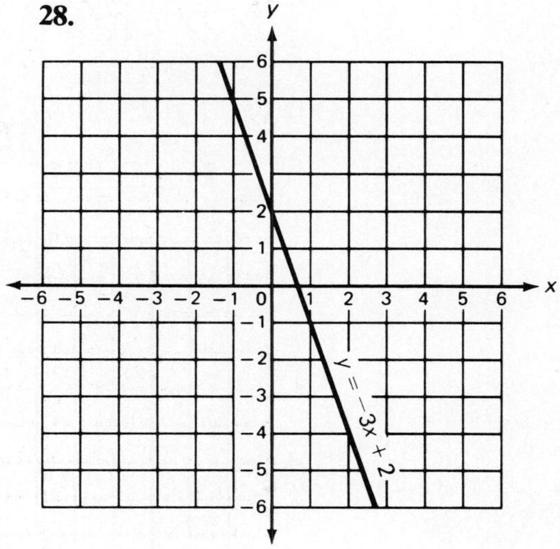

29. $\dfrac{axc^2 + bc + dx}{x^3c^2}$ **30.** $\dfrac{am + k + m}{km + m^2}$ **31.** $-3a^{-2}b$

32. $a^{-6}b^{-2} + 2a^{-4}b^2x^{-1}$ **33.** $b^3 + 2a^8$ **34.** $a^{-2}xp^{-1}m^2$ **35.** $\dfrac{x^2}{a^2}$

36. 7 **37.** $x = \frac{1}{6}$ **38.** $y = -1$ **39.** -6 **40.** 10 **41.** $\dfrac{x + 5}{x - 7}$

42. $\dfrac{x - 3}{x + 1}$ **43.** -16 **44.** True **45.** Rational, reals **46.** $p = \pm\sqrt{41}$

47. $k = \pm\sqrt{13}$ **48.** $p = \pm 2$

problem set 25

1. $N_D + N_Q = 50$, $10N_D + 25N_Q = 1100$, $N_D = 10$, $N_Q = 40$

2. $S + L = 535$, $S + 15 = L$, $S = 260$, $L = 275$

3. $N_A + N_C = 1300$, $1N_A + .5N_C = 1000$, $N_A = 700$, $N_C = 600$

4. $N_{13} + N_{20} = 40$, $13N_{13} + 20N_{20} = 660$, $N_{13} = 20$, $N_{20} = 20$

5. $6800 = (.17)(CH)$, $CH = 40{,}000$ **6.** $DP = (.2)(6400)$, $DP = 1280$

7. $10(2N + 1) = 4(2N + 5) + 2$, $N = 1$, integers = 3, 5, 7

8. $3N = 2(N + 2) + 17$, $N = 21$, integers = 21, 22, 23

9. $\sqrt{33}$ **10.** $\sqrt{89}$ **11.** $D = \sqrt{65}$ **12.** $xyz = (xy)z = (yx)z = yxz$

13. $x + y + z = x + (y + z) = x + (z + y) = (x + z) + y$
$= (z + x) + y = z + (x + y) = z + (y + x) = z + y + x$

14. Additive property of zero, $0 + a = a$; multiplicative property of zero, $0 \cdot a = 0$; multiplicative property of one, $1 \cdot a = a$.

15. See body of text. **16.** Distributive property **17.** Reciprocal

18. Opposite of **19.** $a + b = b + a$ **20.** $(4 \cdot 7)8 = 4(7 \cdot 8)$

21. $30 - 12\sqrt{5}$ **22.** $42\sqrt{6}$ **23.** $18\sqrt{5} + 3\sqrt{15}$

24. $x^3 + 2x^2 + 3x + 8 + \dfrac{19}{x-2}$ **25.** $x = +\frac{7}{3}, -\frac{7}{3}$ **26.** $x = 4, 1$

27. $x = 5, -7$ **28.** $T_E = 5, \quad T_W = 4$ **29.** 2×10^{-5} **30.** $(5, -4)$

31. $y = -2$ **32.** $y = -2x$ **33.** Addition **34.** $x = \frac{16}{25}$

35.

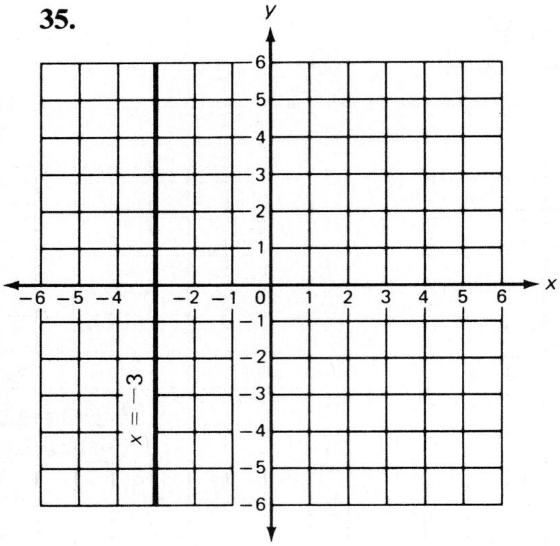

36.

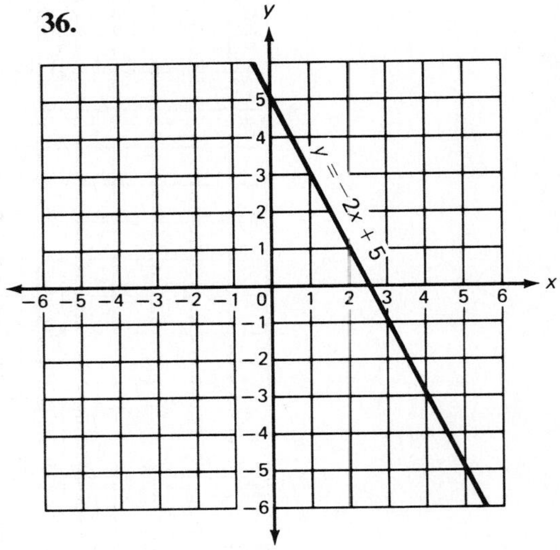

37. $\dfrac{xb}{a+b}$ **38.** $\dfrac{ab+a}{xb-1}$ **39.** $\dfrac{mc+md+ab}{a^2c+a^2d}$ **40.** $\dfrac{kb+kc+b^4L}{b^5+b^4c}$

41. $2x^{-2}a + 2x^3a^{-2}$ **42.** $\dfrac{x+3}{x+5}$ **43.** p^4 **44.** x^2y^3 **45.** $\frac{30}{11}$

46. 0 **47.** 9 **48.** False. Would be true if $\subset$ had been used instead of $\in$.

49. Whole, integers, rational, real.

50.

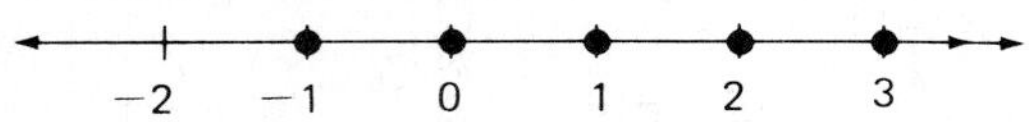

problem set 26

1. $N_C + N_{CO} = 40, \quad 150N_C + 15N_{CO} = 4650, \quad N_C = 30, \quad N_{CO} = 10$

2. $P_1 + P_2 = 98, \quad 2P_1 = P_2 + 10, \quad P_1 = 36, \quad P_2 = 62$

3. $300N_P + 150N_B = 3000, \quad N_P = 2N_B, \quad N_B = 4, \quad N_P = 8$

4. $25N_Q + 5N_N = 2750, \quad N_Q = 2N_N, \quad N_Q = 100, \quad N_N = 50$

5. $10.80 = (.06)(TVP), \quad TVP = \$180.00$

6. $90 = (WP)(1500), \quad WP = .06$ (6 percent)

7. $5N = 3(N+2) + 10, \quad N = 8,$ integers = 8, 9, 10

8. $8(2N+1) = 3(2N+5) + 3, \quad N = 1,$ integers = 3, 5, 7

9. $100\sqrt{5}$ **10.** 600 **11.** $(2.03961) \times 10^1$ **12.** $(2.29565) \times 10^{-2}$

13. $(2.29565)(3.16228) \times 10^{-3}$ **14.** $(2.14912)(3.1622) \times 10^{-7}$

15. $m = 2\sqrt{6}$ **16.** $\sqrt{65}$

17. $xyzk = xy(zk) = xy(kz) = x(yk)z = x(ky)z = xkyz$

18. $3 \cdot 4 = 4 \cdot 3$ **19.** See section 24.7.

20. $\frac{2}{3}x = 5$ **21.** $28\sqrt{6}$ **22.** $120\sqrt{2}$ **23.** $-20\sqrt{2}$

24. $4x - 10 + \dfrac{25}{x + 2}$ **25.** $t = +5, -5$ **26.** $p = 6, -7$

27. $k = 9, -5$ **28.** $T_L = 14, T_S = 6$ **29.** 6×10^{-22}

30. $(2, -2)$ **31.** $y = -\frac{9}{7}x - 1$ **32.** $y = 3$

33. Addition, multiplication, division. **34.** $x = \frac{5}{4}$

35.

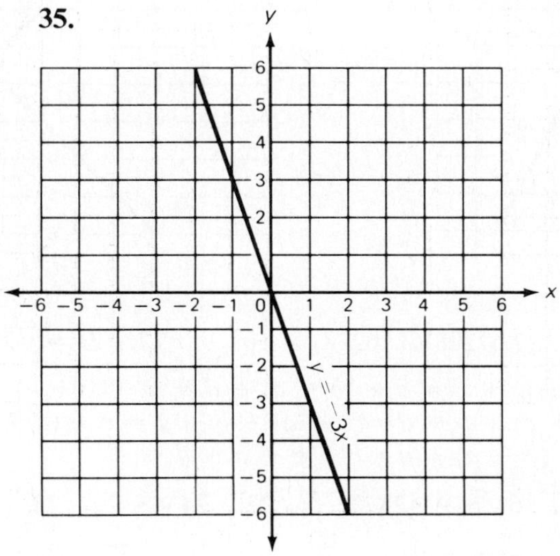

36.

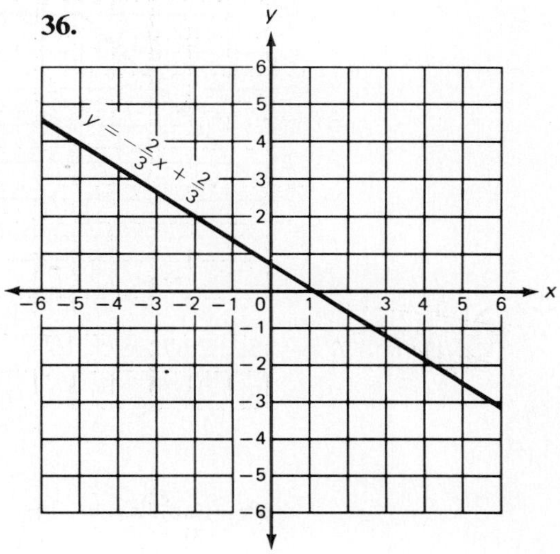

37. $\dfrac{1}{ca + bc}$ **38.** $\dfrac{mc - 4}{k + mc}$ **39.** $\dfrac{b^2m + k + b}{b^2(k + b)}$

40. $\dfrac{a^2x + abx + ma + mb + a^2k}{a^2(a + b)}$ **41.** $4a^2x^{-2} - 4a^6x^2$ **42.** $\dfrac{x + 7}{x + 3}$

43. x^6m^{-2} **44.** x^2 **45.** $p = -20$ **46.** -10 **47.** $\frac{1}{9}$

48. False **49.** Integers, rational, reals. **50.** $\dfrac{6x - px - 3p}{x^2 - x - 12}$

51. $\dfrac{2x - 5}{x(x - 6)(x + 1)}$

problem set 27

1. $N_S + N_D = 75$, $50N_S + 20N_D = 3300$, $N_S = 60$, $N_D = 15$

2. $25N_S + 35N_B = 680$, $2N_B = N_S$, $N_S = 16$, $N_B = 8$

3. $L + S = 50$, $3S + 10 = L$, $L = 40$, $S = 10$

4. $125N_W + 75N_H = 3500$, $N_W + 20 = N_H$, $N_W = 10$, $N_H = 30$

5. $340 = (.20)(B_R)$, $B_R = 1700$ **6.** $R_P = (.15)(800)$, $R_P = 120$

7. $5(2N + 1) + 6 = 4(2N + 5) + 1$, $N = 5$, integers $= 11, 13, 15$

8. $4(2N) = 3(2N + 2)$, $N = 11$, integers $= 22, 24$

9.

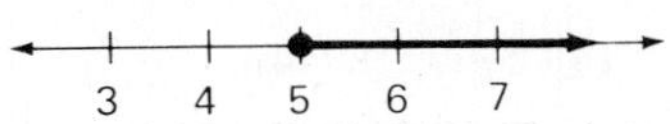

10. {integers}

11.

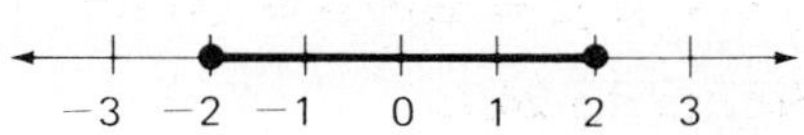

12.

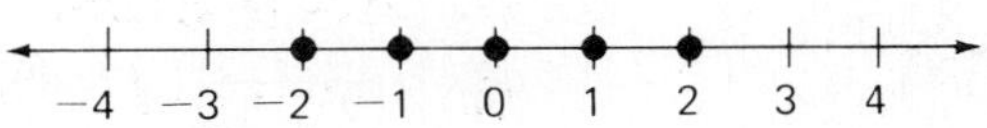

13.

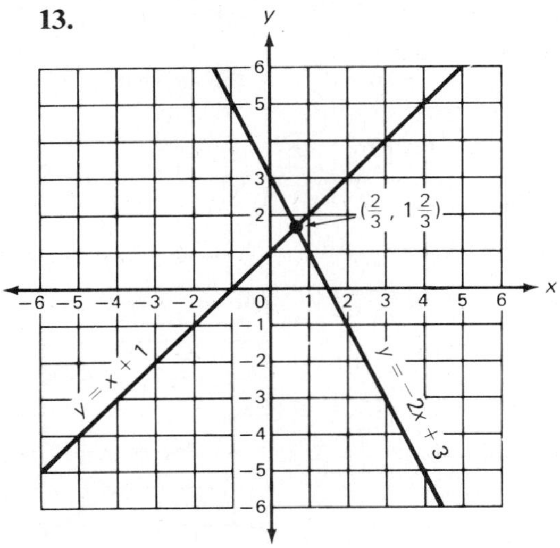

14.

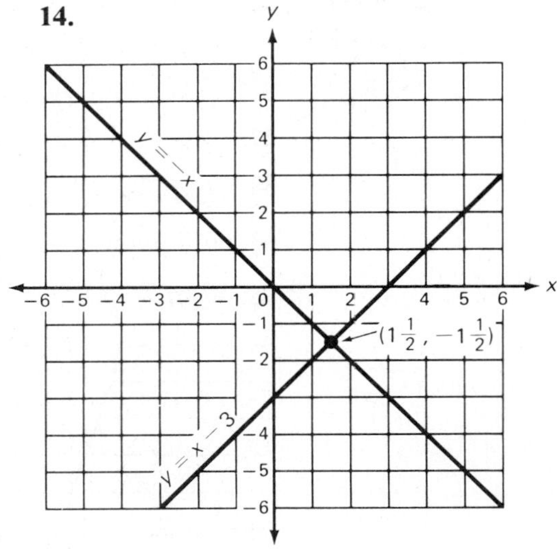

15. 2.67208×10^3 **16.** $(2.67208)(3.16228) \times 10^3$ **17.** $H = \sqrt{29}$

18. $7\sqrt{2}$ **19.** $a + b + c + d = a + (b + c) + d = a + (c + b) + d$
$= (a + c) + b + d = (c + a) + b + d = c + (a + b) + d$
$= c + (b + a) + d = c + b + a + d$

20. See lesson 24. **21.** $6\sqrt{6} + 12\sqrt{2}$ **22.** $24\sqrt{3}$ **23.** $15\sqrt{3} - 30$

24. $x^2 - 3x + 1$ **25.** $x = 3, -10$ **26.** $x = 7$ **27.** $R_1 = 20$, $R_2 = 50$

28. 4×10^1 **29.** $x = 1, y = 2$ **30.** $y = x - 1$

31. $x = -4.5$ **32.** Addition and multiplication **33.** $x = \frac{14}{17}$

34.

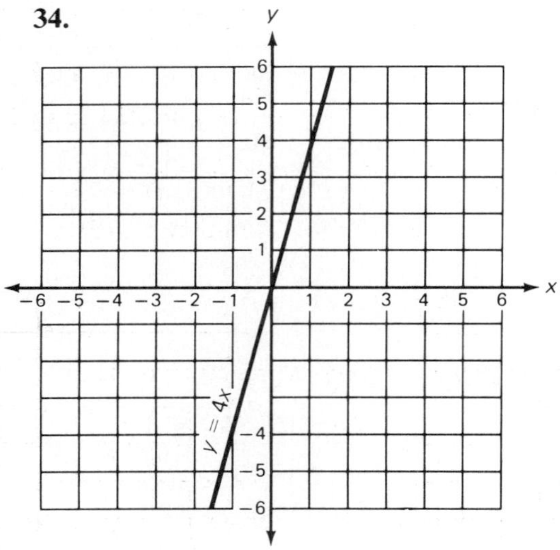

35.

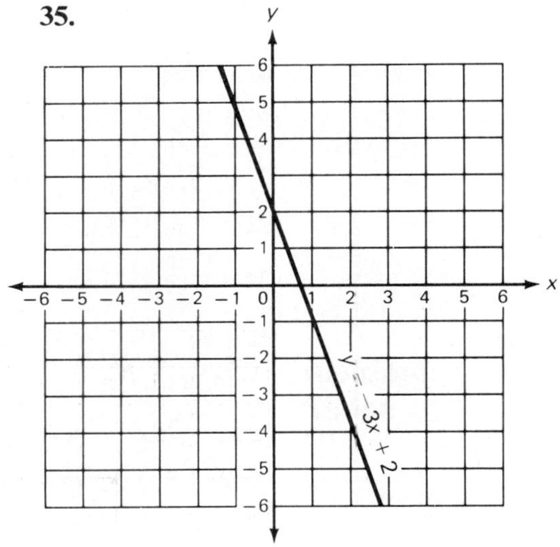

36. $\dfrac{b^2 + ab}{a}$ **37.** $\dfrac{x - y}{x - a}$ **38.** $\dfrac{x^2c^2a + dxc + c}{x^3c^2}$ **39.** $\dfrac{ky + x + y}{y(x + y)}$

40. $a^5c^{-5} + 3a^5b^{-1}c^{-1}$ **41.** $3a^{-2}b^{-1} + 2ab^3 + ab$ **42.** $x^2a^{-1}m^{-1}p^2$

43. $\dfrac{x + 8}{x - 4}$ **44.** $x = -6$ **45.** -7 **46.** $-\frac{1}{81}$ **47.** Irrational, reals

48. $\dfrac{-5}{x^2 - 3x}$ **49.** $\dfrac{-4p}{x^2 - 5x - 6}$

problem set 28

1. $2N_P + 3N_C = 200$, $N_C = 2N_P$, $N_C = 50$, $N_P = 25$
2. $S + L = 90$, $L = 5S + 18$, $S = 12$, $L = 78$
3. $10N_S + 15N_H = 625$, $N_S + N_H = 55$, $N_H = 15$
4. $50N_T + 300N_I = 9500$, $N_T + N_I = 40$, $N_T = 10$
5. $I_C = (.60)(1230)$, $I_C = 738$ **6.** $72 = (.80)(WN)$, $WN = 90$
7. $3[N + (N + 1)] = 2[(N + 2) + (N + 3)] + 1$, $N = 4$, integers = 4, 5, 6, 7
8. $6(2N + 5) = 5(2N + 4) + 6$, $N = -2$, integers = $-4, -2, 0$
9. $t = 5$ **10.** $x = -\frac{7}{9}$ **11.** $x = \frac{12}{55}$ **12.** $n = 2$
13. C is not a function; 1 has two images.
14. 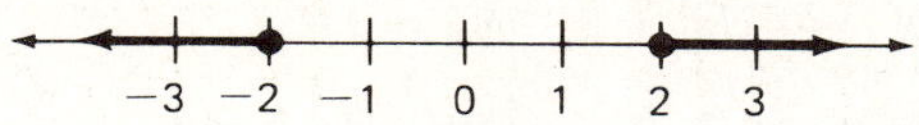

15. 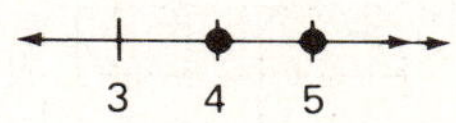

16. 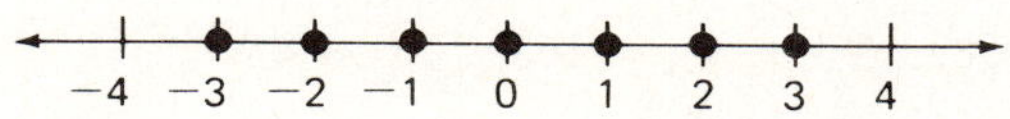

17. 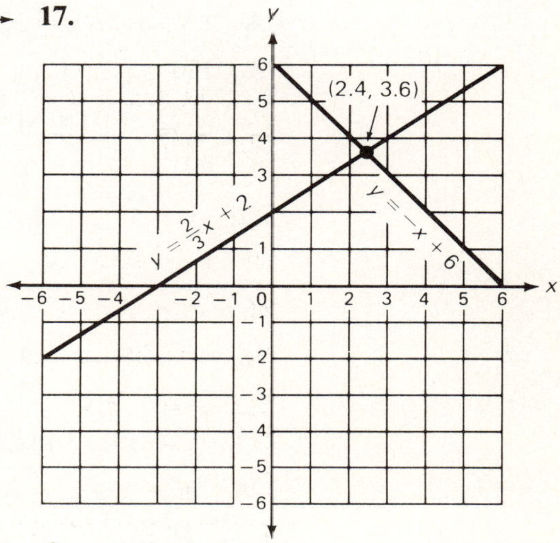

18. 1.15326×10^{-4}
19. $(1.15326)(3.16228) \times 10^{-5}$
20. $m = \sqrt{21}$ **21.** $\sqrt{58}$
22. $$\begin{aligned} x + y + m &= x + (y + m) \\ &= x + (m + y) \\ &= x + m + y \end{aligned}$$
23. $60 + 3\sqrt{6}$
24. $48\sqrt{3}$
25. $18\sqrt{3} + 4\sqrt{13}$ **26.** $2x^2 + x - 3 - \dfrac{9}{x - 2}$ **27.** $x = 2, 5$ **28.** 3
29. $R_1 = 23$, $R_2 = 66$ **30.** 2×10^2 **31.** $(2, -3)$
32. $y = -\frac{3}{4}x + 3$ **33.** $y = -3$ **34.** Addition **35.** $x = 12$
36. 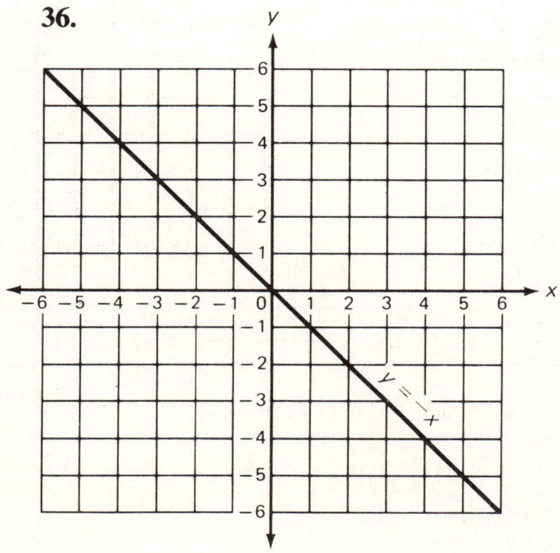

37. 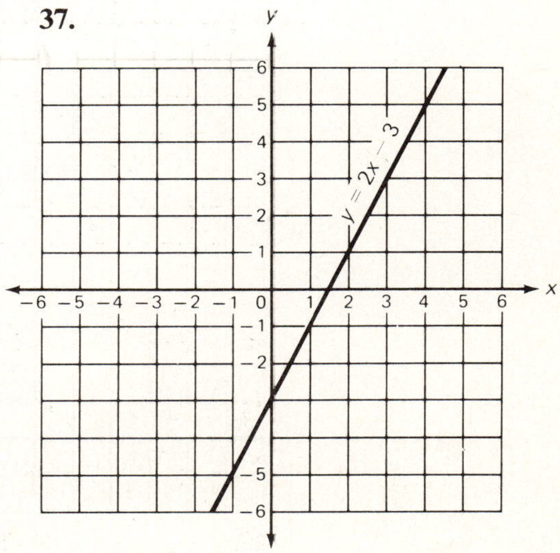

38. $\dfrac{a}{x + a}$ **39.** $\dfrac{1}{ab + b^2}$ **40.** $\dfrac{ab - bc - c^2}{b^2 + bc}$ **41.** $\dfrac{x^2y^2c^2 + zyc + 1}{y^3c^2}$

42. $x^{-6}y^{-2} + 2x^{-4}y^2b^{-1}$ **43.** $-2x^{-4}y + 3x^4y - 2b^{-3}y$

44. $\dfrac{b^2}{a^2y}$ **45.** $\dfrac{x + 6}{x - 2}$ **46.** $x = -10$ **47.** -8 **48.** $-\frac{1}{81}$

49. False. It's the other way around. **50.** $\dfrac{10p - 6k}{x^2 - 9}$ **51.** $\dfrac{8}{x^2 + 6x + 8}$

problem set 29

1. $3N_S + 5N_L = 559$, $N_S + N_L = 153$, $N_L = 50$

2. $N_W + N_B = 143$, $N_B = 4N_W + 8$, $N_B = 116$

3. $60N_L + 20N_S = 4000$, $N_L + N_S = 140$, $N_L = 30$

4. $1N_S + 3N_Y = 258$, $N_S + N_O = 168$, $N_O = 45$

5. $180 = (.2)(WN)$, $WN = 900$ **6.** $238 = (WP)(1400)$, $WP = .17$ (17 percent)

7. $5[(2N + 1) + 7] = -(10)(2N + 5)$, $N = -3$, integers $= -5, -3, -1$

8. $f(2) = 10$ **9.** $g(4) = 9$ **10.** $\phi(-2) = -1$ **11.** (a), (e), (f)

12. $x = 5$ **13.** $x = \frac{41}{7}$ **14.**

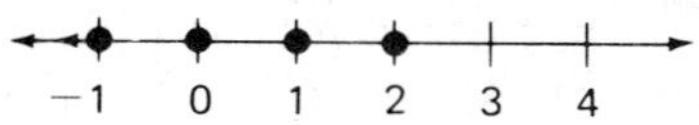

15.

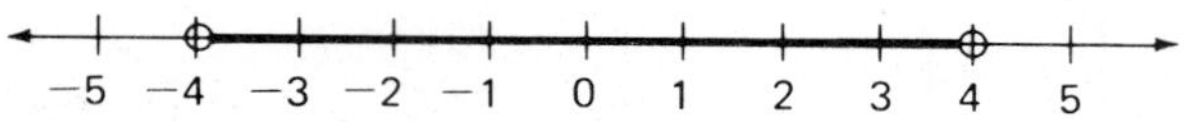

16.

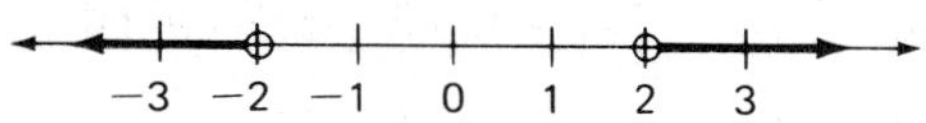

17. $(2.48193)(3.16228) \times 10^4$ **18.** 2.48193×10^4 **19.** $\sqrt{106}$

20. $m + x + y = (m + x) + y = (x + m) + y = x + m + y$

21. $4 + 8\sqrt{2}$ **22.** 0 **23.** 72 **24.** $x^2 + x + \dfrac{2}{x - 1}$

25. $x = -6, -2$ **26.** $2x^2 - 2x$ **27.** 4.9×10^{-15} **28.** $x = 4$, $y = -2$

29. $y \approx -\frac{8}{9}x - \frac{3}{2}$ **30.** $x \approx -3.6$ **31.** Multiplication

32.

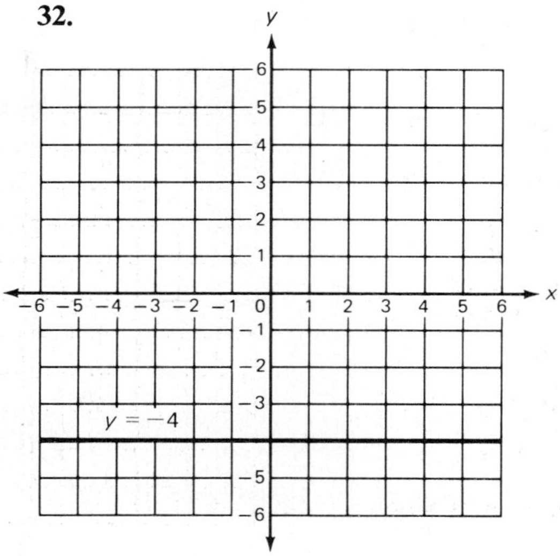

33.

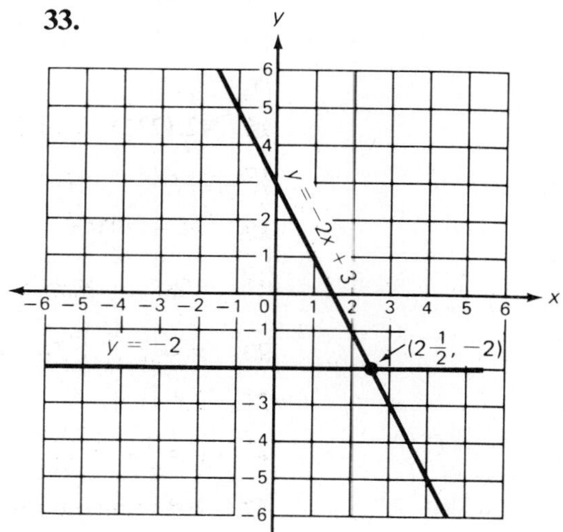

34. $\frac{1}{a}$ **35.** $\frac{x}{a^2}$ **36.** $\frac{ax + cx + a^2b}{a^3 + a^2c}$ **37.** $\frac{ayc + by + dc}{y^2c^2}$

38. $6a^2cx - 2acx^2$ **39.** $\frac{x - 5}{x + 7}$ **40.** $a^2 - 4$

41. $\frac{x^4}{m^2p^9}$ **42.** $\frac{p^2}{ax}$ **43.** $x = 5$ **44.** -9 **45.** 1

46. False. Wrong symbol used. **47.** Rational, Reals

48. $\frac{-12k}{x^2 - x - 20}$ **49.** $\frac{3p - 2x + 2}{x^2 - x}$ **50.** $T_M = 10,\ T_P = 20$

problem set 30

1. 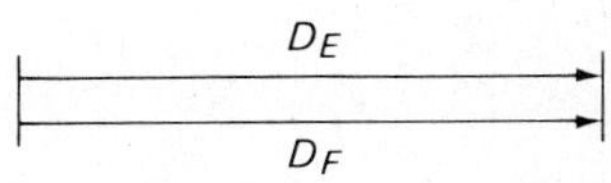

$R_ET_E = R_FT_F,\ R_E = R_F + 15,\ T_E = 12,\ T_F = 16,\ R_E = 60,\ R_F = 45$

2. 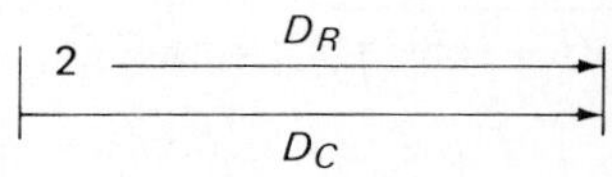

$R_RT_R + 2 = R_CT_C,\ R_R = 70,\ R_C = 100,\ T_R = T_C,\ T = \frac{1}{15}\text{hr} = 4\text{ min}$

3. 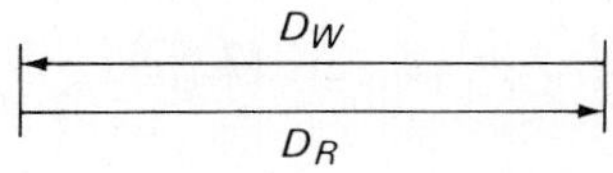

$R_WT_W = R_RT_R,\ T_W + T_R = 10,\ R_W = 2,\ R_R = 18,\ D = 18$

4. 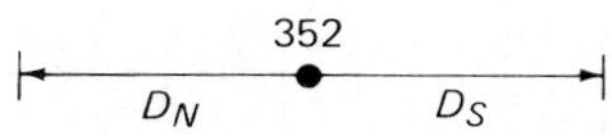

$R_ST_S + R_NT_N = 352,\ R_S = R_N,\ T_S = 5,\ T_N = 3,\ R_S = R_N = 44$

5. $L + S = 98,\ L + 10 = 2S,\ L = 62,\ S = 36$

6. $N_B + N_N = 142,\ N_B = 3N_N + 10,\ N_N = 33$

7. $(WN) = (.04)(12{,}000),\ WN = 480$

8. $12(2N + 1) = 5[(2N + 3) + 3] + 10,\ N = 2,$ integers = 5, 7, 9

9. $f(-2) = 2$ **10.** $g(3) = 8$ **11.** (a), (b), (d) **12.** $x = \frac{268}{15}$

13. $x = -\frac{8}{5}$ **14.** 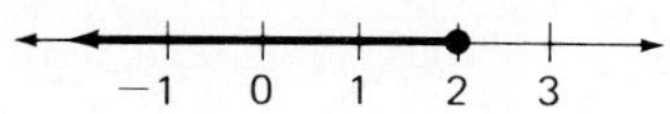

15. 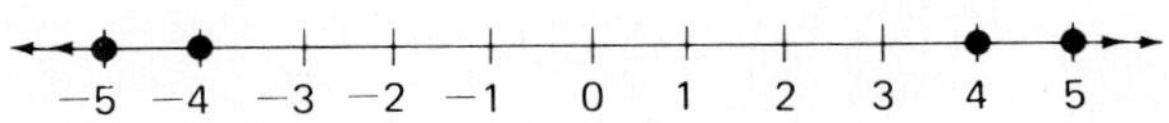

16. $(2.12368)(3.16228) \times 10^{-4}$ **17.** $\sqrt{58}$

18. $xyp = x(yp) = x(py) = xpy$

19. $24 - 20\sqrt{3}$ **20.** $4\sqrt{2}$

21. 108 **22.** $x - 5 + \frac{21}{x + 5}$

23. $x = 6$ **24.** $4z + 6$

25. 2.4×10^{-2} **26.** $x = 1,\ y = 4$

27. $y \approx \frac{1}{2}x + 4.5$ **28.** $x = 3$

29. Addition, subtraction, multiplication.

30. 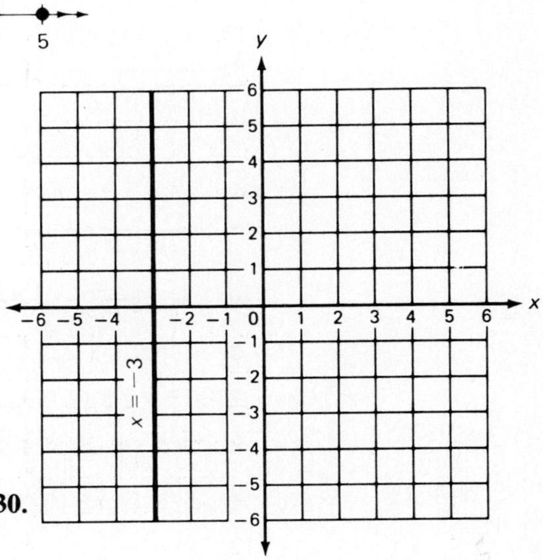

31.

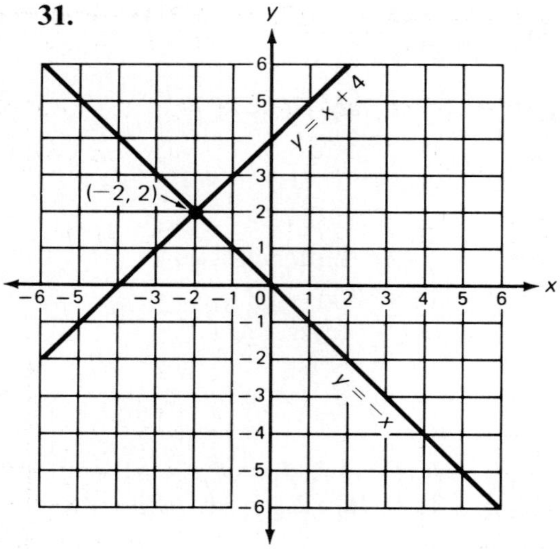

32. y **33.** $\frac{ab}{1 + b}$ **34.** $\frac{ax + ay + c^2x^2}{x^3 + x^2y}$ **35.** $\frac{b^2m + bx + ak}{a^2b^2}$

36. $-3ax^2 + 5ax^{-2}$ **37.** $-3x^2am^{-2} + 5x^2a^{-1}m^{-2}$ **38.** $\frac{x - 3}{x + 5}$

39. $\frac{-4y^2x^3}{z}$ **40.** $\frac{m^4y^2}{x^3}$ **41.** $y = 25$ **42.** 4 **43.** $-\frac{1}{16}$

44. No way! **45.** Natural, wholes, integers, rationals, reals.

46. $\frac{px}{x^2 - 25}$ **47.** $\frac{3mx}{x^2 + 2x - 8}$

problem set 31

1.

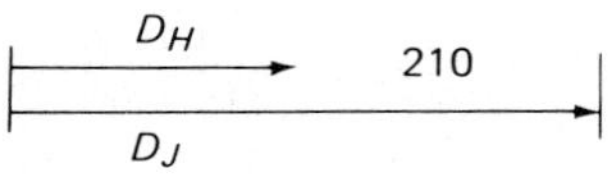

$R_JT_J = R_HT_H + 210$, $R_J = 230$, $R_H = 200$, $T_J = T_H$, $T = 7$

2.

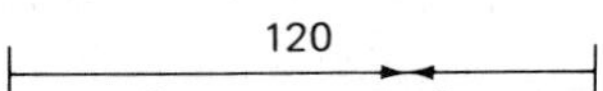

$R_FT_F + R_ST_S = 120$, $R_F = R_S + 4$, $T_F = 3$, $T_S = 3$, $R_F = 22$, $R_S = 18$

3.

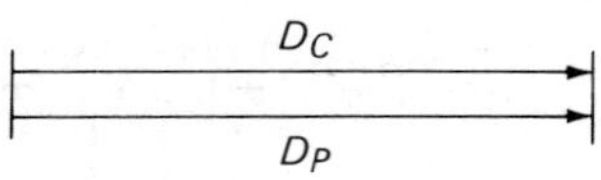

$R_CT_C = R_PT_P$, $R_C = 250$, $R_P = 350$, $T_C = T_P + 2$, $T_P = 5$ hr

4.

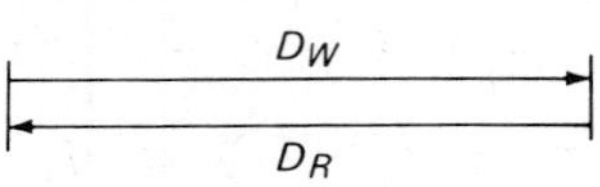

$R_WT_W = R_RT_R$, $R_W = 5$, $R_R = 20$, $T_W + T_R = 1$, $D = 4$

5. $N_E + N_{MC} = 752$, $N_E = 15N_{MC} + 144$, $N_E = 714$

6. $175N_C + 225N_S = 4075$, $N_C + N_S = 23$, $N_C = 22$

7. $12 = (.8)(LVO)$, $LVO = 15¢$

8. $5(N) = 2[(N + 2) + 4] + 6$, $N = 6$, integers = 6, 7, 8

9. $y \approx -\frac{1}{2}x + 2$ **10.** $y = \frac{5}{9}x - \frac{2}{9}$ **11.** $y = -\frac{4}{3}x - \frac{13}{3}$

12. $y = -\frac{3}{4}x + \frac{25}{4}$ **13.** $\phi(0) = 2$ **14.** (b), (c), (d), (e) **15.** $x = 5$

16. $t = \frac{276}{5}$ **17.** 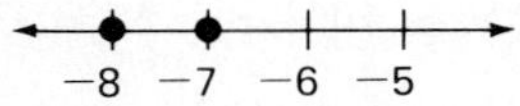

18. 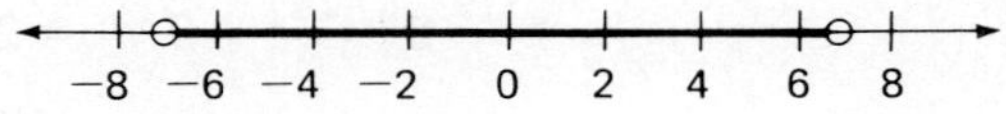

19. $p = 3\sqrt{3}$ **20.** $xypm = xy(pm) = xy(mp) = x(ym)p = x(my)p = xmyp$

21. $-5 - \sqrt{3}$ **22.** $12 - 18\sqrt{3}$ **23.** 28 **24.** $x^2 + 7x + 43 + \dfrac{172}{x - 4}$

25. $x = 1, 8$ **26.** $\dfrac{a - b}{a - 4}$ **27.** 1×10^{-15} **28.** (3, 4)

29. Addition

30. 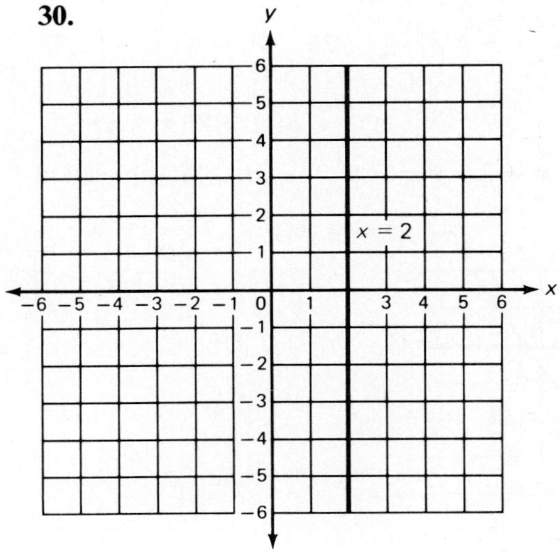

31. 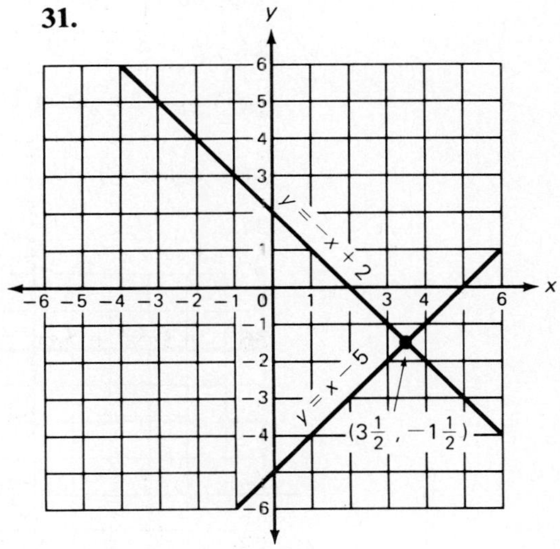

32. $\dfrac{xy - 1}{y^2 - x}$ **33.** $\dfrac{cy}{x}$ **34.** $\dfrac{a + b - ax}{a^2 + ab}$ **35.** $\dfrac{ab + mb^2 - cx}{x^2b^2}$

36. $abc^2 + 5abc^{-2}$ **37.** $5x^2ym^2$ **38.** $x^5a^4y^{-1} + 5x^2ay$

39. $\dfrac{a^4c^4x}{b^3}$ **40.** $\dfrac{xk}{a^4p^2m^3}$ **41.** $x = \frac{39}{8}$ **42.** 8 **43.** $-\frac{1}{4}$

44. True **45.** Irrational, real **46.** $\dfrac{xp}{4 - x^2}$ **47.** $\dfrac{3m}{x^2 + 5x + 4}$

problem set 32

1. 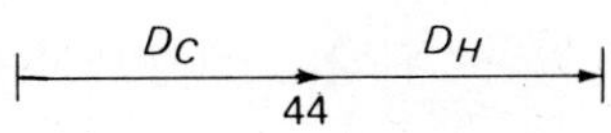

$R_CT_C + R_HT_H = 44$, $R_C = 5$, $R_H = 8$, $T_C + T_H = 7$, $\mathbf{T_C = 4}$, $\mathbf{T_H = 3}$

2. 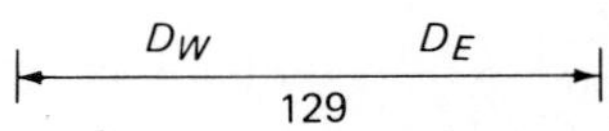

$T_W = 3$, $T_E = 9$, $R_W = 10$, $R_WT_W + R_ET_E = 129$, $\mathbf{R_E = 11}$

3. 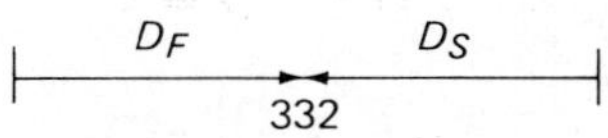

$R_FT_F + R_ST_S = 332$, $T_F = 2$, $T_S = 2$, $R_S = 65$, $\mathbf{R_F = 101}$

4. 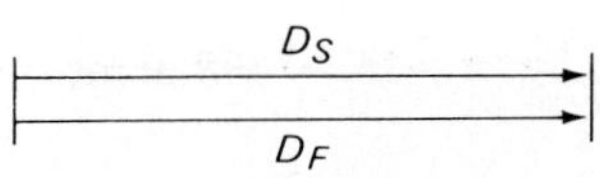

$R_ST_S = R_FT_F$, $R_S = 4$, $R_F = 5$, $T_S = T_F + 1$, **distance = 20**

5. $N_P + N_Q = 742$, $N_P + 25N_Q = 1702$, $N_P = 702$

6. $93{,}480 = (.82)(WN)$, $WN = \$114{,}000$

7. $4 = (WP)(20)$, $WP = 20\%$

8. $3(4 + 2N) = 2[2 + (2N + 4)] + 6$, $N = 3$, integers = 6, 8, 10

9. $x = 38$ **10.** $x = 4, -4$ **11.** $x = 2$ **12.** $x = 5$

13. $y \approx \frac{4.3}{1.8}x + 3.6$ **14.** $y = -\frac{3}{4}x + \frac{1}{4}$ **15.** $y = 2x - 5$

16. $y = -\frac{4}{3}x + \frac{1}{3}$ **17.** $g(2) = 0$ **18.** (b), (c), (d), (e) **19.** $y = 6$

20. $x = \frac{78}{7}$ **21.**

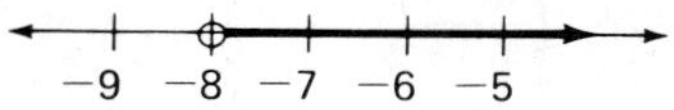

22.

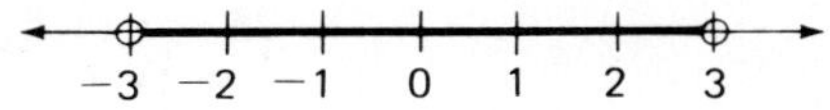

23. $abcy = (ab)(cy) = (cy)(ab) = (yc)(ba) = ycba$

24. $p = 4$ **25.** $5 - 2\sqrt{3} + 4\sqrt{2}$ **26.** $18 - 8\sqrt{6}$ **27.** $10\sqrt{10}$

28. $3x^2 + 5x - 3$ **29.** $x = 8, 1$ **30.** $\frac{1}{(x - 5)(x + y)}$

31. 4.9×10^{-21} **32.** $x = 3$, $y = -2$ **33.** Multiplication

34.

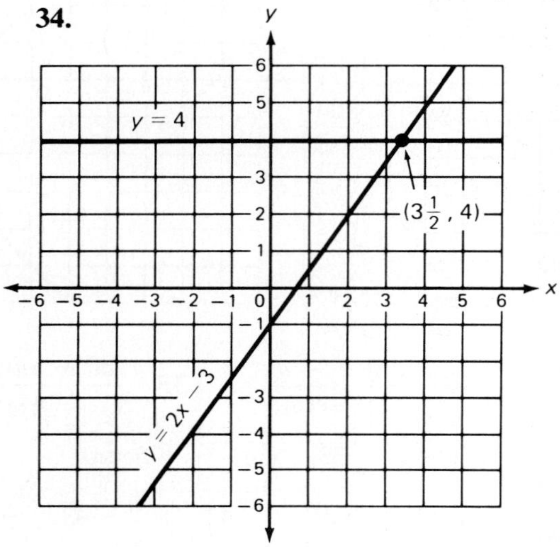

35. $\frac{yx + x}{y^2 - 1}$ **36.** $\frac{by}{x + y}$ **37.** $\frac{2x^2 + 2y - cx}{x^3 + xy}$ **38.** $\frac{c + ax^2 - mb^2x^2}{b^2x^2}$

39. $9x^2y^3z - 2x^2y^{-3}z$ **40.** $\frac{x + 2}{x + 4}$ **41.** $b^4a^5y^{-1} + 7a^2y$ **42.** $\frac{b^2x^4c^4}{a^2}$

43. $\frac{m^5k^4p}{y^7}$ **44.** $x = 1$ **45.** -1 **46.** -2 **47.** False

48. Rational, reals **49.** $\frac{-6ky}{x^2 - 36}$

problem set 33

1. $W = 4$, $B = 5$, $T = 9$

$\frac{B}{T} = \frac{5}{9}$ $\frac{B}{423} = \frac{5}{9}$ $\longrightarrow$ **$B = 235$ acres**

$\frac{W}{T} = \frac{4}{9}$ $\frac{W}{423} = \frac{4}{9}$ $\longrightarrow$ **$W = 188$ acres**

2. $M = 5$, $W = 3$, $T = 8$

$\frac{M}{T} = \frac{5}{8}$ $\frac{M}{184} = \frac{5}{8}$ $\longrightarrow$ **$M = 115$**

3. Rate $= \frac{4}{3}\frac{\text{liter}}{\text{rooms}}$, $\frac{4}{3}\frac{\text{liter}}{~~\text{rooms}~~} \cdot 15$ ~~rooms~~ $\longrightarrow$ **20 liter**

4. $\frac{6 \text{ ft-man}}{4 \text{ ft-shadow}}$, $\frac{6}{4} = \frac{30}{S_T} \longrightarrow \boldsymbol{S_T = 20}$ **ft**

5. $1800 = (.12)(WN)$, $WN = 15{,}000$

6. 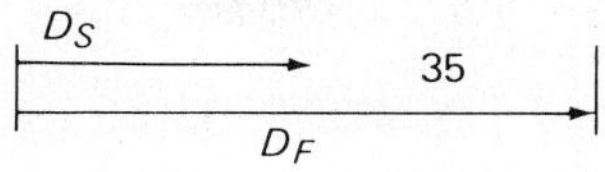

$R_F = 15$, $R_S = 8$, $T_F = T_S$, $R_S T_S + 35 = R_F T_F$, $\boldsymbol{T_F = 5}$ **hr**

7. $N_D + N_N = 123$, $N_D = N_N + 77 \longrightarrow N_N = N_D - 77$, $\boldsymbol{N_D = 100}$

8. $4(2N + 1) = 2[(2N + 5) + 5] + 4$, $N = 5$, integers = **11, 13, 15**

9. 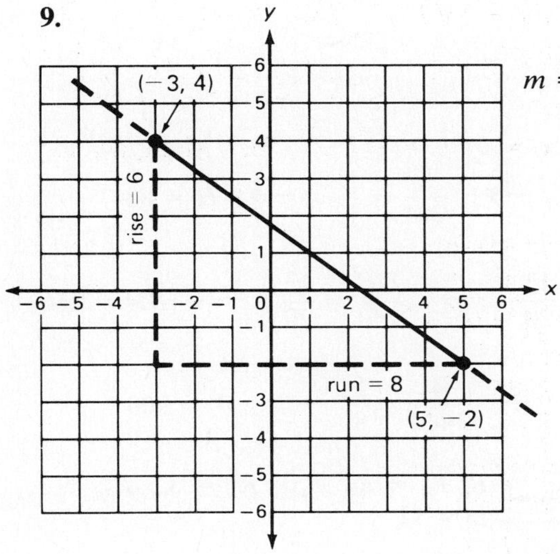

$m = -\frac{3}{4}$

10. $m = \frac{y_2 - y_1}{x_2 - x_1} = \frac{4 - (-2)}{-3 - 5} = -\frac{6}{8} = -\frac{3}{4}$ **11.** $x = 6, -1$ **12.** $x = 37$

13. $y \approx \frac{1}{2}x + \frac{3}{2}$

14. $y = \frac{5}{9}x - \frac{43}{9}$ **15.** $y = -x - 1$ **16.** $y = \frac{3}{2}x + \frac{11}{2}$

17. (a) $f(-2) = -8$ (b) $g(-2) = 4$ (c) $h(-2) = 4$

18. (b), (e), (f) **19.** $x = 18$ **20.** $x = 3$ **21.** 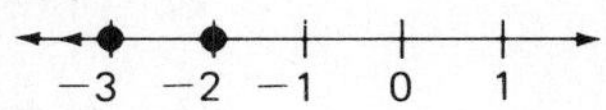

22. 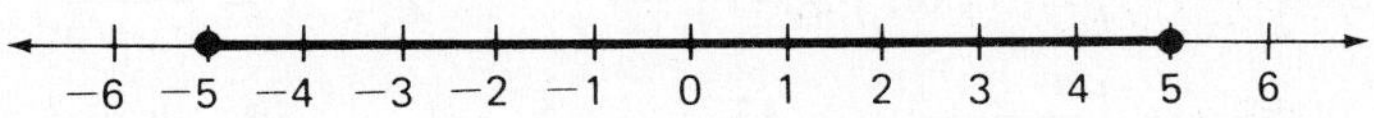

23. $x = 2\sqrt{10}$ **24.** $D = 5\sqrt{2}$

25.
$$\begin{aligned} x + y + z &= x + (y + z) \\ &= x + (z + y) \\ &= x + z + y \end{aligned}$$

26. $-8\sqrt{3}$ **27.** $24 - 8\sqrt{3}$

28. $4\sqrt{6}$ **29.** $x^2 + 4x + 15 + \frac{62}{x - 4}$

30. 1.6×10^{-22} **31.** $x = 2, y = -4$

32. $x = 6, -4$ **33.** $\frac{x - 1}{x - 7}$

34. Addition, multiplication, division.

35. 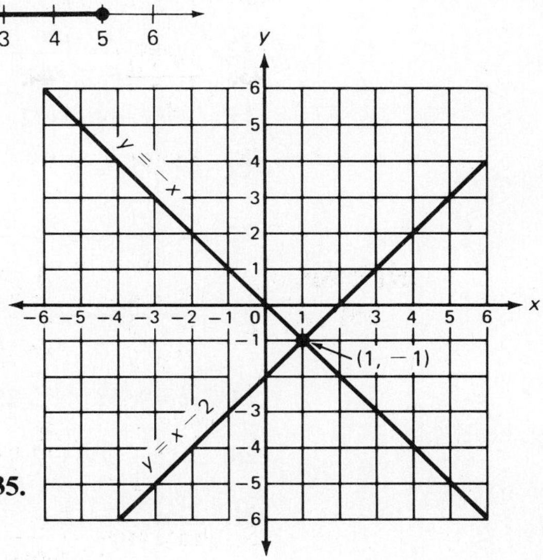

36. $\dfrac{ac - b}{ac + b}$ **37.** $\dfrac{1}{xy}$ **38.** $\dfrac{p + a + p^2}{p^2(p + a)}$ **39.** $\dfrac{ax + bx + cx^2 - d}{x^3}$

40. $2ax^2m^{-2} - 3a^{-1}x^2m^{-2}$ **41.** $5abc^{-2} - 2ac^2b$ **42.** $\dfrac{x - 5}{x + 3}$ **43.** $\dfrac{b^2y^4}{a^{12}x^4}$

44. $\dfrac{y^{13}}{m^9x^4}$ **45.** $p = \frac{1}{3}$ **46.** -2 **47.** -32

48. False, it's the other way around. **49.** Rational, reals

50. $\dfrac{x - 15}{x^2 - 16}$

problem set 34

1. $C = 2$, $H = 11$, $T = 13$ $\quad \dfrac{H}{T} = \dfrac{11}{13} \rightarrow \dfrac{H}{104} = \dfrac{11}{13} \rightarrow$ $\mathbf{H = 88}$

2. $G = 7$, $DH = 13$, $T = 20$ $\quad \dfrac{DH}{T} = \dfrac{13}{20} \rightarrow \dfrac{DH}{240} = \dfrac{13}{20} \rightarrow$ $\mathbf{DH = 156}$

3. $L = 3$, $W = 5$, $T = 8$ $\quad \dfrac{W}{T} = \dfrac{5}{8} \rightarrow \dfrac{W}{1440} = \dfrac{5}{8} \rightarrow$ $\mathbf{W = 900}$

4. 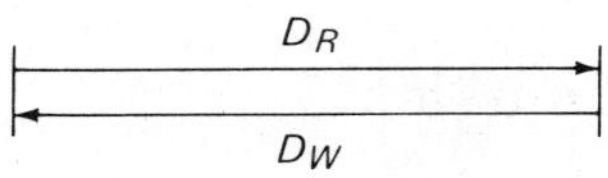

$R_RT_R = R_WT_W$, $R_R = 6$, $R_W = 3$, $T_R + T_W = 6$, $\mathbf{D = 12}$

5. 

$R_NT_N + R_ST_S = 320$, $R_N = 35$, $R_S = 45$, $T_N = T_S$, $\mathbf{T = 4\ hr}$

6. $N_H + N_C = 100$, $4N_H + 6N_C = 540$, $N_H = 30$

7. $910 = (.52)(WN)$, voters = 1750.

8. $(4)(2N) = 16[(2N + 4) + 2]$, $N = -4$, integers $= -8, -6, -4$

9. $m = \dfrac{bx}{cx - 1}$ **10.** $b = \dfrac{ad}{dx - c}$ **11.** $x = \dfrac{bd}{a - bc}$

12. $x = \dfrac{an}{kn + yn - m}$ **13.** $x = 5, -4$ **14.** $x = 31$

15. $y \approx -\frac{1}{6}x + 2$ **16.** $y = -x + 2$ **17.** $y = -2x + 1$

18. $g(-2) = 6$ **19.** (a), (c), (f) **20.** $a = 10$ **21.** $x = 4$

22. 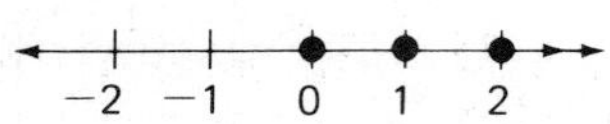

23. 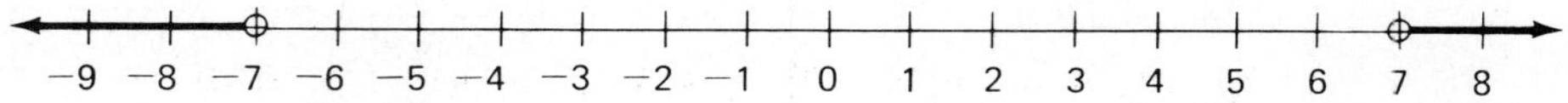

24. $h = \sqrt{52}$ **25.** $2\sqrt{3} - \sqrt{21} + 7$ **26.** $30 - 2\sqrt{10}$ **27.** $18\sqrt{3}$

28. $7x^2 - 3x + 2 + \dfrac{1}{x + 3}$ **29.** $x = 7, -2$ **30.** $\dfrac{x - 2}{a + b}$

31. 5×10^{-11} **32.** $(2, -3)$ **33.** Addition, multiplication

34.

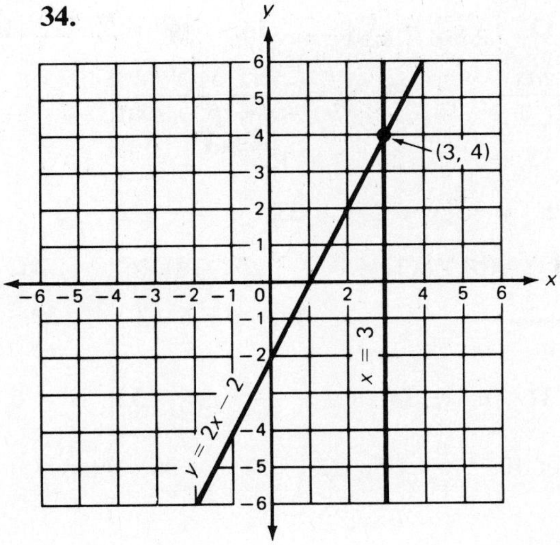

35. $\dfrac{a + b}{b^2 - a}$ **36.** $\dfrac{y + y^2}{x + y^2}$ **37.** $\dfrac{2b^2x - 3}{3b^2}$ **38.** $\dfrac{a + b^2x - a^2bm}{a^2b^2}$

39. $x^2yz^3 + 2x^{-2}y^{-1}z^3$ **40.** $\dfrac{x + 2}{x - 2}$ **41.** $a^2y^3 - 3x^2$

42. $\dfrac{y^3m^2k}{p}$ **43.** a^3bc^2 **44.** 3 **45.** $-\frac{1}{3}$ **46.** True

47. Rational, reals **48.** $\dfrac{7x^2 - 14x - 4}{x^2 - x - 6}$

problem set 35

1. $P = 7$, $S = 5$, $T = 12$ $\quad \dfrac{P}{T} = \dfrac{7}{12} \rightarrow \dfrac{P}{204} = \dfrac{7}{12} \rightarrow$ $\boldsymbol{P = 119}$

2. $H = 7$, $C = 11$, $T = 18$ $\quad \dfrac{C}{T} = \dfrac{11}{18} \rightarrow \dfrac{C}{12{,}600} = \dfrac{11}{18} \rightarrow$ $\boldsymbol{C = 7700}$

3. $F = 5$, $B = 3$, $T = 8$ $\quad \dfrac{B}{T} = \dfrac{3}{8} \rightarrow \dfrac{B}{32} = \dfrac{3}{8} \rightarrow$ $\boldsymbol{B = 12}$ **min**

4.

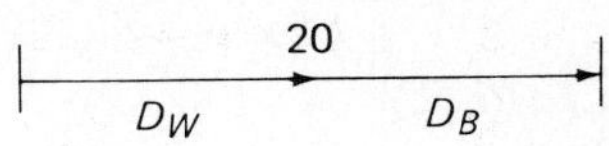

$R_WT_W + R_BT_B = 20$, $R_W = 5$, $R_B = 15$, $T_W + T_B = 2$, $\boldsymbol{D_B = 15}$

5.

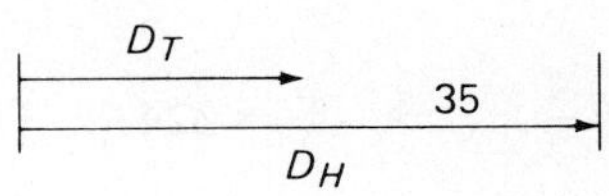

$R_TT_T + 35 = R_HT_H$, $R_H = 12$, $R_T = 2$, $T_H = T_T$, $\boldsymbol{T_H = 3\frac{1}{2}}$ **hr**

6. $N_S + N_R = 15$, $75N_S + 35N_R = 965$, $\boldsymbol{N_R = 4}$

7. $2N_B + 3N_C = 65$, $N_C + 15 = N_B$, $\boldsymbol{N_C = 7}$

8. $5N = 6[(N + 1) + (-3)] + 5$, $N = 7$, integers = **7, 8**

9. $700 = (.2)(C)$, $C = \dfrac{700}{.2} \rightarrow$ **C = 3,500, thus 2800 remain**

10. 37 **11.** $\frac{29}{110}$ **12.** $\frac{134}{100,000} = \frac{67}{50,000}$ **13.** $\frac{14,325}{100} = \frac{573}{4}$ **14.** $\frac{412}{99}$

15. $\frac{14,307}{9990}$ **16.** $p = \frac{mx}{dx + c}$ **17.** $c = \frac{x}{kx + p - nx}$

18. $x = \frac{md}{c + pd}$ **19.** 3, −5 **20.** $t = 15$

21. $y \approx -\frac{1}{2}x - 3$ **22.** $y = \frac{9}{5}x + 2$ **23.** $y = -\frac{2}{3}x - \frac{31}{3}$

24. (a) $m(-5) = -25$, (b) $k(3) = 3$ **25.** (d), (e) **26.** $w = 10$

27. 

28. $s = 6\sqrt{2}$ **29.** $9\sqrt{5}$

30. $12\sqrt{3} - 36\sqrt{2}$ **31.** $m^2 + 3m + 9 + \frac{12}{m - 3}$ **32.** $x = 6, \quad -4$

33. $\frac{x - 1}{x + 4}$ **34.** 3.5×10^{-3} **35.** (3, −3) **36.** None

37.

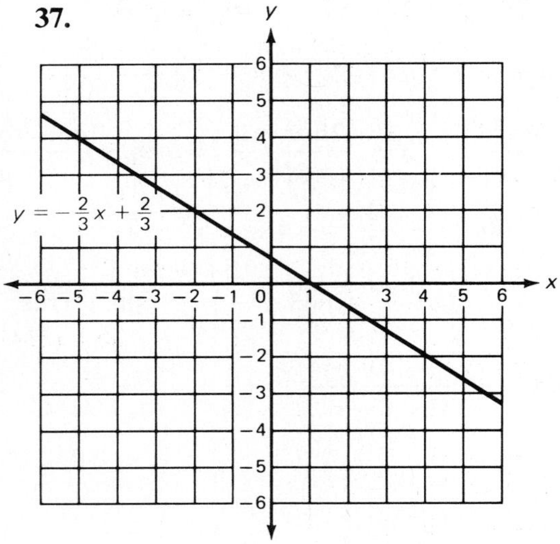

38. $-\frac{a^2 + ab - 1}{b}$ **39.** $\frac{xc}{y}$ **40.** $\frac{mx + 2ax + 3m^2}{m^3 + 2am^2}$

41. $2a^{-2}bc^{-1} + 6a^{-2}b^{-1}c$ **42.** $4x^2 + \frac{32x^3}{a^4}$ **43.** $\frac{4x^6}{a^6y^5}$

44. $x = \frac{31}{6}$ **45.** −173 **46.** $-\frac{1}{16}$

47. False **48.** Rational, reals **49.** $\frac{12x^2 + 4x - 7}{2x^2 - x}$

problem set 36

1. $S = 3$, $L = 4$, $T = 7$ $\quad \frac{L}{T} = \frac{4}{7} \longrightarrow \frac{L}{567} = \frac{4}{7} \longrightarrow$ $\mathbf{L = 324}$

2. $L + S = 24$, $\frac{S}{L} = \frac{3}{5}$, $\mathbf{L = 15}$, $\mathbf{S = 9}$

3. $E = 5$, $S = 3$, $T = 8$ $\quad \frac{S}{T} = \frac{3}{8} \longrightarrow \frac{S}{272} = \frac{3}{8} \longrightarrow$ $\mathbf{S = 102}$

4. 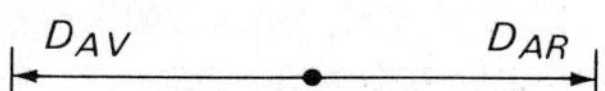

$R_{AR} = 3$, $R_{AV} = 5$, $T_{AV} + 2 = T_{AR}$,
$R_{AR}T_{AR} + R_{AV}T_{AV} = 18$, **T = 3:30 pm**

5. 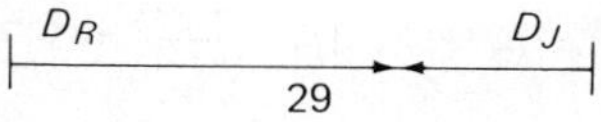

$R_RT_R + R_JT_J = 29$, $T_R = 5$, $T_J = 2$, $R_R = 4$, $\boldsymbol{R_J = 4\frac{1}{2}}$

6. $N_P + 5N_N = 67$, $N_P = N_N + 1$, $\boldsymbol{N_P = 12}$

7. $N_P + 5N_N = 67$, $N_P + N_N = 19$, $\boldsymbol{N_P = 7}$

8. $9(2N) = 2[(2N + 4) + (3)] + 14$, $N = 2$, integers = **4, 6, 8**

9. $27 = (.15)(WN)$, $\boldsymbol{WN = 180}$

10. $N_P = 5$, $N_N = 10$, $N_D = 15$ **11.** $N_P = 4$, $N_E = 8$, $N_C = 52$

12. 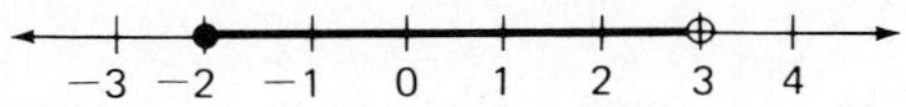

13. 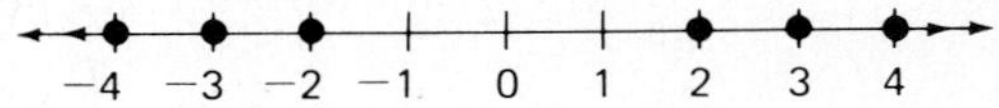

14. $\frac{44}{105}$ **15.** $\frac{1312}{990{,}000}$ **16.** $n = \dfrac{ym}{x - py}$ **17.** $x = \dfrac{1}{m - a - c}$

18. $c = \dfrac{p}{mp - x + kp}$ **19.** $p = 5, \; -1$ **20.** $x = 20$

21. $y = \frac{5}{3}x - \frac{9}{10}$ **22.** $y = x + 5$ **23.** $y = -\frac{3}{2}x + \frac{7}{2}$

24. $f(2) = 1$ **25.** (c), (d), (f) **26.** $y = -3$

27. 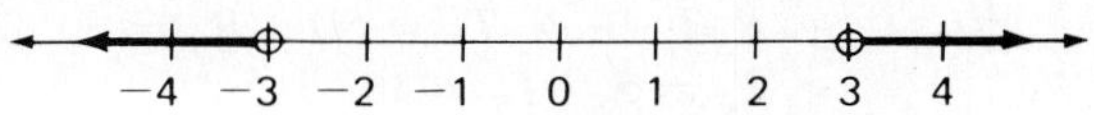

28. $\sqrt{41}$

29. $3\sqrt{2} - 2\sqrt{3}$ **30.** $18\sqrt{2} - 18$ **31.** $2x^2 + 3x - 2 - \dfrac{1}{x - 2}$

32. $x = 1, \; -\frac{5}{3}$ **33.** $\dfrac{3a + 5}{x + y}$ **34.** 5×10^{-10} **35.** $a(bc) = (ab)c$

36.

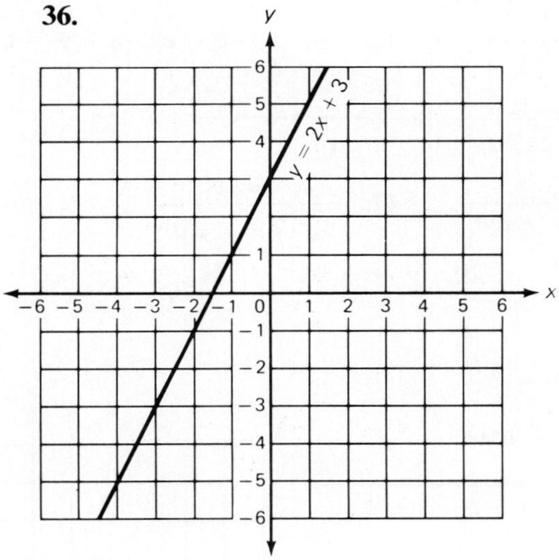

37. $\dfrac{1 + xy}{y^2 - x}$ **38.** $\dfrac{a}{bc}$ **39.** $\dfrac{x + bm - cm^2x^2}{m^2x^2}$ **40.** $ab^2c + 2ab^2c^{-1}$

41. $2b^4cx - 3$ **42.** $\dfrac{a^4}{c^3x^3}$ **43.** 1 **44.** $x = -13$

45. 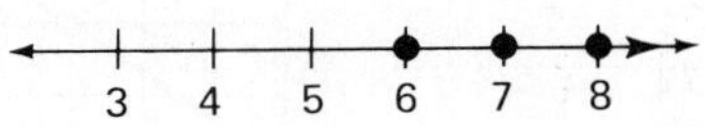

46. -8

47. True **48.** Natural, whole, integers, rational, real. **49.** $\dfrac{3x + 17}{x^2 - 25}$

problem set 37

1. (a) $W = kV$; (b) $10 = k(2) \longrightarrow k = 5$;
(c) $W = 5V$; (d) $35 = 5V \longrightarrow \mathbf{V = 7}$

2. (a) $D = kT$; (b) $75 = k(5) \longrightarrow k = 15$;
(c) $D = 15T$; (d) $D = 15(7) \longrightarrow \mathbf{D = 105}$

3. (a) $P = \frac{k}{V}$; (b) $7 = \frac{k}{75} \longrightarrow k = 525$;
(c) $P = \frac{525}{V}$; (d) $15 = \frac{525}{V} \longrightarrow \mathbf{V = 35}$;

4. (a) $V = \frac{k}{T}$; (b) $60 = \frac{k}{4} \longrightarrow k = 240$;
(c) $V = \frac{240}{T}$; (d) $80 = \frac{240}{T} \longrightarrow \mathbf{T = 3\ hr}$

5. $P = 7$, $TU = 2$, $T = 9$ $\qquad \frac{P}{TU} = \frac{7}{2}$

$\frac{P}{T} = \frac{7}{9} \longrightarrow \frac{P}{207} = \frac{7}{9} \longrightarrow \mathbf{P = 161}$

$\frac{TU}{T} = \frac{2}{9} \longrightarrow \frac{TU}{207} = \frac{2}{9} \longrightarrow \mathbf{TU = 46}$

6. 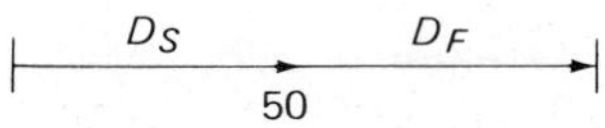

$R_S T_S + R_F T_F = 50$, $R_S = 3$, $R_F = 4$, $T_S + T_F = 15$, $\mathbf{T_F = 5}$, $\mathbf{T_S = 10}$

7. $422N_G + 4N_C = 2122$, $N_G + N_C = 8$, $\mathbf{N_G = 5}$, $\mathbf{N_C = 3}$

8. $4(2N + 1) + 14 = 2(2 + 2N + 5)$, $N = -1$, integers $= -1, 1, 3$

9. $81{,}150 = (.3)(P)$, $P = 270{,}500$

10. $N_Q = 5$, $N_N = 3$, $N_P = 15$ $\qquad$ 11. 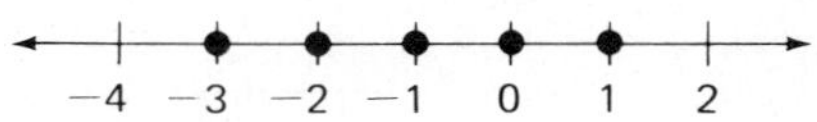

12. 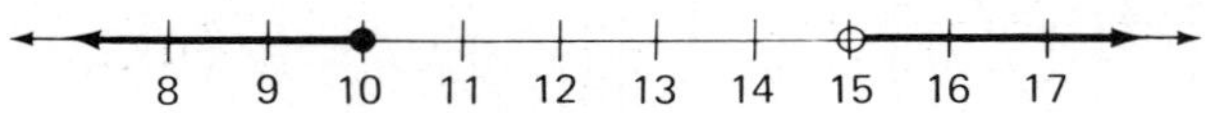

13. $\frac{11}{15}$ $\qquad$ 14. $\frac{1{,}219{,}369}{99{,}000}$

15. $\frac{99{,}323}{99{,}000}$ $\qquad$ 16. $\frac{cx}{1 + pc - cm}$ $\qquad$ 17. $\frac{ay}{6y - 3 - my}$

18. $-2, \ -4$ $\qquad$ 19. 58 $\qquad$ 20. $y = -3x$ $\qquad$ 21. $y = -\frac{1}{3}x - 4$

22. $y = -3x - 18$

23. (a) $p(5) = 5$; (b) $\phi(0) = 2$

24. (a), (b), (d), (e)

25. $p = 8$ $\qquad$ 26. 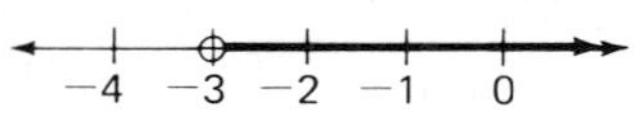

27. $2\sqrt{6}$

28. $-\sqrt{3}$

29. $30\sqrt{3} - 12\sqrt{5}$

30. $3x - 10 + \frac{31}{x + 2}$

31. $x = 11, \ -2$

32. $\frac{x - 1}{x + 2}$ $\qquad$ 33. 1×10^{31}

34. Addition

35. 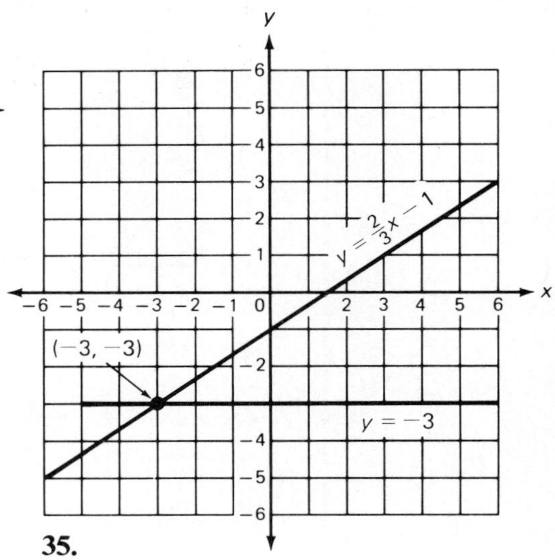

36. $\dfrac{ay - x}{b - ay}$ **37.** $\dfrac{1}{ac}$ **38.** $\dfrac{px - pa - 3x^4}{x^5 - xa}$ **39.** $-x^{-1}a^2b - 3a^2bx$

40. $8a^{-4} + 6a^2b^2$ **41.** $m^{-5}x^{-2}$ **42.** $x = \frac{10}{7}$ **43.** -20

44. $\dfrac{xa + a^2y - zb}{a^2b^2}$

45. -3 **46.** False. Wrong symbol used.

47. Irrational, real **48.** $\dfrac{-2x - 3}{x^2 + 2x + 1}$

problem set 38

1. (a) $D_S = kV^2$; (b) $200 = k(20)^2 \longrightarrow k = \frac{1}{2}$;
(c) $D_S = \frac{1}{2}V^2$; (d) $D_S = \frac{1}{2}(28)^2 \longrightarrow \mathbf{D_S = 392}$

2. (a) $W = \dfrac{k}{D^2}$; (b) $10{,}000 = \dfrac{k}{(5000)^2} \longrightarrow k = 25 \times 10^{10}$;
(c) $W = \dfrac{25 \times 10^{10}}{D^2}$; (d) $W = \dfrac{25 \times 10^{10}}{(50{,}000)^2} \longrightarrow \mathbf{W = 100}$

3. $N_L + N_S = 42$, $5N_L + 3N_S = 186$, $\mathbf{N_S = 12}$, $\mathbf{N_L = 30}$

4. $P = 7$, $A = 4$, $T = 11$ $\qquad \dfrac{P}{A} = \frac{7}{4}$

$\dfrac{A}{T} = \frac{4}{11} \longrightarrow \dfrac{A}{3850} = \frac{4}{11} \longrightarrow \mathbf{A = 1400}$

$\dfrac{P}{T} = \frac{7}{11} \longrightarrow \dfrac{P}{3850} = \frac{7}{11} \longrightarrow \mathbf{P = 2450}$

5. 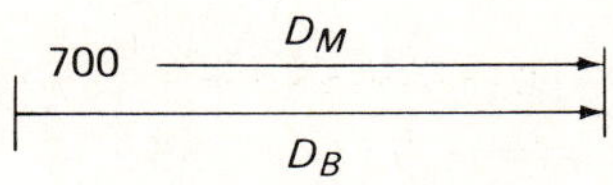

$R_MT_M + 700 = R_BT_B$, $R_M = 7$, $R_B = 11$,
$T_M = T_B$, $\mathbf{T_B = 175}$ **sec**

6. 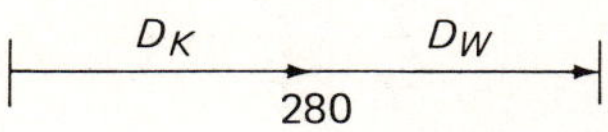

$R_KT_K + R_WT_W = 280$, $T_W + 1 = T_K$,
$R_K = 40$, $R_W = 40$, $T_W = 3$, $T_K = 4$,
12 o'clock noon

7. $N_N + N_D = 35$, $5N_N + 10(35 - N_N) = 200$, $\mathbf{N_N = 30}$, $\mathbf{N_D = 5}$

8. $N + 100 = 3(N + 2 + 5) + 69$, integers = 5, 6, 7

9. $\frac{40}{100} = \dfrac{SP_R}{T_R}$, $\frac{40}{100} = \dfrac{240}{T_R}$, $\mathbf{T_R = 600}$

10. 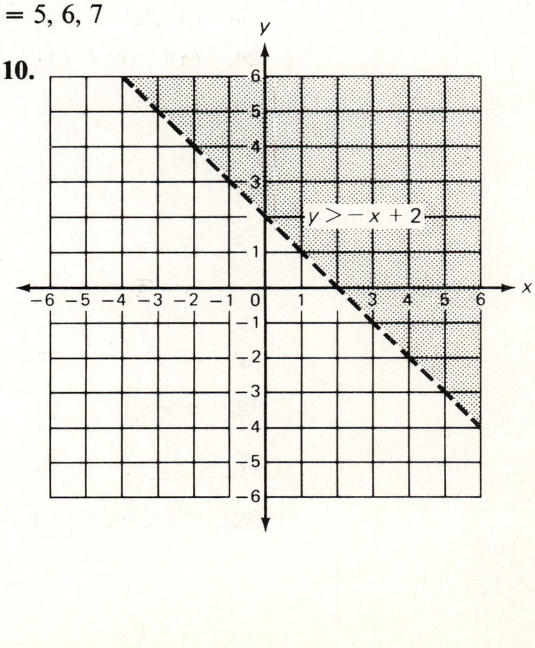

11.

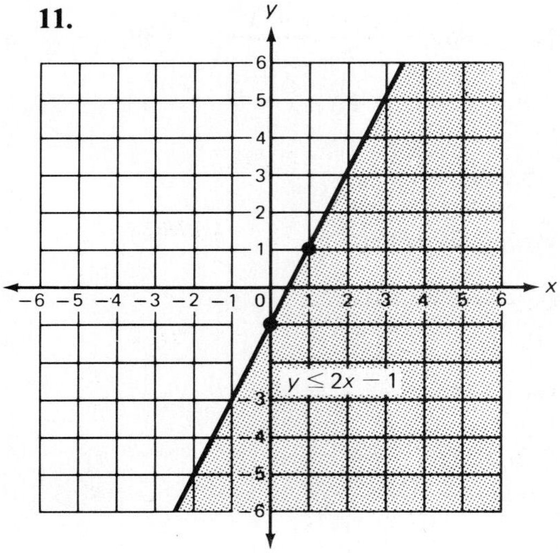

12.

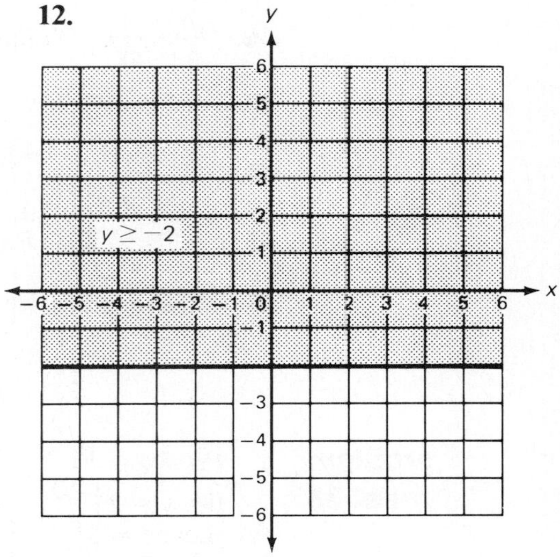

13. $x = -1, 25$ **14.** $x = 9, -1$ **15.** $N_D = 4$, $N_N = 8$, $N_P = 3$

16.

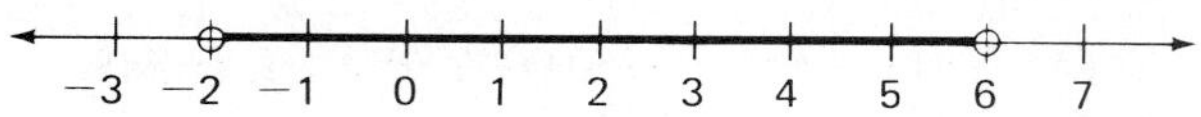

17. $\frac{11}{20}$ **18.** $\frac{10,242}{9900}$ **19.** $a = \dfrac{km}{mp + cm - x}$ **20.** $x = 28$

21. $y = \frac{1}{3}x + 2$ **22.** $y = -\frac{4}{3}x + \frac{1}{3}$ **23.** $y = -\frac{2}{3}x + \frac{11}{3}$

24. (a) $f(0) = 2$, (b) $\phi(-5) = -18$ **25.** (b), (c), (d), (e), (f)

26.

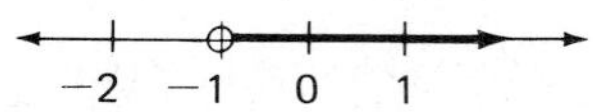

27. All real numbers **28.** $m = \sqrt{65}$

29. $c = \sqrt{82}$ **30.** $21\sqrt{2} - 7\sqrt{3}$ **31.** $14\sqrt{3} + 18\sqrt{7}$ **32.** $4 - 8\sqrt{2}$

33. $x^2 - 5x + 27 - \dfrac{140}{x + 5}$ **34.** $x = -7, 3$ **35.** $\dfrac{3}{k}$ **36.** 1.5×10^{46}

37. $x(y + z) = xy + xz$ **38.**

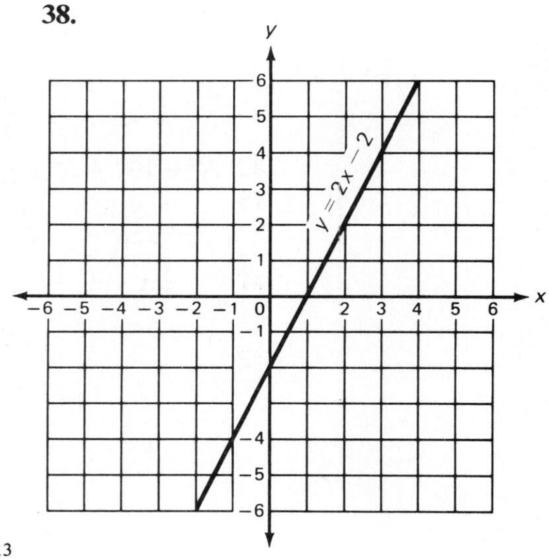

39. $\dfrac{ax + ay + 1}{a + x - y}$

40. $\dfrac{a}{a + b}$ **41.** $\dfrac{mx^2 + c^2(b + y) + kx^3}{x^3c^3}$

42. $8a^2b^{-4}c^{-2} - 3b^{-4}c^{-2}$ **43.** $2 - \dfrac{6}{x^8}$ **44.** $a^3x^2y^{-1}$ **45.** $x = -\frac{16}{5}$

46. -3 **47.** False. It's the other way around.

48. Integers, rationals, reals. **49.** $\dfrac{x^2 - x + 1}{x^2 - 2x}$

problem set 39

1. (a) $N_y = \dfrac{k}{N_m}$; (b) $8 = \dfrac{k}{6{,}000} \longrightarrow k = 48{,}000$;
(c) $N_y = \dfrac{48{,}000}{N_m}$; (d) $3 = \dfrac{48{,}000}{N_m} \longrightarrow \mathbf{N_m = 16{,}000}$

2. (a) $V_E = kW^2$ (b) $3000 = k(2)^2 \longrightarrow k = 750$;
(c) $V_E = 750W^2$; (d) $V_E = 750(3)^2 \longrightarrow \mathbf{V_E = \$6750}$

3. (a) $S_W = \dfrac{k}{D_W}$; (b) $60 = \dfrac{k}{4} \longrightarrow k = 240$; (c) $S_W = \dfrac{240}{D_W}$;
(d) $S_W = \frac{240}{.5} \longrightarrow \mathbf{S_W = 480}$

4. Fats = 2
Leans = 5
$T = 7$

$\dfrac{F}{L} = \frac{2}{5}$

$\dfrac{F}{T} = \frac{2}{7} \longrightarrow \dfrac{F}{343} = \frac{2}{7} \longrightarrow \mathbf{F = 98}$

$\dfrac{L}{T} = \frac{5}{7} \longrightarrow \dfrac{L}{343} = \frac{5}{7} \longrightarrow \mathbf{L = 245}$

5. 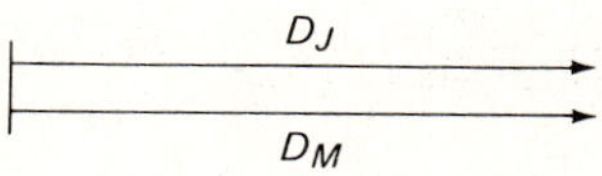

$R_J T_J = R_M T_M$, $R_J = 45$, $R_M = 60$,
$T_J = T_M + 1$, $\underline{T_M = 3}$, **1 o'clock p.m.**

6. 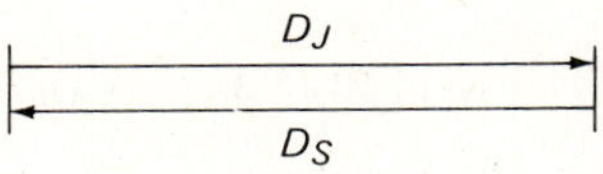

$R_J T_J = R_S T_S$, $R_J = 10$, $R_S = 40$, $T_J + T_S = 3$,
$\mathbf{D_J = 24}$ **km**

7. $N_L + N_{VL} = 30$, $5N_L + 15N_{VL} = 250$, $\mathbf{N_L = 20}$, $\mathbf{N_{VL} = 10}$

8. $14 + 2(2N) = \frac{1}{2}[3(2N + 4 + 2)]$, even integers $= -10, -8, -6$

9. Green = 2
Blue = 5
$T = 7$

$\dfrac{G}{B} = \frac{2}{5}$

$\dfrac{G}{T} = \frac{2}{7} \longrightarrow \dfrac{G}{154} = \frac{2}{7} \longrightarrow \mathbf{G = 44}$

$\dfrac{B}{T} = \frac{5}{7} \longrightarrow \dfrac{B}{154} = \frac{5}{7} \longrightarrow \mathbf{B = 110}$

10. $x = -\frac{3}{2} \pm \dfrac{\sqrt{29}}{2}$

11. $-\frac{5}{2} \pm \dfrac{\sqrt{57}}{2}$

12. $x = -\frac{1}{12} \pm \dfrac{\sqrt{433}}{12}$

13.

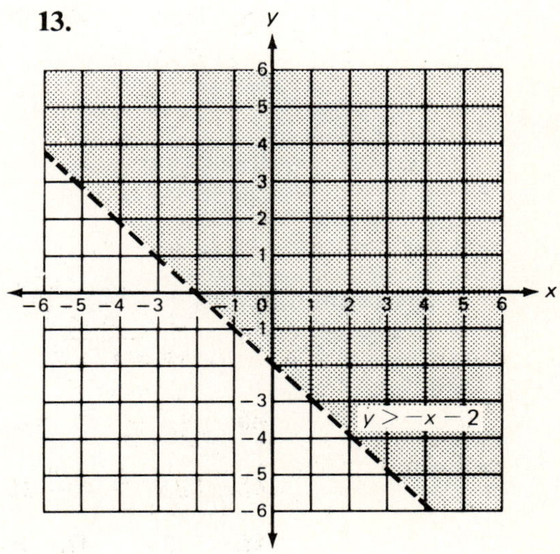

14. $x = 16, -1$ **15.** $x = 4, -1$ **16.** $N_D = 4$, $N_P = 8$, $N_N = 3$

17. 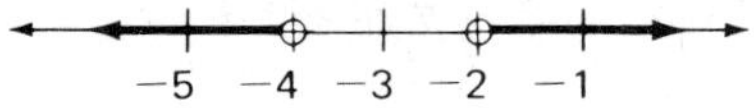

18. $\frac{192}{225}$ **19.** $\frac{1}{b - c} = a$

20. $k = \dfrac{yc}{x - my}$ **21.** $x = 7, -4$ **22.** $y = -x + 1$

23. $y = -\frac{2}{3}x$ **24.** $y = \frac{9}{5}x + \frac{2}{5}$ **25.** $f(4) = 13$ **26.** (a), (b), (c), (f)

27. 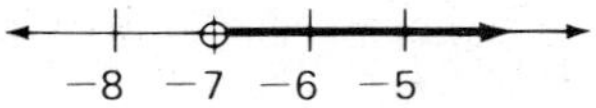

28. $p = \sqrt{13^2 - 5^2} = 12$ **29.** $5 - 3\sqrt{2}$

30. $24 - 48\sqrt{6}$ **31.** $2x^2 + 3x - 5 - \dfrac{3}{x + 2}$ **32.** $x = \frac{1}{2}, -3$ **33.** 1

34. 1×10^{-15} **35.** Multiplication

36.

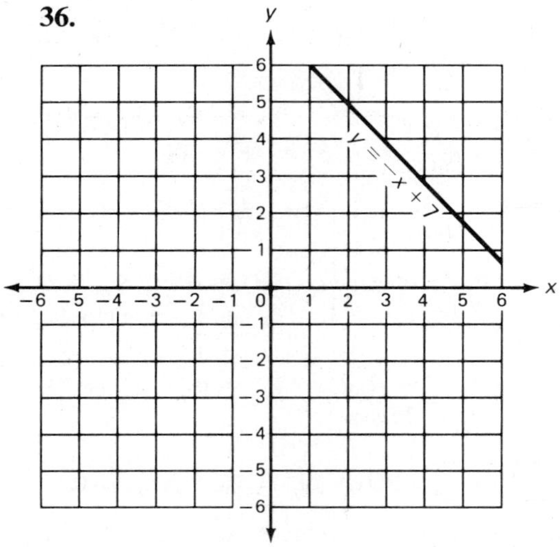

37. $\dfrac{x + 1}{7x}$ **38.** $\dfrac{a + b}{x + 2b}$ **39.** $\dfrac{x + b - ax}{x(x + b)}$ **40.** $-2xyz^2$

41. $2a^4y - 10yb^2$ **42.** $\dfrac{y^3k}{tx^6p^4b^2}$ **43.** $x = -\frac{4}{5}$

44. -3 **45.** $y = -\frac{2}{3}x + \frac{8}{3}$ **46.** $\dfrac{-1}{324}$

47. True **48.** Natural, whole, integers, rational, real

49. $\dfrac{-6x^2 + 12x + 4}{x^2 - 5x + 6}$

problem set 40

1. (a) $P = \dfrac{k}{V}$; (b) $40 = \dfrac{k}{60} \longrightarrow k = 2400$;

(c) $P = \dfrac{2400}{V}$; (d) $30 = \dfrac{2400}{V} \longrightarrow \mathbf{V = 80}$

2. (a) $C_1 = \dfrac{k}{N_P}$; (b) $48 = \dfrac{k}{5} \longrightarrow k = 240$;

(c) $C_1 = \dfrac{240}{N_P}$; (d) $C_1 = \frac{240}{8} \longrightarrow \mathbf{C_1 = \$30}$

3. (a) $T = kP$; (b) $400 = k(4) \longrightarrow k = 100$;
(c) $T = 100P$; (d) $T = 100(6) \longrightarrow$ $\mathbf{T = 600°K}$

4. Hardware = 5
Software = 2
Total = 7

$\frac{H}{S} = \frac{5}{2}$

$\frac{H}{T} = \frac{5}{7} \longrightarrow \frac{H}{14{,}700} = \frac{5}{7} \longrightarrow$ $\mathbf{H = 10{,}500}$

$\frac{S}{T} = \frac{2}{7} \longrightarrow \frac{S}{14{,}700} = \frac{2}{7} \longrightarrow$ $\mathbf{S = 4{,}200}$

5. 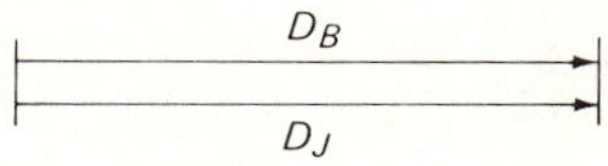

$R_BT_B = R_JT_J$, $R_B = 30$, $R_J = 60$,
$T_B = T_J + 2$, $\mathbf{T_J = 2}$ **hr**

6. 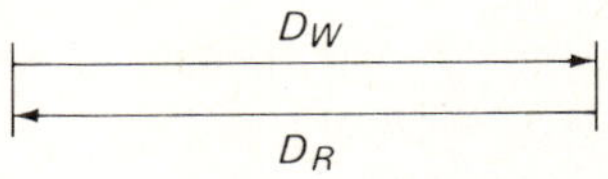

$R_WT_W = R_RT_R$, $R_W = 4$, $R_R = 20$,
$T_W + T_R = 6$, $\mathbf{T_W = 5}$, $\mathbf{T_R = 1}$, $\mathbf{D_W = 20}$

7. $150N_P + 125N_C = 5750$, $N_P + N_C = 40$, $\mathbf{N_P = 30}$, $\mathbf{N_C = 10}$

8. $4(N + 12) = 13 + 5(N + 3)$, integers = 20, 21, 22, 23

9. $1258 = .17(N_P)$, $N_P = 7400$

10. See page 383. **11.** $x = -\frac{1}{3} \pm \frac{\sqrt{21}}{3}$

12. $x = \frac{1}{2} \pm \frac{5}{2} \longrightarrow$ $\mathbf{x = 3, \ -2}$ **13.** $x = \frac{1}{5} \pm \frac{\sqrt{6}}{5}$ **14.** $x = 2, -1$

15.

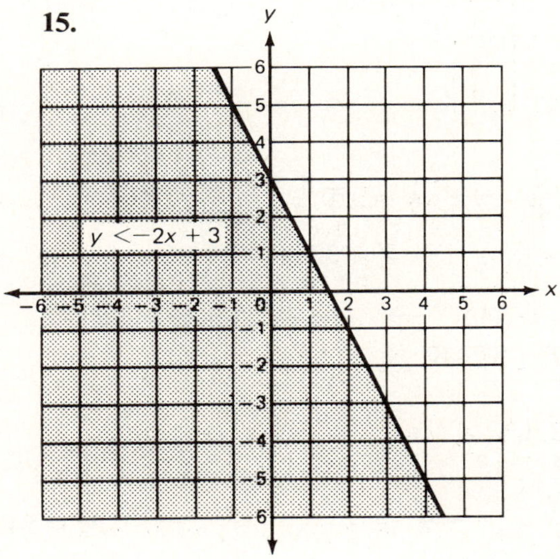

16. $x = 18, -2$ **17.** $N_D = 5$, $N_N = 10$, $N_P = 4$

18. 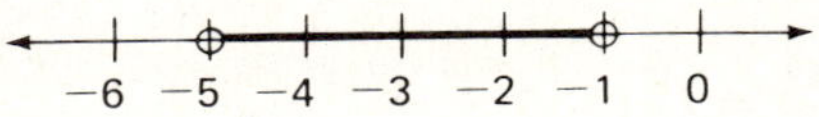

19. $\frac{47}{90}$ **20.** $\frac{1013}{990}$ **21.** $a = \frac{bx}{cb + db - y}$

22. $x = 5, -3$ **23.** $y = \frac{1}{4}x + \frac{7}{4}$ **24.** $y = \frac{1}{2}x - 4$ **25.** $y = -\frac{1}{2}x + 5$

26. $g(3) = 2$ **27.** (a), (b), (c), (d), (e) **28.** 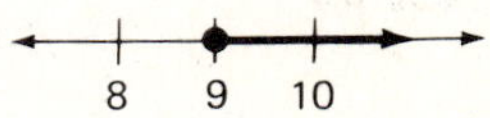

29. 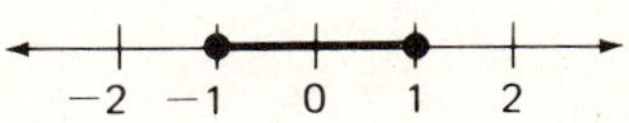

30. $\sqrt{61}$ **31.** $6\sqrt{2}$ **32.** $20\sqrt{3} - 18\sqrt{2}$

33. $3x^2 + 20x + 52 + \frac{162}{x - 3}$ **34.** $x = -\frac{2}{3}, \frac{3}{2}$ **35.** $\frac{y - 3}{y + 2}$

36. 1×10^{-16} **37.** Addition, subtraction, multiplication.

38.

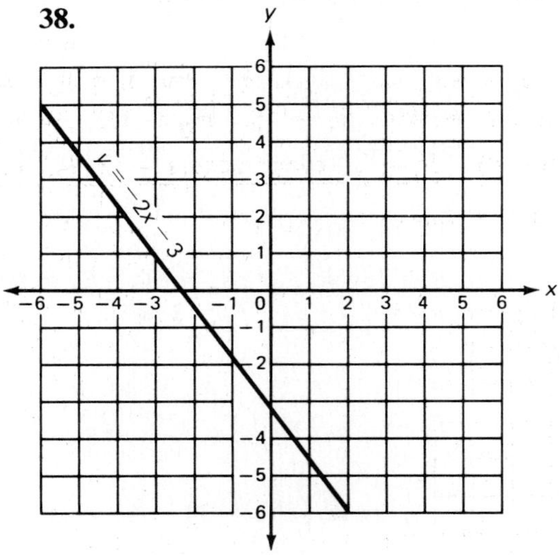

39. $\dfrac{t(t-4)}{t-2}$ **40.** $\dfrac{y}{x^2}$ **41.** $\dfrac{x(a+b)+ya}{a(a+b)}$

42. $4a^2by^3 - 2a^2b^{-1}y^3$ **43.** $2x^{-1}y^6 + 3x$ **44.** $x^{-2}yb^3$ **45.** $x = \frac{20}{11}$

46. -4 **47.** True **48.** Irrational, real

49. $\dfrac{4x^2 + 7x + 6}{x^2 - 9}$

problem set 41

1. (a) $V = \dfrac{k}{P}$; (b) $80 = \dfrac{k}{4} \longrightarrow k = 320$;
(c) $V = \dfrac{320}{P}$; (d) $V = \frac{320}{10} \longrightarrow$ $\mathbf{V = 32}$

2. (a) $V = kT$; (b) $80 = k(700) \longrightarrow k = \frac{4}{35}$; (c) $V = \frac{4}{35}T$
(d) $100 = \frac{4}{35}T \longrightarrow$ $\mathbf{T = 875°K}$

3. Mosquitoes = 2
Other bugs = 7
Total = 9

$\dfrac{M}{B} = \frac{2}{7}$

$\dfrac{M}{T} = \frac{2}{9} \longrightarrow \dfrac{M}{2{,}430{,}000} = \frac{2}{9} \longrightarrow$ $\mathbf{M = 540{,}000}$

$\dfrac{B}{T} = \frac{7}{9} \longrightarrow \dfrac{B}{2{,}430{,}000} = \frac{7}{9} \longrightarrow$ $\mathbf{B = 1{,}890{,}000}$

4. Field goals = 2
Missed goals = 5
Total = 7

$\dfrac{FG}{MG} = \frac{2}{5}$

$\dfrac{FG}{T} = \frac{2}{7} \longrightarrow \dfrac{FG}{84} = \frac{2}{7} \longrightarrow$ $\mathbf{FG = 24}$

$\dfrac{MG}{T} = \frac{5}{7} \longrightarrow \dfrac{MG}{84} = \frac{5}{7} \longrightarrow$ $\mathbf{MG = 60}$

5.

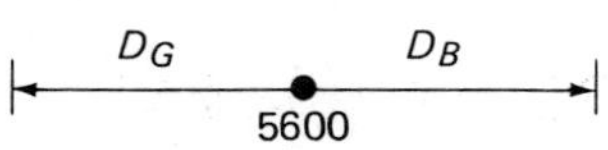

$R_GT_G + R_BT_B = 5600$, $R_G = 400$, $R_B = 700$,
$T_G = T_B + 3$, $\underline{T_B = 4}$, $\underline{T_G = 7}$, **1 o'clock p.m.**

6. 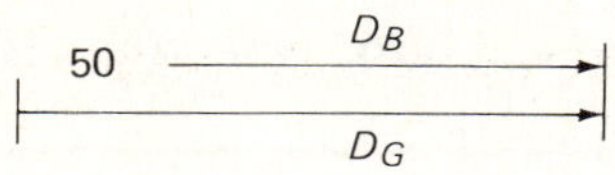

$R_B T_B + 50 = R_G T_G$, $R_B = 6$, $R_G = 11$,
$T_B = T_G$, $\boldsymbol{T_G = 10}$ **sec**

7. $N_P + N_N + N_Q = 52$, $N_P + 5N_N + 25N_Q = 800$, $N_Q = 2N_P$, $N_N = 52 - 3N_P$,
$N_P + 5(52 - 3N_P) + 25(2N_P) = 800$, $\boldsymbol{N_P = 15}$, $\boldsymbol{N_Q = 30}$, $\boldsymbol{N_N = 7}$

8. $5(2N) = 16 + 2(2N + 6 + 4)$, $\underline{N = 6, \text{ even integers } = 12, 14, 16, 18}$

9. $WN = (1.06)(12{,}500)$, $WN = 750$

10.

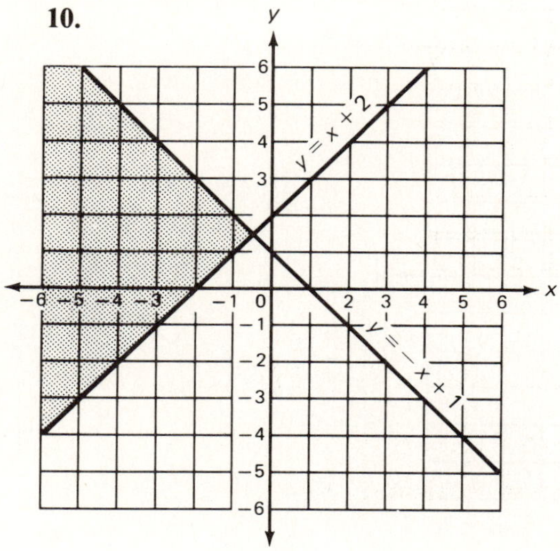

11. $x = \frac{1}{8} \pm \frac{\sqrt{129}}{8}$ **12.** $x = -1 \pm 2\sqrt{2}$ **13.** $x = \frac{1}{4} \pm \frac{\sqrt{41}}{4}$

14.

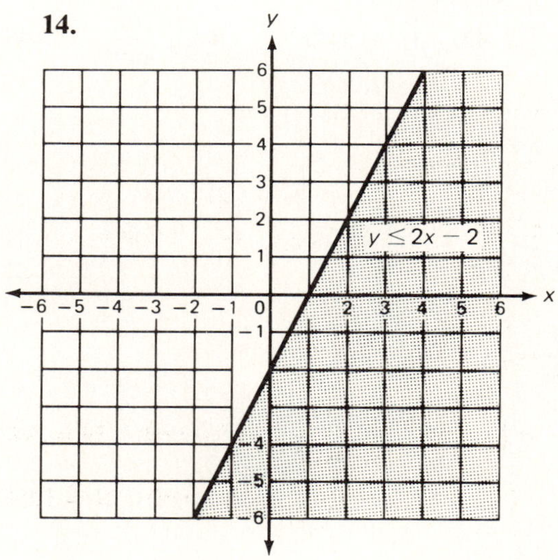

15. 16, −1 **16.** $N_P + N_N + 3N_P = 22$
$N_N = 22 - 4N_P$
$N_P = 5$, $N_D = 15$, $N_N = 2$

17. 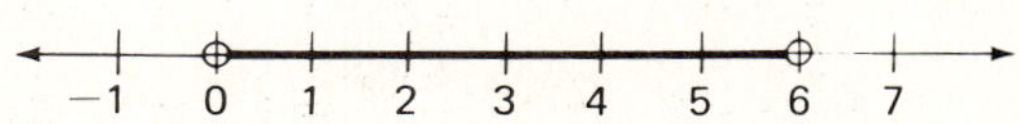

18. $\frac{147}{180} \longrightarrow \underline{\frac{49}{60}}$ **19.** $\frac{6117}{999{,}000}$

20. $c = \dfrac{a}{y + ma - xa}$

21. $x = 52$ **22.** $y = \frac{1}{2}x - 2$ **23.** $y = -\frac{1}{2}x + 3$

24. $y = -\frac{2}{3}x + 1$ **25.** $g(1) = -4$ **26.** (a), (c), (d), (e)

27. 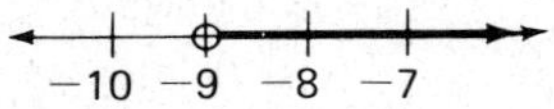

28. 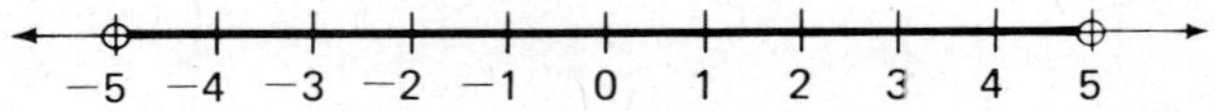

29. $\sqrt{21}$ **30.** $4\sqrt{3} - 8\sqrt{6}$ **31.** $6\sqrt{6}$ **32.** $3x^2 + 20x + 52 + \dfrac{165}{x - 3}$

33. $-3, 1$ **34.** $\dfrac{x - 2}{y + 1}$ **35.** 2×10^{17} **36.** $a + b = b + a$

37.

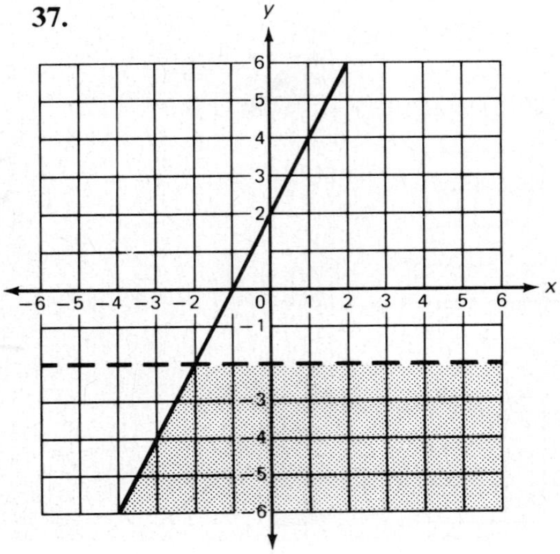

38. $\dfrac{ab + 1}{yb - a}$ **39.** $\dfrac{c - x}{c + x}$ **40.** $\dfrac{3(a + b) - a^2}{a^2(a + b)}$ **41.** $\dfrac{1 - ma + ba^2x}{a^2x}$

42. $3ax^{-1}y^{-2}z - a^{-1}xy^2z$ **43.** $3y + 6x^4z^4$ **44.** $\dfrac{b^2}{acx}$ **45.** 4

46. $\dfrac{x^2 + 11x + 36}{x^2 - 36}$

Index